Tropical Rainforest Responses to Climatic Change

Mark B. Bush and John R. Flenley

Tropical Rainforest Responses to Climatic Change

Springer

Published in association with
Praxis Publishing
Chichester, UK

Professor Mark B. Bush
Chair Ecology Program
Department of Biological Sciences
Florida Institute of Technology
Melbourne
Florida
USA

Professor John R. Flenley
Geography Programme
School of People, Environment and Planning
Massey University
New Zealand

SPRINGER–PRAXIS BOOKS IN ENVIRONMENTAL SCIENCES
SUBJECT *ADVISORY EDITOR*: John Mason B.Sc., M.Sc., Ph.D.

ISBN 3-540-23908-1 Springer Berlin Heidelberg New York

Springer is part of Springer-Science + Business Media (springer.com)

Bibliographic information published by Die Deutsche Bibliothek

Die Deutsche Bibliothek lists this publication in the Deutsche Nationalbibliografie; detailed bibliographic data are available from the Internet at http://dnb.ddb.de

Library of Congress Control Number: 2006928274

Cover design: Jim Wilkie
Project management: Originator Publishing Services, Gt Yarmouth, Norfolk, UK

Printed on acid-free paper

Contents

Preface

Never before in human history has the need for an understanding of climatic change been so great. Nowhere in the world is that need so serious as in the tropics, where deforestation and extinction are at their most rapid, biodiversity greatest, and human lifestyles at their most precarious. We therefore hope and believe that this book will be timely and useful, as it attempts to describe and explain in scientific terms the past, present, and future changes in Earth's most complex terrestrial ecosystem, the Tropical Rain Forest.

The project grew from a discussion between Clive Horwood of Praxis and Mark Bush on the status of climate change research in Tropical Rain Forest settings. Mark's own involvement in attempting to apply lessons learned from the past to the formulation of conservation theory and practice led to a desire to move beyond a simple review of paleoclimatic data. The text aims to build upon and update the foundation of John Flenley's (1979) *The Equatorial Rain Forest: A Geological History* (Butterworth, London). In the intervening period our understanding of individualistic species migration, of potential interactions between climate and physical process, phylogenies, and of the looming impact of global climate change has revolutionized community ecology. In that same period the coverage of tropical paleoecological data has exploded—for example, there was not a single datum from Amazonia when John wrote his book.

John Flenley was called in to help when the sheer enormity of the task became evident to Mark. John had recently moved onto part time so was able to bring his experience of tropical regions fully into play, especially in the area of vegetational history.

We hope that the book will be used by scholars and senior students throughout the world, but especially in the developing countries of the Tropics, where climatic change may spell ecological and economic disaster very soon indeed. Perhaps it is not too much to hope that our book may contribute to influencing world policies in relation to technology and economics, before the climatic changes become irreversible.

We are deeply indebted to all our contributors, a varied selection of excellent researchers, who have given their time and effort unstintingly to make this book possible. We are also grateful to those who have helped with the editing, especially Olive Harris. John Flenley wishes particularly to thank his wife, Helen, for her understanding and support. Our publishers, especially Clive Horwood, deserve exceptional thanks for their patience, tolerance and skill. Any remaining errors are of course our responsibility.

Mark Bush   *John Flenley*
30 April 2006

To
VHB and HCF

Figures

Tables

Abbreviations and acronyms

ABRACOS	Anglo BRazilian Amazon Climate Observational Study
AGCM	Atmospheric General Circulation Model
ALLJ	American Low-Level Jet
AMIP	Atmospheric Model Inter-comparison Program
AMS	Accelerator Mass Spectrometry
ANEEL	Brazilian National Water Agency
ANN	Artificial Neural Network
a.s.l.	Above sea level
ATDF	Amazon Tree Diversity Network
BIOCLIM	BIOlogical CLIMate model
BIOME-3	BIOsphere Model
CAM	Crassulacean Acid Metabolism
CARAIB	CARbon Assimilation In the Biosphere
CBD	Convention on Biological Diversity
CCCma	Canadian Center for Climate Modelling and Analysis
CCM3	Community Climate Model 3
CCSR/NIES	Center for Climate Systems Research/National Institute for Environmental Sciences (Japan)
CLIMAP	CLImate/Long Range Investigation Mappings And Predictions Report
CLIMBER	CLIMate model of intermediate complexity
CMAP	Climate Prediction Center Merged Analysis of Precipitation
CPRM	*Companhia de Pesquisa de Recursos Mineirais* (Brazilian Geological Survey climate model)
CRU	Climate Research Unit
CSM	National Center for Atmospheric Research (U.S.A.) Climate System Model
d.b.h.	Diameter at breast height

DCA	Detrended Correspondence Analysis
DGVM	Dynamic Global Vegetation Model
DIF	Differentials
DJF	December, January, February
DO2	Dansgaard–Oeschger event 2
DOMAIN	A computer model to delimit the potential range of plants and animals
ECMWF	European Center for Medium Range Weather Forecast
ED	Ecosystem Demography model
ELA	Equilibrium Line Altitude
ENSO	El Niño–Southern Oscillation
ET	EvapoTranspiration
FATE	Functional Attributes in Terrestrial Ecosystems Model
FOAM	Fast Ocean Atmosphere Model
FORCLIM	FORests in a Changing CLIMate Model
FORET	FORests of Eastern Tennessee Model
FORMIND	INDividual-based Mixed RainFORest Growth Simulator
GAM	Generalized Additive Modeling; Generalized Additive Model
GARP	Genetic Algorithm for Rule-set Prediction
GCM	General Circulation Model; Global Climate Model
GEN2	Global Environmental and Ecological Simulation of Interactive Systems Version 2
GENESIS-IBIS	Coupled Global Biosphere–Atmosphere Model
GHCN	Global Historical Climatology Network
GHG	GreenHouse Gas
GISP	Greenland Ice Sheet Project
GLM	Generalized Linear Modeling
GPCP	Global Precipitation Climatology Project
HadCM3LC	A Hadley Centre model
He-1	Heinrich Event 1
HYDRA	Terrestrial hydrology model
IAI	Inter American Institute for Global Change
IBIS	Integrated BIosphere Simulator
IGBP-PAGES	International Geosphere–Biosphere Programme—PAst Global ChangES
INGENMET	*Instituto Geológico Minero y Metalúrgico* (Peru)
IPCC	Intergovernmental Panel on Climate Change
IS92a	A "business as usual" scenario for the CCCma (above)
ITCZ	Inter Tropical Convergence Zone
kcal.yr BP	Kilo calibrated years before present
LAI	Leaf Area Index
LBA	Large Scale Atmosphere Biosphere Experiment in Amazonia; Large Scale Biosphere Atmosphere experiment
LBA-WET AMC	Large Scale Biosphere–Atmosphere Experiment in

	Amazonia—WET Season Atmospheric Mesoscale Campaign
LGM	Last Glacial Maximum
LPJ	Lund–Potsdam–Jena Research Group
LW	Legates–Wilmott
MAT	Mean Annual Temperature
MC1	Corvallis Dynamic Vegetation Model
MIS	Marine Isotope Stage
MJO	Madden Julian Oscillation
MTCO	Mean Temperature of the COldest month
MWP	Medieval Warm Period
NAP	Non-Arboreal Pollen
NASA GEOS	National Aeronautics and Space Administration Geodetic and Earth Orbiting Satellite
NCEP	National Center for Environmental Prediction
NEE	Net Ecosystem Exchange
NPP	Net Primary Productivity
OSL	Optically-Stimulated Luminescence
PDO	Pacific Decadal Oscillation
PEP	PhosphoEnolPyruvate
PET	Potential EvapoTranspiration
PFT	Plant-Functional Type
PMIP	Palaeoclimate Model Inter-comparison Program
p.p.m.V	Parts per million by Volume
R30	Geophysical Fluid Dynamics Laboratory Coupled Climate Model
RAINFOR	*Red Amazónica de Inventarios FORestales* (Amazon Forest-Inventory Network)
SACZ	South Atlantic Convergence Zone
SALLJEX	South American Low-Level Jet field EXperiment
SAMS	South American Monsoon System
SASM	South American Summer Monsoon
SDGVM	Sheffield Dynamic Global Vegetation Model
SO	Southern Oscillation
SRES	Special Report on Emissions Scenarios
SST	Sea Surface Temperature
TDF	Tropical Deciduous Forest
TEF	Tropical Evergreen Forest
TRIFFID	Terrestrial carbon cycle model
TRMM	Tropical Rainfall Measuring Mission
TSEF	Tropical Semi-Evergreen Forest
UGAMP	United Kingdom Universities Global Atmospheric Modelling Programme
UKMO	United Kingdom Meteorological Office
UMRF	Upper Montane Rain Forest

UNFCCC	United Nations Framework Convention on Climate Change
UV	UltraViolet
VECODE	VEgetation COntinuous DEscription Model
VEMAP	Vegetation/Ecosystem Modeling and Analysis Project
VPD	Vapor Pressure Deficit
WUE	Water-Use Efficiency

Authors

Mark B. Bush
Department of Biological Sciences, Florida Institute of Technology, 150 W. University Blvd.,
Melbourne, FL 32901, U.S.A.
Email: mbush@fit.edu

John R. Flenley
Geography Programme, School of People, Environment and Planning, Massey University,
Palmerston North, New Zealand.
Email: J.Flenley@massey.ac.nz

Contributors

Timothy R. Baker
Earth and Biosphere Institute, Geography, University of Leeds, Leeds LS2 9JT, U.K. and Max-Planck-Institut für Biogeochemie, Postfach 100164, 07701 Jena, Germany
Email: geotrb@leeds.ac.uk

Richard A. Betts
Met Office, Hadley Centre for Climate Prediction and Research, Fitzroy Road, Exeter EX1 3PB, U.K.
Email: richard.betts@metoffice.com

Raymonde Bonnefille
Cerege, CNRS, Route Leon Lachamp, BP 80, 13545 Aix-en-Provence, Cedex 04 France.
Email: rbonnefi@cerege.fr

Mark B. Bush
Department of Biological Sciences, Florida Institute of Technology, 150 W. University Blvd., Melbourne, FL 32901, U.S.A.
Email: mbush@fit.edu

Paul A. Colinvaux
Marine Biological Laboratory, Woods Hole, MA 02543, U.S.A.
Email: pcolinva@mbl.edu

Sharon A. Cowling
Department of Geography, University of Toronto, 100 St. George Street, Toronto, Ontario, Canada, M5S 3G3.
Email: cowling@geog.utoronto.ca

John R. Flenley
Geography Programme, School of People, Environment and Planning, Massey University, Palmerston North, New Zealand.
Email: J.Flenley@massey.ac.nz

William D. Gosling
Department of Earth Sciences, Open University, Walton Hall, Milton Keynes MK7 6AA, U.K.
Email: W.D.Gosling@open.ac.uk

Lee Hannah
Senior Fellow, Climate Change Biology, Center for Applied Biodiversity Science, Conservation International, 1919 M Street NW, Washington, DC 20036. Current address: Bren School of Environmental Science and Management, UCSB, Santa Barbara, CA 93106, U.S.A.
Email: lhannah@conservation.org

Jennifer A. Hanselman
Department of Biological Sciences, Florida Institute of Technology, 150 W. University Boulevard, Melbourne, FL 32901, U.S.A.
Email: jhanselm@fit.edu

Henry Hooghiemstra
Palynology and Quaternary Ecology, Institute for Biodiversity and Ecosystem Dynamics (IBED), Palynology and Paleo/Actuo-ecology, Faculty of Science, University of Amsterdam, Kruislaan 318, 1098 SM Amsterdam, The Netherlands.
Email: hooghiemstra@science.uva.nl

Peter A. Kershaw
School of Geography and Environmental Science, P.O. Box Number 11A, Monash University, Victoria 3800, Australia.
Email: Peter.Kershaw@arts.monash.edu.au

Simon L. Lewis
Earth and Biosphere Institute, Geography, University of Leeds, Leeds LS2 9JT, U.K.
Email: s.l.lewis@leeds.ac.uk

Jon Lovett
Environment Department, University of York, Heslington, York YO10 5DD, U.K.
Email: jl15@york.ac.uk

Tom Lovejoy
The H. John Heinz III Center for Science, Economics and the Environment, 1001 Pennsylvania Ave., NW, Suite 735 South, Washington, DC 20004, U.S.A.
Email: lovejoy@heinzctr.org

Yadvinder Malhi
School of Geography and the Environment, Oxford University, Oxford, U.K.
Email: yadvinder.malhi@ouce.ox.ac.uk

Rob Marchant
Environment Department, University of York, Heslington, York YO10 5DD, U.K.
Email: rm524@york.au.uk

José Marengo
CPTEC/INPE-Center for Weather Forecasts and Climate Studies/National Institute for Space Research, Rodovia Presidente Dutra, km 40, Caixa Postal 01, Cachoeira Paulista, SP 12630-000, Brazil.
Email: Marengo@cptec.inpe.br

Megan E. McGroddy
Dept. of Biology, West Virginia University, P.O. Box 6057, Morgantown, WV 26506, U.S.A.
Email: Megan.McGroddy@mail.wvu.edu

Robert J. Morley
School of Geography and Environmental Science, Monash University, Victoria 3800, Australia and Dept. of Geology, Royal Holloway University, Egham, Surrey TW20 0EX, U.K. Mailing address: Palynova/PT Eksindo Pratama, Vila Indah Pajajaran, Jl Kertarajasa No. 12A, Bogor, Indonesia 16153.
Email: pollenpower@indo.net.id

Oliver L. Phillips
Earth and Biosphere Institute, Geography, University of Leeds, Leeds LS2 9JT, U.K.
Email: o.phillips@leeds.ac.uk

Dolores R. Piperno
Smithsonian Tropical Research Institute, Balboa, Panama and Department of Anthropology, National Museum of Natural History, Washington, DC 20013, U.S.A.
Email: PIPERNOD@si.edu

Herman H. Shugart
Department of Environmental Sciences, University of Virginia, Charlottesville, VA 22903, U.S.A.
Email: hhs@virginia.edu

Miles R. Silman
Department of Biology, Wake Forest University, P.O. Box 7325, Reynolda Station, Winston-Salem, NC 27109-7325, U.S.A.
Email: silmanmr@wfu.edu

Whendee L. Silver
Ecosystem Sciences Division, Department of Environmental Science, Policy, and Management, 137 Mulford Hall #3114, University of California, Berkeley, CA 94720, U.S.A.
Email: wsilver@nature.berkeley.edu

Sander van der Kaars
School of Geography and Environmental Science, P.O. Box Number 11A, Monash University, Victoria 3800, Australia.
Email: Sander.vanderKaars@arts.monash.edu.au

1

Cretaceous and Tertiary climate change and the past distribution of megathermal rainforests

R. J. Morley

1.1 INTRODUCTION

The history of megathermal (currently "tropical") rainforests over the last 30 kyr is now becoming relatively well-understood, as demonstrated by the many contributions in this volume. However, our perception of their longer-term history remains highly fragmentary. There is a real need for a better understanding of rainforest history on an *evolutionary* timescale, not only to have a better idea of the biological, geological, and climatic factors which have led to the development of the most diverse ecosystem ever to have developed on planet Earth, but also since the implications of rainforest history on an evolutionary timescale are inextricably linked to a plethora of other issues currently receiving wide attention. Determining the place and time of origin and/or radiation of angiosperms (which overwhelmingly dominate present day megathermal rainforests), establishing patterns of global climate change, clarifying the nature of global temperature gradients through time, understanding the successive switching from greenhouse to icehouse climates, global warming, patterns of dispersal of megathermal plants and animals, higher rank (ordinal) taxonomy and the nature of controls on global diversity gradients are but some issues which are being clarified with the better understanding of the long-term history of megathermal rainforests.

This chapter attempts to examine climatic controls on megathermal rainforests since their first appearance in the Cretaceous Period up until the end Pliocene, a period more than 60 times longer than that considered by the other contributions in this book. The first part summarizes the pattern of initial radiation of angiosperms, and the first physiognomic evidence for closed multistratal forests during the Late Cretaceous. For the earlier Tertiary, the pattern of changing rainforest climates is viewed on a very broad scale, through the construction of rainforest maps, each representing periods of perhaps 5 Myr. These maps follow those of Morley (2000a) but have been substantially improved by integrating the comprehensive database on the global distribution of climatically sensitive lithologies (primarily evaporites, bauxites and coals) compiled by Boucot *et al.* (in press). Bauxites were given little consideration in the Morley (2000a) maps due to difficulties regarding age determination, but are critical in evaluating past megathermal climates since they are generated under hot and wet climates that are strongly seasonal. They therefore reflect the former occurrence of monsoonal climates. Boucot's comprehensive database allows bauxites to be placed within an appropriate perspective, despite difficulties of precise dating.

For the mid- and younger Tertiary, in addition to presenting generalized global maps, the approach followed allows climate change over this period to be viewed more from the perspective of the Quaternary. For the later Quaternary, radiometric dates and oxygen isotope signals provide a precise time framework within which scenarios of climate change can be established and regionally correlated. The pattern is of astronomically driven climate cycles each comprising (a) an initial period of rapid warming, followed by (b) warm, everwet climates, and (c) by a period of gradual, sometimes intermittent temperature decline with reduced moisture availability, culminating in (d) a period during which everwet tropical climates were of much more restricted distribution (Flenley, 1979; Morley, 2000a). In synchronization with climate fluctuations, global sea levels have risen and fallen following shedding from and subsequent sequestration of seawater in polar ice caps.

The precision of dating which can be applied in the younger Quaternary is rarely available for Tertiary sediments. By applying a *sequence stratigraphic* approach in the Tertiary, which is widely used in the petroleum industry, and emphasizes patterns of sediment deposition in relation to fluctuating sea levels (e.g., Wilgus *et al.*, 1988; Posamentier and Allen, 1999) by equating periods of sea level lowstand with "glacials" and highstand with "interglacials", patterns of climate change from fossil data relative to sea level change can be viewed in the same perspective as Quaternary fluctuations even where independent dating is of relatively low precision (Morley, 2000a). Such an approach is applicable for the post Middle Eocene, during which time ice accumulation has been taking place in polar areas, and most sea level changes are thought to reflect the sequestration of seawater into polar ice caps (Abreu and Anderson, 1998; Bartek *et al.*, 1991; Zachos *et al.*, 2001). Consequently, over the period from the Late Eocene to Pliocene global sea level change may be used as a proxy for global climate change. However, there remains debate about the nature of sea level fluctuations during earlier "greenhouse" phases (e.g., Hallam, 1992; Miller *et al.*, 2004).

Indications that many Tertiary sea level changes parallel periods of climate change are illustrated by palynological analyses through successive transgressive/regressive cycles, especially in areas of high rates of sedimentation, as may occur in Tertiary deltas, such as the Niger (Nigeria) or Mahakam (Indonesia). Palynological signals from such sections, albeit on a different timescale, can be compared with those seen in Late Quaternary deep-sea cores, such as the Lombok Ridge core G6-4 from Indonesia reported by van der Kaars (1991), Papalang-10 core offshore Mahakam Delta by Morley *et al.* (2004) Niger Delta core GIK 16856 by Dupont and Weinelt (1996), Amazon Fan ODP Leg 155 cores by Haberle (1997), Haberle and Malin (1998), and Hoorn (1997). The current phase of active, deep-water hydrocarbon exploration in these areas provides a rich source of (mainly unpublished) data which emphasizes periods of lowest sea levels (and coolest climate), since the main exploration targets in these settings are sands which would have been swept down the continental slope when sea levels dropped below the level of continental shelves.

1.2   DIFFERENCES BETWEEN QUATERNARY AND TERTIARY MEGATHERMAL FORESTS

Using Quaternary analogs to interpret ecological and climatic successions from the Tertiary raises two main issues; (1) was there a fundamental difference between Quaternary and Tertiary rainforests and (2) were Tertiary species compositions so different from the Quaternary as to make comparison fruitless?

With respect to the first of these issues, it is now clear that there was one major difference between Quaternary rainforests, and those from the Miocene and Early Pliocene. Over the last 2.8 Myr, equatorial climates were, at least intermittently, significantly cooler than at any time since the Oligocene, as indicated by the sudden dispersal of numerous microthermal taxa into equatorial montane forests of each rainforest block in the mid-Pliocene (Morley, 2000a, 2003; Van der Hammen and Hooghiemstra, 2000 and Figure 1.1). This also implies that pre Late Pliocene equatorial climates were a few degrees warmer than today. Quaternary rainforests in the equatorial zone are therefore likely to have exhibited greater altitudinal stratification than in the Neogene since lowland rainforests, which exhibit little internal altitudinal stratification, would have extended to higher altitudes, giving less room for montane forests. During the Early Pliocene, and most of the Miocene, microthermal taxa were essentially missing at equatorial latitudes, or so poorly represented as to go virtually unrecorded in palynological analyses. During the Oligocene, cooler climates resulted in the intermittent expansion of frost-tolerant vegetation into the equatorial zone in a manner not even seen in the Quaternary, clearly shown for the Southeast Asian region (Morley *et al.*, 2003).

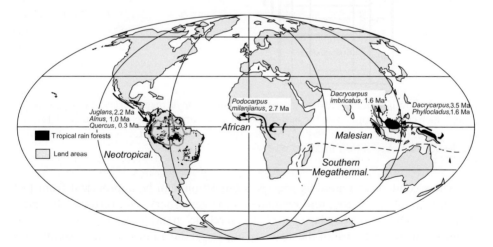

Figure 1.1. Ice-age distributions of closed canopy megathermal rainforests has been a subject of debate. The refugial hypothesis depicted substantial replacement of forest by savanna during ice ages (e.g., Whitmore and Prance, 1987) (*shaded areas*) and for Amazonia (Van der Hammen and Hooghiemstra, 2000) (*circled by gray line*; figure after Morley, 2000a). While historically important, this view of forest fragmentation has now been replaced by paleoecologically-based reconstructions that show much less change in forest cover. Also shown are noteworthy instances of Pliocene and Pleistocene dispersal directions of microthermal taxa into low latitudes.

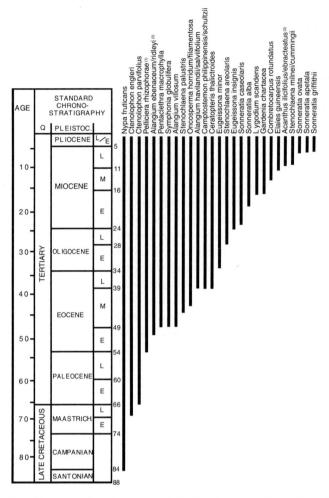

Figure 1.2. Stratigraphic range of angiosperm and pteridophyte megathermal species, or species pairs, which can be identified on the basis of pollen and spores, and have a well-defined Tertiary fossil record: (1) from Rull (1999); (2) Morley (unpublished); (3) Morley (1991); others from Morley (2000a). Age shown in Myr.

The second issue, regarding species composition, can be approached from two angles: rates of speciation, and comparison of Quaternary and Tertiary ecological successions. During the heyday of the "glacial refuge" hypothesis, the suggestion was frequently made that most of the diversity of present day rainforests was essentially a Quaternary phenomenon, with new species being generated by successive isolation and subsequent expansion of populations (Haffer, 1969; Prance, 1982). This scenario always seemed at odds with the pollen record of those megathermal *species* that can be differentiated on the basis of pollen (Figure 1.2), most of which show very long histories. It is thus comforting that this theory is now discredited on paleoecological grounds (Colinvaux et al., 2000; Bennett, 2004); also, current molecular studies of rainforest trees demonstrate species longevity of the same order as the pollen record (e.g., Dick *et al.*, 2003), emphasizing that rainforests contain many species of great antiquity; Kutschera and Niklas (2004) estimate that shrubs and hardwoods have mean species durations of 27–34 Myr. With respect to comparing Quaternary and Tertiary ecological successions, the classic study of a Middle Miocene coal from Brunei (Anderson and Muller, 1975), that showed a peat swamp succession with phasic communities with close similarities to those seen in present day peat swamps (Morley, in press), demonstrates close ecological parallels between Neogene and Quaternary vegetation. There is therefore a just case for using Quaternary analogs to interpret Tertiary vegetational scenarios, particularly back as far as the Oligocene.

1.3   LATE CRETACEOUS EXPANSION OF MEGATHERMAL FORESTS

Angiosperms, which dominate megathermal rainforests today, first radiated during the Early Cretaceous from mid- to low latitudes (Crane *et al.*, 1995; Hickey and Doyle, 1977) in response to climatic stress (Stebbins, 1974; Doyle and Donaghue, 1987). They are unlikely to have become initially established in a closed, rainforest setting as previously inferred by Takhtajan (1969) and Thorne (1976) on the assumption that "primitive" angiosperms such as members of Winteraceae, *Trochodendron* and *Tetra-centron* (with vesseless wood which require a mesic climate) evolved in such areas. The vesseless habit in these angiosperms is now considered a derived character (Doyle and Endress, 1997). They came to dominate over other plant groups in the Albian and Cenomanian (Crane, 1987). The equatorial zone at this time was likely to have been hot (Barron and Washington, 1985; Pearson *et al.*, 2001) and strongly monsoonal (Parrish *et al.*, 1982; Morley, 2000a), but not necessarily "semi-arid" as suggested by Herngreen and Duenas-Jimenez (1990) and Herngreen *et al.* (1996). The equatorial zone was therefore an unlikely zone for the establishment of the first megathermal, mesic forests. The paucity of mesic low-latitude settings in the Turonian is emphasized by the particularly low diversity of fern spores from the equatorial regions at this time, but their diverse representation in mid-latitudes continued (Crane and Lidgard, 1990).

It was in mid-Cretaceous mid-latitudes, which were in part characterized by perhumid, frost-free climates, that mesic forests first became an important setting for angiosperms in both hemispheres, and by the Cenomanian most of the physiognomic leaf types characteristic of megathermal forests—including simple entire leaves with drip tips, compound and palmate leaves—were already in place (Upchurch and Wolfe, 1987). From the Turonian to the Maastrichtian, many groups that we consider as strictly "tropical" have their first records from these areas, with families such as Bombacaceae, Clusiaceae, Cunoniaceae, Icacinaceae, Menispermaceae, Rutaceae, Sabiaceae, Saurauiaceae, Theaceae, and Zingiberaceae (Mai, 1991; Morley, 2000a; Davis *et al.*, 2005) first appearing within northern hemisphere mid-latitudes, whereas southern hemisphere mid-latitudes saw the appearance of Aquifoliaceae and Proteaceae, and became a harbour for Winteraceae and Chloranthaceae (Dettmann, 1994).

Within the equatorial zone, mesic angiosperm-dominated forests did not appear until some time after their appearance in mid-latitudes (Morley, 2000a). The first evidence for the development of everwet equatorial climates is probably from Nigeria, where coal deposits are represented from the Campanian to Maastrichtian (Reyment, 1965; Salami, 1991; Mebradu *et al.*, 1986), suggesting an everwet climate. Groups that show their initial radiation in the Cretaceous of the equatorial zone are Annonaceae, Arecaceae, Ctenolophonaceae, Gunneraceae, Fabaceae, Myrtaceae, Restionaceae, and Sapindaceae (Morley, 2000a).

Molecular studies sometimes help to determine which taxonomic groups originated as Northern Megathermal (or Boreotropical) elements, and which have always been equatorial lineages; thus, Davis *et al.* (2002) indicate that Malphigiaceae are likely to be Boreotropical. Doyle and Le Thomas (1997) show that Anonaceae are an equatorial group, as did Givnish *et al.* (2000) for Rapateaceae. However, care needs to be exercised in assessing the often geographically biased and scattered fossil record of groups being assessed by molecular analyses since the macrofossil record is strongly biased to Europe and North America where most collecting has been done (Morley and Dick, 2003).

The biogeographical histories of the major groups of megathermal angiosperms for the remainder of the Cretaceous and Tertiary periods can be divided into two main phases. During the first phase, from the latest Cretaceous to Middle Eocene, the Earth was characterized by greenhouse climates, and predominantly by plate tectonic disassembly (Morley, 2000a, 2003). This was a period of widespread range expansion and diversification of megathermal plants. The post-Middle Eocene, on the other hand, was a period essentially of global cooling and the successive expansion of icehouse climates, coupled with plate tectonic collision, and was mainly a period of range retraction of megathermal taxa.

The time from which mesic megathermal forests can be visualized as closed, multi-storeyed forests, and thus resemble modern rainforests in terms of physiognomy, is debatable. Upchurch and Wolfe (1987) suggested that leaf morphologies from the Cenomanian Dakota Formation reflect such a setting, but at this time angiosperm wood fossils are generally small-dimensioned, and seed sizes small (Wing and Tiffney, 1987), militating against the presence of modern aspect rainforests at this time. A re-examination of leaf assemblages from the same Dakota Formation locality by Johnson (2003, and pers. commun.) show that this locality was dominated by large, lobed angiosperm leaves, not reminiscent of rainforest physiognomy. However, Davis *et al.* (2005) have used molecular evidence to show that the clade Malpighiales, which constitute a large percentage of species in the shaded, shrub and small-tree layer in tropical rainforests worldwide, radiated rapidly in the Albian–Cenomanian, and suggest that this radiation was a response to adaptations to survive and reproduce under a closed forest canopy.

The first evidence for typical closed, multi-stratal forest synusiae based on fossils comes from the latest Cretaceous of Senegal and Nigeria in West Africa. Evidence includes the presence of casts of large seeds from the Campanian of Senegal (Monteillet and Lappartient, 1981), a large supply of endosperm in an enlarged seed allowing successful germination under a forest canopy (Grime, 1979). A molecular link between life form and seed size has recently been established (Moles *et al.*, 2005) with large seeds being linked with tropical trees. The presence of seeds or fruit attributable to climbers from Nigeria (Chesters, 1955) and the presence of large-girth angiosperm wood (Duperon-Ladouneix, 1991) also suggests the presence of tall canopy trees. The oldest locality for multi-storeyed forests is therefore likely to have been in the

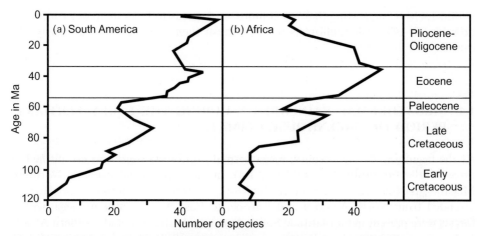

Figure 1.3. Numbers of stratigraphically useful angiosperm pollen types per epoch, for: (a) South America (data from Muller *et al.*, 1987) and (b) West Africa (data from Salard-Chebaldaeff, 1990), providing a rough proxy for angiosperm diversity through time (from Morley, 2000a).

equatorial zone. Subsequently, large-dimensioned seeds are widespread from the Paleocene onward in North America (Wing and Tiffney, 1987) suggesting that following the demise of the dinosaurs, closed multi-storeyed forests became widespread, perhaps coinciding with the radiation of frugiverous mammals. The appearance of evidence for multi-storeyed forests in West Africa coincides with a distinct diversity increase of fossil angiosperm pollen (Figure 1.3, from Morley, 2000a), which was thought to reflect evolutionary adaptations associated with the development of the forest canopy by Niklas *et al.* (1980).

Kubitski (2005) considers the development of the rainforest canopy in the Late Cretaceous of Africa and South America to be one of the major stages in the development of all land plants. The presence of the rainforest canopy not only facilitated the diversification of most angiosperm families in a manner not seen previously, but also provided a setting for the renewed diversification of pteridophytes, under its shadow, as suggested both from molecular data (Schneider *et al.*, 2004), and also from changes in pteridophyte spore assemblages in the low-latitude palynological record from the latest Cretaceous onward, with the increased representation and diversification of monolete, as opposed to trilete spores from this time.

The K-T meteorite impact probably had a major effect on rainforests globally (Figure 1.3) but did not substantially affect the main angiosperm *lineages* that characterized each area. Gymnosperms, however, fared particularly poorly in the low latitudes following the K-T event. In the earliest Tertiary gymnosperms were virtually absent from each of the equatorial rainforest blocks. Recovery of rainforest diversity after the K-T event is generally acknowledged to have taken some 10 Myr (Fredriksen, 1994), but a recently discovered leaf fossil flora from the Paleocene of Colorado (Johnson and Ellis, 2002) suggests much more rapid recovery, perhaps within 1.4 Myr, suggesting that much more work needs to be done to determine just how long it takes for rainforests to re-establish their diversity after a cataclysmic event.

1.4   MEGATHERMAL RAINFORESTS DURING THE EARLY TERTIARY PERIOD OF GREENHOUSE CLIMATE

At the beginning of the Tertiary, megathermal rainforests were thus established in three parallel latitudinal zones (Figure 1.4). In the northern hemisphere, Northern Megathermal (termed "Boreotropical" in Morley, 2000a) mesic and monsoonal forests extended from North America and Europe, to East Asia, Southern Megathermal forests were present in mid-latitude South America, Australasia and southern Africa, and equatorial forests of the Palmae province were well-developed in northern South America, Africa, India and probably Southeast Asia (Morley, 2000a). The Paleocene saw global temperatures rise dramatically (Figure 1.5), due to increased atmospheric CO_2 (Pearson and Palmer, 2000). At the Paleocene–Eocene boundary, megathermal forests were thus at their most extensive (Figure 1.6), more or less reaching the polar regions, as far as 60°N in Alaska (Wolfe, 1985), and with *Nypa* swamps at 57°S in Tasmania (Pole and McPhail, 1996). At this time, intermittent land connections from North America to Europe via Greenland, from South America to Australasia via Antarctica and with a filter dispersal route between the Americas (Hallam, 1994; Morley, 2003) megathermal plants were able to disperse globally in a manner seen at

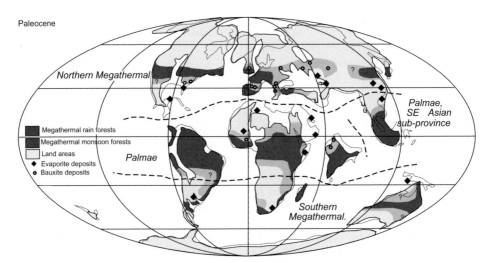

Figure 1.4. Closed canopy megathermal rainforests first became widespread during the Paleocene (Morley, 2000a). Paleogeography and paleocoastlines from Smith *et al.* (1994). Occurrences of evaporites and bauxites from Boucot *et al.* (in press). *Dotted lines* are floristic province boundaries.

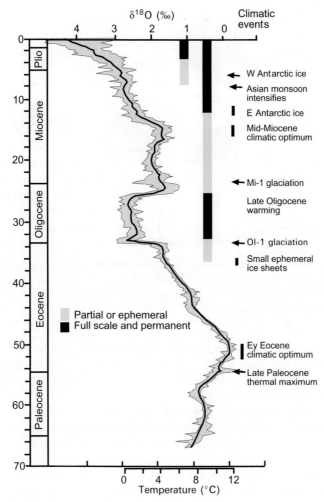

Figure 1.5. Generalised oxygen isotope curve for benthonic (bottom dwelling) foraminifera through the Cenozoic (from Zachos *et al.*, 2001). The ratio of [16]O to [18]O for benthonic foraminifera provides a proxy for high-latitude surface marine temperatures (Hudson and Anderson, 1989), and therefore is a guide to global temperature trends: Oi = glacial interval at beginning of Oligocene; Mi = glacial at beginning of Miocene. The temperature scale was computed for an ice-free ocean, and thus applies only to the pre-Oligocene period of greenhouse climates.

no other time—with, for instance, members of the family Bombacaceae spreading from North America to Europe, on the one hand, and via South America and presumably Antarctica to Australia and New Zealand, on the other (Morley, 2000a; 2003). In the Middle and Late Eocene, subsequent to the thermal maximum, climate oscillations resulted in the successive expansion and contraction of megathermal forests in mid-latitudes, as recorded for North America by Wolfe (1977).

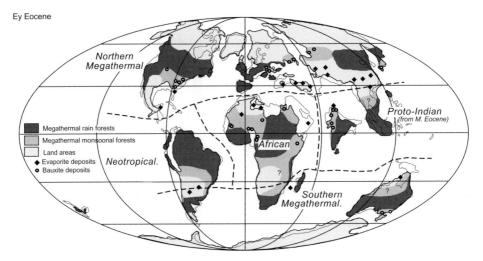

Figure 1.6. Distribution of closed canopy megathermal rainforests during the Late Paleocene/ Early Eocene thermal maximum (Morley, 2000a). Paleogeography and paleocoastlines from Smith *et al.* (1994). Occurrences of evaporites and bauxites from Boucot *et al.* (in press). *Dotted lines* are floristic province boundaries.

The nature of the vegetation that characterized mid-latitudes at the time of the thermal maximum has been widely studied, with classic fossil localities in Europe, such as the London Clay (e.g., Reid and Chandler, 1933; Chandler, 1964; Collinson, 1983) and Messel in Germany (Collinson, 1988), North America (Wolfe, 1977; Manchester, 1994, 1999), South America (Wilf *et al.*, 2003) and Australia (Christophel, 1994; Greenwood, 1994), but the character of equatorial vegetation at the time of the thermal maximum remains unclear. There has been some discussion as to whether mid-latitude areas experienced a "tropical" climate at this time (Daley, 1972; Martin, 1992). Most authors logically conclude that climates at this time were different from any present day climates. The critical factors were lack of frosts and absence of a water deficit. Summer-wet climates in Indochina and Mexico are probably the closest modern analogs, not surprisingly, in areas where many Boreotropical elements are relict (Morley, 2000a, Figure 11.9).

Very few studies demonstrating ecological succession from low latitudes from this critical period of the Paleocene–Eocene thermal maximum have been published. Reference has been made to "reduced global climate gradients" based on oxygen isotope analysis of calcareous foraminiferal tests (e.g., Shackleton and Boersma, 1983), but current evidence shows that low-temperature estimates from the equatorial zone are erroneous and due to diagenetic effects. Recent sea surface estimates based on very well-preserved microfossils from the equatorial zone suggest Eocene sea surface temperatures were at least 28–32°C (Pearson *et al.*, 2001; Zachos *et al.*, 2003). Evidence from paleofloras suggests that there was a marked vegetational zonation from the equator to mid-latitudes (Morley, 2000a)—for instance, equatorial and South Africa were characterized by very different floras at this time, indicating a climatic zonation from mid- to low latitudes and current Eocene sea surface estimates are in line with those expected by modelling climates from vegetational data.

A study of the palynological succession through the Venezuelan Guasare, Mirador and Misoa formations by Rull (1999) provides a glimpse of the evolutionary and ecological changes that characterized the Late Paleocene to Early Eocene thermal maximum onset in northern South America. A conspicuous ecological change took place at the Paleocene–Eocene boundary. The Late Paleocene flora is similar to other low-latitude pollen floras of similar age, such as that from Pakistan (Frederiksen, 1994), emphasizing its pantropical character, whereas the Early Eocene palynoflora is geographically more differentiated, owing to a high proportion of restricted elements caused by the extinction of Paleocene taxa and the incoming of new components. The incoming of new Eocene taxa was gradual (or possibly stepped), and diversities increase in a manner that parallels global temperature estimates. At a detailed level several palynocycles could be defined, both in terms of assemblage and diversity changes, suggesting cyclic forcing mechanisms controlling vegetation changes. This study clearly suggests that vegetation change at low latitudes at the beginning of the thermal maximum was as pronounced as at mid-latitudes. A substantial temperature increase most likely accounted for the vegetation change recorded.

Some recent studies suggest that Early and Middle Eocene low-latitude climates were moisture-deficient or strongly seasonal in some areas. A well-dated Middle Eocene leaf flora from Tanzania, about 15°S paleolatitude, suggests the presence of wooded, rather than forest vegetation with near-modern precipitation estimates for this area (Jacobs and Heerenden, 2004). The plant community was dominated by caesalpinoid legumes and was physiognomically comparable to miombo woodland. Data from a very thick Early and Middle Eocene succession from southwest Sulawesi in Indonesia indicates alternating phases of dry climate (possibly reflecting periods of low sea level), in which Restionaceae were prominent members, and wetter climate, dominated by palms (Morley, unpublished).

1.5   MIDDLE EOCENE TO OLIGOCENE CLIMATES

1.5.1   General trends

From the Middle Eocene through to Late Eocene global climates show an overall cooling, with a further rapid temperature decline at the end of the Eocene (Miller *et al.*, 1987; Zachos *et al.*, 2001) following which mid-latitude northern hemisphere climates mostly became too cold to support megathermal vegetation. The decline in global temperatures is associated with a major build-up of polar ice, initially over Antarctica; and consequently sea levels fell globally, especially during the Oligocene.

With cooler temperatures in mid-latitudes, megathermal rainforests underwent a major retraction to low latitudes (Figure 1.7). This retraction was particularly pro-nounced in the northern hemisphere, with megathermal forests virtually disappearing from most of the North American continent (Wolfe, 1985) and becoming much more restricted in Europe, some elements possibly being maintained along the Atlantic

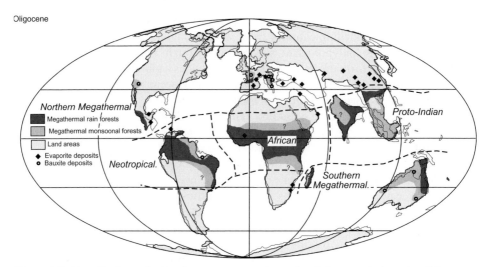

Figure 1.7. Distribution of closed canopy megathermal rainforests during the Oligocene, following the terminal Eocene cooling event (Morley, 2000a). Paleogeography and paleo-coastlines from Smith *et al.* (1994). Occurrences of evaporites and bauxites from Boucot *et al.* (in press). *Dotted lines* are floristic province boundaries.

coast as a result of warm currents. Northern hemisphere megathermal forest species had to disperse equatorward or face extinction. Their success at southward dispersal was related to the different tectonic setting in each of the three main areas. Because there was a continuous land connection from East Asia to the equatorial zone, many Boreotropical elements were able to find refuge in the forests of Southeast Asia. The Boreotropical relicts included many so-called primitive angiosperms, and as a result there is a concentration of such taxa in that area, especially in the rainforest refugia of southern China and Vietnam (e.g., Magnoliaceae, *Trochodendron*). This area has also provided a refuge for many Boreotropical gymnosperms, such as *Cunninghamia*, *Glyptostrobus*, and *Metasequoia*.

With respect to North America, Northern Megathermal elements may have been able to find refuge along the southern margin of the North American Plate, but could not disperse to the equatorial zone until the formation of the Isthmus of Panama in the Pliocene (Burnham and Graham, 1999). As a result, many more Northern Mega-thermal elements are likely to have become extinct in the Americas than in Southeast Asia. Many of those that did survive, and have parallel occurrences in Southeast Asian forests, are now extant as the amphi-Pacific element of van Steenis (1962).

For Europe, the east–west barriers of Tethys, the Alps, and the Sahara combined to limit equatorward dispersal to Africa to just a few taxa; hence, there are barely any true Northern Megathermal elements in present day African rainforests (Tiffney, 1985; Morley, 2001).

In the southern hemisphere, the end Eocene cooling event had a negative impact on the Southern Megathermal forests of South Africa and southern South America. However, the northward drift of the Australian Plate at the time of the period of major mid-Tertiary climate decline, allowed most Australian Southern Megathermal elements to survive this event. Today, the concentration of primitive angiosperm elements in the rainforests of northeast Australia is testament to reduced Australasian climate stress during the period of mid-Tertiary global cooling. The isolation of Australia and associated continental fragments has resulted in opportunities for many primitive elements to survive in this area compared with elsewhere. The concentration of primitive angiosperms in the area from "Assam to Fiji", which Takhtajan (1969) termed his "cradle of the angiosperms", has nothing to do with angiosperm origins, but is the response of these groups to finding refugia in a tectonically active global plate-tectonic setting during the period of mid- to Late Tertiary climate decline.

1.5.2   Climate change in low latitudes

Coinciding with the end Eocene cooling event, low-latitude climates also changed substantially, becoming significantly cooler and drier. This was particularly the case in Southeast Asia, where both palynological and lithological evidence suggests that everwet climates became of very limited extent, except, perhaps in the areas of Assam and Myanmar, where Oligocene coals yield rainforest leaf floras (Awasthi and Mehrota, 1995). There may also have been small refugia in other areas, such as the southeast margin of Sundaland (Morley, unpublished). The terminal Eocene event resulted in numerous extinctions across the tropics—for example, of *Nypa* from Africa and South America (Germeraad *et al.*, 1968). However, the impact of this event was probably felt less in South America than other areas, since several taxa persisted there into the Neogene, such as mauritioid and other palm lineages. In general, equatorial floras began to take on an increasingly modern aspect during the course of the Oligocene.

A detailed pattern of climate change for the Oligocene is forthcoming from the Indonesian West Natuna Basin, which contains thick deposits of latest Eocene to Oligocene freshwater lacustrine and brackish lagoonal, followed by Neogene paralic, deposits that yield a rich palynomorph succession (Morley *et al.*, 2003). Sediments were sourced primarily from the paleo Chao Phraya/Pahang catchments (Figure 1.8) and pollen data probably reflect vegetation change on a catchment rather than local scale. The latest Eocene and earlier Oligocene are characterized by pollen assemblages rich in Gramineae and with the very limited representation of "wet climate" elements, such as pollen of peat swamp trees, suggesting a warm, but seasonally dry climate (Figure 1.9). However, the mid- and Late Oligocene contains four maxima of temperate gymnosperms, which include *Abies*, *Picea*, and *Tsuga*, associated with *Alnus*, and also with *Pinus* and Poaceae (seasonal climate elements) and some pollen of rainforest taxa, followed by acmes with rainforest elements correlating with periods of higher relative sea level. These assemblages suggest alternating cool, seasonal, followed by warm, seasonal climates. The maxima of temperate gymnosperms suggest that cool climate oscillations brought freezing temperatures to tropical mountains, and consequently relatively cool lowland climates must also have been present.

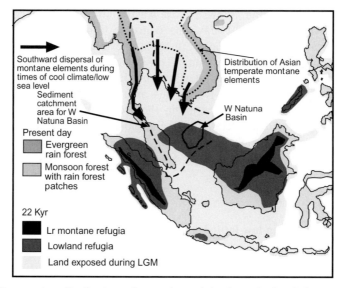

Figure 1.8. Present day distribution of megathermal (and tropical) rainforests in Southeast Asia, and probable distribution at c. 22 cal. yr BP showing positions of rainforest refugia. The shoreline at c. 22 cal. yr BP is also shown, together with the position of the catchment that fed the Malay/West Natuna Basins, and Natuna Basin, modified from Morley (2000a) using current palynological data (especially Morley *et al.*, 2004), and taking account of the mammalian data of Meijaard (2003) and Bornean generic diversity data of Slik *et al.* (2003).

Previously, the high representation of montane gymnosperms in the Southeast Asian area has been interpreted as reflecting a source from high mountains (Muller, 1966, 1972), but geological data suggest an inverse relationship between phases of mountain building and the general abundance of temperate elements (Morley, 2000b), emphasizing that most abundance variation within montane gymnosperms is climatic. The four cool climate intervals in Natuna coincide roughly, but not precisely, with the Oligocene cooler climate episode indicated from benthic isotope data (Miller *et al.*, 1987; Zachos *et al.*, 2001), and can be approximately tied to positive oxygen isotope excursions in the high-resolution oxygen isotope curve of Abreu and Anderson (1998).

Grass pollen also shows a series of maxima through the Oligocene of West Africa (Morley, 2000a, p. 140), reflecting similar drier and wetter periods, but without evidence for temperature change.

1.6   EARLY AND EARLIEST MIDDLE MIOCENE, RETURN OF GREENHOUSE CLIMATES

1.6.1   General trends

The latest Oligocene/earliest Miocene was characterized by globally warmer climates but with some cooler episodes (Zachos *et al.*, 2001). The highest global temperatures

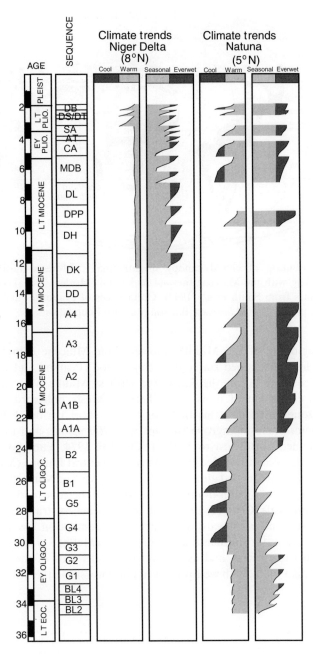

Figure 1.9. Summary of Oligocene to Pliocene climatic change in relation to sea level change, suggested from Natuna Basin palynological studies, together with Middle Miocene to Pliocene climate cycles for the Niger Delta. Sequence nomenclature follows Morley *et al.* (2003) for Oligocene to Early Miocene and Morley (2000a, Figure 7.13) for Middle Miocene to Pliocene. Timescale used is that of Berggren *et al.* (1995).

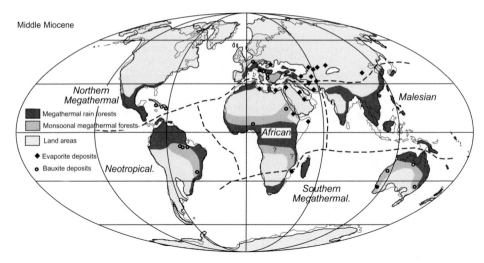

Figure 1.10. Distribution of closed canopy megathermal rainforests during the Middle Miocene, coinciding with the Miocene thermal maximum (Morley, 2000a). Paleogeography and paleocoastlines from Smith *et al.* (1994). Occurrences of evaporites and bauxites from Boucot *et al.* (in press). *Dotted lines* are floristic province boundaries.

were at the beginning of the Middle Miocene (mid-Miocene climatic optimum), although CO_2 levels remained stable over this period (Pearson and Palmer, 2000).

The renewed warming in the Early and earliest Middle Miocene once again resulted in the expansion of moist megathermal forests poleward of subtropical high-pressure zones, although this time for only a short period (Figure 1.10). In the northern hemisphere, mangrove swamps with *Rhizophora* and rainforests with *Dacrydium* extended northward to Japan (Yamanoi, 1974; Yamanoi *et al.*, 1980), *Symplocos* and *Mastixia* diversified in southern and central Europe (Mai, 1970), and megathermal elements extended along the eastern seaboard of North America (Wolfe, 1985). In South Africa, palm-dominated vegetation became widespread at two successive time intervals (Coetzee, 1978), and in southeast Australia the combination of warmer climates and northward drift once again resulted in the development of megathermal forests as far south as the Murray Basin (McPhail *et al.*, 1994). Climates in India again became moist, and as a result many elements of the Malesian flora spread to the Indian Plate, with well-preserved macrofossils in the Siwaliks (Awasthi, 1992).

1.6.2   Climate change in low latitudes

The most pronounced climate change in low latitudes occurred in the Southeast Asian region, where climates change from seasonally dry (monsoonal) to everwet at about the Oligo-Miocene boundary, 23.6 Myr (Morley *et al.*, 2003; previously estimated at about 20 Myr in Morley 1998, 2000a). This dramatic change is reflected both by pollen floras (disappearance of Gramineae pollen, dramatic increase in pollen from peat swamps) and the sudden appearance of coals in the lithological record. It is from the Oligo-Miocene boundary that the Southeast Asian region has an essentially "modern" flora ("Malesian" flora in Morley, 2000a), with only minor subsequent modifications.

This change coincides almost precisely with the time of collision of the Australian Plate with the Philippine and Asian Plates (Hall, 1996, 2002), and I suggest that the dramatic climate change may relate to this collision, which is likely to have caused major disruption to Indonesian throughflow (Morley, 2003), with the result that warm moist air from the Pacific Warm Pool probably shed its moisture content in Sundaland, rather than farther to the west, from this time onward. This suggestion has significant implications. First, it is likely that a single climatic scenario, with mainly warm, wet phases followed by cooler, locally drier phases, for the major part of the Sunda region had been in place for the subsequent 23 Myr. Second, it is likely that the El Niño oscillation, which provides the main trigger for Sundanian rainforest regeneration (Ashton et al., 1988; Curran et al., 2004) by cueing all trees to fruit at the same time (mast-fruiting), thus reducing total seed predation (Janzen, 1974, 1976), may also have been in place since this time. This may explain why so many Southeast Asian rainforest taxa depend on El Niño for their reproductive success. In this respect, it is noteworthy that Dipterocarpaceae pollen (together with Poaceae) was very common over a wide area of the Southeast Asian region just prior to the change to wetter climates, with dipterocarps presumably being seasonal forest elements (Morley, 1991, 2000a, 2003). This raises the possibility that dipterocarps were able to take advantage of the new everwet climate scenario that came about following the disruption of Indonesian throughflow by establishing a rhythmic flowering pattern along with El Niño, and become dominant elements in Southeast Asian rainforests.

Detailed trends of climate change from the Natuna Basin (Morley et al., 2003) indicate that over this period climates oscillated from cool and wet, during periods of low sea level, to warm and wet following sea level rise. Low sea level "glacial" settings were often characterized by thick coal deposits, often containing common Casuarina and Dacrydium pollen, suggesting Kerapah peats (Morley, 2000a, 2004) rather than basinal peats of the type which are currently characteristic of the coastal areas of Borneo and Sumatra today. Low sea level intervals are also characterized by common temperate elements—such as Abies, Alnus, Picea, and Tsuga—but in lower abundance than in the Oligocene. High sea level periods, on the other hand, were characterized by the expansion of basinal peats and mangroves, and the disappearance of temperate elements (Figure 1.9).

In equatorial Africa and South America, there is less evidence for a sudden change of climate at the beginning of the Miocene. African forests are thought to have gone through a period of decline at this time in a manner not seen in either Southeast Asia or South America, for many extinctions are recorded in the pollen record (Legoux, 1978; Morley, 2000a). Also, grasslands increased in representation, especially during periods of low sea level, from about 21 Myr onward, and the first evidence for burning of grasslands, from the occurrence of charred grass cuticle, is from about 15 Myr (Morley and Richards, 1993).

1.7 LATER MIDDLE MIOCENE TO PLIOCENE, GLOBAL COOLING AND RETRACTION OF MEGATHERMAL RAINFORESTS TO THE TROPICS

The phase of global cooling—starting at about 15 Myr in the Middle Miocene, and subsequently from about 2.8 Myr in the mid-Pliocene—resulted in the restriction of moist megathermal vegetation to the tropical zone, and coincided with the expansion of grasslands and deserts across much of the lower to mid-latitudes.

Megathermal elements disappeared from the mid-latitudes, with the exception of the Australasian region; here the drift of the Australian Plate into the southern hemisphere mid-latitude high pressure zone accentuated the effect of late Neogene desiccation in Australia, and rainforests became restricted to tiny pockets along the east coast, but its northerly drift maintained frost-free climates allowing mesic megathermal elements to survive in areas of everwet climate. This was not the case for New Zealand, being positioned a little farther to the south, which lost its megathermal elements in the Pliocene. The northern drift of the Australian Plate, coupled with global climate deterioration, resulted in the expansion of high pressure over northern Australia and this was primarily responsible for Pliocene development of the Javanese monsoon and the establishment of seasonal climates across Nusa Tenggara (Pribatini and Morley, 1999). With the drift of India into the northern hemisphere high-pressure zone, it also lost most of its moist vegetation.

The effect of global climate change on equatorial floras is illustrated by comparing histories over the last 15 Myr from Africa, where climatic perturbations have had a particularly deleterious effect, and Southeast Asia, where everwet climates have been much more the rule.

The record from the late Neogene of the Natuna Basin (5°N) is intermittent compared with that for the Oligocene and Early Miocene (Figure 1.9). However, some critical trends are clear. Climates have been predominantly moist for most of the late Neogene even during periods of low sea level; however, seasonal climate elements, such as Gramineae pollen, are much more persistently present than in the Early Miocene. It is likely that the Natuna area was characterized by wet climates, but seasonal, open vegetation may have been well-developed to the north in the mid- and upper region of the catchment that fed the basin.

For Africa, pollen data are available from the Niger Delta (Morley, 2000a) with macrofossil records from Tanzania (Jacobs, 2002). In the Niger Delta the pattern is of much drier "glacial" intervals throughout the late Neogene during low sea level periods with relatively little evidence for temperature change until about 2.7 Myr, corresponding closely to the time of cooler climates in Borneo (Figure 1.8). The Late Quaternary pollen diagram from offshore the Niger Delta by Dupont and Wienelt (1996), which spans the last interglacial/glacial cycle, shows that rainforests covered the delta during the Holocene and last interglacial, and savanna during the last glacial, and provides an excellent analog for climate changes in the delta region over the remainder of the Quaternary as well as the late Neogene back to at least 13 Myr. Over the late Neogene many oscillations of grass and rainforest pollen can be seen in sections studied from Niger Delta well sections (e.g., Morley, 2000a, Figures 7.11,

7.13); during periods of low sea level, climates were substantially drier, with wide-spread grasslands, whereas rainforest elements expanded during high sea level periods. Grasslands were associated with burning, for charred Gramineae cuticular debris, suggesting widespread savanna fires, is a persistent feature of low sea level periods at least since the beginning of the Late Miocene (Morley and Richards, 1993). Both grass pollen and charred cuticle become more common after about 7 Myr, suggesting that grasslands and burning were more widespread after that time, a date that fits well with the retraction of forest and expansion of C4 grasses in other areas (e.g., in Pakistan, Quade *et al.*, 1989; and Siwaliks, Nepal, Hoorne, 2000). Grasslands were probably less extensive during the Early Pliocene, but expanded further after about 3.0 Myr and into the Quaternary.

Leaf floras recently reported from the Late Miocene of Tanzania indicate different degrees of drying (Jacobs, 1999, 2002). One leaf flora (Waril), dated at 9–10 Myr, suggests an open vegetation and a climate with a pronounced dry season, whereas a leaf flora dated 6.6 Myr (Kapturo) suggests a woodland or dry forest setting. Jacobs suggests from these data that there was not a unidirectional change from forested to open environments in the Kenya rift valley during the Miocene (which is often proposed to explain the evolution of hominids in Africa). As with the Niger Delta area, it is more likely that a succession of alternately wetter and drier phases occurred, but with an overall trend toward cooler and drier climates, in line with global models. The Neogene vegetational history of Amazonia has recently been reviewed by Hooghiemstra and Van der Hammen (1998), and Van der Hammen and Hooghiemstra (2000). They emphasize that temperature oscillations took place over the entire Neogene, with cooler phases interrupting a climate that was mostly warmer than today. During the Pliocene the climate seems to have been generally cooler than during the Miocene, and between 3.0 and 2.5 Myr a strong cooling produced the first glacial period, closely paralleling the pattern seen in Southeast Asia and West Africa.

The likelihood that grass pollen maxima indicate the successive expansion of savanna has been downplayed by Hoorn (1994), Hooghiemstra and Van der Hammen (1998) and Bush (2002), who emphasize that grass pollen may be sourced from a variety of vegetation types, including swamp forest (where grasses are often found as a component of floating vegetation communities). However, in cases where grass pollen acmes occur in association with charred grass cuticle the likelihood of a derivation from more seasonal climate sources is much greater (Morley and Richards, 1993). For Amazonia such assemblages are probably derived from "cerrado" (wooded grassland) or semi-deciduous woodland. Maxima of charred grass cuticle associated with Gram-ineae pollen maxima have been recorded from the Late Pliocene of the Amazon Fan (Richards, 2000; Richards and Lowe, 2003), suggesting that during Late Pliocene times, Amazon climates were substantially drier than either the Miocene/Early Pliocene or the Pleistocene. This has implications regarding the long-term history of Amazonian vegetation and the "refuge" theory. Whereas the Amazon may have existed under continuous forest cover during the Pleistocene (Colinvaux *et al.*, 2000), this may not have been the case during the Late Pliocene, during which time fragmentation of Amazonian rainforests may have been a real possibility.

The climate oscillations discussed here from West Africa and Southeast Asia based on petroleum exploration data must be considered generalized compared with the high-resolution patterns seen in the later Quaternary. A study of a deep marine Pliocene profile from ODP 658, offshore northwest Africa by Leroy and Dupont (1994) shows high-resolution oscillations of grassland and desert elements and emphasizes that Milankovich scale climate changes, driven by astronomical cycles, were the rule just as in the Quaternary. The Pliocene section of the Sabana de Bogotá core from Colombia shows similar scale oscillations (Hooghiemstra, 1984; Hooghiemstra and Ran, 1994). It is therefore not unrealistic to suggest that astronomical cycles were also the driving force behind cyclical climate change throughout the Miocene and Oligocene.

1.8   TRENDS IN RAINFOREST DIVERSITY BASED ON THE PALYNOLOGICAL RECORD

Obtaining meaningful data regarding palynomorph diversity, which can be interpreted in terms of species diversity of vegetation, is fraught with difficulties, and so is rarely attempted (Birks and Line, 1990). Differences in depositional environment, as well as taphonomic factors, have a significant effect on such estimates, making "number of pollen types", or pollen diversity indices difficult to interpret. Also, individual analysts may have different concepts of what constitutes a "pollen type" (and how to deal with "undetermined" pollen) with the result that data from different analysts sometimes cannot be directly compared.

Trends in taxon richness from the Paleocene into the Eocene have been demonstrated by Rull (1999) in Venezuela and by Jaramillo and Dilcher (2000) in Colombia. Rull (1999) calculated diversity indices (Shannon–Weaver index) and pollen richness (number of palynomorph taxa per sample analyzed), and showed that there is a clear trend to increasing species richness into the Paleocene–Eocene thermal maximum from about 15 pollen types per sample in the Paleocene to about 20 in the Middle Eocene (in a count of 200 grains). Jaramillo and Dilcher (2000) compared diversities in the Late Paleocene and Middle Eocene principally using rarefaction (Raup, 1975) and the "range through" method of Boltovskoy (1988) to estimate standing diversity, with both methods indicating an increase in diversity from the Late Paleocene into the Middle Eocene. Late Paleocene samples yielded an average of 28 types and Middle Eocene samples averaged 54 pollen types, while estimates of standing diversity averaged 38 for the Late Paleocene and 73 for the Middle Eocene. Eocene taxon richness in both Venezuela and Colombia is low compared with palynomorph richness in modern low-latitude assemblages, where 60–70 pollen types in a count of 250 are more the rule.

Palynomorph assemblages from the Middle Eocene from the southern margin of Sundaland (Java) also show higher diversities than those from South America with typically 70–80 types in counts of 250 (Lelono, 2000), and standing diversities of 115–140 depending on the calculation method (Morley, unpublished). The diversity of the

Southeast Asian flora increased following collision of the Indian and Asian Plates and the mixing of Indian and Southeast Asian elements, resulting in the formation of the Proto-Indian Flora (Morley, 2000a, b). A parallel diversity increase within the palyno-flora of the Malay Basin (Malesian Flora) is noted in the Middle Miocene (Jaizan Md Jais, 1997; Morley and Jaizan Md Jais, new data) following collision of the Australian and Asian Plates. This raises the possibility that floristic interchange following plate collision may be a general feature in promoting high levels of species diversity in tropical floras.

The trend of gradually increasing floristic diversity through the Paleocene and Eocene in South America and Africa (Figure 1.3) has previously been brought to attention by Morley (2000a). This trend comes to an abrupt halt at the end of the Eocene. Following the end Eocene cooling event, low-latitude floras show a sudden reduction in diversity in South America and Africa and also Southeast Asia (Morley, unpublished), coinciding with cooler and drier low-latitude climates.

A large Mio-Pliocene database is currently being generated from the Makassar Straits, east of Borneo, by analysis of boreholes on the continental slope and basin floor offshore the Mahakam Delta, all in very uniform, deep marine (1,000 m+) depositional settings. The Mahakam River catchment occupies a rainforest refuge area (Morley *et al.*, 2004). Preliminary results show very uniform numbers of pollen types per sample from the Middle Miocene (typically 60 types in a count of 300) up to the mid-Pliocene (typically 70–80 types per sample), with the possibility of a minor reduction in numbers per sample in the Late Pliocene. Pollen floras yield about 170–200 determinable types per stratigraphic section of 60–100 samples (Morley, new data). Trends closely parallel those seen in Miocene Mahakam Delta plain sediments (Morley, 2000a, Figure 7.13). The conclusion from these studies is that Bornean rainforests slowly increase in diversity over time from the Early Miocene to mid-Pliocene; data are currently insufficient to confidently demonstrate any real diversity reduction after 2.8 Myr.

For Amazonia, however, Van der Hammen and Hooghiemstra (2000) note that Hoorn (1994) found 280 pollen types in the Rio Caquetá area in Miocene river valley sediments, but note that Holocene river sediments from the same area yield only 140 pollen types, despite the present day vegetation in the area being very diverse, with 140 species per 0.1 ha (Ureggo, 1997). On the basis of these data they suggest that present day Amazonian vegetation is less diverse than that of the Miocene, a proposition also discussed by Flenley (2005) on the basis of the same data. Many questions need to be answered before reaching such conclusions: (1) Were the depositional settings directly comparable, and the same facies/subfacies represented? (2) Could taphonomic factors be in play to account for these differences? (3) Did the Miocene and Holocene river systems have the same vegetation types growing in the upper catchment? (4) How did the Holocene sediments recruit pollen from the surrounding vegetation, and were the same taxonomic concepts applied to both Miocene and Holocene sediments? Experi-ence from working in fluvial sediments in Southeast Asia suggests that the number of pollen types preserved may vary considerably from one depositional locality to another, and that a large database from different depositional facies is needed to assess the richness of the pollen flora on a regional basis.

Palynomorph richness data from the Niger Delta (Morley, 2000a, Figure 7.13) based on analyses of petroleum exploration boreholes suggests that floristic diversity underwent several sudden reductions following phases of sea level fall and expansion of seasonal climate vegetation, especially at the beginning of the Late Miocene (about 11.7 Myr using the timescale of Berggren *et al.*, 1995), at about 7.0 Myr and following 2.8 Myr. There were also numerous extinctions during the Early Miocene as noted above. I suggest that the present day low diversity of African rainforest flora is a result of the successive expansion of seasonally dry climates in a manner not seen in either Southeast Asia or South America. These dry climate episodes occurred from the Early Miocene onward and were not just a Quaternary phenomenon. Successive dry climate episodes go some way to help explain why present day African equatorial flora is more species-poor than elsewhere.

1.9   SCENARIO FOR RAINFOREST EVOLUTION AND DIVERSIFICATION

The evolution and diversification of megathermal rainforests has been dependent on, and proceeded parallel with, a succession of geological and climatic and dispersal events, controlled largely by plate-tectonic and astronomical processes, in parallel with evolutionary pressures for plants to reproduce and colonize all available land space. These events have occurred in a unique time sequence. The result is that today each geographically separated rainforest area contains its own association of species, largely descended from ancestors that were established perhaps over 70 Myr ago, and subsequently became modified, and diversified, so as to occupy the available niches within each region.

From the perspective of Quaternary studies in relation to the explanation of rainforest diversity, processes of evolution of rainforest taxa have mostly focused on the "refuge hypothesis", which maintains that the successive isolation of populations in relation to the expansion and contraction of forested areas following Quaternary Milankovich cycle driven climate changes acted as a "species pump", triggering speciation. Such an approach has paid little attention to the antiquity of rainforest species as suggested from the Tertiary fossil pollen record, and now being substantiated by molecular studies, or to the high Tertiary floristic diversities suggested by pollen diversity data. For instance, a molecular analysis of one of the fastest-evolving rainforest taxa, the species-rich Neotropical genus *Inga* (Fabaceae), shows that its radiation is thought to have been promoted by the later phases of Andean orogeny and the bridging of the Panama Isthmus, perhaps coupled with climatic fluctuations, (Richardson *et al.*, 2001), but as noted by Bermingham and Dick (2001) provides little support for the idea that Pleistocene ice ages played a grandiose part in generating tropical species diversity.

This chapter attempts to show that—to understand the diversity of tropical rainforests—their development must be viewed on a much longer timescale. The climate changes which characterize the Quaternary were also taking place over much of the Tertiary period, the only difference between Quaternary and later Tertiary climate changes being one of degree of change, since "glacial" climates from the equatorial zone were clearly cooler from 2.8 Myr onward, and the vertical vegetational migration on tropical mountains over this period was likely to be more pronounced from this time onward.

From the Early Miocene to the mid-Pliocene, rainforest diversity in Southeast Asia (based on pollen-type richness) has gradually increased. A slight reduction in pollen-type richness after 2.8 Myr is not reliable in reflecting a diversity reduction in rainforest flora. Data from the Niger Delta regarding pollen-type richness suggests that West African flora has reduced significantly in diversity over the same period, with taxon losses throughout the Miocene, and with significant reductions at 11.7 and 7.0 Myr, and a particularly sharp reduction at about 2.8 Myr in the mid-Pliocene.

Palynological data from the Southeast Asian Neogene also demonstrates that the diversity of Southeast Asian flora may well have become accentuated as a result of the successive reformation of lowland vegetation on the continental shelves following periods of sea level fall over a period of at least 20 Myr (Morley, 2000a). Over this period major areas of the region have experienced everwet climates during both high and low sea level periods, with wet/dry oscillations being restricted to the Oligocene, and to some degree the Late Miocene.

In West Africa the pattern was of the alternating expansion and contraction of rainforests in relation to more open vegetation with grasslands over a period of some 30 Myr, with dry episodes—which included burning of savanna—becoming more pronounced, particularly after 7 Myr, and then again in the Late Pliocene. The depauperate nature of African rainforest flora compared with other areas is thought to be due to the decimating effect of these dry climate events, not on a Quaternary timescale, but over some 20 Myr, as emphasized by the higher number of extinctions seen in the West African Miocene pollen record than in other areas.

The scene for evolution of rainforest species is thus of gradual differentiation over a long time period with different forcing mechanisms inhibiting dispersal and isolating populations. The high diversities seen in rainforest refugia—or hot spots—are likely to relate to areas of long-term continuity of moist climates within those areas rather than to allopatric speciation driven by habitat fragmentation. The highest diversities, however, are seen where climatic stability coincides with areas which have experienced phases of orogeny and especially of plate collision, as seen from pollen data for Java following the Middle/Early Eocene collision of the Indian and Asian Plates, and for the Middle Miocene of the Sunda region following the collision of the Australian and Asian Plates. From the neotropics, molecular and biogeographical data suggest that high diversities may relate to the uplift of the Andes in the Miocene and the formation of the Panamanian Isthmus in the Pliocene.

Low equatorial floristic diversities may follow periods of cool, and particularly dry climates, as was the case following the end Eocene cooling event, when in Southeast Asia cool climate oscillations brought freezing temperatures to tropical mountains with corresponding seasonally dry lowland climates. Similarly, for equatorial Africa increased seasonality of climate from the Early Miocene onward accounts for the current depauperate nature of African rainforest flora.

1.10   REFERENCES

Abreu, V. S. and Anderson, J. B. (1998) Glacial eustacy during the Cenozoic: Sequence stratigraphic implications. *AAPG Bulletin* **82**, 1385–1400.

Anderson, J. A. R. and Muller, J. (1975) Palynological study of a Holocene peat and a Miocene coal deposit from N.W. Borneo. *Review of Palaeobotany and Palynology* **19**, 291–351.

Ashton, P., Givnish, T. and Appanah, S. (1988) Staggered flowering in the Dipterocarpaceae: New insights into floral induction and the evolution of mast fruiting in the aseasonal tropics. *American Naturalist* **132**, 44–66.

Awasthi, N. (1992) Changing patterns of vegetation succession through Siwalik succession. *Palaeobotanist* **40**, 312–327.

Awasthi, N. and Mehrota, R. C. (1995) Oligocene flora from Makum Coalfield, Assam, India. *Palaeobotanist* **44**, 157–188.

Barron, E. J. and Washington, W. M. (1985) Warm Cretaceous climates: High atmospheric CO_2 as a plausible mechanism. In: Sundquist *et al.* (eds.), *The Carbon Cycle and Atmospheric CO_2: Natural Variations, Archaean to the Present* (pp. 546–553). American Geophysical Union, Washington, D.C.

Bartek, L. R., Vail, P. R., Anderson, J. B., Emmet, P. A., and Wu, S. (1991) Effect of Cenozoic ice sheet fluctuations in Antarctica on the stratigraphic signature of the Neogene. *Journal of Geophysical Research* **96**, 6753–6778.

Bennett, K. D. (2004) Continuing the debate on the role of Quaternary environmental change on macroevolution. *Philosophical Transactions of the Royal Society B* **359**, 295–303.

Berggren, W. A., Kent, V. D., Swisher III, C. C., and Aubrey, M. -P. (1995) A revised Cenozoic geochronology and chronostratigraphy. In: *Time Scales and Global Stratigraphic Correlation* (SEPM Special Publication 54, pp. 129–212). Society of Economic Palaeontologists and Mineralogists, Tulsa, OK.

Bermingham, E. and Dick, C. (2001) The *Inga*, newcomer or museum antiquity. *Science* **293**, 2214–2216.

Birks, H. J. B. and Line, J. M. (1990) The use of rarefaction analysis for estimating palynological richness from Quaternary pollen-analytical data. *The Holocene* **2**, 1–10.

Boltovskoy, D. (1988) The range-through method and first–last appearance data in palaeontological surveys. *Journal of Palaeontology* **62**, 157–159.

Boucot, A. J., Chen Xu, and Scotese, C. R. with contributions by Morley, R. J. (2004, in press). *Preliminary Compilation of Cambrian through Miocene Climatically Sensitive Deposits* (SEPM Special Publication Series). Society of Economic Palaeontologists and Mineralogists, Tulsa, OK.

Burnham, R. J. and Graham, A. (1999) The history of Neotropical vegetation: New developments and status. *Annals of the Missouri Botanic Garden* **86**, 546–589.

Bush, M. B. (2002) On the interpretation of fossil Poaceae pollen in the lowland humid neotropics. *Palaeogeography, Palaeoclimatology, Palaeoecology* **177**, 5–17.

Chandler, M. E. J. (1964) *The Lower Tertiary Floras of Southern England: A Summary and Survey of Findings in the Light of Recent Botanical Observations* (151 pp.). British Museum, London.

Chesters, K. I. M. (1955) Some plant remains from the Upper Cretaceous and Tertiary of West Africa. *Annals and Magazine of Natural History* **12**, 498–504.

Christophel, D. C. (1994) The Early Tertiary macrofloras of continental Australia. In: R.S. Hill (ed.), *History of Australian Vegetation, Cretaceous to Recent* (pp. 262–275). Cambridge University Press, Cambridge, U.K.

Coetzee, J. A. (1978) Climatic and biological changes in southwestern Africa during the late Cainozoic. *Palaeoecology of Africa* **10**, 13–29.

Colinvaux, P. A., De Oliviera, P. E., and Bush, M. B. (2000) Amazonian and Neotropical plant communities on glacial time-scales: The failure of the aridity and refuge hypotheses. *Quaternary Science Reviews* **19**, 141–169.

Collinson, M. E. (1983) *Fossil Plants of the London Clay* (Palaeontological Association Field Guides to Fossils No 1, 121 pp.). Palaeontological Association, London.

Collinson, M. E. (1988) The special significance of the Middle Eocene fruit and seed flora from Messel, Western Germany. *Courier Forschunginst. Senkenberg* **107**, 187–197.

Crane, P. R. (1987) Vegetational consequences of the angiosperm diversification. In: E. M. Friis, W. G. Chaloner, and P. R. Crane (eds.), *The Origins of Angiosperms and Their Biological Consequences* (pp. 107–144). Cambridge University Press, Cambridge, U.K.

Crane, P. R. and Lidgard, S. (1990) Angiosperm radiation and patterns of Cretaceous palynological diversity. In: P. D. Taylor and G. P. Larwood (eds.), *Major Evolutionary Radiations* (Systematics Association Special Volume 42, pp. 377–407). Systematics Association, London.

Crane, P. R., Friis, E. M., and Pedersen, K. R. (1995) The origin and early diversification of angiosperms. *Nature* **374**, 27–34.

Curran, L. M., Trigg, S., McDonald, A. K., Astiani, D., Hardiono, Y. M., Siregar, P., Caniago, I., and Kasischke, E. (2004) Forest loss in protected areas of Indonesian Borneo. *Science* **303**, 1000–1003.

Daley, B. (1972) Some problems concerning the Early Tertiary climate of southern Britain. *Palaeogeography, Palaeoclimatology, Palaeoecology* **11**, 177–190.

Davis, C. C., Bell, C. D., Matthews, S., and Donaghue M. J. (2002) Laurasian migration explains Gondwanan disjunctions: Evidence from Malpighiaceae. *Proceedings of the National Academy of Sciences U.S.A.* **99**, 6933–6937.

Davis, C. C., Webb, C. O., Wurdack, K. J., Jaramillo, C. A., and Donaghue, M. J. (2005) Explosive radiation of Malpighiales supports a mid-Cretaceous origin of modern tropical rain forests. *The American Naturalist* **165**, E36–E65.

Dettmann. M. E. (1994) Cretaceous vegetation: The microfossil record. In: R. S. Hill (ed.), *History of Australian Vegetation: Cretaceous to Recent* (pp. 143–170). Cambridge University Press, Cambridge, U.K.

Dick, C. W., Abdul-Salim, K., and Bermingham, E. (2003) Molecular systematics reveals cryptic Tertiary diversification of a widespread tropical rainforest tree. *American Naturalist* **160**, 691–703.

Doyle, J. A. and Donaghue, M. J. (1987) The origin of angiosperms: A cladistic approach. In: E. M. Friis, W. G. Chaloner, and P. R. Crane (eds.), *Introduction to Angiosperms: The Origins of Angiosperms and Their Biological Consequences* (pp. 17–49). Cambridge University Press, Cambridge, U.K.

Doyle, J. A. and Endress, P. K. (1997) Morphological phylogenetic analysis of basal angiosperms: Comparison and combination with molecular data. *International Journal of Plant Science* **161**, S121–S153.

Doyle, J. A. and Le Thomas, A. (1997) Phylogeny and geographic history of Annonaceae. *Geographie Physique et Quaternaire* **51**, 353–351.

Duperon-Laudoueneix, M. (1991) Importance of fossil woods (conifers and angiosperms) discovered in continental Mesozoic sediments of Northern Equatorial Africa. *Journal of African Earth Sciences* **12**, 391–396.

Dupont, L. M. and Wienelt, M. (1996) Vegetation history of the savanna corridor between the Guinean and the Congolian rain forest during the last 150,000 years. *Veget. Hist. Archaeobot.* **5**, 273–292.

Flenley, J. R. (1979) *The Equatorial Rain Forest: A Geological History* (162 pp.). Butterworths, London.

Flenley, J. R. (2005) Palynological richness and the tropical rain forest. In: E. Bermingham, C. Dick, and C. Moritz (eds.), *Tropical Rainforests: Past, Present, and Future* (pp. 73–77). Chicago University Press, Chicago.

Frederiksen, N. O. (1994) Paleocene floral diversities and turnover events in eastern North America and their relation to diversity models. *Review of Palaeobotany and Palynology* **82**, 225–238.

Germeraad, J. H., Hopping, C. A., and Muller, J. (1968) Palynology of Tertiary sediments from tropical areas. *Review of Palaeobotany and Palynology* **6**, 189–348.

Givnish, T. J., Evans, T. M., Zihra, M. L., Patterson, T. B., Berry, P. E., and Systma, K.J. (2000) Molecular evolution, adaptive radiation, and geographic diversification in the amphi-Atlantic family Rapateaceae: Evidence from ndhF sequences and morphology. *Evolution* **54**, 1915–1937.

Greenwood, D. R., 1994. Palaeobotanical evidence for Tertiary climates. In: R. S. Hill (ed.), *History of Australian Vegetation: Cretaceous to Recent* (pp. 44–59). Cambridge University Press, Cambridge, U.K.

Grime, J. B. (1979) *Plant Strategies and Vegetation Process*. John Wiley & Sons, Chichester, U.K.

Haberle, S. (1997) Upper Quaternary vegetation and climate history of the Amazon Basin: Correlating marine and terrestrial records. *Proceedings of the Ocean Drilling Program, Scientific Results* **155**, 381–396.

Haberle, S. G and Malin, M. (1998) Late Quaternary vegetation and climate change in the Amazon Basin based on a 50000-year pollen record from the Amazon Fan, ODP Site 932. *Quaternary Research* **51**, 27–38.

Haffer, J. (1969) Speciation in Amazonian birds. *Science* **165**, 131–137.

Hall, R. (1996) Reconstructing Cenozoic SE Asia. In: R. Hall and D. J. Blundell (eds.), *Tectonic Evolution of Southeast Asia* (Geological Society Special Publication 106, pp. 152–184). Geological Society, London.

Hall, R. (2002) Cenozoic geological and plate tectonic evolution of SE Asia and the SW Pacific: Computer-based reconstructions, model and animations. *Journal of Asian Earth Sciences* **20**, 353–431.

Hallam, A. (1992) *Phanerozoic Sea Level Changes* (265 pp.). Colombia University Press, New York.

Hallam, A. (1994) *An Outline of Phanerozoic Biogeography* (246 pp.). Oxford University Press, New York.

Herngreen, G. F. W. and Duenas-Jimenez, H. (1990) Dating of the Cretaceous Une Formation, Colombia, and the relationship with the Albian–Cenomanian African–South American microfloral province. *Review of Palaeobotany and Palynology* **66**, 345–359.

Herngreen, G. F. W., Kedves, M., Rovnina, L. V., and Smirnova, S. B. (1996) Cretaceous palynological provinces: A review. In: J. Jansonius and D. C. McGregor (eds.), *Palynology: Principles and Applications* (Vol. 3, pp. 1157–1188). American Association of Stratigraphic Palynologists Foundation, Houston, TX.

Hickey, L. J. and Doyle, J. A. (1977) Early Cretaceous fossil evidence for angiosperm evolution. *The Botanical Review* **43**, 1–183.

Hooghiemstra, H. (1984) *Vegetational and Climatic History of the High Plain of Bogotá, Colombia: A Continuous Record of the Last 3.5 Million Years* (Dissertationes Botanicae 79, 368 pp.). J. Cramer, Vaduz, Germany.

Hooghiemstra, H. and Ran, E. T. H. (1994) Late Pliocene–Pleistocene high resolution pollen sequence of Colombia: An overview of climatic change. *Quaternary International* **21**, 63–80.

Hooghiemstra, H. and Van der Hammen, T. (1998) Neogene and Quaternary development of the Neotropical rain forest. *Earth Science Reviews* **44**, 147–183.

Hoorn, C. (1994) An environmental reconstruction of the palaeo-Amazon River system (Middle–Late Miocene, NW Amazonia). *Palaeogeography, Palaeoclimatology, Palaeoecology* **112**, 187–238.

Hoorn, C. (1997) Palynology of the Pleistocene glacial/interglacial cycles of the Amazon Fan (holes 940A, 944A and 946A). *Proceedings of the Ocean Drilling Program, Initial Results* **155**, 397–407.

Hoorn, C. (2000) Palynological evidence for vegetation development and climatic change in the Sub-Himalayan Zone (Neogene, Central Nepal). *Palaeogeography, Palaeoclimatology, Palaeoecology* **163**, 133–161.

Hudson, J. D. and Anderson, T. F. (1989) Ocean temperatures and isotopic compositions through time. *Transactions Royal Society of Edinburgh, Earth Sciences* **80**, 183–192.

Jacobs, B. F. (1999) Estimation of rainfall variables from leaf characters in tropical Africa. *Palaeogeography, Palaeoclimatology, Palaeoecology* **145**, 231–250.

Jacobs, B. F. (2002) Estimation of low-latitude climates using fossil angiosperm leaves: Examples from the Miocene Tugen Hills, Kenya. *Palaeobiology* **28**, 399–421.

Jacobs, B. F. and Heerenden, P. S. (2004) Eocene dry climate and woodland vegetation reconstructed from fossil leaves from northern Tanzania. *Palaeogeography, Palaeoclimatology, Palaeoecology* **213**, 115–123.

Jaizan Md Jais (1997) Oligocene to Pliocene quantitative stratigraphic palynology of the southern Malay Basin, offshore Malaysia. Unpublished Ph.D. thesis, University of Sheffield (321 pp. + 98 plates).

Janzen, D. (1974) Tropical blackwater rivers, animals, and mast fruiting by the Dipterocarpaceae. *Biotropica* **6**, 69–103.

Janzen, D. (1976) Why bamboos wait so long to flower. *Annual Review of Ecology and Systematics* **7**, 347–391.

Jaramillo, C. and Dilcher, D. L. (2000) Microfloral diversity patterns of the last Paleocene–Eocene interval in Colombia, northern South America. *Geology* **28**, 815–818.

Johnson, K. R. and Ellis, B. (2002) A tropical rain forest in Colorado 1.4 million years after the Cretaceous–Tertiary boundary. *Science* **296**, 2379–2383.

Kubitski, K. (2005) Major evolutionary advances in the history of green plants. *Acta Phytotaxonomica et Geobotanica* **56**, 1–10.

Kutschera, U. and Niklas, K. J. (2004) The modern theory of biological evolution: An expanded synthesis. *Naturwissenschaften*, online publication. Visit *www.uni-kassel.de/fb19/plantphysiology/niklas.pdf*

Legoux, O. (1978) Quelques espèces de pollen caractéristiques du Néogène du Nigeria. *Bulletin Centre Recherche Exploration-Production Elf-Aquitaine* **2**, 265–317.

Lelono, E. B. (2000) Palynological study of the Eocene Nanggulan Formation, Central Java, Indonesia. Ph.D. thesis, Royal Holloway, University of London (413 pp.).

Leroy, S. and Dupont, L. (1994) Development of vegetation and continental aridity in northwestern Africa during the Late Pliocene: The pollen record of ODP Site 658. *Palaeogeography, Palaeoclimatology, Palaeoecology* **109**, 295–316.

McPhail, M. K., Alley, N. F., Truswell, E. M. and Sluiter, R. K. (1994) Early Tertiary vegetation: Evidence from spores and pollen. In: R. S. Hill (ed.), *History of Australian vegetation: Cretaceous to Recent* (pp. 262–275). Cambridge University Press, Cambridge, U.K.

Mai, D. H. (1970) Subtropische Elemente im europaischen Tertiare. *Palaeontol. Abh. Abt. Palaobot.* **3**, 441–503.

Mai, D. H. (1991) Palaeofloristic changes in Europe and the confirmation of the arctotertiary palaeofloral geofloral concept. *Review of Palaeobotany and Palynology* **68**, 28–36.

Manchester, S. R. (1994) Fruits and seeds of the Middle Eocene Nut Beds Flora, Clarno Formation, Oregon. *Palaeontographica Americana* **58**, 1–205.

Manchester, S. R. (1999) Biogeographical relationships of North American Tertiary floras. *Annals of the Missouri Botanic Garden* **86**, 472–522.

Mebradu, S., Imnanobe, J., and Kpandei, L.Z. (1986) Palynostratigraphy of the Ahoko sediments from the Nupe Basin, N.W. Nigeria. *Review of Palaeobotany and Palynology* **48**, 303–310.

Meijaard, E. (2003) Mammals of south-east Asian islands and their Late Pleistocene environments. *Journal of Biogeography* **30**, 1245–1257.

Martin, H. A. (1992) The Tertiary of southeastern Australia: Was it tropical? *Palaeobotanist* **39**, 270–280.

Miller, K. G., Fairbanks, R. G., and Mountain, G. S. (1987) Tertiary oxygen isotope synthesis, sea level history and continental margin erosion. *Paleoceanography* **2**, 1–19.

Miller, K. G., Sugarman, P. J., Browning, J. V., Kominz, M. A., Ollson, R. K., Feigenssen, M. D., and Hernandez, J. C. (2004) Upper Cretaceous sequences and sea-level history. *New Jersey Coastal Plain Geological Society of America Bull.* **32**, 368–393.

Moles, A. T., Ackerly, D. D., Webb, C. O., Tweddle, J. C., Dickie, J. B., and Westoby, M. (2005) A brief history of seed size. *Science* **307**, 576–580.

Monteillet, J. and Lappartient, J.-R. (1981) Fruits et graines du Crétace supérieur des Carrières de Paki (Senegal). *Review of Palaeobotany and Palynology* **34**, 331–344.

Morley, R. J. (1991) Tertiary stratigraphic palynology in Southeast Asia: Current status and new directions. *Geol. Soc. Malaysia. Bull.* **28**, 1–36.

Morley, R. J. (1998) Palynological evidence for Tertiary plant dispersals in the Southeast Asian region in relation to plate tectonics and climate. In: R. Hall and J. D. Holloway (eds.), *Biogeography and Geological Evolution of SE Asia* (pp. 211–234). Backhuys, Leiden, The Netherlands.

Morley, R. J. (2000a) *Geological Evolution of Tropical Rain Forests* (362 pp.). John Wiley & Sons, London.

Morley, R. J. (2000b) The Tertiary history of the Malesian Flora. In: L. G. Saw *et al.* (eds.), *Proceedings of the IVth Flora Malesiana Symposium, Kuala Lumpur* (pp. 197–210). Forest Research Institute, Kepong, Malaysia.

Morley, R. J. (2001) Why are there so many primitive angiosperms in the rain forests of Asia–Australia? In: I. Metcalfe, J. M. B. Smith, M. Morwood, and I. Davidson (eds.), *Floral and Faunal Migrations and Evolution in SE Asia–Australia* (pp. 185–200). Swetz & Zeitliner, Lisse, The Netherlands.

Morley, R. J. (2003) Interplate dispersal routes for megathermal angiosperms. *Perspectives in Plant Ecology, Evolution and Systematics* **6**, 5–20.

Morley, R. J. (in press) Ecology of Tertiary coals in SE Asia. In: T. A. Moore (ed.), *Coal Geology of Indonesia: From Peat Formation to Oil Generation* (Advances in Sedimentology Series). Elsevier, North-Holland, The Netherlands.

Morley, R. J. and Dick, C. W. (2003) Missing fossils, molecular clocks and the origin of the Melastomataceae. *American Journal of Botany* **90**, 1638–1644.

Morley, R. J. and Richards, K. (1993) Gramineae cuticle: A key indicator of late Cenozoic climatic change in the Niger Delta. *Review of Palaeobotany and Palynology* **77**, 119–127.

Morley, R. J., Morley, H. P., and Restrepo-Pace, P. (2003) Unravelling the tectonically controlled stratigraphy of the West Natuna Basin by means of palaeo-derived Mid Tertiary climate changes. *29th IPA Proceedings* (Vol. 1).

Morley, R. J., Morley, H. P., Wonders, A. A., Sukarno, H. W., and Van Der Kaars, S. (2004) Biostratigraphy of Modern (Holocene and Late Pleistocene) Sediment cores from Makassar Straits. In: *Deepwater and Frontier Exploration in Asia & Australasia Proceedings, Jakarta, December 2004.*

Muller, J. (1966) Montane pollen from the Tertiary of N.W. Borneo. *Blumea* **14**, 231–235.

Muller, J. (1972) Palynological evidence for change in geomorphology, climate and vegetation in the Mio-Pliocene of Malesia. In: P. S. Ashton and M. Ashton (eds.), *The Quaternary Era in Malesia* (Geogr. Dept, University of Hull, Misc. Ser 13, pp. 6–34). University of Hull, Hull, U.K.

Muller, J., De di Giacomo, E., and Van Erve, A.W. (1987) A palynological zonation for the Cretaceous, Tertiary and Quaternary of northern South America. *American Association of Stratigraphic Palynologists, Contributions Series* **16**, 7–76.

Nicklas, K. J., Tiffney, B. H., and Knoll, A. (1980) Apparent changes in the diversity of fossil plants. *Evolutionary Biology* **12**, 1–89.

Parrish, J. T., Ziegler, A. M. and Scotese, C. R. (1982) Rainfall patterns and the distribution of coals and evaporites in the Mesozoic and Cenozoic. *Palaeogeography, Palaeoclimatology, Palaeoecology* **40**, 67–101.

Pearson, P. H. and Palmer, M. R. (2000) Atmospheric carbon dioxide concentrations over the past 60 million years. *Nature* **406**, 695–699.

Pearson, P. H., Ditchfield, P. W., Singano, J., Harcourt-Brown, J. C., Nicholas, C. J., Olsson, K. R., Shackleton, N. J., and Hall, M. A. (2001) Warm tropical sea surface temperatures in the Late Cretaceous and Eocene epochs. *Nature* **413**, 481–487.

Pole, M. S. and McPhail, M. K. (1996) Eocene *Nypa* from Regatta Point, Tasmania. *Review of Palaeobotany and Palynology* **92**, 55–67.

Posamentier, H. W. and Allen, G. P. (1999) *Siliciclastic Sequence Stratigraphy: Concepts and Applications* (SEPM Special Publication). Society of Economic Palaeontologists and Mineralogists, Tulsa, OK.

Prance, G.T. (1982) *Biological Diversification in the Tropics*. Columbia University Press, New York.

Pribatini, H. and Morley, R. J. (1999) Palynology of the Pliocene Kalibiuk and Kaliglagah Formations, near Bumiayu, Central Java. In: *Tectonics and Sedimentation of Indonesia, Indonesian Sedimentologists Forum* (Special Publication No 1, p, 53, Abstract).

Quade, J. J., Cerling, T. E. and Bowman, J. R. (1989) Development of Asian monsoon revealed by marked ecological shift during the latest Miocene in northern Pakistan. *Nature* **342**, 163–166.

Raup, D.M. (1975) Taxonomic diversity estimation using rarefaction. *Paleobiology* **1**, 333–342.

Reid, E. M. and Chandler, M. E. J. (1933) *The Flora of the London Clay* (561 pp.). British Museum (Natural History), London.

Reyment, R. A. (1965) *Aspects of the Geology of Nigeria* (145 pp.). University Press, Ibadan, Nigeria.

Richards, K. (2000) Grass pollen and charred cuticle in the Amazon Basin. *Linnean Society Palynology Special Interest Group Meeting, October 2000* (Abstract).

Richards, K. and Lowe, S. (2003) Plio-Pleistocene palynostratigraphy of the Amazon Fan, offshore Brazil: Insights into vegetation history, palaeoclimate and sequence stratigraphy. *Conference of American Association of Stratigraphic Palynologists (AASP), London, England, September 2002* (Abstract).

Richardson, J. E., Pennington, R. T., Pennington, T. D., and Hollingsworth, P. M. (2001) Rapid diversification of a species-rich genus of Neotropical rain forest trees. *Science* **293**, 2242–2245.

Rull, V. (1999) Palaeofloristic and palaeovegetational changes across the Paleocene/Eocene boundary in northern South America. *Review of Palaeobotany and Palynology* **107**, 83–95.

Salard-Cheboldaeff, M. (1990) Intertropical African palynostratigraphy from Cretaceous to Late Quaternary times. *Journal of African Earth Sciences* **11**, 1–24.

Salami, M. B. (1991) Palynomorph taxa from the "Lower Coal Measures" deposits (?Campanian–Maastrichtian) of Anambra Trough, Southeastern Nigeria. *Journal of African Earth Sciences* **11**, 135–150.

Schneider, H., Schuettpelz, E., Pryer, K. M., Cranfill, R., Magallon, S., and Lupia, R. (2004) Ferns diversified in the shadow of angiosperms. *Nature* **428**, 553–557.

Shackleton, N. and Boersma, A. (1983) The climate of the Eocene ocean. *Journal of the Geological Society* **138**, 153–157.

Slik, J. W. F., Poulsen, A. D., Ashton, P. S., Cannon, C. H., Eichhorn, K. A. O., Kartawinata, K., Lanniari, I., Nagamasu, H., Nakagawa, M., van Nieuwstadt, M. G. L. *et al.* (2003) A floristic analysis of the lowland dipterocarp forests of Borneo. *Journal of Biogeography* **30**, 1517–1531.

Smith, A. G., Smith, D. G., and Funnell, B. M. (1994) *Atlas of Mesozoic and Cenozoic coastlines* (99 pp.). Cambridge University Press, Cambridge, U.K.

Stebbins, G. L. (1974) *Flowering Plants: Evolution above the Species Level.* Belknap Press, Cambridge, MA.

Takhtajan, A. (1969) *Flowering Plants, Origin and Dispersal* (transl. C. Jeffrey, 300 pp.). Oliver & Boyd, Edinburgh/Smithsonian Institution, Washington, D.C.

Thorne, R. F. (1976) When and where might the tropical angiospermous flora have originated? In: D. J. Mabberley and Chang Kiaw Lan (eds.), *Tropical Botany* (Vol. 29, pp. 183–189). Gardens Bulletin, Singapore.

Tiffney, B. H. (1985b) Perspectives on the origin of the floristic similarity between eastern Asia and Eastern North America. *Journal of the Arnold Arboretum* **66**, 73–94.

Upchurch, G. R. and Wolfe, J. A. (1987) Mid-Cretaceous to Early Tertiary vegetation and climate: evidence from fossil leaves and woods. In: E. M. Friis, W. G. Chaloner, and P. H. Crane (eds.), *The Origins of Angiosperms and Their Biological Consequences* (pp. 75–105). Cambridge University Press, Cambridge, U.K.

Urrego, L. E. (1997) *Los Bosques Inundables del Medio Caqueta: Caracterización y Sucesión* (Estudios en la Amazonia Colombiana 14, pp. 1–133). Tropenbos, Bogotá [in Spanish].

Van der Hammen, T. and Hooghiemstra, H. (2000) Neogene and Quaternary history of vegetation, climate and plant diversity in Amazonia. *Quaternary Science Reviews* **19**, 725–742.

Van der Kaas, W. A. (1991) Palynology of eastern Indonesian marine piston-cores: A Late Quaternary vegetational and climatic history for Australasia. *Palaeogeography, Palaeoclimatology, Palaeoecology* **85**, 239–302.

Van Steenis, C. G. G. J. (1962) The land-bridge theory in botany. *Blumea* **11**, 235–372.

Whitmore, T. C. and Prance, G. T. (1987) *Biogeography and Quaternary History in Tropical America.* Clarendon Press, Oxford, U.K.

Wilf, P., Ruben Cuneo, M., Johnson, K. R., Hicks, J. F., Wing, S. L., and Obradovich, J. D. (2003) High plant diversity from Eocene South America: Evidence from Patagonia. *Science* **300**, 122–125.

Wilgus, C. K., Hastings, B. S., Kendall, C. G. St. C., Posamentier, W. H., Ross, C. A., and van Wagoner, J. C. (1988) *Sea-level Changes: An Integrated Approach* (SEPM Special Publication 42, 407 pp.). Society of Economic Palaeontologists and Mineralogists, Tulsa, OK.

Wing, S. L. and Tiffney, B. H. (1987) Interactions of angiosperms and herbivorous tetrapods through time. In: E. M. Friis, W. G. Chaloner, and P. H. Crane (eds.), *The Origins of Angiosperms and Their Biological Consequences* (pp 203–224). Cambridge University Press, Cambridge, U.K.

Wolfe, J. A. (1977) *Palaeogene Floras from the Gulf of Alaska Region* (U.S. Geological Survey Professional Paper 997, 208 pp.). U.S. Geological Survey, Reston, VA.

Wolfe, J. A. (1985) distributions of major vegetation types during the Tertiary. In: E. T. Sundquist and W. S. Broekner (eds.), *The Carbon Cycle and Atmospheric CO_2: Natural Variations, Archean to Present* (American Geophysical Union Monograph 32, pp. 357–376). American Geophysical Union, Washington, D.C.

Yamanoi, T. (1974) Note on the first fossil record of genus *Dacrydium* from the Japanese Tertiary. *Journal of the Geological Society of Japan* **80**, 421–423.

Yamanoi, T., Tsuda, K., Itoigawa, J., Okamoto, K., and Tacuchi, K. (1980) On the mangrove community discovered from the Middle Miocene formations in southwest Japan. *Journal of the Geological Society of Japan* **86**, 635–638.

Zachos, J. C., Pagini, M., Sloan, L., Thomas, E., and Billups, K. (2001) Trends, rhythms and aberrations in global climate 65 Ma to Present. *Science* **292**, 686–693.

Zachos, J. C., Wara, W. M., Bohaty, S., Delaney, M. L., Pettrizzo, M. R., Brill, A., Bralower, T. J. and Premoli-Silva, I. (2003) A transient rise in tropical sea surface temperature during the Paleocene–Eocene thermal maximum. *Sciencexpress* 23 October 2003, pp. 1–4.

2

Andean montane forests and climate change

M. B. Bush, J. A. Hanselman, and H. Hooghiemstra

2.1 INTRODUCTION

The montane forest habitats of the Andes support exceptionally high biodiversity, with many species occupying narrow elevational ranges (e.g., Terborgh, 1977). These attributes, combined with the short migratory distances (often <30 km separates the lowlands from the upper forest line) allow montane forests to be extremely sensitive monitors of climatic change.

Andean montane forests, which we define to encompass temperate and montane rainforests within the tropical zone (after Huber and Riina, 1997), range from about 1,300 m up to about 3,300 km elevation. The mean annual temperature at the lower limit of the montane forest is about 20°C, with minima of c. 7°C (Colinvaux *et al.*, 1997). Annual precipitation generally exceeds c. 1,000–1,200 mm, and ground-level cloud is frequent. Some caution is needed in grouping all montane forests together and assuming that they will respond similarly to a common forcing as species composition of montane forests varies significantly according to latitude, altitude, aspect, local precipitation, and soil type (Gentry, 1988). A further variable that must be included is that humans have occupied and modified these landscapes for millennia (Erickson 1999; Kolata *et al.*, 2000), and there is uncertainty over the elevation of the upper forest limit in many parts of the Andes (Erickson, 1999; Wille *et al.*, 2002).

In this chapter we will address some of the larger scale issues—for example, the migration of species in response to tectonic and climatic change, the stability of systems but instability of communities through time, and whether there is an out-of-phase climatic influence on southern and northern Andean sites during the last glacial maximum (LGM).

2.2   TECTONIC CHANGES AND THE RISE OF THE ANDES

For the last 20 million years the Andes have been rising as a result of the subduction of several oceanic plates beneath the South American Plate. The uplift transformed a rather flat continent into one with strong physical separation of lowlands and a host of new habitats ranging from humid foothills to ice-covered summits. The rise of the Andes had no less radical an effect on the biogeography of the continent. Drainages of great rivers were reversed (Damuth and Kumar, 1975; Hoorn *et al.*, 1995), and the related orogeny in Central America provided, first, stepping stones and, ultimately, a landbridge connecting a Gondwanan to a Laurasian flora and fauna (Terborgh, 1992; Webb and Rancy, 1996). The great American faunal interchange in which successive waves of taxa moved north and south and then underwent adaptive radiation began as early as 16 million years ago and reached its peak (Webb, 1997) following the closure of the Isthmus of Panama, a progressive process in which the final phase took place between 5 and 4 million years ago.

The arrival of such animals as monkeys, sloths, elephantids, camelids, rats, and cats left a lasting impression on these systems. Many entered unoccupied niches, while others may have gone into direct competition with marsupial counterparts, or the indigenous array of flightless predatory birds. The net result was rather lop-sided with relatively few genera moving into North America, though Glyptodont, a re-radiation of sloth species, possum, armadillos, and porcupines were clear exceptions. While only the latter three have surviving representatives in North America, >50% of mammal genera in South America are derived from Laurasian immigrants (Terborgh, 1992).

This pattern obeys the basic biogeographic rule that the flora of larger source areas will outcompete those of smaller source areas. Consequently, lowland rainforest taxa from South America surged up into Central America, and became the dominant vegetation of the lowland tropics. Contrastingly, Laurasian elements swept south along mountain chains occupying the climatically temperate zone of Central and South American mountains.

Many modern genera were extant and clearly recognizable in the pollen of Miocene sediments (23 to 6 million years ago) (Jaramillo and Dilcher, 2002). During this time the Andes were rising, attaining about half their modern height reaching c. 2,500–3,000 m about 10 million years ago (Hoorn *et al.*, 1995). Thus, these were forest-shrouded systems. Additionally, low passes—such as the Guayaquil gap and the Maracaibo area—maintained lowland connectivity from the Pacific to the interior of the continent until the mid- to Late Miocene (Hoorn *et al.*, 1995). Only in later stages of uplift did the Andes rise above elevations supporting diverse montane forests (i.e., above 3,300–3,500 m).

Montane-dwelling migrants into this setting from North America had to island-hop—either literally, or from hilltop to hilltop—including passage across a broad lowland plain in central Panama. This gap without highlands over 1,000 m was at least 130 km and may have acted as a severe filter to large seeded species, such as *Quercus*. Indeed, *Quercus* diversity in western Panama is about 13 species, whereas only one species is present in eastern Panama (Gentry, 1985).

The southward migrations of Laurasian taxa such as Annonaceae, *Hedyosmum*,

Salix and *Rumex* are inferred rather than observed, but the arrival of *Myrica*, *Alnus*, and *Quercus* are apparent in the paleoecological records from the high plain of Bogotá (Hooghiemstra, 1984, 1989; Van der Hammen, 1985; Van der Hammen *et al.*, 1992; Van 't Veer and Hooghiemstra, 2000). *Myrica* arrived in the Middle Pliocene, *Alnus* first occurs in the Colombian pollen record about 1 million years ago. *Quercus* first occurs about 478 kyr BP (Van 't Veer and Hooghiemstra, 2000) but probably only attained its modern dominance between 1,000 m and 2,500 m elevation about 200 kyr BP (Hooghiemstra *et al.*, 2002). Since the first arrival of these species *Alnus* has spread as far south as Chile, whereas the southernmost distribution of *Quercus* coincides with the Colombian–Ecuadorian border (Gentry 1993). *Alnus*, a pioneer species, thrives in disturbed settings, whereas *Quercus humboldtii* is a dominant of mature Andean forests. The arrival of *Quercus* in Colombia clearly impacted previously established taxa such as *Hedyosmum*, *Vallea*, and *Weinmannia* (Hooghiemstra, 1984)—species that are still the common components of upper Andean forest in Peru and Ecuador.

Progressive cooling during the Quaternary led the upper limit of diverse forest to move downslope ranging between 3,600 m and 2,800 m during warm periods and probably as low as 1,900 m during peak glacial conditions. The consequent expansion of montane grasslands, through a combination of uplift and cooling, provided habitat for newly arriving holarctic species that enriched Puna and Paramo floras.

The new arrivals to forest and grassland settings created novel communities. The concept of no-analog communities is usually used in reference to suggest that communities of the past differed from those of the present (e.g., Overpeck *et al.*, 1985), but ecologically a no-analog community is most significant if it is the formation of a novel community compared with those that preceded it. The sequential arrival of *Myrica*, *Alnus*, and *Quercus* established such novel communities. Furthermore, the faunal interchange between the Americas altered predator–prey relationships, seed-dispersal and plant recruitment (Janzen and Martin, 1982; Wille *et al.*, 2002). While many of the megaherbivores died out in the terminal Pleistocene, the surviving camelids exert a significant grazing influence on montane grasslands. Such a basic observation is an important reminder that the loss of some of the other megafauna could have had a substantial impact on the openness of all Neotropical settings (Janzen and Martin, 1982). Thus, during the Quaternary there has been a major re-shuffling of community composition, both in plants and animals, that resulted from plate tectonics. Glacial activity added additional layers of migrational change.

2.3 SENSITIVITY AND QUANTIFYING COOLING

Modern pollen studies are the backbone of any attempt to quantify past forest changes. Over the past 20 years a series of studies in Colombia (Grabandt, 1985), Ecuador (Bush, 1991), and Peru (Weng *et al.*, 2004), have provided considerable data against which past migrations can be calibrated. The most recent of these included a blind study to test the sensitivity of pollen as a tool to predict modern forest community. Moss polsters (collected from a 20 m^2 area of forest) at 50–100 m vertical

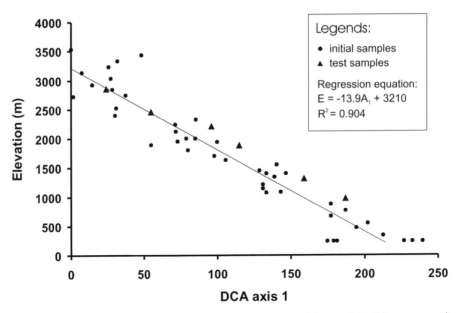

Figure 2.1. Modern pollen rain and elevation. Regression of first axis DCA scores against elevation for log-transformed modern pollen data. All data from a line transect from Amazonia into the Andes in eastern Peru (Weng *et al.*, 2004). *Circles* represent samples of known elevation. The six *triangles* represent a blind study in which the analyst did not know sample elevation.

increments were collected along a transect in the Kosñipata Valley, Peru. The elevational range of the samples was from 300 m to 3,400 m. Despite high pollen diversity (more than 400 types identified) in these samples, >90% of pollen was identified to family level and 30–50% to genus level (Weng *et al.*, 2004). Log-transformed pollen data were analyzed with detrended correspondence analysis (Hill, 1979; McCune and Mefford, 1999) and the resulting Axis 1 scores regressed against elevation (Figure 2.1). In 2001 we collected six further samples and conducted a blind study. The six unlabelled samples were given to the analyst (Weng) to determine the accuracy with which they could be placed into the data set (Figure 2.1). From these results we determined that our accuracy in assigning an elevation to an unknown sample is about ±260 m. Local moist air adiabatic lapse rates are almost exactly 5.5°C per 1,000 m of ascent (Weng *et al.*, 2004). From this we infer that our error in assessing temperature based on palynology is c. ±1.5°C. The Kosñipata transect was mature second-growth forest, disturbed by road construction. As disturbance-tolerant species tend to produce a lot of pollen and are often generalist species, ongoing study of less disturbed transects may provide even narrower error ranges in temperature estimates.

It will be noted that the samples in Figure 2.1 from 3,350 and 3,400 m do not fall close to the regression line. Both of these samples were collected from sheltered gullies that contained shrubs of *Weinmannia*, woody Asteraceae and *Polylepis*, giving these samples a "low" signature in the analysis.

2.4 SITES IN SPACE AND TIME

Sites that provide paleoecological records from within modern montane forest settings are thinly scattered. The cause of this paucity lies in the geography of the Andes themselves (Figure 2.2). The flanks of the Andes are so steep that the vertical elevation occupied by montane forest is often spanned by just 10–30 km laterally. In the inter-Andean plateaus, montane forests are restricted to the wetter and somewhat lower sections of the northern Andes. The small area and lack of glacially-formed lakes within the elevations occupied by modern montane forest, combined with frequent rockslides and active tectonism, contribute to a landscape in which few ancient lakes formed and even fewer have survived.

The list of montane forest sites is expanded when we include those that have supported montane forest in the past. During the thermal optima of previous

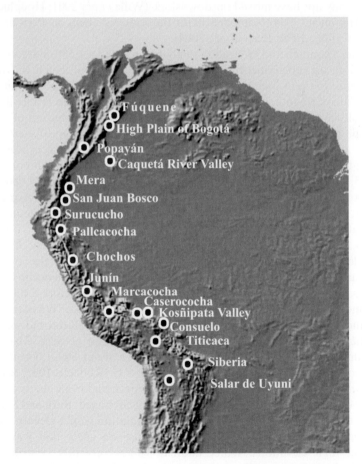

Figure 2.2. The location of sites of paleoecological importance mentioned in the text relative to topography.

interglacials—such as marine isotope stages (MIS) 5e, 7, 9, and 11—it appears that montane forest may have extended upslope by as much as 200 m of its present location. The influence on the lower limit of montane forest during these episodes is more difficult to establish. Bush (2002) hypothesized that—as climates warm—the elevation of cloud formation on the flank of the Andes will increase. In the southern Andes at the peaks of these interglacial events, conditions appear to be drier-than-modern, consistent with an upslope movement of the cloudbase. Under such warm conditions the change in the elevation of cloudbase may have been greater than the upslope expansion of the montane forest, creating a narrower total elevational range supporting montane forest. Contrastingly, during the glacial periods montane forest species invaded downslope in response to cooling and the lower formation of cloud. Although the descent of montane taxa and the lowering of the upper forest line appear broadly similar (c. 1,500 m) along the Andes, the movement of the lower limit of the cloudbase may be more variable regionally. In the drier lowlands of Colombia this cloudbase may not have moved far downslope (Wille et al., 2001; Hooghiemstra and van der Hammen, 2004), compared with the wetter systems of Peru and Ecuador (Colinvaux et al., 1996; Bush et al., 2004).

Van der Hammen and González (1960) documented a 1,500 m descent of vegetation based on the replacement of forest with grasslands and then a widening downslope distance to the estimated position of upper forest line. Since that initial study of the high plain of Bogotá, virtually every Andean record from the last ice age indicates at least a 1,000-m descent of vegetation and often a 1,500-m descent of some pollen taxa at the LGM. The moist air adiabatic lapse rate (Chapter 10)—evident on the Andean flank—provides a means to translate this vegetational movement into a change in temperature. Modern lapse rates vary according to local humidity, ranging between 5.5°C and 6.2°C (Witte, 1994) in Colombia, and c. 5.2°C per 1,000 m of ascent in Peru and Ecuador (Colinvaux et al., 1997; Bush and Silman, 2004). Accordingly, for a 1,000–1,500 m descent of vegetation the inferred change in paleotemperature is a cooling relative to modern of 5–8.5°C.

Most Andean LGM pollen records are consistent with a cooling of c. 8°C in the highest elevations tapering down to a cooling of c. 4–5°C in the lowlands. This temperature differential suggests a steeper-than-modern temperature gradient. As there is no suggestion that the Andean slopes were ever without forest, it is improbable that the moist air adiabatic lapse rate would change very much (Webster and Streten, 1978; Rind and Peteet, 1985). Evidence from studies of glacial moraines leads to reconstructions of the equilibrium line altitude (ELA) for glaciers. Glaciers in Peru and Ecuador are generally inferred to have ELAs about 800–1,000 m lower than modern counterparts, suggesting a cooling of 4–5°C (Rodbell, 1992; Seltzer, 1992; Smith et al., 2005). Hence, the inferred temperature signal from plants at high elevations may contain a more complex signal than first envisaged. Bush and Silman (2004) proposed one such effect in which black body radiation would elevate sensible heat loss under low atmospheric CO_2 concentrations—an effect that would be more extreme at high elevations. Other additive effects probably contributed to the observed high elevation cooling.

2.5   QUATERNARY INTERGLACIALS

Within the dating resolution available to us, Neotropical interglacials appear to coincide in timing and general character with those documented elsewhere. The interglacials are known by their marine isotope stages (MIS): 5e (c. 130–116 kyr BP), MIS 7 (c. 240–200 kyr BP), MIS 9 (330–300 kyr BP) and MIS 11 (425–390 kyr BP). They generally last about 15–40 kyr. While a 100-kyr cycle appears to underlie the glaciations of the last half million years, the intensity of interglacial periods appears to be related to precessional amplitude (Broecker, 2006).

  Three records exist that provide insights into multiple glacial cycles in the Andes. The High Plain of Bogotá, Lake Titicaca (Hanselman *et al.*, 2005) and the Salar de Uyuni, although only the MIS 6 to MIS 1 portion of this record has been published so far (Fritz *et al.*, 2004). The amount of ecological change in three fossil pollen records that cover either or both of MIS 1 and MIS 5e from Lake Titicaca reveals that—true to the precessional prediction—MIS 5e is the most extreme of these interglacials (Figure 2.3). The onset of MIS 5e is taken to be at 136 kyr BP based on the chronology used in Hanselman *et al.* (2005) and 11 kcal. yr BP is taken as the start of the Holocene.

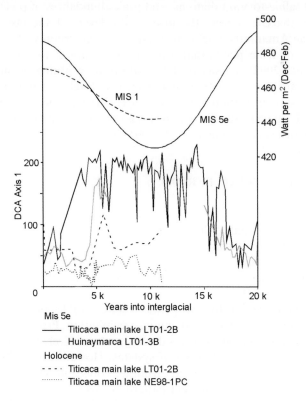

Figure 2.3. A comparison of MIS 5e and the Holocene based on insolation and changes in community composition revealed through DCA. Data are from Hanselman *et al.* (2005) and insolation curves from Analyseries 1.2 (Berger, 1992; Paillard *et al.*, 1996)}.

The data are drawn from a deep water core from Lake Titicaca LT01-2B (240-m water depth), a shallower water core (40-m water depth) from the Huinaymarka sub-basin (Core LT01-3B), and a piston core from 130-m water depth that provides a detailed Holocene record from the main basin (core NE98-1PC; Paduano *et al.*, 2003). The fossil pollen data were combined into a single matrix and ordinated using detrended correspondence analysis (DCA) (Hill, 1979; McCune and Mefford, 1999). The scores for Axis 1 are plotted against time since the start of the relevant interglacial, and therefore provide a comparative trajectory of the amount of ecological change that took place. Core LT01-3B has a hiatus in the middle of the interglacial, but shows a very similar pattern of community change leading into and out of the event as found in the deep water core LT01-2B. That hiatus is entirely consistent with our hypothesis of increased evaporation and reduced precipitation compared with that of the Holocene that marked the peak of MIS 5e.

During 5e, and to a slightly lesser extent in previous interglacials, warming caused the upper Andean forest limit to move about 200 m above its modern elevation. Both the Colombian and Bolivian records indicate that the peak of MIS 5e may have been relatively dry. This drying is especially evident in Lake Titicaca, where the abundance of benthic and saline-tolerant diatoms, and peak abundances of pollen of Chenopodiaceae/Amaranthaceae suggest the lowest lake levels of the last 340,000 years. Chenopodiaceae/Amaranthaceae pollen types are commonly derived from salt-tolerant plants, or from plants that grow in areas subject to irregular inundation (Marchant *et al.*, 2002). Parsing out the effects of warming versus reduction in precipitation is not easy, as both would lower lake level. In Colombia the estimate of warming based on migration of tree line appears to have been about 1°C (Van der Hammen and Hooghiemstra, 2003), and this is consistent with the tentative estimate of 1–2°C for Titicaca (Hanselman *et al.*, 2005).

2.6   THE LAST GLACIAL PERIOD

If the pattern of temperature change is traced from the last interglacial to the present, there was a substantial rapid cooling—perhaps 3°C—at the end of the last interglacial that marked the onset of glacial conditions in the Andes (Van 't Veer and Hooghiemstra, 2000). Following this cooling, temperatures bumped up and down, tracking the Milankovitch cycles, but gradually declining to the coldest time at the LGM (Hooghiemstra *et al.*, 1993).

The precipitation record for this period is harder to decipher, and inferred lake depth is a major proxy for changes in annual precipitation. Precipitation patterns are often highly localized, and when one is dealing with relatively few sites it is possible that such local effects skew our view of systems. However, if we look outside the montane forest region and include data from ice cores and high Andean lakes, and from the Amazonian plain, a coherent pattern begins to emerge (Table 2.1).

In Colombia the Funza-2 record terminates about 30,000 cal yr BP when the lake dried out. The Fúquene-3 record suggests a progressive lowering of lake level beginning around 60,000 years ago and culminating in a depositional hiatus between

Table 2.1. Inferred LGM moisture from described sites in the northern and southern Andes.

Latitude (°S, unless otherwise stated)	Elevation	LGM (wet/dry)	Onset of deglaciation	Timing of mid-Holocene dry event	Literature source
5°N	2,580	Dry	c. 25,000	—	van der Hammen and Hooghiemstra (2003)
4–5°N	2,600	Dry	c. 24,000	—	Hooghiemstra (1984)
2°N	1,750	Dry	c. 24,000	—	Wille et al. (2000)
1	c. 400	Dry	—	—	van der Hammen et al. (1992)
2	4,060	—	—	7,500–4,000	Hansen et al. (2003)
3	3,180	Wet*	—	—	Colinvaux et al. (1997)
7	3,300	—	c. 15,000	9,000–3,800	Bush et al. (this volume)
7	3,575	—	—	c. 6,000	Hansen and Rodbell (1995)
9	6,048	Dry*	—	8,400–5,200	Thompson et al. (1995)
11	4,100	—	—	c. 6,000	Hansen et al. (1994)
11	4,100	Wet	c. 22,000	—	Seltzer et al. (2002)
13	1,360	Wet	c. 21,000	8,200–4,000	Bush et al. (2004)
16	2,350–2,750	—	—	Wet mid-Holocene	Holmgren et al. (2001)
16–17	3,810	Wet	c. 21,000	6,000–4,000	Paduano et al. (2003)
16–17	3,810	Wet	c, 21,000	8,000–5,500	Baker et al. (2001)
16–17	3,810	Wet	c. 22,000	—	Seltzer et al. (2002)
16–17	3,810	Wet	—	6,000–3,500	Tapia et al. (2003)
16	4,300	—	—	8,500–2,500	Abbott et al. (2000, 2003)
17	c. 3,800	Wet*	—	—	Mourguiart et al. (1997, 1998)
17	2,920	Wet*	c. 21,000	11,000–4,000	Mourguiart and Ledru (2003)
18	6,542	Wet	c. 21,000	9,000–3,000	Thompson et al. (1998)
20	3,653	Wet	—	—	Chepstow-Lusty (2005), Fritz et al. (2004)
20	3,653	Wet	—	—	Baker et al. (2001)

* As interpreted by the authors of this chapter.

c. 22 kcal.yr BP and 12 kcal.yr BP (Van der Hammen and Hooghiemstra, 2003). The Altiplano of Peru and Bolivia appears to have become wetter after c. 60 kyr BP (Fritz *et al.*, 2004); given the uncertainties in dating, this may or may not be related to the beginning of the drier conditions in Colombia. However, the LGM does provide support for asynchrony in wet episodes, as this was a time of flooding in the Altiplano, and low lake level in Colombia.

At least three giant paleolakes occupied the Altiplano at various times during the Quaternary (Servant, 1977; Placzek *et al.*, 2006). A paleolake appears to have formed about 26 kcal.yr BP (Baker *et al.*, 2001), coincidental with the onset of ice accumulation at Sajama (Thompson *et al.*, 1998). This wet event appears to have lasted until c. 16 kcal.yr BP. Some recent evidence suggests even greater complexity in this sequence of high and low stands suggesting strong swings in precipitation and lake level between 26 and 16 kcal.yr BP (Placzek *et al.*, 2006). The combination of extreme cold and wet conditions during the LGM caused ice lobes to advance to within 100 m vertically of the modern Titicaca shoreline (a vertical descent of about 1,300 m; Seltzer *et al.*, 1995, 2002). Baker *et al.* (2001) suggest that lake level in the Salar de Uyuni followed the precessional cycle for the last 50,000 years. Highstands corresponded to maxima of insolation occurring during the wet season (December–January–February), and lowstands during the corresponding minima.

Thus, the period centered on 22 kcal.yr BP was wet and the period centered on 33 kcal.yr BP was dry. Indeed, during the 33 kcal.yr BP dry event, lake level in Titicaca dropped by as much as 130 m, certainly the lowest lake levels of the last glacial cycle. This dry event is also documented in the northern Amazon lowlands, though not in southeastern Amazonia (Chapter 3). In Colombia, the Fuquene-2 and Fuquene-7 pollen records (Van Geel and Van der Hammen, 1973; Mommersteeg, 1998) demonstrate that this period was exceptionally wet, providing another good example of antiphasing between the Altiplano and the northern Andes.

While the evidence of precessional oscillations has a long history in Colombia (Hooghiemstra *et al.*, 1993) on the Altiplano this synchrony is only evident in the last two glacial cycles. Prior to c. 60 kyr BP the Salar de Uyuni was predominantly dry, with only sporadic flooded episodes (Fritz *et al.*, 2004; Chepstow-Lusty *et al.*, 2005). Two plausible scenarios have yet to be tested: one is that the climate was significantly drier prior to 60 kyr BP, and the other is that tectonic change altered the hydrology of the basin at this time, making it more probable that it would hold water.

The critical question in a discussion of montane forest is: Can these upslope systems provide proxy data about what was happening to the montane forests? The answer to that appears to be a qualified "yes".

In Colombia the Caquetá River valley (Van der Hammen *et al.*, 1992) documents a relatively wet time between c. 50 kyr BP and 30 kyr BP and a drier LGM, consistent with the records from the high plain of Bogotá. A record from Popayán (1,700 m; Wille *et al.*, 2000) reveals the presence of either a cool open forest or closed montane forest throughout the last 30,000 years. The data from this site suggest a cooling of 5–7.5°C at the LGM. In Ecuador the premontane sites of Mera and San Juan Bosco (1,100 m and 970 m, respectively; Bush *et al.* 1990) match this interpretation closely, suggesting synchrony at least as far south as the equator.

Lake Consuelo in southern Peru provides the most detailed view yet of the lower Andes during the last glacial maximum. At 1,360 m elevation the modern lake lies at exactly the elevation of cloud formation in this section of the Andes. The modern flora is dominated by lowland elements—for example, *Alchornea, Brosimum, Euterpe, Ficus, Guatteria, Maquira, Unionopsis,* and *Wettinia.* Premontane elements such as *Dictyocaryum, Myrsine, Alsophila,* and *Cyathea* are also present (M. Silman, pers. commun.). The pollen types of the Holocene reflect this lowland mixture of species, but those of the glacial clearly indicate the presence of a montane forest. *Podocarpus, Alnus, Hedyosmum, Weinmannia, Bocconia, Vallea,* Ericaceae, and *Polylepis/Acaena* replaced the lowland flora.

Range data for 24 abundant taxa that could be identified to genus in the pollen record were collected from the TROPICOS database (Bush *et al.,* 2004). A Bayesian analysis of these data in combination with the fossil pollen data generated a probability distribution for the most likely elevation at which the combination of species in a fossil spectrum could occur. Note this analysis does not rely on community structure, as it is based on total documented ranges of species, not on co-occurrence. The outcome of this analysis—the first complete deglacial sequence from the premontane forest region of the southern tropical Andes—revealed a remarkably stable system during MIS 2 and 3 (c. 43–20 kcal. yr BP). These data are particularly interesting as two upslope sites—Siberia (Mourguiart and Ledru, 2003) and the Altiplano (Baker *et al.,* 2001)—both reveal the period prior to the LGM, centered on 33,000 cal yr BP to be dry. Apparently the moisture flow into the montane forests at the base of the Andes was not interrupted, demonstrating continuity of habitat availability for these systems, through one of the driest times of the last glacial cycle.

2.7   DEGLACIATION

The timing and rate of Andean deglaciation is somewhat contentious, as it has been suggested that the southern Andes mirrors the Vostok record from Antarctica, while the northern Andes mirrors the GISP record from Greenland (Seltzer *et al.,* 2002). It appears that the more southern tropical Andes entered a deglacial phase between 21 kcal. yr BP and 19 kcal. yr BP (within the classic LGM of the northern hemisphere), while the northern Andes may not have warmed until c. 16 kcal. yr BP. This relatively early deglaciation is manifested in most Central Andean records (Figure 2.4).

The deglaciational path had some bumps in it, though most of the apparent hiccups were probably abrupt changes in precipitation rather than deviations from a steady change in temperature (Bush *et al.,* 2004). A dry event at c. 16.5 kcal. yr BP is recorded strongly in Lake Titicaca, and was followed by a cool, wet event that centers on c. 15.1 kcal. yr BP. This latter event is consistently manifested in Central Andean records, though it is more evident in the records between c. 11°S and 7°S rather than those at higher latitudes.

Thereafter, the trend out of the last ice age is relatively constant in the southern tropical Andes, whereas the northern Andes appears to reflect the Caribbean and northern hemispheric episodes of abrupt warming and cooling. Consequently, in Peru

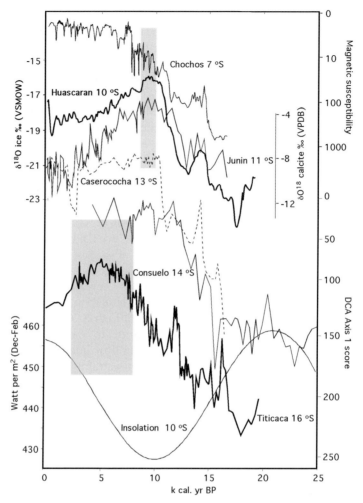

Figure 2.4. Central Andean insolation, and the extent of physical and community change during deglaciation and the Holocene. Datasets are Lake Chochos magnetic susceptibility (note inverted log scale; Bush *et al.*, 2005); Huascaran δ^{18}O ice core (Thompson *et al.*, 1995); Lake Junin δ^{18}O calcite (Seltzer *et al.*, 2000); Lake Caserococha fossil pollen DCA Axis 1 (Paduano, 2001); Lake Consuelo fossil pollen DCA Axis 1 (Bush *et al.*, 2004 and new data from D. Urrego); Lake Titicaca fossil pollen DCA Axis 1 (Paduano *et al.*, 2003); insolation (DJF) for 10°S from Analyseries 1.2 (Berger, 1992; Paillard *et al.*, 1996). VSMOW = Vienna standard marine ocean water; VPDB = Vienna peedee belemnite.

and Bolivia the deglaciational warming appears to have been on average <1°C per millennium, whereas in the northern Andes a relatively large jump in temperatures at the onset of the Holocene, perhaps 4°C within the space of a few hundred years, is thought to have occurred. Thus, these systems have responded to warming events whose rates differed by about an order of magnitude.

Evidence for the presence, or absence, of the Younger Dryas event in South America has engendered considerable debate (Heine, 1993; Hansen, 1995; Van der Hammen and Hooghiemstra, 1995; Rodbell and Seltzer, 2000; Van 't Veer *et al.*, 2000; Bush *et al.*, 2005). Some records reveal an oscillation that fits well with the chronology of the North Atlantic event (e.g., Van 't Veer *et al.*, 2000); however, other records—such as that of Titicaca and glacial advances in Ecuador and Peru—reveal an oscillation that pre-dates the Younger Dryas by 500 years (Rodbell and Seltzer, 2000; Paduano *et al.*, 2003). In summary, it appears that the Younger Dryas is better represented in the northern section of the Neotropics than south of the equator. Furthermore, in most settings if a change is contemporaneous with that of the North Atlantic it is manifested in precipitation change rather than in temperature shifts resulting in some glacial re-advance or retreat (Clapperton, 1993; Rodbell and Seltzer, 2000; Smith *et al.*, 2005).

2.8   THE HOLOCENE

Farther south—in Ecuador—the related sites of Sacurococha (3,180 m: Colinvaux *et al.*, 1997) and Pallcacocha (4,200 m; Moy *et al.*, 2002) begin their sedimentary record at c. 15 kcal . yr BP. These two sites lie in the same drainage and each has a markedly laminated stratigraphy. The laminations have been suggested to reflect El Niño related storm intensity (Rodbell *et al.*, 1999; Moy *et al.*, 2002). While these sites cannot inform us of climate change in the Pleistocene, they do suggest an affinity with the Colombian sites rather than sites of southern Peru and Bolivia, which show a very marked dry event in the mid-Holocene (Wirrmann *et al.*, 1992; Ybert, 1992; Paduano *et al.*, 2003; Rowe *et al.*, 2003). Again the southern and northern sites appear to be asynchronous in their precipitation signals, with all sites north of Junin exhibiting a dry start to the Holocene followed by rising lake levels between 10 kcal . yr BP and 8 kcal . yr BP. Sites in the southern tropical Andes are generally entering a dry phase at that time and experience low lake levels until c. 4 kcal . yr BP (Bradbury *et al.*, 2001). The only record from the southern tropical Andes that spans a portion of this event is Lake Siberia (Mourguiart and Ledru, 2003). This record terminates at c. 5.1 kcal . yr BP, but the period from 10 kcal . yr BP to 5 kcal . yr BP shows the expansion of grassland, consistent with more open conditions, but the return of some forest taxa in the uppermost samples.

When records resume regionally, human impacts are evident in many sites—for example, Marcacocha (Chepstow-Lusty *et al.*, 2002), Titicaca (Paduano *et al.*, 2003), and Junin (Hansen and Rodbell, 1995); the uplands were being transformed by burning and deforestation. The modern upper forest line may be a baseline that was created at this time. How different a truly natural upper forest line would be is a matter of ongoing debate. Almost 50 years ago Ellenberg (1958) suggested that *Polylepis* woodland could have been extensive up to elevations of 4,000 m on the wetter slopes and 5,000 m on the drier slopes of the Andes. Though falling from favor for many years, his ideas have been resurrected (e.g., Fjeldså, 1992; Kessler, 1995; Wille *et al.*, 2002). No resolution has been reached regarding either the natural elevation of upper forest lines, or the past importance of *Polylepis* in Andean floras, but it certainly seems likely that the often stark separation of forest and grasslands is an artifact of millennia of human landuse.

While humans altered the highland landscape, it is also probable that climate influenced human populations. The mid-Holocene drought on the Altiplano induced a period termed the "Silencio Arqueológico" in which there was widespread abandonment (Núñez *et al.*, 2002). Where did these populations go? Into the montane forest? Perhaps. The Lake Siberia record shows an increase in charcoal coincident with the peak of this drought (Mourguiart and Ledru, 2003). Whether these fires resulted from human occupation of a moister site than could be found in the highlands, or whether this area was merely more drought-prone has yet to be resolved. Later droughts are implicated in the cultural collapse of civilizations such as the Huari, Tiwanaku, and Chiripó (Brenner *et al.*, 2001; Chepstow-Lusty *et al.*, 2002). Too few records exist to document the effect of these Late Holocene droughts on montane forests, and these are data that are badly needed.

2.9   THE PAST AS A KEY TO THE FUTURE

The potential for previous interglacials to serve as a guide to the climatic future of the Holocene has attracted considerable attention (e.g., Ruddiman, 2003; Broecker, 2006). That the full biodiversity of the Andean system appears to have survived the intensity of MIS 5e offers some hope that systems will be able to adjust to the next 50–100 years of projected climate change. Most climate simulations project the Amazon Basin to become warmer and drier over the next century, and for a warming of tropical mountains to be about 2–3°C (IPCC, 2001). Estimates of species' migrational responses to such climate change suggest that the tropical Andes will be one of the most sensitive regions to biome-level change; that is, the Andes have a high proportion of pixels representing the region that change from one biome type to another (Malcolm *et al.*, 2006).

Melting tropical ice caps (Thompson *et al.*, 2002) and the upslope migration of species (Pounds *et al.*, 1999) are evidence that these changes are already taking effect. The stress of warming may induce complex interactions—for example, between droughts, chytrid fungae, and frogs—that may lead to extinctions (Pounds, 2001; Pounds *et al.*, 2006).

The rate of response of communities to climate change has been tested in temperate northern latitudes by rapid warming events such as the termination of the Younger Dryas. That particular warming was similar in its rate of change to the anticipated warming of the next century. If the tropics were similarly exposed to rapid warming, and there was no corresponding wave of extinction, we might be able to predict a sturdy migrational response that would accommodate climate change. However, such a clear, sharp warming is seldom evident in tropical records.

The flat spot in the ^{14}C record that provides relatively large possible calibration solutions between c. 10,000 and 11,000 ^{14}C years often frustrates efforts to provide a definitive chronology. From the available records, it appears that there was no rapid warming at the onset of the Holocene in much of Amazonia and the tropical Andes. Species in the biodiversity hot spots of the Peruvian Andes have not contended with change faster than c. 1°C of warming per millennium (Bush *et al.*, 2004), and therefore while the range of temperatures projected for the next 50–100 years may be within their Quaternary experience, the rate of climate change probably is not.

2.10 CONCLUSIONS

Paleoecological research in the Andes has provided some exciting insights into the both long-term migrations of species and also responses to rapid climatic oscillations. In Europe and North America the accumulation of thousands of pollen records allowed Holocene migrations to be mapped in great detail. From those studies emerged the understanding that temperate communities are ephemeral, perhaps the most important ecological insight to arise from Quaternary paleoecology. However, simply applying the rules of temperate ecology to the tropics has been shown repeatedly to be unwise. The Andes offer a very different migratory environment to the great plains of Europe and eastern North America. The Amazonian lowlands are often separated from Andean snows by <50 km. The complex topography of Andean valleys, ridges, and streambeds offer a mass of microhabitats that can range from xeric scrub to lush forest in a few tens of meters. The consequence of this heterogeneity is that migration could have been nearly instantaneous rather than lagged by thousands of years. Under these circumstances continuity of habitat availability, rather than ability to migrate in and out of refugia, may be the key to diversity.

Paleoecological records from the Andes show a remarkable continuity of montane forest availability for species. Although the area with ground-level cloud moved up and down a mountain, it appears probable that this niche has been a continuous feature of the environment since the Andean orogeny created uplands high enough to induce cloud formation. Where it can be measured, rates of community change are low for tens of millennia, though communities are changing throughout that time. Novel assemblages arose due to continental-scale as well as local migrations, but the overall niche of living within a montane forest may have changed less than its cloud-free counterparts up and downslope.

Regional asynchrony is a feature of the paleoeclimatic literature with Lake Junin, Peru (11°S) cited as the southernmost record that had a full glacial precipitational pattern common to sites south of Mexico (Bradbury *et al.*, 2001); farther south tropical systems were somewhat out of phase with this northern group of sites. However, Seltzer *et al.* (2000) argue that moisture change between Lake Junin and sites in the Caribbean were asymmetric in the Holocene. This latter argument is based on the apparent fit of moisture availability and regional wet season insolation. These apparently contradictory assessments are not necessarily inconsistent as we have yet to discover the proximate causes of glacial age precipitation change in the Andes. The important points that can be derived from the paleoecological data are that precipitation and temperature patterns varied substantially with latitude along the tropical

Andes, and that regions exhibiting synchronous changes in one period could be asynchronous in another.

The paleoecological record needs to be incorporated in conservation thinking to devise appropriate strategies to avert an imminent loss of biodiversity. However, for paleoecology to become genuinely integrated with conservation science we will need to provide more detailed records, especially increasing our taxonomic precision. Furthermore, new paleoecological records from the montane forest region are desperately needed to expand our spatial data set and test the many emerging theories relating to this fascinating ecosystem.

Acknowledgements

This work was funded by National Science Foundation grants ATM 9906107 and 0317539. Two anonymous reviewers are thanked for their help and comments.

2.11   REFERENCES

Abbott, M. B., Wolfe, B. B., Aravena, R., Wolfe, A. P., and Seltzer, G. O. (2000) Holocene hydrological reconstructions from stable isotopes and paleolimnology, Cordillera Real, Bolivia. *Quaternary Science Reviews* **19**, 1801–1820.

Abbott, M. B., Wolfe, B. B., Wolfe, A. P., Seltzer, G. O., Aravena, R., Mark, B. G., Polissar, P. J., Rodbell, D. T., Rowe, H. D., and Vuille, M. (2003) Holocene paleohydrology and glacial history of the central Andes using multiproxy lake sediment studies. *Palaeogeography, Palaeoclimatology, Palaeoecology* **194**, 123–138.

Baker, P. A., Seltzer, G. O., Fritz, S. C., Dunbar, R. B., Grove, M. J., Tapia, P. M., Cross, S. L., Rowe, H. D., and Broda, J. P. (2001) The history of South American tropical precipitation for the past 25,000 years. *Science* **291**, 640–643.

Berger, A. (1992) Astronomical theory of paleoclimates and the last glacial–interglacial cycle. *Quaternary Science Reviews* **11**, 571–581.

Bradbury, J. P., Grosjean, M., Stine, S., and Sylvestre, F. (2001) Full and late glacial lake records along the PEP1 transect: Their role in developing interhemispheric paleoclimate interactions. In: V. Markgraf (ed.), *Interhemispheric Climate Linkages* (pp. 265–291). Academic Press, San Diego.

Brenner, M., Hodell, D. A., Curtis, J. H., Rosenmeier, M. F., Binford, M. W., and Abbott, M. B. (2001) Abrupt climate change and Pre-Columbian cultural collapse. In: V. Markgraf (ed.), *Interhemispheric Climate Linkages* (pp. 87–104). Academic Press, San Diego.

Broecker, W. (2006) The Holocene CO_2 rise: Anthropogenic or natural? *EOS Supplement* **87**, 27–29.

Bush, M. B. (1991) Modern pollen-rain data from South and Central America: A test of the feasibility of fine-resolution lowland tropical palynology. *The Holocene* **1**, 162–167.

Bush M. B. (2002) Distributional change and conservation on the Andean flank: A palaeoecology perspective. *Global Ecology and Biogeography* **11**, 463–473.

Bush, M. B. and Silman, M. R. (2004) Observations on Late Pleistocene cooling and precipitation in the lowland Neotropics. *Journal of Quaternary Science* **19**, 677–684.

Bush, M. B., Weimann, M., Piperno, D. R., Liu, K.-b., and Colinvaux, P. A. (1990) Pleistocene temperature depression and vegetation change in Ecuadorian Amazonia. *Quaternary Research* **34**, 330–345.

Bush, M. B., Silman, M. R., and Urrego, D. H. (2004) 48,000 years of climate and forest change from a biodiversity hotspot. *Science* **303**, 827–829.

Bush, M. B., Hansen, B. C. S., Rodbell, D., Seltzer, G. O., Young, K. R., León, B., Silman, M. R., Abbott, M. B., and Gosling, W. D. (2005) A 17,000 year history of Andean climatic and vegetation change from Laguna de Chochos, Peru. *Journal of Quaternary Science* **20**, 703–714.

Bush, M. B., Listopad, M. C. S., and Silman, M. R. (in press) Climate change and human occupation in Peruvian Amazonia. *Journal of Biogeography.*

Chepstow-Lusty, A., Frogley, M. R., Bauer, B. S., Bennett, K. D., Bush, M. B., Chutas, T. A., Goldsworthy, S., Tupayachi Herrera, A., Leng, M., Rousseau, D.-D. *et al.* (2002) A tale of two lakes: Droughts, El Niños and major cultural change during the last 5000 years in the Cuzco region of Peru. In: S. Leroy and I. S. Stewart (eds.), *Environmental Catastrophes and Recovery in the Holocene* (Abstracts Volume). Brunel University, London.

Chepstow-Lusty, A. J., Bush, M. B., Frogley, M. R., Baker, P. A., Fritz, S. C., and Aronson, J. (2005) Vegetation and climate change on the Bolivian Altiplano between 108,000 and 18,000 yr ago. *Quaternary Research* **102**, 90–98.

Clapperton, C. W. (1993) *Quaternary Geology and Geomorphology of South America.* Elsevier, Amsterdam.

Colinvaux, P. A., De Oliveira, P. E., Moreno, J. E., Miller, M. C., and Bush, M. B. (1996) A long pollen record from lowland Amazonia: Forest and cooling in glacial times. *Science* **274**, 85–88.

Colinvaux, P. A., Bush, M. B., Steinitz-Kannan, M., and Miller, M. C. (1997) Glacial and postglacial pollen records from the Ecuadorian Andes and Amazon. *Quaternary Research* **48**, 69–78.

Damuth, J. E. and Kumar, N. (1975) Amazon cone: Morphology, sediments, age, and growth pattern. *Geological Society of America Bulletin* **86**, 863–878.

Erickson, C. L. (1999) Neo-environmental determinism and agrarian "collapse" in Andean prehistory. *Antiquity* **73**, 634.

Fjeldså, J. (1992) Biogeographic patterns and evolution of the avifauna of relict high-altitude woodlands of the Andes. *Streenstrupia* **18**, 9–62.

Fritz, S. C., Baker, P. A., Lowenstein, T. K., Seltzer, G. O., Rigsby, C. A., Dwyer, G. S., Tapia, P. M., Arnold, K. K., Ku, T. L. and Luo, S. (2004) Hydrologic variation during the last 170,000 years in the southern hemisphere tropics of South America. *Quaternary Research* **61**, 95–104.

Gentry, A. H. (1985) Contrasting phytogeographic patterns of upland and lowland Panamanian plants. In: W. G. D'Arcy and M. D. Correa A. (eds.), *The Botany and Natural History of Panama* (pp. 147–160). Missouri Botanical Garden, St. Louis.

Gentry, A. H. (1988) Changes in plant community diversity and floristic composition on environmental and geographical gradients. *Annals of the Missouri Botanical Garden* **75**, 1–34.

Gentry, A. H. (1993) *A Field Guide to the Families and Genera of Woody Plants of Northwest South America (Colombia, Ecuador, Peru) with Supplementary Notes on Herbaceous Taxa* (895 pp.). Conservation International, Washington, D.C.

Grabandt, R. A. J. (1985) *Pollen Rain in Relation to Vegetation in the Colombian Cordillera Oriental.* University of Amsterdam, Amsterdam.

Hanselman, J. A., Gosling, W. D., Paduano, G. M., and Bush, M. B. (2005) Contrasting pollen histories of MIS 5e and the Holocene from Lake Titicaca (Bolivia/Peru). *Journal of Quaternary Science* **20**, 663–670.

Hansen, B. C. S. (1995) A review of lateglacial pollen records from Ecuador and Peru with reference to the Younger Dryas Event. *Quaternary Science Reviews* **14**, 853–865.

Hansen, B. C. S. and Rodbell, D. T. (1995) A late-glacial/Holocene pollen record from the eastern Andes of Northern Peru. *Quaternary Research* **44**, 216–227.

Hansen, B. C. S., Seltzer, G. O., and Wright, H. E. (1994) Late Quaternary vegetational change in the central Peruvian Andes. *Palaeogeography, Palaeoclimatology, Palaeoecology* **109**, 263–285.

Hansen, B. C. S., Rodbell, D. T., Seltzer, G. O., Leon, B., Young, K. R., and Abbott, M. (2003) Late-glacial and Holocene vegetational history from two sites in the western Cordillera of southwestern Ecuador. *Palaeogeography, Palaeoclimatology, Palaeoecology* **194**, 79–108.

Heine, J. T. (1993) A reevaluation of the evidence for a Younger Dryas climatic reversal in the tropical Andes. *Quaternary Science Reviews* **12**, 769–779.

Hill, M. O. (1979) DECORANA: A FORTRAN program for detrended correspondence analysis and reciprocal averaging. *Ecology and Systematics*. Cornell University, New York.

Holmgren, C. A., Betancourt, J. L., Rylander, K. A., Roque, J., Tovar, O., Zeballos, H., Linares, E., and Quade, J. (2001) Holocene vegetation history from fossil rodent middens near Arequipa, Peru. *Quaternary Research* **56**, 242–251.

Hooghiemstra, H. (1984) *Vegetational and Climatic History of the High Plain of Bogotá, Colombia* (Dissertaciones Botanicae 79, 368 pp.). J. Cramer, Vaduz, Liechtenstein.

Hooghiemstra, H. (1989) Quaternary and upper-Pliocene glaciations and forest development in the tropical Andes: Evidence from a long high-resolution pollen record from the sedimentary basin of Bogotá, Colombia. *Palaeogeography, Palaeoclimatology, Palaeoecology* **11**, 26.

Hooghiemstra, H. and van der Hammen, T. (2004) Quaternary ice-age in the Colombian Andes: Developing an understanding of our legacy. *Philosophical Transactions of the Royal Society of London (B)* **359**, 173–181.

Hooghiemstra, H., Melice, J. L., Berger, A., and Shackleton, N. J. (1993) Frequency spectra and paleoclimatic variability of the high-resolution 30–1450 ka Funza I pollen record (Eastern Cordillera, Colombia). *Quaternary Science Reviews* **12**, 141–156.

Hooghiemstra, H., van der Hammen, T., and Cleef, A. (2002) Paleoecología de la flora boscosa. In: M. Guariguata and G. Kattan (eds.), *Ecología y Conservaciœn de Bosques Neotropicales* (pp. 43–58). Libro Universitario Regional, Cartago, Costa Rica.

Hoorn, C., Guerrero, J., Sarmiento, G. A., and Lorente, M. A. (1995) Andean tectonics as a cause for changing drainage patterns in Miocene northern South America. *Geology* **23**, 237–240.

Huber, O. and Riina, R. E. (1997) *Glosario Fitoecológico de las Américas. Vol. 1. América del Sur: Países Hispanoparlantes* (500 pp.). Fundación Instituto Botánico de Venezuela y UNESCO, Caracas.

IPCC (Intergovernmental Panel on Climate Change) (2001) *IPCC Summary for Policymakers. Climate Change 2001: Impacts, Adaptation and Vulnerability* (A report of Working Group II, 83 pp.). Cambridge University Press, Cambridge, U.K.

Janzen, D. H. and Martin, P. S. (1982) Neotropical anachronisms: The fruits the gomphotheres ate. *Science* **215**, 19–27.

Jaramillo, C. A. and Dilcher, D. L. (2002) Middle Paleogene palynology of central Colombia, South America: A study of pollen and spores from tropical latitudes. *Palaeontographica Abteiling B* **258**, 87–213.

Kessler M. (1995) Present and potential distribution of *Polylepis* (Rosaceae) forests in Bolivia. In: S. P. Churchill, H. Balslev, E. Forero, and J. L. Luteyn (eds.), *Biodiversity and Conservation of Neotropical Montane Forests* (pp. 281–294). New York Botanical Garden, New York.

Kolata, A. L., Binford ,M. W., and Ortloff, C. (2000) Environmental thresholds and the empirical reality of state collapse: A response to Erickson (1999). *Antiquity* **74**, 424.

Malcolm, J. R., Markham, A., Neilson, R. P., and Garaci, M. (2002) Estimated migration rates under scenarios of global climate change. *Journal of Biogeography* **29**, 835–849.

Malcolm, J. R., Liu, C., Neilson, R. P., Hansen, L., and Hannah, L. (2006) Global warming and extinctions of endemic species from biodiversity hotspots. *Conservation Biology* **20**, 538–548.

Marchant, R. *et al.* (2002) Distribution and ecology of parent taxa of pollen lodged within the Latin American Database. *Review of Palaeobotany and Palynology* **121**, 1–75.

McCune, B. and Mefford, M. J. (1999) *PC_ORD: Multivariate Analysis of Ecological Data.* MJM Software Design, Gleneden Beach, OR.

Mommersteeg, H. (1998) *Vegetation Development and Cyclic and Abrupt Climatic Changes during the Late Quaternary: Palynological Evidence from the Colombian Eastern Cordillera* (pp. 131–191). University of Amsterdam, Amsterdam.

Mourguiart, P. and Ledru, M. P. (2003) Last Glacial Maximum in an Andean cloud forest environment (Eastern Cordillera, Bolivia). *Geology* **31**, 195–198.

Mourguiart, P., Argollo, J., Correge, T., Martin, L., Montenegro, M. E., Sifeddine, A., and Wirrmann, D. (1997) Changements limnologiques et climatologiques dans le bassin du lac Titicaca (Bolivie), depuis 30 000 ans. *Comptes rendus de l'Académie des sciences, series 2* **325**, 139–146.

Mourguiart, P., Correge, T., Wirrmann, D., Argollo, J., Montenegro, M. E., Pourchet, M., and Carbonel, P. (1998) Holocene palaeohydrology of Lake Titicaca estimated from an ostracod-based transfer function. *Palaeogeography, Palaeoclimatology, Palaeoecology* **143**, 51–72.

Moy, C. M., Seltzer, G. O., Rodbell, D. T., and Anderson D. M. (2002) Variability of El Nino/Southern Oscillation activity at millennial timescales during the Holocene epoch. *Nature* **420**, 162–164.

Núñez, L., Grosjean, M., and Cartajena, I. (2002) Human occupations and climate change in the Puna de Atacama, Chile. *Science* **298**, 821–824.

Overpeck, J. T., Webb, T. I., and Prentice, I. C. (1985) Quantitative interpretation of fossil pollen spectra: Dissimilarity coefficients and the method of modern analogs. *Quaternary Research* **23**, 87–108.

Paduano, G. (2001) *Vegetation and Fire History of Two Tropical Andean Lakes, Titicaca (Peru/Bolivia), and Caserochocha (Peru) with Special Emphasis on the Younger Dryas Chronozone.* Department of Biological Sciences. Florida Institute of Technology, Melbourne, FL.

Paduano, G. M., Bush, M. B., Baker, P. A., Fritz, S. C., and Seltzer, G. O. (2003) A vegetation and fire history of Lake Titicaca since the Last Glacial Maximum. *Palaeogeography, Palaeoclimatology, Palaeoecology* **194**, 259–279.

Paillard, D., Labeyrie, L. and Yiou, P. (1996) Macintosh program performs time-series analysis. *EOS Trans. AGU* **77**, 379.

Placzek, C., Quade, J., and Patchett, P. J. (2006) Geochronology and stratigraphy of late Pleistocene lake cycles on the southern Bolivian Altiplano: Implications for causes of tropical climate change. *Geol. Soc. Am. Bull. 10.1130/B25770.1* **118**, 515–532.

Pounds, J. A. (2001) Climate and amphibian declines. *Nature* **410**, 639–640.

Pounds, J. A., Fogden, M. P. L., and Campbell, J. H. (1999) Biological response to climate change on a tropical mountain. *Nature* **398**, 611–615.

Pounds, J. A., Bustamante, M. R., Coloma, L. A., Consuegra, J. A., Fogden, M. P. L., Foster, P. N., La Marca, E., Masters, K. L., Merino-Viteri, A., Puschendorf, R. *et al.* (2006) Widespread amphibian extinctions from epidemic disease driven by global warming. *Nature* **439**, 161.

Rind, D. and Peteet, D. (1985) Terrestrial conditions at the last glacial maximum and CLIMAP sea-surface temperature estimates: Are they consistent? *Quaternary Research* **24**, 1–22.

Rodbell, D. T. (1992) Late Pleistocene equilibrium-line reconstructions in the Northern Peruvian Andes. *Boreas* **21**, 43–52.

Rodbell, D. T. and Seltzer, G. O. (2000) Rapid ice margin fluctuations during the Younger Dryas in the tropical Andes. *Quaternary Research* **54**, 328–338.

Rodbell, D. T., Seltzer, G. O., Anderson, D. M., Abbott, M. B., Enfield, D. B. and Newman, J. H. (1999) An ~15,000-year record of El Niño-driven alluviation in southwestern Ecuador. *Science* **283**, 516–520.

Rowe, H. D., Guilderson, T. P., Dunbar, R. B., Southon, J. R., Seltzer, G. O., Mucciarone, D. A., Fritz, S. C. and Baker, P. A. (2003) Late Quaternary lake-level changes constrained by radiocarbon and stable isotope studies on sediment cores from Lake Titicaca, South America. *Global and Planetary Change* **38**, 273–290.

Ruddiman, W. F. (2003) The anthropogenic greenhouse era began thousands of years ago. *Climatic Change* **61**, 261–293.

Seltzer, G. O. (1992) Late Quaternary glaciation of the Cordillera Real, Bolivia. *Journal of Quaternary Science* **7**, 87–98.

Seltzer, G. O., Rodbell, D. T., and Abbott, M. B. (1995) Andean glacial lakes and climate variability since the last glacial maximum. *Bulletin de l'Institut Français d'Etudes Andines* **24**, 539–550.

Seltzer, G. O., Rodbell, D. T., and Burns, S. (2000) Isotopic evidence for late Quaternary climatic change in tropical South America. *Geology* **28**, 35–38.

Seltzer, G. O., Rodbell, D. T., Baker, P. A., Fritz, S. C., Tapia, P. M., Rowe, H. D. and Dunbar, R. B. (2002) Early deglaciation in the tropical Andes. Science 298, 1685–1686.

Servant, M. (1977) La cadre stratigraphique du Plio-Quaternaire de l'Altiplano des Andes tropicales en Bolivie. *Res. Franç. Quat., INQUA 1977, Suppl. Bull. AFEQ, 1977-1* **50**, 323–327.

Smith, J. A., Seltzer, G. O., Farber, D. L., Rodbell, D. T., and Finkel, R. C. (2005) Early local last glacial maximum in the tropical Andes. *Science* 308, 678–681.

Tapia, P. M., Fritz, S. C., Baker, P. A., Seltzer, G. O., and Dunbar, R. B. (2003) A late Quaternary diatom record of tropical climatic history from Lake Titicaca (Peru and Bolivia). *Palaeogeography, Palaeoclimatology, Palaeoecology* **194**, 139–164.

Terborgh, J. (1977) Bird species diversity on an Andean elevational gradient. *Ecology* **58**, 1007–1019.

Terborgh, J. (1992) *Diversity and the Tropical Rain Forest* (242 pp.). Freeman, New York.

Thompson, L. G., Mosley-Thompson, E., Davis, M. E., Lin, P.-N., Henderson, K. A., Cole-Dai, J., Bolsan, J. F. and Liu, K. G. (1995) Late glacial stage and Holocene tropical ice core records from Huascaran, Peru. *Science* **269**, 46–50.

Thompson, L. G., Davis, M. E., Mosley-Thompson, E., Sowers, T. A., Henderson, K. A., Zagorodnov, V. S., Lin, P.-N., Mikhalenko, V. N., Campen, R. K., Bolzan, J. F. *et al.*. (1998) A 25,000-year tropical climate history from Bolivian ice cores. *Science* **282**, 1858–1864.

Thompson, L. G., Mosley-Thompson, E., Davis, M. E., Henderson, K. A., Brecher, H. H., Zagorodnov, V. S., Mashiotta, T. A., Lin, P.-N., Mikhalenko, V. N., Hardy, D. R. *et al.* (2002) Kilimanjaro ice core records: Evidence of Holocene climate change in tropical Africa. *Science* **298**, 589–593.

Van 't Veer, R. and Hooghiemstra, H. (2000) Montane forest evolution during the last 650 000 yr in Colombia: A multivariate approach based on pollen record Funza-I. *Journal of Quaternary Science* **15**, 329–346.

Van 't Veer, R., Islebe, G. A. and Hooghiemstra, H. (2000) Climatic change during the Younger Dryas chron in northern south America: A test of the evidence. *Quaternary Science Reviews* **19**, 1821–1835.

Van der Hammen, T. (1985) The Plio-Pleistocene climatic record of the tropical Andes. *Journal of the Geological Society* **142**, 483–489(487).

Van der Hammen, T. and González, E. (1960) Upper Pleistocene and Holocene climate and vegetation of the Sabana de Bogotá (Colombia, South America). *Leidse Geologische Mededelingen* **25**, 261–315.

Van der Hammen, T. and Hooghiemstra, H. (1995) The El Abra stadial: A Younger Dryas equivalent in Colombia. *Quaternary Science Reviews* **14**, 841–851.

Van der Hammen, T. and Hooghiemstra, H (2003) Interglacial–glacial Fuquene-3 pollen record from Colombia: An Eemian to Holocene climate record. *Global and Planetary Change* **36**, 181–199.

Van der Hammen, T., Duivenvoorden, J. F., Lips, J. M., Urrego, L. E., and Espejo, N. (1992) Late Quaternary of the middle Caquetá River area (Colombian, Amazonia). *Journal of Quaternary Science* **7**, 45–55.

Van Geel, B. and Van der Hammen, T. (1973) Upper Quaternary vegetational climate sequence of the Fuquene area (Eastern Cordillera Colombia). *Palaeogeography, Palaeoclimatology, Palaeoecology* **4**, 9–92.

Webb, S. D. (1997) The great American faunal interchange. In: A. G. Coates (ed.), *Central America: A Natural and Cultural History* (pp. 97–122). Yale University Press, New Haven, CT.

Webb, S. D. and Rancy A. (1996) Late Cenozoic evolution of the Neotropical mammal fauna. In: Jackson, J. B. C., Budd, A. F. and Coates, A. G. (eds.), Evolution and environment in tropical America, pp. 335–358. University of Chicago, Chicago.

Webster P. and Streten, N. (1978) Late Quaternary ice-age climates of tropical Australasia, interpretation and reconstruction. *Quaternary Research* **10**, 279–309.

Weng, C., Bush, M. B., and Silman, M. R. (2004) An analysis of modern pollen rain on an elevational gradient in southern Peru. *Journal of Tropical Ecology* **20**, 113–124.

Wille, M., Negret, J. A., and Hooghiemstra, H. (2000) Paleoenvironmental history of the Popayan area since 27000 yr BP at Timbio, Southern Colombia. *Review of Palaeobotany and Palynology* **109**, 45–63.

Wille, M., Hooghiemstra, H., Hofstede, R., Fehse, J., and Sevink, J. (2002) Upper forest line reconstruction in a deforested area in northern Ecuador based on pollen and vegetation analysis. *Journal of Tropical Ecology* **18**, 409–440.

Wirrmann, D., Ybert, J.-P., and Mourguiart, P. (1992) A 20,000 years paleohydrological record from Lake Titicaca. In: C. Dejoux and A. Iltis (eds.), *Lake Titicaca: A Synthesis of Limnological Knowledge* (pp. 40–48). Kluwer Academic Pess, Boston.

Witte, H. J. L. (1994) *Present and Past Vegetation and Climate in the Northern Andes (Cordillera Central, Colombia): A Quantitative Approach* (pp. 269). University of Amsterdam, Amsterdam.

Ybert, J. P. (1992) Ancient lake environments as deduced from pollen analysis. In: C. DeJoux and A. Iltis (eds.), *Lake Titicaca: A Synthesis of Limnological Knowledge* (pp. 49–62). Kluwer Academic Publishers, Boston.

3

Climate change in the lowlands of the Amazon Basin

M. B. Bush, W. D. Gosling, and P. A. Colinvaux

3.1 INTRODUCTION

Data from palynology, taxonomy and isotopic analyses, allied to climate models, reveal the complexity of the history of Amazon ecosystems. Evidence from these records suggests that Pleistocene climatic change was neither uniform nor synchronous across the basin, but that its effects were pervasive.

A major obstacle to Amazon paleoecology is paucity of lakes containing uninterrupted sedimentary sequences spanning one or more glacial cycles. To date, the only fossil pollen records from lowland Amazon lake sediments to span the last glacial maximum (LGM) are those of Carajas (Absy *et al.*, 1991), Maicuru (Colinvaux *et al.*, 2001), three from the Hill of Six Lakes (Colinvaux *et al.*, 1996; Bush *et al.*, 2004a), together with Lakes Chaplin and Bella Vista on the southwestern forest–savanna ecotone in Bolivia (Mayle *et al.*, 2000; Burbridge *et al.*, 2004) (Figure 3.1). All of these are relatively small, shallow bodies of water that are vulnerable to desiccation. That these lakes retained water for most of their >50,000-year histories is testament to the relative constancy of Amazonian precipitational regimes. However, none of these records presents an ideal archive, as they all contain sedimentary gaps or extremely slow rates of sediment accretion, are not necessarily located in optimal locations for studying past climate change, and only represent a small portion of the Quaternary.

These lake records are supplemented by the pollen history of the Amazon lowlands from sediments of the Amazon fan (Haberle, 1997; Haberle and Maslin, 1999) and records near the upper limits of wet Amazon forest on the flanks of the Andes. These lower montane records include Lake Consuelo at 1,360-m elevation in Peru (Bush *et al.*, 2004b), and Mera (1,100 m) and San Juan Bosco (970 m) in Ecuador (Liu and Colinvaux, 1985; Bush *et al.*, 1990) (Figure 3.1).

A combination of new paleoecological records, molecular phylogenies and climatic understanding has contributed to a revised view of the evolutionary origins of Amazonian biodiversity. In the past decade the suggestion that arid glacials led to

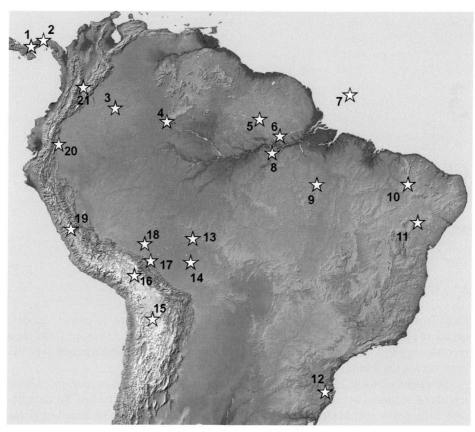

Figure 3.1. The location of paleoecological sites mentioned in the text in relation to topography. Citations are given for locations not specifically identified in text: (1) Lake Gatun (Bartlett and Barghoorn, 1973); (2) El Valle; (3) Loma Lindo (Behling and Hooghiemstra, 1999); (4) Hill of Six Lakes (Pata, Verde, and Dragao; (5) Maicuru; (6) Geral, Santa Maria, and Saracuri (De Toledo, 2004); (7) Amazon fan; (8) Tapajos; (9) Carajas; (10) sites in northeastern Brazil (Stute *et al.*, 1995); (11) Lapa dos Brejões and Toca da Barriguda caves (Wang *et al.*, 2004); (12) Botuverá cave; (13) Acre transect (Pessenda *et al.*, 1998); (14) Lakes Chaplin and Bella Vista (Mayle *et al.*, 2000); (15) Salar de Uyuni; (16) Titicaca (Paduano *et al.*, 2003); (17) Lake Consuelo; (18) Gentry, Werth, Parker, Vargas (Bush *et al.*, in press); (19) Junin; (20) Mera and San Juan Bosco; (21) High Plain of Bogotá.

the contraction of rainforests into relatively small and isolated areas prompting waves of allopatric vicariance—the refugial hypothesis (Haffer, 1969)—has lost traction. Testing the veracity of this hypothesis was the focus of Amazon paleoecological research for three decades and has been exhaustively covered in recent reviews (see Colinvaux *et al.*, 2001; see Haffer and Prance, 2001). In its most recent manifestation (Haffer and Prance, 2001), the refugial hypothesis no longer describes forests replaced by savanna, but by "intermediate forest". These ill-defined vegetation types could include any combination of elements from woody savanna to mesic forest. Haffer and

Prance (2001) also suggest that mesic forest may have persisted in the landscape along riparian corridors. According to this hypothesis not all species would have found this new landscape a barrier to dispersal, whereas others would. Furthermore, the time-frame for these changes has shifted from the last glacial period to some time in the Tertiary. The clearly defined set of predictions that provided the appeal of the original refugial hypothesis has been replaced by a nebulous wraith that defies testing. Despite these flaws, some researchers are not yet prepared to discard the refugial hypothesis. Rather than re-hash that argument in detail, we will make the following observations:

1. Molecular and genetic data for plants, birds and mammals, suggest that specia-tion was not concentrated within the ice ages, but has probably been a continuous process with most modern species appearing since the beginning of the Miocene (e.g., Zink and Slowinski, 1995; Moritz *et al.*, 2000; Pennington *et al.*, 2004).
2. None of the late Quaternary paleoecological records from the center of the Amazon basin provide evidence of widespread, long-term aridity sufficient to cause the fragmentation of rainforests, although there is evidence of expansion of forests in the last few thousand years at the margins (Bush, 2005).
3. Simple contrasts of wet versus dry, cold versus warm, cannot capture the vari-ability of the glacial–interglacial cycles and the communities without modern analogs that were generated (Bush and Silman, 2004).
4. Dry forest species in modern "refugia" have not speciated and so presumably 19,000 years since deglaciation is not long enough for speciation to occur (Pennington *et al.*, 2004).

3.2 EVIDENCE OF TEMPERATURE CHANGE

The first evidence that Amazonia experienced substantially cooler-than-modern con-ditions during the last ice age came from the discovery of *Podocarpus* timbers in an exposure of silty peat near the town of Mera, Ecuador, at 1,100-m elevation (Liu and Colinvaux, 1985). *Podocarpus* spp. are generally found in cloud forests above 1,800 m, and this observation was used to draw the inference of a c. 800-m descent of *Podo-carpus* populations at c. 30–36 kcal. yr BP. Indeed, because *Podocarpus* is not restricted to modern montane forests, Gentry (1993) lists four species that occur above 1,800 m and one that is found in the lowlands—the use of this genus as a paleoecological indicator of cooling has been criticized (van der Hammen and Hooghiemstra, 2000). However, further work on Mera and the site of San Juan Bosco that lies about 160 km to the south of Mera (Bush *et al.*, 1990) revealed a suite of macro- and microfossils of additional taxa that are similarly most abundant in modern montane forests. *Drimys, Alnus, Weinmannia*, and *Hedyosmum*, were found to be abundantly repre-sented in glacial age samples that also were depauperate in the taxa currently associated with this elevation—for example, *Cecropia*, Urticaceae/Moraceae, *Iriartea*.

This finding reinforced the probability that these sites supported cold-adapted elements at the peak of the last ice age. Every Pleistocene-aged pollen record recovered from Amazonia provides similar evidence of cool-tolerant populations moving into the lowland forests. In each case, floral elements became abundant 800 m to 1,500 m below the modern centers of their population. Similar patterns have been found in regions adjacent to the Amazon Basin—for example, southern and central Brazil (de Oliveira, 1992; Ledru, 1993; Salgado-Labouriau, 1997), the Andes (reviewed in Chapter 2), and Central America (Bartlett and Barghoorn, 1973; Bush and Colinvaux, 1990).

Quantifying this cooling has relied on translating the altitudinal descent of species into temperature change via the moist air adiabatic lapse rate, generally taken to be c. 5.5–6°C. Hence, the observed 800-m to 1,500-m descent of thermally sensitive populations translates into a 4°C to 7°C cooling. A similar estimate of cooling was developed from the isotopic analysis of groundwater in eastern Brazil. Stute *et al.* (1995) found the temperature of "fossil" groundwater to be c. 5°C cooler than that of groundwater formed under modern conditions.

Van der Hammen and Hooghiemstra (2000) and Wille *et al.* (2000) have argued for a flexible lapse rate during the ice ages that would significantly steepen the temperature gradient from the lowlands to the highlands. Their contention is that the Colombian Andes at 2,580 m cooled by c. 8°C while the lowlands cooled only between 2.5 and 6°C (according to the data source). If these values are taken at face value, one interpretation is that the moist air adiabatic lapse rate must have steepened. The implied lapse rate to accommodate the difference in montane versus lowland temperature increases for a modern rate of c. 6°C in Colombia (Wille *et al.*, 2000) to unrealistically high values of between 6.7 and 8.1°C. Such high lapse rates are unlikely as they imply very dry air, and—given that no mid-elevation Andean setting with that kind of aridity has been documented so far—a flexible lapse rate is not the solution to the observed variability in data.

The moist air adiabatic lapse rate is controlled by atmospheric humidity and is not seen to vary greatly from one tropical setting to another despite differences in precipitation and seasonality—that is, it is almost always 5.8 ± 0.5°C per 1,000 m of elevation. While narrow fluctuations can be expected through time, lapse rates are unlikely to vary beyond a constrained range (Rind and Peteet 1985).

On first principles it is difficult to envisage a very strong change in lapse rate in humid sections of the Andes, and yet the foothill regions consistently provide a slightly lower (typically 5°C change) temperature reconstruction than the highlands (typically 8°C change). Given that the LGM in Colombia was dry (Hooghiemstra and van der Hammen, 2004), while it was wet in Peru and Bolivia (Baker *et al.*, 2001), we can assume that lapse rates may have risen close to c. 6.3°C in Colombia and been near modern (i.e., 5.5°C) in the central Andes. However, these changes are inadequate to describe the c. 3–6°C discrepancy between the lowlands and the uplands. We advocate taking a step back and considering other mechanisms than lapse rate change to account for the observed data.

We observe that paleovegetation response in mountains does not provide a pure temperature signal and is likely to be exacerbated by factors correlated with elevation.

If not parsed out, these factors can lead to an exaggerated paleotemperature change estimate. For example, changes in black body radiation due to lowered atmospheric CO_2 content (Bush and Silman, 2004 and Chapter 10) and feedback mechanisms involving ultraviolet radiation (Chapter 8) are more extreme with increasing elevation. With a thinner atmosphere—that is, LGM conditions of 170 ppm CO_2 and 350 ppb CH_4 compared with pre-industrial Holocene concentrations of 280 ppm CO_2 and 650 ppb CH_4—more heat is lost during nighttime re-radiation of stored heat than under modern conditions. This black body radiation effect would increase with elevation, be strongest under cloudless skies, and might add an apparent c. 2.3°C of cooling to the true change in temperature (Bush and Silman, 2004). As it is the coldest nighttime temperatures that a plant must survive, the black body radiation imposes an additive thermal stress that can cause mortality as physiological thresholds are exceeded. This mechanism provides one example; more probably exist, contributing to the differential migration distances in montane and lowland settings.

Although the seminal work on the High Plains of Bogotá set a new path for Neotropical paleoecology, many other records now exist that suggest somewhat more modest temperature departures both in the lowlands and in the mountains than the reported 7–9°C. Thus, we suggest that the migration of tree line as inferred from the pollen records is accurate, but that there may be more mechanisms at work than simply temperature change contributing to those shifts. If estimates of glacial descent and ELA (Seltzer, 1990; Rodbell, 1992; Seltzer *et al.*, 2003; Smith *et al.*, 2005) are also considered, the actual LGM cooling at all elevations may have been c. 4–6°C, making it more consistent with data obtained from marine paleotemperature reconstructions (Ballantyne *et al.*, 2005).

It is important to note that this degree of cooling was not temporally uniform throughout the last ice age and that there were other high-magnitude changes (comparable with the Pleistocene/Holocene transition) during this period. That all the lowland records contain gaps in sedimentation makes it difficult to put a firm timeline on when Amazonia was coldest. At the Hill of Six Lakes, montane taxa are clearly abundant in samples that are radiocarbon-infinite in age, and they have their peak occurrence between 21 kcal.yr BP and 18 kcal.yr BP (Bush *et al.*, 2004a). Looking farther afield, Lake Titicaca and Lago Junin in the Peruvian Andes were probably coldest between c. 35 kcal.yr BP and 21 kcal.yr BP (Hansen *et al.* 1984; Seltzer *et al.*, 2000; Smith *et al.*, 2005). But, at the lowest elevation of modern cloud formation, the record from Lake Consuelo (1,360-m elevation) indicates a protracted, steady cooling of about 6°C between 40,000 and 22,000 cal yr BP (Urrego *et al.*, 2005).

In some records, particularly those in eastern Amazonia and coastal Brazil (de Oliveira, 1992; Ledru, 1993; Behling, 1996; Behling and Lichte, 1997; Haberle and Maslin, 1999; Ledru *et al.*, 2001; Sifeddine *et al.*, 2003), cold-tolerant taxa persist and in some cases reach their peak abundance as late as 14 kcal.yr BP, when western Amazonian and selected Andean records are showing considerable, steady, warming (Paduano *et al.*, 2003; Bush *et al.*, 2004b). Such short-term variability has yet to be explained, but it is a marked characteristic of these records that climatic events are neither synchronous nor basin-wide.

3.3   EVIDENCE OF PRECIPITATION CHANGE

The topic of Amazonian precipitation has been divisive, and positions for or against Amazonian aridity relatively entrenched. We have been proponents of an Amazonian system that remained relatively moist during the last ice age, a view contrasted by those who see Amazonia as having been much drier than present (e.g., Haffer, 1969). At first, it was easy to argue either position as the field of debate was uncluttered by empirical data. Now both sides are drifting toward the middle as we recognize that Amazonia did not dry out sufficiently to fragment its forests, but that relatively dry periods that lasted for perhaps as much as 11,000 years were a reality. Furthermore, some migration of the forest boundary, and its replacement by semi-deciduous dry forest or even savanna close to ecotonal boundaries is indicated both by paleo-ecological studies and models (Absy *et al.*, 1991; Behling and Hooghiemstra, 1999; Cowling, 2004; Cowling *et al.*, 2004). To understand why paleoprecipitation was such a hot-button issue for those involved it has to be appreciated that this debate harks back to whether, or not, refugia genuinely drove Amazonian speciation. More was at stake than a simple history of rainfall. The underlying issue was an explanatory mechanism of modern species distributions, the role of late Quaternary climatic change as an agent of evolution and, hence, the age of Amazonian species diversity.

We now have proxy data for past lake level and inferred precipitation balance. We also have global data for changes in atmospheric CO_2 concentrations that may complicate the issue, and phylogenetic histories (both traditional and molecular) that help to simplify it.

The only long, continuous, sedimentary records that span a significant portion of the last glacial cycle come from a string of massifs that arc from northwestern Amazonia across eastern Amazonia down to the south. The Hill of Six Lakes (three records), Maicuru, and Carajas are the five longest Amazonian paleoecological records, and all come from lakes that sit atop edaphically dry massifs or inselbergs. All five records contain a sedimentary hiatus, but the timing of these hiatuses differs between the sites. It is thought likely that these hiatuses are the product of dry periods. At the Hill of Six Lakes the dry event is most intense between c. 40 kcal. yr BP and 27 kcal. yr BP (Bush *et al.*, 2004a), at Maicuru it is most intense between 33 kcal. yr BP and 19 kcal. yr BP (Colinvaux *et al.*, 2001), while Carajas is driest between 28 kcal. yr BP and 16 kcal. yr BP (Absy *et al.*, 1991; Ledru *et al.*, 2001). These data demonstrate that during the late Pleistocene relatively short dry events occurred asynchronously across the basin.

The records of lake level suggest both a west-to-east and south-to-north variation in climate history within the basin. Global circulation models of Amazonian climatic responses at the LGM are similarly regionalized with opposing signatures of precipitation change evident across the basin (Hostetler and Mix, 1999). Indeed, climatic responses to the modern El Niño/Southern Oscillation reiterates this regionalization with markedly different climatic responses across the basin (Bush and Silman, 2004). The over-arching conclusions are that scenarios calling for a uniform basin-wide climate change are inherently implausible, and that when dry events occurred they were relatively short-lived. Equally, it is unlikely—in view of the variation in hydrology, topography, soils and species richness of Amazonia—that the highly diverse vegetation responded uniformly to a given forcing.

3.4 CHANGES IN ATMOSPHERIC CO_2

Simulations of the response of ecosystems to reduced CO_2 (Mayle and Beerling, 2004; Mayle et al., 2004) suggest that dry forest and savannas occupied modern ecotonal areas, especially on the southern rainforest margin. Lowering atmospheric concentrations of CO_2 is generally considered to induce drought stress in plants with a C3 photosynthetic pathway, as they must have their stomates open longer than C4 plants to assimilate enough CO_2 for growth. In very dry systems plants that have C4 or CAM pathways, which can open and close stomata to retain water and photosynthesize more efficiently under intense light and temperatures, may displace C3 species. However, such a change is not detected in any of the paleoecological records during the period of relatively lower atmospheric CO_2 concentrations. These observations are consistent with the lack of observed changes in $\delta^{13}C : {}^{12}C$ ratios across a transect of ecotonal sites in southern Brazil that straddle the Pleistocene–Holocene boundary (Pessenda et al., 1998).

Cowling has suggested that lower CO_2 at the LGM induced plants to support reduced leaf area indices and forests to be structurally simpler (Cowling et al., 2001, 2004; Cowling, 2004). If such systems had more light and higher rates of evaporation at the soil surface the ramifications would be far-reaching for microclimates and soil carbon flux. Clearly, the role of CO_2, and the adaptations that plants show toward it, is an important area of research and provides a substantial variable that is hard to quantify.

Although there has been considerable emphasis on the potential role of CO_2 as a potent limiting factor in the Late Pleistocene, an alternative view is also worth considering. Due to the temporal asymmetry of ice ages it can be argued the rate of change in atmospheric CO_2 concentrations was much slower in the transition into cold episodes than during the warming associated with the onset of interglacials (Figure 3.2). Fast increases in CO_2 concentration may have acted as a fertilizer and influenced community structure, but the declines in CO_2 were so slow there may have been little negative effect on productivity. The Vostok ice-core data reveal rates of change seldom exceeding 0.001 ppm and averaging 0.0001 ppm when CO_2 concentrations were falling (Figure 3.2). Given that modern seasonal oscillations in CO_2 are about 5 ppm, the changes imposed on plants drifting towards an ice age were imperceptibly slow and may have been within the range accommodated by natural selection, thereby minimizing its effect.

3.5 THE PERIODICITY OF CHANGE

For substantial periods western Amazonian and southern Andean sites do appear to have similar histories. Sedimentary records in the Colombian Andes, the Bolivian

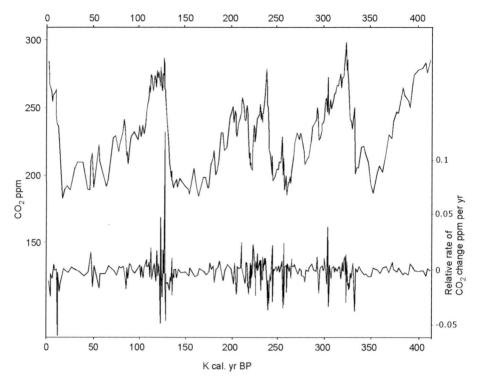

Figure 3.2. CO$_2$ concentrations from the Vostok core (Petit *et al.*, 1999) (*upper line*) and the relative rate of change in CO$_2$ concentrations between samples (*lower line*).

Andes, the Chilean pampas, and the Cariaco Basin all reflect precessional forcing during the last 50,000 years, and so too does the paleoecological record of Lake Pata in the Hill of Six Lakes (Bush *et al.*, 2002). Records from speleothems collected in southern and northeastern Brazil also document a precessional pattern in moisture supply (Wang *et al.*, 2004; Cruz *et al.*, 2005).

However, upon closer inspection, some other patterns emerge. Although precessional forcing is evident in all these records, the Colombian Andes are out-of-phase with the southern Andes (Bush and Silman, 2004), which is not surprising as the wet season for each occurs 6 months apart. Consequently, precipitation in Colombia is in phase with July insolation, whereas that of Titicaca correlates with December insolation. Interestingly, the highstands and lowstands of Lake Pata (0° latitude) are in phase with those of Lake Titicaca (17°S) and so a simple geographic placement of the site is not enough to predict the orbital forcing. The clue to the connection comes from the speleothem record from Botuverá Cave near Rio de Janeiro in southeastern Brazil. This record reveals an oscillation in the strength and position of the South American Summer Monsoon (SASM) resulting from changes in insolation intensity (Cruz *et al.*, 2005) (Figure 3.3). SASM is driven by the convective activity over Amazonia and fed moisture via the South American low-level jet (SALLJ). The SALLJ is strengthened as

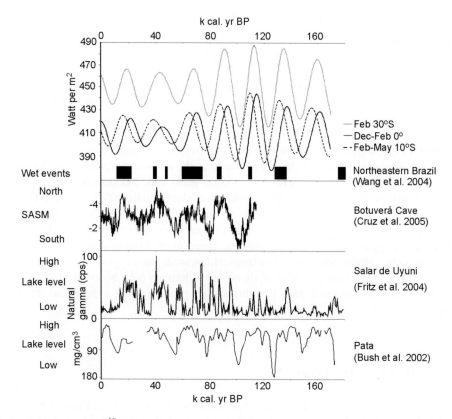

Figure 3.3. Data for $\delta^{18}O$ from Brazilian speleothem records, downcore gamma radiation Salar de Uyuni and K^+ concentration from Lake Pata, compared with mean insolation calculated in Analyseries 1.2 for 0°, 10°S, and 30°S (Berger, 1978). Periods selected are those used by authors in original descriptions relating data to insolation.

convection intensifies with the net result of drawing more moisture from the northern tropical Atlantic into Amazonia. As most of the basin heats most strongly in December, this sets the precessional rhythm for sites receiving moisture from the ALLJ (Bush, 2005). The position of SASM is influenced by convection, so that during strong convection SASM expands farther south, bringing rain to Botuverá Cave. During weaker convection Amazonia is drier and SASM is more restricted in its southerly range. SASM expands progressively southward during the austral summer and so the best correlation with orbital forcing is obtained by tracking peaks in February (late summer) insolation (Figure 3.3). SASM also transports Amazonian moisture to the Altiplano, and hence lake levels in the Titicaca and the Salar de Uyuni records, but here it is the entire wet season that is the time of critical insolation (December–February).

Another tantalizing speleothem data set from eastern Brazil shows a precessional pattern with wet peaks aligning to austral autumn (February–May) peaks in insolation, as opposed to the December cycles of the other sites (Wang *et al.*, 2004). As we gain more high-resolution paleoclimatic records it may be possible to test whether past dry events can truly be predicted based on precessional influences on the prevailing moisture source (Figure 3.3).

Interestingly, the long sedimentary record from the Salar de Uyuni reveals that precession is not strong enough to counteract some other drivers of precipitation change. For example, Fritz *et al.* (2004) suggest that global ice volume is another significant variable in Pleistocene Andean precipitation and that—until ice volume reaches a critical point—precession does not emerge as a significant factor. A similar argument was made that the climate of Panama shows poor correlation with climatic events in the North Atlantic prior to c. 45,000 years ago, but between that time and about 14,000 years ago a closer relationship is evident (Bush, 2002). Thus, in Panama at c. 7°N and in the Altiplano it appears that prevailing controls on climate were modified by global ice volume. Indeed, in Colombia the record from the High Plains of Bogotá appears to reflect faithfully the three main Milankovitch cycles (Hooghiemstra *et al.*, 1993), whereas in the tropical lowlands farther south, precession is by far the most important pacemaker of climate change. An observation that corresponds well with modeled data that highlights precession as a potential driver of tropical paleoclimates (Clement *et al.*, 1999, 2001).

3.6   THE TYPE OF FOREST

Almost all Amazonian pollen records that extend back into the Late Pleistocene reveal higher inputs of pollen from montane forest taxa—for example, *Podocarpus*, *Alnus*, and *Hedyosmum*. Available data suggest that these forests lack exact modern counterparts. Lowland species persisted alongside montane species during the coldest episodes, but for much of the glacial period there may have been minimal penetration of the lowlands by montane taxa. These data argue that lowland Amazon forests of the Pleistocene were mesic systems periodically stressed by strong cooling and drought.

An emerging view supports the presence of forest cover across Amazonia throughout the Quaternary, but raises the possibility that the forests were much drier than those of today and may have supported less biomass (Prado and Gibbs, 1993; Pennington *et al.*, 2000, 2004; Cowling *et al.*, 2004; Mayle and Beerling, 2004). A suggestion for a dry arc of vegetation connecting southern Amazonian savannas to those of Colombia (Pennington *et al.*, 2000) has been revised in the light of new paleoecological data from lowland Bolivia (Mayle *et al.* 2000; Burbridge *et al.* 2004; Pennington *et al.*, 2004).

The finding that dry forest did not expand as predicted in the Bolivian ecotonal region investigated by Mayle's team was a major obstacle to the corridor concept. Similarly, misidentified stone lines within the Belterra clays that were taken as a signal of aridity, have been re-evaluated and are now seen to support continuous Pleistocene edaphic moisture (Colinvaux *et al.*, 2001).

Pennington *et al.* (2000) suggested that the Hill of Six Lakes data could equally represent dry forest as represent wet forest. However, this suggestion was based on the abbreviated list of species presented in the first reporting of the Lake Pata record (Colinvaux *et al.*, 1996). The full species list of >300 pollen and spore types includes many species that are strong indicators of cool mesic forest—such as *Cedrela*, *Cyathea*, *Podocarpus*, *Ilex*, *Brosimum*, and *Myrsine*. Indeed, the most plausible connection from south to north across Amazonia is via the eastern corridor of low precipitation that extends from southern Guiana to eastern Brazil (Figure 3.3). Given the large river barriers, complex orographic features of western Amazonia and the overall climatic heterogeneity of the basin, it is more appropriate to consider Amazonia as a series of patches and connectivity as a shuffling within the patchwork rather than migration in corridors.

Although the dry forest flora is a minority component of the lowland Neotropical system, it is worth exploring the potential for connection between isolated sites as the lessons learned are applicable to the wetter systems as well. A first observation is that many of the sand savannas peripheral to, and within Amazonia contain endemic species and probably have not been completely connected recently, if ever (Pennington, pers. commun.). However, migration through the Amazonian landscape may not have been through white sand soils, as many of the dry forest species demand better soils (Pennington *et al.*, 2004), and their migrational corridors may have been through riparian settings.

A significant problem in this arena is the iconography of biome-watching. Concepts such as "caatinga", "semi-deciduous tropical forest", or "wet forest" are at best the abstractions of mappers. These broad vegetation classifications can help us to think about a system, but we should not believe that they are sharply distinct in their ecology. Many of the same species that occur in a lowland semi-deciduous forest are found at the edge of their ecological tolerance within mesic forest. Nuances of a month more or less of dry season may substantially alter the dominant taxa, though perhaps resulting in much less change in the long tail of rare species (*sensu* Pitman *et al.*, 1999). Tuomisto *et al.* (1995) recognize >100 biotopes of *tierra firme* forest types in Peruvian Amazonia alone, and Duivenvoorden and Lips (1994) some 20 plant communities within a small rainforest near Araracuara, Colombia, and Olson *et al.* (2001) recognize 35 types of dry forest. Even within this wealth of forest types a given species may be facultatively deciduous in one setting, or in one year.

The dry valleys scattered along the eastern flank of the Andes are probably the result of interactions between prevailing winds and topographic blocking (Killeen, in press). As such these features are probably relatively fixed and have offered the potential for "island-hopping" around western Amazonia for several million years. The most probable path for migration through Amazonia would be where vegetation is potentially susceptible to fluctuations in precipitation.

A corridor of low precipitation crosses north–south across Amazonia at c. 50°W (Figure 3.4). Based on the seasonality and totals of precipitation, Nepstad *et al.* (1994) suggest that in this region evergreen forest trees are reliant on deep soil water when seasonal drought deficits occur in the surface soils. If this reliance on hydrology is true, any factor that lowers water tables beyond rooting depth could lead to a substantial

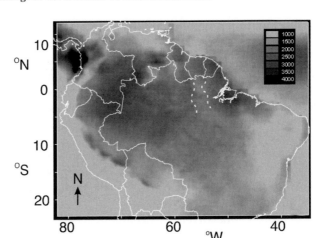

Figure 3.4. Precipitation data for Amazonia based on satellite monitoring (TRMM) showing the corridor (*dotted lines*) identified by Nepstad *et al.* (1994) where evergreen trees are dependent on deep soil moisture.

change in the flora. This same corridor was suggested by Bush (1994) to have been the section of Amazonia most sensitive to reduced precipitation. Whether this region experienced drought synchronously—or whether portions of it were wet while other portions were dry—would not inhibit the eventual migration of a species north to south through this region.

Finally, the distribution of dry forest arc species show some disjunctions across the Andes. As it is implausible that continuous habitat for these species spanned the mountains during the Quaternary, we are left with three possibilities. Congeners might have been mistaken for conspecifics. Or the species in question might be capable of dispersing across a major biogeographic barrier such as the Andes, in which case dispersal across lesser barriers within Amazonia obviates the need for any kind of habitat continuity. The third possibility is that these biogeographic patterns are relictual and derived from the pre-Andean biogeography of South America—that is, at least mid-Miocene in age (Hoorn *et al.*, 1995). We consider the latter two of these options to be the most probable.

The message emerging from this analysis is that simplistic notions requiring the whole of the vast Amazon lowlands to have a single climate subject to uniform change are bound to fail. Pleistocene forests were shaped by cooling, low atmospheric CO_2 concentrations, and at differing times by lessened rainfall. There is a strong suggestion of no-analog communities based on the prevalence of montane taxa interspersed with a full suite of lowland taxa. Some species that we now consider to be dry forest species may also have been able to survive within this forest, especially if it was structurally somewhat more open than modern forests. Thus, asking if a "dry forest arc" once existed may be much less relevant as a question, than "to what extent (if any) would conditions have to change to allow species to migrate through or around Amazonia?" The answer to this latter question may be "surprisingly little".

3.7   PHYLOGENIES

The empirical paleoenvironmental data discussed above have provided an improved understanding of the response of the vegetation in the Amazon Basin to known global climate change. New phylogenetic studies question some long-held assumptions about the nature and pace of Amazonian speciation and offer some critical insights into when speciation occurred.

Examples of biological insights arising from phylogenies come from a variety of organisms. *Heliconius* butterflies (Brown, 1987), frogs, and primates (Vanzolini, 1970) were all advanced as exemplars of refugial evolution, and each group has been revisited using modern molecular cladistic techniques.

Heliconius butterflies provide an excellent example of how reliance on modern biogeographic patterns can result in false assumptions about past evolution. *H. erato* and *H. melpomene* are co-mimetics, whose close co-evolution has been taken to indicate a similar biogeographic history (Brown, 1987), but a genetic analysis reveals that their patterns of divergence are markedly different (Flanagan *et al.* 2004). Sub-populations of *H. erato* show little geographic structure in rapidly evolving alleles, telling us that there is a relatively ancient divergence among a species with fairly high rates of gene flow. *H. erato* is the model, and the initial mimic, *H. melpomene*, has discrete sub-populations consistent with a history of local population isolation and expansion. These data—rather than suggesting environmental change as a cause of phenotypic variation—demonstrate the power of positive feedback in Mullerian mimicry when an abundant model (*H. erato*) is already present (Flanagan et al,. 2004).

Studies of the molecular clock suggest that speciation of birds and *Ateles* (spider monkeys) was more rapid in the Miocene or Pliocene than the Pleistocene, and that many of the species purported to reflect allopatry in an arid Amazon were already formed prior to the Quaternary (Zink and Slowinski, 1995; Collins and Dubach, 2000; Moritz *et al.*, 2000; Zink *et al.*, 2004). The important messages to emerge from these studies are that different environmental factors serve as barriers to different lifeforms, that common biogeographic patterns do not imply similar evolutionary demographies, that speciation is—and has been—a continuous feature of these systems, and lastly that the scale and ecological complexity of Amazonia can serve to induce species to specialize and perhaps even speciate without a vicariant event.

Simple assumptions about the compositions of Amazonian clades were also challenged by an analysis of small mammals that found an overlap of species— and genetic similarity—between specimens from the central Brazilian dry forests and those from adjacent rainforests (Costa, 2003). In some clades, samples collected from the dry forest were more similar to those from rainforests than other rainforest samples. This study highlighted the integrated, polyphyletic history of clades and that no simple model of vicariance is likely to explain their biogeography.

Phylogenetics has played an important role in demonstrating the temporal pattern of speciation and the development of biogeographic regions within Amazonia. Reviews by Moritz *et al.* (2000) and Hall and Harvey (2002) reveal that the geographic divergences contain broadly similar themes between groups. Within the Amazon basin, many clades show a basal split that separates northern and western clades

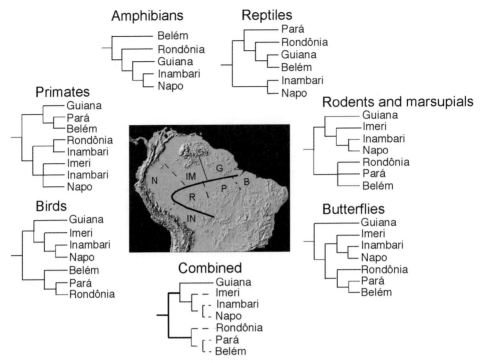

Figure 3.5. Summary phylogenies for a variety of Amazonian animal taxa (after Hall and Harvey, 2002) compared with their biogeographic relationship. The overall combined phylogeny is reproduced on the map. Levels of the phylogeny are reflected in the weight and dash patterns of lines on both the tree and the map: B = Belém; G = Guiana; Im = Imeri; In = Inambari; N = Napo; P = Pará; R = Rondônia.

from southeastern ones (Figure 3.5), but there is no fine-scale pattern indicating distinct centers of endemism (Bates *et al.*, 1998; Collins and Dubach, 2000; Patton *et al.*, 2000; Hall and Harvey, 2002; Symula *et al.*, 2003). Ideally, the dating of the divisions between clades should be underpinned by multiple dated calibration points derived from the fossil record (Near *et al.*, 2005). However, in Amazonia the fossil record for many taxa is so depauperate that many of these studies rely on previously estimated rates of molecular evolution to convert observed genetic distances to absolute age estimates. As these estimates are consistent with records derived from those in phylogenies supported by fossil data, it is probable that their age estimates are correct within a factor of 2.

In general, for clades that extend into the wider lowland Neotropics (e.g., the Chocó), there is a deeper divergence consistent with division of the clade by Miocene Andean orogeny (Hoorn *et al.*, 1995). However, even in Amazonian clades there is increasing evidence that the basal splits took place within the Miocene (Moritz *et al.*, 2000).

Nores (1999) and Grau *et al.* (2005) have suggested that high sea levels during the Pliocene or Quaternary may have fragmented bird populations and account for their patterns of vicariance. The presence of higher Miocene sea levels has been suggested to be linked to the isolation of the northern and southern massifs within the Guianan highlands and allowed divergence of geographic amphibian clades (Noonan and Gaucher, 2005). Considerable uncertainty exists regarding Miocene and Pliocene sea levels; Miller *et al.* (2005) suggest that they may have been only 30–60 m higher than present, whereas other estimates place them 110–140 m higher (Räsänen *et al.*, 1995; Naish, 1997). The rise of the Andes caused forebasin subsidence in western Amazonia. Although sea-level rise was only c. 30–60 m higher than present, when coupled with tectonic subsidence the effect was a c. 100 m rise in relative sea level. Repeated highstands approached 100 m above modern levels—as suggested by Räsänen *et al.* (1995) and Nores (1999)—the basal splits between northern and southeastern Amazonian clades, and the Guiana clades separating from other northern clades, are explicable (Figure 3.6). Nevertheless, considerable debate surrounds the extent

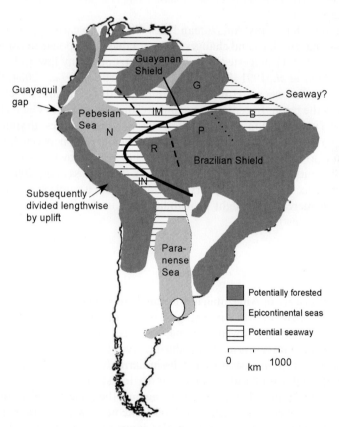

Figure 3.6. Summary diagram showing the relationship between flooding caused by a 100-m marine highstand and known epicontinental seas in Amazonia, and biogeographic patterns. *Darkest area* is postulated Miocene seaway (after Räsänen *et al.*, 1995). The biogeographic divisions and phylogenies are derived from Figure 3.5: B = Belém; G = Guiana; Im = Imeri; In = Inambari; N = Napo; P = Pará; R = Rondônia.

and duration of epicontinental seas within South America during the Miocene (e.g., Hoorn 1993; Hoorn *et al.*, 1995; Räsänen *et al.*, 1995), and before strong conclusions can be drawn regarding their influence on the biogeography of South America, more complete geological research and improved phylogenies are clearly needed.

3.8   HOLOCENE CLIMATE CHANGE

With so many questions regarding the ice age climates of Amazonia, those of the Holocene have received less attention than they deserve. Servant *et al.* (1981) wrote of a major Holocene drought and Meggers (1994) discussed the potential impact of climate change on human societies. The role of human occupation and its impact on the landscape of Amazonia and Central America is a matter of active debate (Denevan, 2003; Heckenberger *et al.*, 2003; Meggers, 2003) and is addressed by Piperno (Chapter 7).

Here we note that many Amazonian lakes filled in the early Holocene, possibly due to precessional rhythms and changes in water tables in response to rising sea level. As Servant *et al.* (1981) suggested, a dry event interrupts many lowland sedimentary records (e.g., Absy *et al.*, 1991; Mayle *et al.*, 2000; Burbridge *et al.*, 2004; Bush *et al.*, 2004a; Bush *et al.*, in review). However, the timing of this event and its severity is not uniform across the basin. In central Amazonia records from Lakes Geral (Bush *et al.*, 2000) and the Tapajós (Irion *et al.*, in press) suggest a very modest drying compared with most other sites. In general, this dry event reaches its peak between 7,000 cal yr BP and 4,000 cal yr BP, and—though no direct link has been shown—this is the period when El Niño/Southern Oscillation was very weak (Sandweiss *et al.*, 2001). However, as more records accumulate, it appears that this was not a single sustained drought, but a period of increased drought probability than in the Early or Late Holocene (Bush *et al.*, in review).

3.9   CONCLUSIONS

The Quaternary history of Amazonia is only partly revealed, and each detailed record adds fresh perspective to what is emerging as a complex history. The weight of evidence suggests that the lowlands remained predominantly forested during the last ice age, and therefore, it is assumed, during preceding Quaternary glacial events. The nature of the forest, its openness, the dissimilarity from that of modern times and the extent to which it was invasible by cool, or dry forest elements are worthy subjects for future research. While advancing the case against the existence of continuous arcs of dry forest within the last ice age, we establish the more general argument that the vast Amazon Basin has a dynamic and complex environment that cannot be treated as a climatic or ecological monolith. The many and various changes in Amazon environments during the Quaternary have combined to produce the modern distribution of plants and animals. However, it seems increasingly likely that these changes were not the driving factors generating the high diversity found within this region. Biogeography has a vital role to play in helping us to ask appropriate questions and to move toward a better understanding of this system. However, to quote the paper stemming from a memorable talk recently given by Russell Coope to the Royal Society of London "reconstructing evolutionary history on the basis of present-day distribution alone is rather like trying to reconstruct the plot of a film from its last few frames". Amazonian paleoecology came of age testing biogeographic hypotheses, now the challenge will be to provide the lead so that biogeographic theory can be based on empirical data, not on assumption.

Acknowledgments

This work was funded by grants from the National Science Foundation Grants DEB 9732951, 0237573 (MBB).

3.10   REFERENCES

Absy, M. L., Clief, A., Fournier, M., Martin, L., Servant, M., Sifeddine, A., Silva, F. D., Soubiès, F., Suguio, K. T., and van der Hammen, T. (1991) Mise en évidence de quatre phases d'ouverture de la forêt dense dans le sud-est de L'Amazonie au cours des 60,000 dernières années. Première comparaison avec d'autres régions tropicales. *Comptes Rendus Academie des Sciences Paris, Series II* **312**, 673–678 [in French].

Baker, P. A., Seltzer, G. O., Fritz, S. C., Dunbar, R. B., Grove, M. J., Tapia, P. M., Cross, S. L., Rowe, H. D., and Broda, J. P. (2001) The history of South American tropical precipitation for the past 25,000 years. *Science* **291**, 640–643.

Ballantyne, A. P., Lavine, M., Crowley, T. J., and Baker, P. A. (2005) Meta-analysis of tropical surface temperatures during the Last Glacial Maximum. *Geophysical Research Letters* **32**, L05712 (05714).

Bartlett, A. S. and Barghoorn, E. S. (1973) Phytogeographic history of the Isthmus of Panama during the past 12,000 years: A history of vegetation, climate, and sea-level change. In: A. Graham (ed.), *Vegetation and Vegetational History of Northern Latin America* (pp. 203–299). Elsevier, Amsterdam.

Bates, J. M., Hackett, S. J., and Cracraft, J. (1998) Area-relationships in the Neotropical lowlands: An hypothesis based on raw distributions of Passerine birds. *Journal of Biogeography* **25**, 783–794.

Behling, H. (1996) First report on new evidence for the occurrence of *Podocarpus* and possible human presence at the mouth of the Amazon during the late-glacial. *Vegetation History and Archaeobotany* **5**, 241–246.

Behling, H. and Hooghiemstra, H. (1999) Environmental history of the Colombian savannas of the Llanos Orientales since the Last Glacial Maximum from lake records El Pinal and Carimagua. *Journal of Paleolimnology* **21**, 461–476.

Behling, H. and Lichte, M. (1997) Evidence of dry and cold climatic conditions at glacial times in tropical Southeastern Brazil. *Quaternary Research* **48**, 348–358.

Berger, A. (1978) Long-term variations of daily insolation and Quaternary climatic change. *Journal of Atmospheric Science* **35**, 2362–2367.

Brown, K. S. R., Jr. (1987) Biogeography and evolution of Neotropical butterflies. In: T. C. Whitmore, and G. T. Prance (eds.), *Biogeography and Quaternary History in Tropical America* (pp. 66–104). Oxford Science Publications, Oxford, U.K.

Burbridge, R. E., Mayle, F. E., and Killeen, T. J. (2004) 50,000 year vegetation and climate history of Noel Kempff Mercado National Park, Bolivian Amazon. *Quaternary Research* **61**, 215–230.

Bush, M. B. (1994) Amazonian speciation: A necessarily complex model. *Journal of Biogeography* **21**, 5–18.

Bush, M. B. (2002) On the interpretation of fossil Poaceae pollen in the humid lowland neotropics. *Palaeogeography, Palaeoclimatology, Palaeoecology* **177**, 5–17.

Bush, M. B. (2005) Of orogeny, precipitation, precession and parrots. *Journal of Biogeography* **32**, 1301–1302.

Bush, M. B. and Colinvaux, P. A. (1990) A long record of climatic and vegetation change in lowland Panama. *Journal of Vegetation Science* **1**, 105–119.

Bush, M. B. and Silman, M. R. (2004) Observations on Late Pleistocene cooling and precipitation in the lowland Neotropics. *Journal of Quaternary Science* **19**, 677–684.

Bush, M. B., Weimann, M., Piperno, D. R., Liu, K.-b., and Colinvaux, P. A. (1990) Pleistocene temperature depression and vegetation change in Ecuadorian Amazonia. *Quaternary Research* **34**, 330–345.

Bush, M. B., Miller, M. C., De Oliveira, P. E., and Colinvaux, P. A. (2000) Two histories of environmental change and human disturbance in eastern lowland Amazonia. *The Holocene* **10**, 543–554.

Bush, M. B., Miller, M. C., de Oliveira, P. E., and Colinvaux, P. A. (2002) Orbital forcing signal in sediments of two Amazonian lakes. *Journal of Paleolimnology* **27**, 341–352.

Bush, M. B., De Oliveira, P. E., Colinvaux, P. A., Miller, M. C., and Moreno, E. (2004a) Amazonian paleoecological histories: One hill, three watersheds. *Palaeogeography, Palaeoclimatology, Palaeoecology* **214**, 359–393.

Bush, M. B., Silman, M. R., and Urrego, D. H. (2004b) 48,000 years of climate and forest change from a biodiversity hotspot. *Science* **303**, 827–829.

Bush, M. B., Silman, M. R., and Listopad, M. C. S. (in press) A regional study of Holocene climate change and human occupation in Peruvian Amazonia. *Journal of Biogeography*.

Clement, A. C., Seager, R., and Cane, M. A. (1999) Orbital controls on the El Niño/Southern Oscillation tropical climate. *Paleoceanography* **14**, 441–456.

Clement, A. C., Cane, M. A., and Seager, R. (2001) An orbitally driven tropical source for abrupt climate change. *Journal of Climate* **14**, 2369–2375.

Colinvaux, P. A., De Oliveira, P. E., Moreno, J. E., Miller, M. C., and Bush, M. B. (1996) A long pollen record from lowland Amazonia: Forest and cooling in glacial times. *Science* **274**, 85–88.

Colinvaux, P. A., Irion, G., Räsänen, M. E., Bush, M. B., and Nunes de Mello J. A. S. (2001) A paradigm to be discarded: Geological and paleoecological data falsify the Haffer and Prance refuge hypothesis of Amazonian speciation. *Amazoniana* **16**, 609–646.

Collins, A. C. and Dubach, J. M. (2000) Biogeographic and ecological forces responsible for speciation in *Ateles*. *International Journal of Primatology* **21**, 421–444.

Costa, L. P. (2003) The historical bridge between the Amazon and the Atlantic Forest of Brazil: A study of molecular phylogeography with small mammals. *Journal of Biogeography* **30**, 71–86.

Cowling, S. A. (2004) Tropical forest structure: A missing dimension to Pleistocene landscapes. *Journal of Quaternary Science* **19**, 733–743.

Cowling, S. A., Maslin, M. A., and Sykes, M. T. (2001) Paleovegetation simulations of lowland Amazonia and implications for neotropical allopatry and speciation. *Quaternary Research* **55**, 140–149.

Cowling, S. A., Betts, R. A., Cox, P. M., Ettwein, V. J., Jones, C. D., Maslin, M. A., and Spall, S. A. (2004) Contrasting simulated past and future responses of the Amazonian forest to atmospheric change. *Philosophical Transactions of the Royal Society of London B* **359**, 539–547.

Cruz, F. W., Jr., Burns, S. J., Karmann, I., Sharp, W. D., Vuille, M., Cardoso, A. O., Ferrari, J. A., Silva Dias, P. L., and Vlana, O., Jr. (2005) Insolation-driven changes in atmospheric circulation over the past 116,000 years in subtropical Brazil. *Nature* **434**, 63–66.

de Oliveira, P. E. (1992) *A Palynological Record of Late Quaternary Vegetational and Climatic Change in Southeastern Brazil*. The Ohio State University, Columbus, OH.

De Toledo, M. B. (2004) *Holocene Vegetation and Climate History of Savanna–Forest Ecotones in Northeastern Amazonia*. Biological Sciences. Florida Institute of Technology, Melbourne, FL.

Denevan, W. M. (2003) The native population of Amazonia in 1492 reconsidered. *Revista De Indias* **62**, 175–188.

Flanagan, N. S., Tobler, A., Davison, A., Pybus, O. G., Kapan, D. D., Planas, S., Linares, M., Heckel, D., and McMillan, W. O. (2004) Historical demography of Müllerian mimicry in the neotropical *Heliconius* butterflies. *Proceedings of the National Academy of Sciences* **101**, 9704–9709.

Fritz, S. C., Baker, P. A., Lowenstein, T. K., Seltzer, G. O., Rigsby, C. A., Dwyer, G. S., Tapia, P. M., Arnold, K. K., Ku, T. L., and Luo, S. (2004) Hydrologic variation during the last 170,000 years in the southern hemisphere tropics of South America. *Quaternary Research* **61**, 95–104.

Gentry, A. H. (1993) *A Field Guide to the Families and Genera of Woody Plants of Northwest South America (Colombia, Ecuador, Peru) with Supplementary Notes on Herbaceous Taxa* (895 pp.). Conservation International, Washington, D.C.

Grau, E. T., Pereira, S. L., Silveira, L. F., Hofling, E., and Wajntal, A. (2005) Molecular phylogenetics and biogeography of Neotropical piping guans (Aves: Galliformes): *Pipile* Bonaparte, 1856 is a synonym of *Aburria* Reichenbach, 1853. *Molecular Phylogenetics and Evolution* **35**, 637–645.

Haberle, S. G. (1997) Late Quaternary vegetation and climate history of the Amazon Basin: Correlating marine and terrestrial pollen records. In: R. D. Flood, D. J. W. Piper, A. Klaus, and L. C. Peterson (eds.), *Proceedings of the Ocean Drilling Program, Scientific Results* (pp. 381–396). Ocean Drilling Program, College Station, TX.

Haberle, S. G. and Maslin, M. A. (1999) Late Quaternary vegetation and climate change in the Amazon basin based on a 50,000 year pollen record from the Amazon fan, PDP site 932. *Quaternary Research* **51**, 27–38.

Haffer, J. (1969) Speciation in Amazonian forest birds. *Science* **165**, 131–137.

Haffer, J. and Prance, G. T. (2001) Climatic forcing of evolution in Amazonia during the Cenozoic: On the refuge theory of biotic differentiation. *Amazoniana* **16**, 579–608.

Hall, J. P. and Harvey, D. J. (2002) The phylogeography of Amazonia revisited: New evidence from Riodinid butterflies. *Evolution* **56**, 1489–1497.

Hansen, B. C. S., Wright, H. E., Jr., and Bradbury, J. P. (1984) Pollen studies in the Junín area, Central Peruvian Andes. *Geological Society of America Bulletin* **95**, 1454–1465.

Heckenberger, M. J., Kuikuro, A., Kuikuro, U. T., Russell, J. C., Schmidt, M., Fausto, C., and Franchetto, B. (2003) Amazonia 1492: Pristine forest or cultural parkland? *Science* **301**, 1710–1714.

Hooghiemstra, H. and van der Hammen, T. (2004) Quaternary ice age in the Colombian Andes: Developing an understanding of our legacy. *Philosophical Transactions of the Royal Society of London B* **359**, 173–181.

Hooghiemstra, H., Melice, J. L., Berger, A., and Shackleton, N. J. (1993) Frequency spectra and paleoclimatic variability of the high-resolution 30–1450 ka Funza I pollen record (Eastern Cordillera, Colombia). *Quaternary Science Reviews* **12**, 141–156.

Hoorn, C. (1993) Marine incursions and the influence of Andean tectonics on the Miocene depositional history of northwestern Amazonia: Results of a palynostratigraphic study. *Palaeogeography, Palaeoclimatology, Palaeoecology* **105**, 267–309.

Hoorn, C., Guerrero, J., Sarmiento, G. A., and Lorente, M. A. (1995) Andean tectonics as a cause for changing drainage patterns in Miocene northern South America. *Geology* **23**, 237–240.

Hostetler, S. W. and Mix, A. C. (1999) Reassessment of ice-age cooling of the tropical ocean and atmosphere. *Nature* **399**, 673–676.

Ledru, M.-P. (1993) Late Quaternary environmental and climatic changes in Central Brazil. *Quaternary Research* **39**, 90–98.

Ledru, M.-P., Cordeiro, R. C., Dominguez, J. M. L., Martin, L., Mourguiart, P., Sifeddine, A., and Turcq, B. (2001) Late-Glacial cooling in Amazonia inferred from pollen at Lagoa do Caco, northern Brazil. *Quaternary Research* **55**, 47–56.

Liu, K. and Colinvaux, P. A. (1985) Forest changes in the Amazon Basin during the last glacial maximum. *Nature* **318**, 556–557.

Mayle, F. E. and Beerling, D. J. (2004) Late Quaternary changes in Amazonian ecosystems and their implications for global carbon cycling. *Palaeogeography, Palaeoclimatology, Palaeoecology* **214**, 11–25.

Mayle, F. E., Burbridge, R., and Killeen, T. J. (2000) Millennial-scale dynamics of southern Amazonian rain forests. *Science* **290**, 2291–2294.

Mayle, F. E., Beerling, D. J., Gosling, W. D., and Bush, M. B. (2004) Responses of Amazonian ecosystems to climatic and atmospheric carbon dioxide changes since the last glacial maximum. *Philosophical Transactions of the Royal Society of London B* **359**, 499–514.

Meggers, B. J. (1994) Archaeological evidence for the impact of mega-Niño events on Amazonia during the past two millennia. *Climate Change* **28**, 321–338.

Meggers, B. J. (2003) Revisiting Amazonia circa 1492. *Science* **301**, 2067.

Miller, K. G., Kominz, M. A., Browning, J. V., Wright, J. D., Mountain, G. S., Katz, M. E., Sugarman, P. J., Cramer, B. S., Christie-Black, N., and Pekar, S. (2005) The Phanerozoic record of global sea-level change. *Science* **310**, 1293–1298.

Moritz, C., Patton, J. L., Schneider, C. J., and Smith, T. B. (2000) Diversification of rainforest faunas: An integrated molecular approach. *Annual Review of Ecology and Systematics* **31**, 533–563.

Naish, T. (1997) Constraints on the amplitude of late Pliocene eustatic sea-level fluctuations: New evidence from New Zealand shallow-marine sediment record. *Geology* **25**, 1139–1142.

Near, T. J., Meylan, P. A. and Shaffer, H. B. (2005) Assessing concordance of fossil calibration points in molecular clock studies: An example using turtles. *American Naturalist* **165**, 137–146.

Nepstad, D. C., de Carvalho, C. R., Davidson, E. A., Jipp, P. H., Lefebvre, P. A., Negreiros, G. H., da Silva, E. D., Stone, T. A., Trumbore, S. E., and Vieira, S. (1994) The role of deep roots in the hydrological and carbon cycles of Amazonian forests an pastures. *Nature* **372**, 666–669.

Noonan, B. P. and Gaucher, P. (2005) Phylogeography and demography of Guianan harlequin toads (*Atelopus*): Diversification within a refuge. *Molecular Ecology* **14**, 3017–3031.

Nores, M. (1999) An alternative hypothesis for the origin of Amazonian bird diversity. *Journal of Biogeography* **26**, 475–485.

Olson, D. M., Dinerstein, E., Wikramanayake, E. D., Burgess, N. D., Powell, G. V. N.,
    Underwood, E. C., D'Amico, J. A., Itoua, I., Strand, H. E., Morrison, J. C. *et al.*
    (2001) Terrestrial ecoregions of the world: A new map of life on earth. *BioScience* **51**,
    933–938.
Paduano, G. M., Bush, M. B., Baker, P. A., Fritz, S. C., and Seltzer, G. O. (2003) A vegetation
    and fire history of Lake Titicaca since the Last Glacial Maximum. *Palaeogeography,
    Palaeoclimatology, Palaeoecology* **194**, 259–279.
Patton, J. L., da Silva, M. N. F., and Malcolm, J. R. (2000) Mammals of the Rio Juruá and the
    evolutionary and ecological diversification of Amazonia. *Bulletin of the American Museum
    of Natural History* **244**, 1–306.
Pennington, R. T., Prado, D. E., and Pendry, C. A. (2000) Neotropical seasonally dry forests
    and Quaternary vegetation changes. *Journal of Biogeography* **27**, 261–273.
Pennington, R. T., Lavin, M., Prado, D. E., Pendry, C. A., Pell, S. K., and Butterworth, C. A.
    (2004) Historical climate change and speciation: Neotropical seasonally dry forest plants
    show patterns of both Tertiary and Quaternary diversification. *Philosophical Transactions
    of the Royal Society of London B* **359**, 515–538.
Pessenda, L. C. R., Gomes, B. M., Aravena, R., Ribeiro, A. S., Boulet, R., and Gouveia, S. E.
    M. (1998) The carbon isotope record in soils along a forest–cerrado ecosystem transect:
    Implications for vegetation changes in the Rondonia state, southwestern Brazilian
    Amazon region. *The Holocene* **8**, 599–603.
Petit, J. R., Jouzel, J., Raynaud, D., Barkov, N. I., Barnola, J.-M., Basile, I., Bender, M., and
    Chappellaz, J. (1999) Climate and atmospheric history of the past 420,000 years from the
    Vostok ice core, Antarctica. *Nature* **399**, 429–436.
Pitman, N. C. A., Terborgh, J., Silman, M. R., and Nuez, P. V. (1999) Tree species distributions
    in an upper Amazonian forest. *Ecology* **80**, 2651–2661.
Prado, D. E. and Gibbs, P. E. (1993) Patterns of species distributions in the seasonal dry forests
    of South America. *Annals of the Missouri Botanical Garden* **80**, 902–927.
Rind, D. and Peteet, D. (1985) Terrestrial conditions at the last glacial maximum and CLIMAP
    sea-surface temperature estimates: Are they consistent? *Quaternary Research* **24**, 1–22.
Rodbell, D. T. (1992) Late Pleistocene equilibrium-line reconstructions in the Northern
    Peruvian Andes. *Boreas* **21**, 43–52.
Räsänen, M. E., Linna, A. M., Santos, J. C. R., and Negri, F. R. (1995) Late Miocene tidal
    deposits in the Amazon foreland basin. *Science* **269**, 386–390.
Salgado-Labouriau, M. L. (1997) Late Quaternary palaeoclimate in savannas of South
    America. *Journal of Quaternary Science* **12**, 371–379.
Sandweiss, D. H., Maasch, K. A., Burger, R. L., Richardson III, J. B., Rollins, H. B., and
    Clement, A. (2001) Variation in Holocene El Niño frequencies: Climate records and
    cultural consequences in ancient Peru. *Geology* **29**, 603–606.
Seltzer, G. O. (1990) Recent glacial history and paleoclimate of the Peruvian–Bolivian Andes.
    Quaternary Science Reviews **9**, 137–152.
Seltzer, G., Rodbell, D. and Burns, S. (2000) Isotopic evidence for late Quaternary climatic
    change in tropical South America. *Geology* **28**, 35–38.
Seltzer, G. O., Rodbell, D. T., and Wright, H. E. (2003) Late-Quaternary paleoclimates of
    the southern tropical Andes and adjacent regions. *Palaeogeography, Palaeoclimatology,
    Palaeoecology* **194**, 1–3.
Servant, M., Fontes, J.-C., Rieu, M., and Saliège, X. (1981) Phases climatiques arides holocènes
    dans le sud-ouest de l'Amazonie (Bolivie). *Comptes Rendus Academie Scientifique Paris,
    Series II* **292**, 1295–1297 [in French].

Sifeddine, A., Albuquerque, A. L. S., Ledru, M.-P., Turcq, B., Knoppers, B., Martin, L., de
    Mello, W. Z., Passenau, H., Dominguez, J. M. L., Cordeiro, R. C. *et al.* (2003) A 21,000 cal
    years paleoclimatic record from Caco Lake, northern Brazil: Evidence from sedimentary
    and pollen analyses. *Palaeogeography, Palaeoclimatology, Palaeoecology* **189**, 25–34.
Smith, J. A., Seltzer, G. O., Farber, D. L., Rodbell, D. T., and Finkel, R. C. (2005) Early local
    last glacial maximum in the tropical Andes. *Science* **308**, 678–681.
Stute, M., Forster, M., Frischkorn, H., Serejo, A., Clark, J. F., Schlosser, P., Broecker, W. S.,
    and Bonani, G. (1995) Cooling of tropical Brazil (5°C) during the last glacial maximum.
    Science **269**, 379–383.
Symula, R., Schulte, R., and Summers, K. (2003) Molecular systematics and phylogeography of
    Amazonian poison frogs of the genus *Dendrobates*. *Molecular Phylogenetics and Evolution*
    26, 452–475.
Tuomisto, H., Ruokolainen, K., Kalliola, R., Linna, A., Danjoy, W., and Rodriguez, Z. (1995)
    Dissecting Amazonian biodiversity. *Science* **269**, 63–66.
Urrego, D. H., Silman, M. R., and Bush, M. B. (2005) The last glacial maximum: Stability and
    change in an Andean cloud forest. *Journal of Quaternary Science* **20**, 693–701.
van der Hammen, T. and Hooghiemstra, H. (2000) Neogene and Quaternary history of
    vegetation, climate and plant diversity in Amazonia. *Quaternary Science Reviews* **19**,
    725–742.
Vanzolini, P. E. (1970) Zoologia sistematica, geografia e a origem das especies. *Instituto
    Geografico São Paulo, Serie Teses e Monografias* **3**, 1–56.
Wang, X., Auler, A. S., Edwards, R. L., Cheng, H., Cristalli, P. S., Smart, P. L., Richards, D. A.,
    and Shen, C.-C. (2004) Wet periods in northeastern Brazil over the past 210 kyr linked to
    distant climate anomalies. *Nature* **432**, 740–743.
Wille, M., Negret, J. A., and Hooghiemstra, H. (2000) Paleoenvironmental history of the
    Popayan area since 27,000 yr BP at Timbio, Southern Colombia. *Review of Palaeobotany
    and Palynology* **109**, 45–63.
Zink, R. M. and Slowinski, J. B. (1995) Evidence from molecular systematics for decreased
    avian diversification in the Pleistocene epoch. *Proceedings of the National Academy of
    Science* **92**, 5832–5835.
Zink, R. M., Klicka, J., and Barber, B. R. (2004) The tempo of avian diversification during the
    Quaternary. *Philosophiocal Transactions of the Royal Society, London B* **359**, 215–220.

4

The Quaternary history of far eastern rainforests

A. P. Kershaw, S. van der Kaars, and J. R. Flenley

4.1 INTRODUCTION

4.1.1 Present setting

This region differs from those supporting tropical rainforest in other parts of the world in that it is less continental and geologically much more dynamic. It incorporates some major pieces of continental plate, but its center—the so-called "Maritime Continent" (Ramage, 1968)—is largely a complex interaction zone between the Asian and Australian Plates resulting from the continued movement of the Australian Plate into Southeast Asia (Metcalfe, 2002). The effects of tectonic and volcanic activity have resulted in mountain uplift, particularly in New Guinea, and formation of the volcanic island chain of Indonesia. Vulcanicity also occurs out into the Pacific beyond the "andesite line" where most "high" islands are volcanic and most "low" islands are coral islands developed on sunken volcanoes.

The extensive areas of continental shelf—particularly the Sunda and Sahul Shelves—but including the shelves along the east coast of northern Australia and around the South China Sea, combined with the impact of the Indonesian throughflow that restricts the movement of warm water from the Pacific to the Indian Ocean, have resulted in the highest sea surface temperatures on Earth in the form of the West Pacific Warm Pool. The enhanced convective activity associated with the warm pool results in high rainfall through much of the year in the heart of the Maritime Continent and dominance of the vegetation by evergreen rainforest. The area also provides the major source of heat release that drives the East Asian–Australasian summer monsoon system reflected in the strong summer rainfall patterns beyond the Intertropical Convergence Zone in each hemisphere, and resulting in the occurrence of seasonal, raingreen or "monsoon" semi-evergreen to deciduous rainforest over much of continental Southeast Asia and the very north of Australia (Figure 4.1). Additional influences on rainforest distribution are the warm northerly and southerly currents

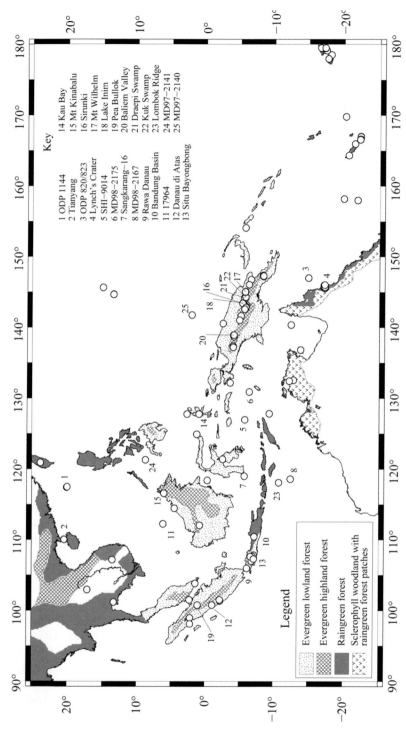

Figure 4.1. Distribution of rainforest vegetation in the far east and pollen-analyzed sites covering at least the last 6,000 years. Rainforest types have been simplified from distributions and descriptions of communities identified by Fedorova *et al.* (1993, 1994).

emanating from the Pacific equatorial current that, in combination with the southeast and northeast trade winds from the Pacific, result in the production of high orographic rainfall and associated rainforest along mountainous eastern coastal areas of Southeast Asia and Australia.

Most of the region is subjected to high interannual rainfall variability that is also, to a large degree, a product of its particular geography and the dynamics of oceanic and atmospheric circulation systems. The energy provided by convective activity within the Maritime Continent is the major contributor to the operation of the east–west Walker circulation that breaks down periodically, resulting in the movement of the warm water banked up against the Indonesian throughflow eastwards and resulting in a substantial reduction in precipitation from all sources over most of the region. These El Niño phases of the so-called "El Niño–Southern Oscillation" (ENSO)—that have also been linked to a weakening of the monsoon (Soman and Slingo, 1997)—can cause severe droughts and fires, even within rainforest, especially where there is disturbance from human activity.

Although tropical influences dominate the climate of the region, the Tibetan Plateau is important in creating a strong winter monsoon influence. The height and extent of this plateau results in the production of cool dry air that exacerbates seasonal contrasts in the northern part of the region and has a push effect on summer monsoon development in the southern hemisphere.

4.1.2 Nature of the evidence

Most of the evidence for past vegetation and climate from the region is derived from palynological studies. Perceived problems of pollen analysis in the lowland tropics—due to the richness of the flora, dominance of effective animal pollination, and lack of strong winds within the core area—resulted in most early research being focused on highland communities (Flenley, 1979). In these per-humid areas, a major interest has been and continues to be on altitudinal variation in the changing position and composition of montane rainforest and alpine zones in relation to global climate influences. Studies have been restricted mainly to swamps and shallow lakes covering the latter part of the last glacial period and Holocene.

Ventures into the terrestrial lowlands have generally not proved particularly successful due not only to original perceptions but also to the dearth of continuous sediment sequences in both perennially and seasonally wet environments, and lack of differentiation of peatland, riparian and dryland forest communities in the extensive peatlands that are otherwise very suitable for pollen analysis. Notable exceptions are deep defined basins of volcanic origin that have revealed detailed records of both vegetation and climate change, sometimes covering long periods of time.

A major feature of the region—that has been exploited in recent years—is the maritime setting whereby ocean basins occur in close proximity to land areas. A number of sediment cores have provided long and fairly continuous regional records of vegetation and climate change, securely dated from associated oxygen isotope records. Nevertheless, none of these records yet covers the whole of the Quaternary and reliance is placed on geologically isolated glimpses of past environments for some indication of the nature of the early part of this period.

4.2   MODERN POLLEN SAMPLING

Some basis for interpretation of Quaternary palynological records is derived from examination of patterns and processes of modern pollen deposition recorded in pollen traps and surface sediments. Such sampling examines deposition both within the rainforest and outside the rainforest in lake, swamp, and marine environments used for reconstructions of vegetation history.

The first quantitative study of pollen deposition in rainforest was by Flenley (1973). In the lowland rainforest of Malaysia he found significant pollen influx (between 800 and 2,020 grains/cm^2/annum) and relatively high pollen diversity (60 to 62 taxa) although representation within taxa through time was very variable. Similar results were found by Kershaw and Strickland (1990) in a north Queensland rainforest. They also found, from a knowledge of the distribution of trees surrounding the traps, that two-thirds of the pollen could have been derived from within 30 m of the traps. An examination of traps situated less than 100 m outside rainforest, in a small crater lake on the Atherton Tableland in north Queensland, demonstrated an enormous reduction in pollen deposition and substantial sifting out of pollen of local producers (Kershaw and Hyland, 1975). Pollen influx values dropped to below 200 grains/cm^2/annum and spectra were dominated by a relatively small number of taxa with significant regional pollen dispersal. It was determined that there was about equal representation of pollen from above canopy and rainout components. Any trunk space component was small and the high degree of correspondence between trap assemblages and those derived from the topmost part of a sediment core from the lake (Kershaw, 1970) suggested also that there was little inwash of pollen, though this component may have been trapped by marginal swamp.

Despite the great variability of pollen deposition within rainforest, patterns of representation appear to reflect systematic vegetation variation on a regional scale. Numerical analysis of a number of surface litter samples from throughout the lowland and sub-montane forests of northeast Queensland (Kershaw, 1973; Kershaw and Bulman, 1994) revealed a similar pattern to floristic analysis of forest plots from which the samples were derived. Although there was little in common between taxon representation and abundance in the two groups, it suggested that pollen assemblages could be used to characterize the broad environmental features of the landscape, including the vegetation. A similar result was achieved with the use of percentages of only those taxa that had been identified from lake-trapping and existing fossil pollen records as regionally important. This finding indicated the potential for analysis of pollen diagrams from tropical rainforest in a similar manner to those from other vegetation types where variation in abundance of a small number of taxa provides the basis for interpretation. Bioclimatic estimates for such "common taxa" in north-eastern Queensland (Moss and Kershaw, 2000) demonstrate their potential for quantitative paleoclimatic reconstruction (Figure 4.2). The presence of numerous other taxa can allow refinement of interpretation (Kershaw and Nix, 1989) although insufficient pollen may be present in samples to allow counts of a sufficient size to demonstrate presence or absence in potential source vegetation. Figure 4.2 also demonstrates the degree of penetration into rainforest of pollen from the dominants of surrounding sclerophyll vegetation, *Eucalyptus* and *Casuarina*, that have generally wider pollen dispersal than rainforest taxa.

The heterogeneous nature of lowland tropical rainforest is an impediment to determination of the actual sources of "common" taxa and, therefore, their relative degree of dispersal. This complication is reduced at higher altitudes where widely dispersed taxa, many of which are clearly wind-pollinated, make up significant and identifiable components of the vegetation. The compilation of Flenley (1979) provides an excellent summary of variation in pollen representation along an altitudinal transect in New Guinea (Figure 4.3). Above the highly human-modified vegetation, clearly recognized by high values of Poaceae or *Casuarina*, the montane zones of oak and beech forest are dominated by pollen of their dominant taxa: *Lithocarpus*/*Castanopsis* and *Nothofagus*, respectively. Upper montane mixed forest is characterized by *Quintinia* while alpine vegetation is recognized by the only occurrences of "alpine pollen taxa". The bare ground on the mountain summit has a unique pollen signature that clearly identifies those taxa, *Nothofagus* and *Casuarina*, which have wide pollen dispersal. Flenley (1979) remarks on the tendency for pollen to be carried uphill and suggests it is due to the fact that pollen is released during the day when anabatic winds are active.

A much broader indication of pollen transport, including a potentially major water-transported component, is provided by recent analyses of suites of core-top pollen samples from the Indonesian–Australian region (van der Kaars, 2001; van der Kaars and De Deckker, 2003; van der Kaars, new data) and the South China Sea (Sun *et al.*, 1999). Isopolls interpolated from samples along the steep precipitation gradient from east Indonesia to northwest Australia are shown for major pollen groupings based on a dryland pollen sum, excluding pteridophytes (Figure 4.4). This gradient is clearly reflected in the pollen with predominantly rainforest taxa including pteridophytes showing high values in the rainforested Indonesian region and then progressively declining relative to the predominantly sclerophyll taxa of Myrtaceae (attributable mainly to *Eucalyptus*) and Poaceae that dominate Australian vegetation. Compared with other pollen types, the pollen of rainforest angiosperms reflect most faithfully the distribution of rainforest. Rainforest conifers are much better represented than angiosperms considering their almost total restriction to montane forests, a feature no doubt due to obligate wind dispersal of pollen and greater opportunity for wind transport from higher altitudes. The major concentration of montane pollen types between Sulawesi and New Guinea reflects also the proximity to mountainous areas within the study area. Pteridophyte spores have a very similar distribution to the rainforest conifers and, although this pattern can be accounted for—to some degree—by the fact that they are most abundant in wet tropical and often montane forest, transport is facilitated also by water. The fact that percentages of pteridophytes are so much higher than those of pollen is probably the result of effective water transport.

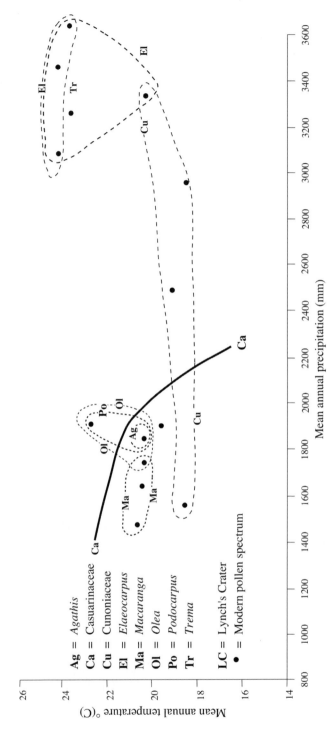

Figure 4.2. Climatic ranges for highest representation of major rainforest taxa in relation to bioclimatic estimates for modern pollen samples from northeast Queensland rainforests. The extent of penetration of high values for the sclerophyll woodland taxon Casuarinaceae is also shown (adapted from Moss and Kershaw, 2000).

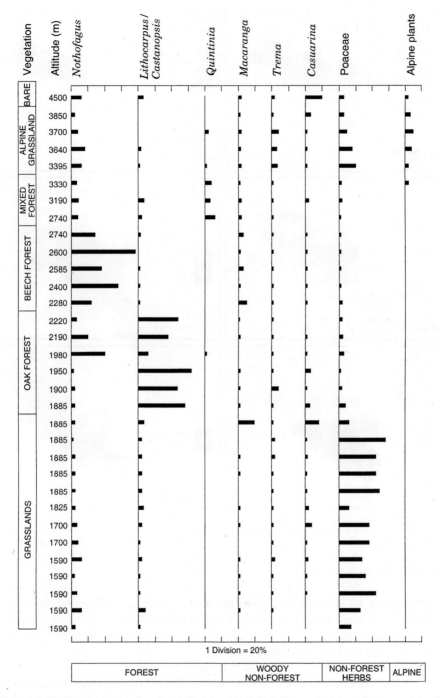

Figure 4.3. Representation of major pollen taxa in relation to vegetation along an altitudinal surface sample transect in Papua–New Guinea (modified from Flenley, 1973).

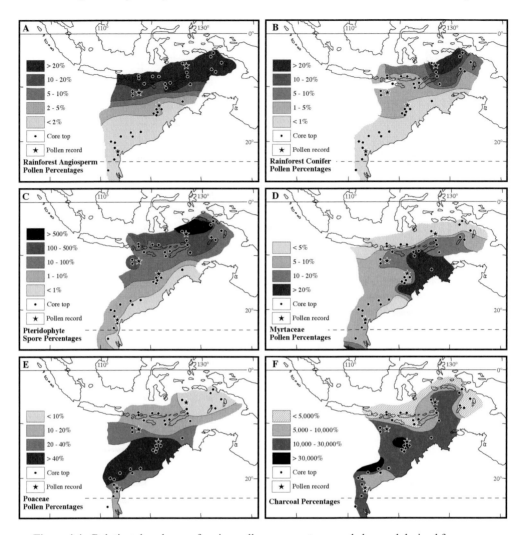

Figure 4.4. Relative abundance of major pollen groups, taxa, and charcoal derived from core-top samples in the northwestern Australian–southern Indonesian region based on a pollen sum of total dryland pollen excluding pteridophytes. Data from van der Kaars and De Deckker (2003) and van der Kaars (2001).

In contrast to pollen—that generally reflects the regional representation of vegetation—charcoal, derived from the same samples, shows a less certain pattern, at least in percentage terms. It is very unlikely that fire activity is highest in the open ocean where charcoal values are highest. This clearly indicates that charcoal particles, on average, are transported farther than pollen. However, it is clear from a general decline northwards that charcoal is, as expected, derived mainly from Australia. The reduction in charcoal percentages in the very northwest of Australia may be realistic as, within this very dry area, a lack of fuel would allow only the occasional burn.

The South China Sea shows largely an inversion of the southern hemisphere pattern (Figure 4.5). Here, tropical rainforest angiosperm pollen is derived largely from the equatorial humid region centred on Borneo and decreases northwards relative to *Pinus* which may be regarded as the equivalent of Australian sclerophyllous trees, in its dominance of drier and more seasonal open forests that cover large areas of Peninsular Southeast Asia and more subtropical forests of southern China. The much higher values of *Pinus* are considered to result from the influence of the strong northerly winter monsoons that blow while *Pinus* trees are still in cone, in addition to the high production and dispersal rates of its pollen (Sun *et al.*, 1999). The rainforest gymnosperms—represented in Figure 4.5 by their most conspicuous genus,

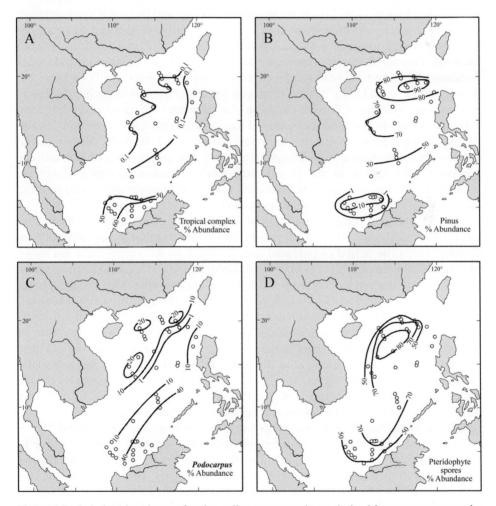

Figure 4.5. Relative abundance of major pollen groups and taxa derived from core-top samples in the South China Sea based on a pollen sum of total dryland pollen excluding pteridophyes. Adapted from Sun *et al.* (1999).

Podocarpus—clearly show much broader pollen dispersal than the angiosperms and sources in both the equatorial tropics and mountains in southern China. Pteridophyte values, when consideration is given to the different basis for calculation of the pollen sum, are similar to those in southern waters, but are demonstrably well dispersed with highest percentages towards the center of the basin. Although not illustrated, mangroves in both data sets show highest values close to coastal locations with a substantial fall away from the coast.

4.3   REGIONAL TAXON REPRESENTATION

Some indications of those taxa that contribute to a regional picture of the vegetation within the tropical rainforest-dominated regions of Southeast Asia and Australia are shown in relation to major ecological groups in Table 4.1. The distinctions between the major taxonomic groups—between essentially evergreen or raingreen and winter deciduous trees, and between rainforest and open forest trees—are fairly clear but those between altitudinally defined rainforest groups are somewhat arbitrary due to the continuous nature of floristic variation, the influences of factors other than temperature on distribution, and the variety of terminologies used for vegetation description in different areas. With the "montane" conifers, for example, Morley (2000) questions the designation of *Podocarpus* and *Dacrydium* as indicative of high-altitude rainforest as they can occur in lowlands, particularly in association with low nutrient-status soils. The inclusion of New Caledonia with its ultra-mafic soils would result in an almost total breakdown of an altitudinal classification.

Despite the long period of isolation of the Australian and Southeast Asian continental plates (Morley, 2000) and apparent limited taxon exchange—apart from New Guinea, within the period of potential contact, the Late Miocene—there are major similarities between pollen floras from the different regions, at least at identified levels. This similarity is most evident with the lowland angiosperms. This group is large, usefully reflecting the floristic diversity of these forests, though low pollen taxonomic resolution disguises much variation in regional representation. Much of this diversity can be accounted for by the lack of dominant wind-dispersed taxa. Although many of the important families and genera in the forest are recognized, there are major biases in representation. For example, the dominant family in Southeast Asian lowland forests, Dipterocarpaceae, is very much under-represented in pollen spectra while the pollen of the dominant family in Australian rainforests, Lauraceae, is hardly recorded. Secondary or successional taxa—such as many Moraceae/Urticaceae, *Macaranga/Mallotus*, *Trema*, and *Celtis*—are, by contrast, over-represented.

The greater differentiation in montane and lower montane elements is due to refined identification of a more limited suite of taxa, many of which contain few species, as well as distinctive northern or southern origins. There is no evidence of taxa—such as *Engelhardia*, *Myrica*, *Altingia*, *Liquidambar*, *Lithocarpus/Castanopsis*, or *Quercus*—reaching Australia although a number have reached New Guinea, while *Dodonaea*, *Nothofagus*, *Quintinia*, and *Araucaria* have not expanded northwards into, or through the whole of, the Southeast Asian region. It is interesting that the southern

Table 4.1. Common pollen taxa of major ecological groups in the far east.

Major ecological groups	Common pollen taxa	China Sea region	Southern Indonesia	New Guinea	Northeast Queensland
Montane conifers	*Dacrycarpus*	×	×	×	
	Dacrydium	×	×	×	
	Podocarpus	×	×	×	×
	Phyllocladus	×	×	×	
Montane angiosperms	*Coprosma*			×	
	Dodonaea		×		(×)
	Drimys	×		×	×
	Engelhardia	×	×		
	Epacridaceae	×			×
	Ericaceae	×	×	×	
	Leptospermum	×			×
	Myrica	×			
	Myrsinaceae	×		×	×
	Nothofagus			×	
	Quintinia			×	×
Lower montane conifers	*Agathis* (Araucariaceae)			×	×
	Araucaria (Araucariaceae)				×
Lower montane angiosperms	*Altingia*	×			
	Cunoniaceae			×	×
	Hammamelidaceae	×			
	Liquidamber	×			
	Lithocarpus/Castanopsis	×	×		
	Quercus	×	×		
Lowland angiosperms	*Acalypha*	×		Insuf.	×
	Anacardiaceae	×	×	data	×
	Barringtonia		×		
	Bischoffia		×		×
	Calamus		×		×
	Celastraceae	×			×
	Celtis	×	×		×
	Calophyllum		×		×
	Dipterocarpaceae		×		
	Elaeocarpaceae	×	×		×
	Euphorbiaceae	×	×		×
	Ilex	×	×		×
	Macaranga/Mallotus	×	×		×
	Melasomataceae	×			×
	Meliaceae	×	×		×
	Moraceae/Urticaceae	×	×		×

(*continued*)

Table 4.1 (*cont.*)

Major ecological groups	Common pollen taxa	China Sea region	Southern Indonesia	New Guinea	Northeast Queensland
Lowland angiosperms	Myrtaceae	×	×		×
	Nauclea		×		
	Oleaceae	×	×		×
	Palmae	×	×		×
	Proteaceae		×		×
	Rubiaceae	×	×		×
	Rutaceae	×	×		×
	Sapindaceae	×	×		×
	Sapotaceae	×	×		×
	Sterculiaceae				×
	Trema	×	×		×
Open forest (savanna)	*Casuarina*		×	×	×
	Eucalyptus		×	×	×
	Melaleuca		×	×	×
	Pinus	×			
Herbs	*Artemisia*	×			
	Cyperaceae	×	×	×	×
	Poaceae	×	×	×	×
Alpine herbs and shrubs	*Astelia*			×	
Pteridophytes		×	×	×	×
Winter deciduous forest		×			

podocarps have their lowest diversity in northeast Queensland despite the fact that they probably dispersed from the Australian region. *Dacrycarpus* and *Phyllocladus* did not arrive in Southeast Asia until the Plio-Pleistocene, but *Dacrydium* and *Podocarpus* have had a much longer residence (Morley, 2002). *Phyllocladus* has extended no further north than Borneo, so its pollen representation in the South China Sea region must derive from this source. *Nothofagus brassospora* has failed to cross into montane Southeast Asia and has also been lost from Australia. Indeed, percentages of *Nothofagus* pollen fall off rapidly with distance from New Guinea. The poor representation of *Nothofagus* pollen in marine sediments is perhaps unexpected considering its proposed high dispersal capacity (Flenley, 1979). Pollen of the open forest regional dominants of the northern and southern subtropics—*Pinus* and *Eucalyptus*/Casuarinaceae, respectively—hardly extend into the other hemisphere despite proposed continuity of monsoon influences across the equator. This pollen is probably explained by limited direct wind connections and the efficiency of pollen removal from the atmosphere within the everwet equatorial zone. Patterns for Casuarinaceae and *Eucalyptus* are regionally complicated: the former by the existence of several component species within rainforest in Southeast Asia and New Guinea, and the latter by problems in consistent separation from rainforest Myrtaceae and, less importantly, from the other major, myrtaceous, sclerophyll genus, *Melaleuca*.

There are no identifiable extratropical elements in pollen assemblages from the Australasian region. Both subtropical and warm temperate rainforest floras are essentially depauperate tropical floras, while sclerophyll elements have a recognizable pollen flora similar to that in the tropics. In contrast, assemblages from the South China Sea contain notable percentages of deciduous temperate taxa—as well as the temperate steppe taxon, *Artemisia*—and no doubt a temperate *Pinus* component transported, at the present day, by the winter monsoon.

Herbs other than *Artemisia* would have been derived mainly from the tropical savannas. However, some component would have been derived from anthropogenic grasslands and perhaps also—in the case of the Cyperaceae, in particular—from coastal rivers and swamps, although under natural conditions peat swamp forests and mangroves would have dominated many coastal communities. Rhizophoraceae is the most conspicuous mangrove taxon, and especially so in pollen assemblages. The percentages of pollen from mangroves fall off rapidly from the coast in marine assemblages.

No differentiation is made within the pteridophyte category that is composed largely of fern spores. It is clear from the surface samples that fern spores have highest values in wetter forests. They are derived mainly from tree ferns and epiphytes, with ground ferns becoming common in higher altitude forests and in some mangroves.

4.4   LONG-TERM PATTERNS OF CHANGE

The general composition and distribution of tropical rainforest was established well before the onset of the Quaternary period (Morley, 2000), with the only recognized major dispersal since this time being the expansion of *Phyllocladus*, probably from the New Guinea highlands into the highlands of the southern part of Southeast Asia.

There are only two records that cover much of the Quaternary period—one from the northern part of the South China Sea, and the other from the Coral Sea off northeastern Australia—and these necessarily provide the framework for establishment of temporal patterns of change during much of this period.

4.4.1   The South China Sea region

Sun *et al.* (2003) provide a near-continuous, high resolution (i.e., 820-year average time interval between samples) record from ODP Site 1144 (20°3″N, 117°25″E) taken from a water depth of 2,037 m equidistant from southern China, Taiwan, and the Philippines (Figures 4.1 and 4.6). The record covers marine isotope stages (MIS) 29 to 1 or the last 1.03 Myr. In terms of major ecological groups represented, there is little

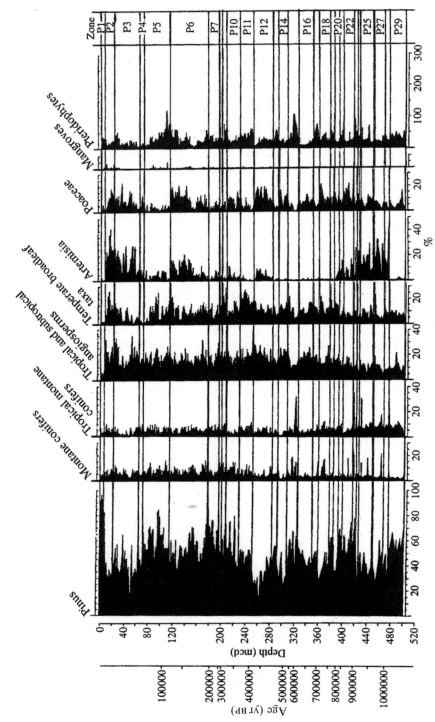

Figure 4.6. Relative abundance of major taxa and taxon groups in the pollen record from ODP Site 1144, South China Sea (after Sun et al., 2003). All taxa are expressed as percentages of the dryland pollen sum, that excludes Pteridophyta and mangroves.

consistent change in representation and variability is generally lowest in the rainforest groups, tropical and lower montane (subtropical) taxa, montane conifers, and, to some degree, pteridophytes. There is some indication that lower altitude rainforest has generally expanded from about 600 kyr, after the Early–Middle Pleistocene transition, while there has been an overall reduction since this time in montane conifers. Throughout the record, the greatest variation is in the dominant groups: the herbs and *Pinus*, with the former showing highest values during the glacial periods and the latter during interglacials. It is proposed that some of this variation is a result of glacials being drier than interglacials, an interpretation supported by generally higher values of pteridophytes during interglacials. It is suggested by Sun *et al.* (2003), however, that it was not a simple replacement of herb vegetation by *Pinus*, but that herbs were largely derived from the exposed continental shelf and *Pinus* from more distant mountain areas. The substantial component of *Artemisia* at times within the herb component might suggest also that temperatures during glacial periods were much reduced, even at sea level. However, the very high values for *Artemisia* steppe vegetation during the Last Glacial Maximum are interpreted not in terms of climate but considered the result of a tectonically-induced broader continental shelf relating to an uplift of the Tibetan Plateau around 150 kyr. This interpretation seems inconsistent, though, with the presence of equally high values of *Artemisia* through much of the period from 1,000 to 900 kyr, where no expansion of the continental shelf area is inferred.

This overview of major pollen components in the ODP Site 1144 record masks some important changes that have taken place in the representation of tropical rainforest taxa over the last million years. From a separate portrayal of relative taxon abundance in relation to a lowland and montane tropical rainforest sum, Sun *et al.* (2003) identify three major periods. The earliest, before 900 kyr, is characterized by relatively high values for tropical montane taxa and *Altingia*, suggesting cool conditions, consistent with a climatic interpretation for the high *Artemisia* values. The period corresponds with minimum variation in the marine isotope record. Increased temperature is inferred for the subsequent period, 900 kyr to 355 kyr, where lowland taxa including Dipterocarpaceae, Celastraceae, *Macaranga/Mallotus*, and *Trema* were more conspicuous. This interpretation appears counter to that from the accompanying isotope record where generally lower temperatures may be inferred. From about 355 kyr, the submontane taxa *Quercus* and *Castanopsis* markedly increase their dominance. The abundance of these taxa in seasonal forests in southern China suggests that the climate was more seasonal as well as cooler. The period also marks the entry into and consistent representation of Moraceae, Oleaceae, and *Symplocos* in the pollen record as well as marked increases in Apocynaceae, Rubiaceae, Sapindaceae, and Sapotaceae, indicating a substantial change in the composition of lower altitude rainforest. The restriction of mangrove pollen to this period may seem surprising but could relate to the achievement of higher temperatures periodically with more pronounced glacial–interglacial cyclicity or to changes in coastal configuration. However, the marine isotope record indicates that the amplitude of glacial oscillations increased about 600 kyr, much before the onset of this period.

A pollen record from the lowland Tianyang volcanic basin of the Leizhou Peninsula on the northern coast of the South China Sea and at a similar latitude to ODP Site 1144 (Zheng and Lei, 1999) provides a useful terrestrial comparison of vegetation changes for the region to the marine core for the Late Quaternary period. OSL and radiocarbon dating combined with paleomagnetic analyses have allowed tentative correlation of the record with the marine isotope record over the last 400 kyr. In contrast to the marine record, the dominant pollen types are the evergreen oaks (*Quercus* and *Castanopsis*) that derive from the mountains surrounding the site. The fact that there is only low representation of lowland rainforest taxa, despite the fact that tropical semi-evergreen rainforest surrounds the site, supports the evidence for relatively low pollen production and dispersal from this vegetation formation within modern pollen studies. However, the much lower values for *Pinus* and pterido-phytes—together with the fact that they tend to peak in glacial rather than interglacial periods—brings into question the regional climatic significance of these taxa in the marine record. It is inferred by Zheng and Lei (1999) that glacial periods generally remained wet, although the last glacial period was an exception, with abnormally high values for Poaceae and the only significant values for *Artemisia* interpreted as indicating much drier conditions than present. Temperatures are estimated, from an inferred lowering of montane forest by at least 600 m, to have been some 4°C lower than today during earlier glacial periods, and even lower during the last glacial period. However, there is little variation in lowland forest elements through the record.

4.4.2   The Coral Sea region

Marine records from the Coral Sea adjacent to the humid tropics region of north-eastern Australia provide a coarse resolution coverage of much of the last 10 Myr (Kershaw *et al.*, 1993, 2005; Martin and McMinn, 1993) (Figures 4.1 and 4.7). Throughout almost all of the period, the dry land pollen assemblages are dominated by the rainforest taxa Araucariaceae (predominantly *Araucaria*) and *Podocarpus* that, in this region, may have been abundantly represented in lowland as well as higher altitude communities, and the predominantly sclerophyll taxon Casuarinaceae. Late Miocene to Early Pliocene assemblages also contain notable percentages of other montane rainforest taxa—especially *Dacrydium guillauminii* type, *Dacrycarpus*, *Phyllocladus*, and *Nothofagus*—while rainforest angiosperms (including lowland taxa, herbs, and mangroves) are poorly represented. The climate was wet and probably substantially cooler than today. There is a gap in the record from the Early Pliocene to the very early Quaternary, but dominance of a Late Pliocene terrestrial sequence on the Atherton Tableland by *Podocarpus*, *Nothofagus*, and Casuarinaceae, with the full complement of southern conifers, suggests a continuation of wet and cool conditions until at least close to Pleistocene times (Kershaw and Sluiter 1982). The reduced representation of *Araucaria* at this site can be explained by the per-humid conditions as araucarian forest is generally confined to drier rainforest margins. Conversely, *Nothofagus* would have thrived under the high rainfall as well as the higher altitude of the Tableland.

At the time of recommencement of pollen preservation in the marine record about 1.6 Myr, values for Casuarinaceae, Araucariaceae, and *Podocarpus* are maintained, but other southern conifers have much reduced percentages, with *Phyllocladus* having

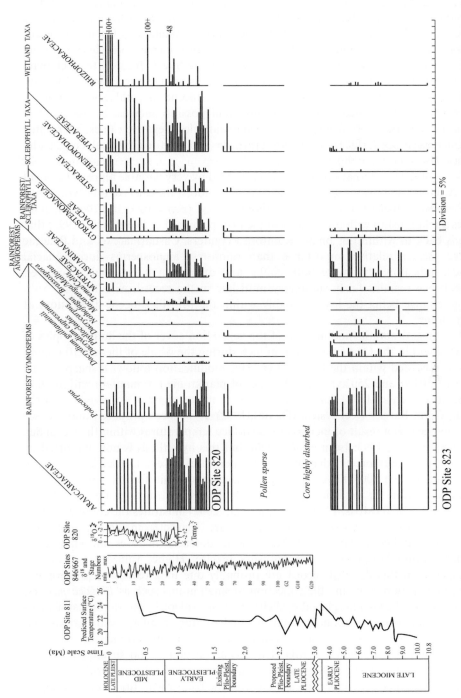

Figure 4.7. Representation of major and indicator taxa in pollen records from ODP Site 823 (Martin and McMinn, 1993) and ODP Site 820 (Kershaw *et al.*, 1993) in relation to the Coral Sea marine isotope record from ODP Site 820 (Peerdeman, 1993) and inferred sea surface temperature records from ODP Sites 820 and 811, and the combined "global" isotope record of Sites 846 and 667 (Shackleton *et al.*, 1995). All terrestrial pollen, excluding mangroves and pteridophyte spores, make up the pollen sum on which all percentages are based.

disappeared from the record. *Nothofagus* also was probably regionally extinct with occasional grains most likely derived by long-distance transport from New Guinea There are marked increases from 1.4 Myr in Poaceae, Asteraceae, Chenopodiaceae, rainforest angiosperms, and mangroves, but—as the record changes at this point from a deep-sea (ODP Site 823) to a continental slope (ODP Site 820) core—these changes may reflect differential pollen transport as much as source vegetation. However, differences between Late Tertiary and Early Pleistocene assemblages do indicate that rainfall had declined and temperatures had possibly increased. The trend in the marine isotope record from the Coral Sea towards less negative values, interpreted as an increase in sea surface temperatures, provides support for a regional temperature increase (Isern *et al.*, 1996).

Most of the Quaternary period is characterized by very variable representation of taxa. Although sample resolution is too coarse to address the cause of this variation, a detailed record from at least 1 Myr to about 950 kyr (Kershaw *et al.*, 2005) demonstrates a relationship with oscillations in the isotope record of Peerdeman (1993) that extends back to within this phase. Rainforest conifers, Casuarinaceae, and Poaceae achieve greater relative importance than rainforest angiosperms during glacial periods, indicating they were both cooler and drier than interglacials. However, the sequence remains fairly stationary until late in the record, with the only substantial change being the loss, probably within the Mid-Pleistocene Transition, of *Dacrycarpus*.

The most dramatic modification of the vegetation cover of the humid tropics of Australia within the Quaternary and, in fact, during the whole of the last 10 million years, is recorded within the last 200 kyr. This modification follows a sharp decline between 350 kyr and 250 kyr in δ^{18}O values of planktonic foraminifera within ODP Site 820 (Peerdeman 1993) that correlated with a major phase of development of the present Great Barrier Reef system (Davies, 1992). It has been proposed that the isotopic change was a result of increased sea surface temperatures within the Coral Sea (Peerdeman *et al.*, 1993; Isern *et al.*, 1996), but this hypothesis is not supported by alkenone paleothermometry that suggests temperatures have not varied by more than 1.5°C over the last 800,000 years and that diagenesis within foraminifera is a more likely explanation for the isotope trend. A detailed record through the last 250 kyr from ODP Site 820 (Moss, 1999; Kershaw *et al.*, 2002) illustrates the complex nature of the vegetation changes (Figure 4.8). Dates are derived from the accompanying isotope record (Peerdeman, 1993), but—due to potential hiatuses and changes in sediment accumulation rates—are not very precise. Higher values for rainforest angiosperms in this detailed record are probably due to a reduction in sieve size during preparation, allowing the collection of small grains such as *Elaeocarpus* and Cunoniaceae. Glacial–interglacial cyclicity is most evident in the rainforest angiosperm and mangrove components that are highest during interglacial periods and during interglacial transgressions, respectively, but is over-ridden by stepwise changes in other major components. There is a substantial and sustained increase in Poaceae around 180 kyr with apparent compensatory decreases in pteridophyes and Arecaceae. Southern conifers decrease through the record with the last major representation of *Podocarpus* and *Dacrydium* about 190 kyr: the latter disappearing from the record around 25 kyr, and sustained decreases in Araucariaceae about 140 kyr and

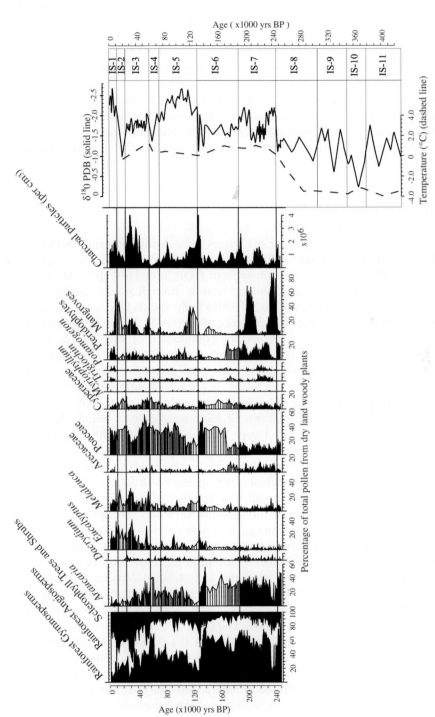

Figure 4.8. Selective features of the detailed Late Quaternary record from ODP Site 820 (Moss, 1999) in relation to the marine isotope record of Peerdeman *et al.* (1993). All taxon abundances are expressed as percentages of the dry land arboreal pollen sum (excluding aquatics, mangroves, and pteridophytes).

30 kyr. By contrast, *Eucalyptus* increases from very low values around 130 kyr and increases again around 40–30 kyr, with rises corresponding to highest charcoal peaks in this record. There is little sustained change in representation of rainforest angiosperms, although the trend towards higher values for Cunoniaceae—resulting in a greater contribution of sub-montane pollen—certainly does not support a general temperature increase. Many of these changes are identified within the later part of the record by those in the adjacent terrestrial record of Lynch's Crater: notably, the initiation of burning around the site, dated to about 45 kyr (Turney *et al.*, 2001), with a sustained increase in *Eucalyptus* and decline in *Araucaria* a few thousand years later, and a similar age for the disappearance of *Dacrydium*. However, there is no evidence for sustained changes before this time, back to the initiation of the record about 220 kyr (Kershaw *et al.*, 2002).

4.5   THE LATER PLEISTOCENE

A more spatially representative picture of changes in the distribution and composition of rainforest can be constructed for the later part of the Pleistocene from a greater number of sites that, in addition to those from ODP Sites 1144 and 820, provide continuous or near-continuous palynological records through at least the last glacial cycle (Figure 4.1). The record from core SHI-9014 in the Banda Sea (van der Kaars *et al.*, 2000) provides the most substantial evidence of vegetation change in the rainforest core of Southeast Asia and a framework for examination of variation within the broader area (Figure 4.9). The regional significance of the Banda Sea record is demonstrated by its remarkable similarity to a recent record from core MD98-2175 in the adjacent Aru Sea (van der Kaars, new data). Rainforest is a prominent component of the pollen spectra throughout, although the substantial sclerophyll component—largely *Eucalyptus*, that would have been derived mainly from the Australian mainland—demonstrates a very broad pollen catchment area. Highest rainforest and pteridophyte values occur during interglacials, indicating that they were much wetter than glacial periods. However, expansion of the Sahul continental shelf during times of low sea level, much of which appears to have been covered largely by grassland (Chivas *et al.*, 2001), would have resulted in excessive Poaceae representation and probable exaggeration of moisture variation through the recorded period. Wet conditions and the maintenance of near-continuous rainforest are certainly evident in some areas—such as highland New Guinea (Walker and Flenley, 1979) and much of the island of Borneo (Anshari *et al.*, 2004; Morley *et al.*, 2004) during the last glacial period, including the Last Glacial Maximum. However, grassland may have disrupted forest growth in more peripheral rainforest (see Section 4.6). The Banda Sea record displays higher values for upper montane taxa during interglacial than glacial periods, and this is surprising considering the abundant evidence from sites in highland parts of the region for much expanded montane vegetation with substantial temperature lowering (see Section 4.6 for discussion of this). This Banda Sea pattern may be a result of an overall reduction in lowland rainforest that is demonstrated to have covered at least parts of the Sunda continental shelf within core rainforest areas during the last glacial period (Morley *et*

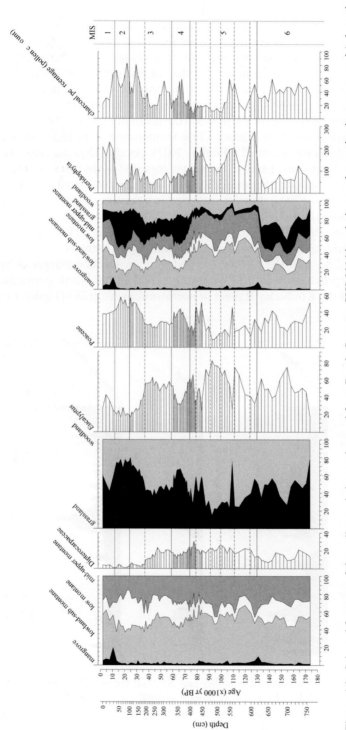

Figure 4.9. Selected features of the pollen and charcoal record from Banda Sea core SHI-9014 in relation to the marine isotope record (adapted from van der Kaars *et al.*, 2000). All taxa are expressed as percentages of the rainforest pollen sum.

al., 2004). It may also be the case that the present day terrestrial area of lowland rainforest was little reduced during glacial periods if there was a smaller degree of temperature lowering at low altitudes. One feature of the Banda Sea record that is shared with those from the Aru Sea and Sangkarang-16, offshore Sulawesi, is a dramatic and sustained reduction in pollen of the dominant lowland rainforest family Dipterocarpaceae about 37 kyr. It is possible that the present pattern of representation of the family—that is lower in abundance and diversity in the eastern than western part of the region—is as much the result of this Late Pleistocene event as it is of the historical barrier of Wallace's Line to migration of the Indo-Malaysian flora westwards as generally assumed (Whiffin, 2002). As this Dipterocarpaceae is associated with an increase in charcoal, burning is regarded as the primary cause, and the impact of early people—rather than climate—has been postulated as its major cause (van der Kaars *et al.*, 2000). Similar sustained increases in charcoal recorded in long marine cores from the Sulu Sea (Beaufort *et al.* 2003) and to the north of New Guinea (Thevenon *et al.*, 2004)—but from a different time, about 52 kyr—have been considered as providing support for the human burning hypothesis.

A rare insight into the history of rainforest on the dry margin of rainforest distribution is provided by marine core MD98-2167 in the North Australian Basin, off the coast of the Kimberley Ranges of northwestern Australia (Figure 4.10). Here,

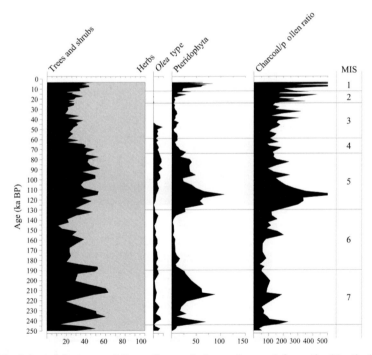

Figure 4.10. Selected features of the pollen and charcoal record from the North Australian Basin core MD98-2167 (Kershaw *et al.*, in press; van der Kaars, unpublished data) in relation to the marine isotope record of Brad Opdyke (unpublished data). All taxa are expressed as percentages of the total dryland pollen sum.

deciduous vine thickets exist in small pockets surrounded by eucalypt-dominated savanna woodland. The major representative of these vine thickets—that have generally poor pollen dispersal—is considered to be *Olea* type. It has low but relatively consistent representation through the record—showing little response to inferred changes in rainfall as indicated by broad glacial–interglacial changes in relative abundance of tree and shrub pollen, that derive largely from the Kimberley region—and particularly the pteridophyte spores that must have been derived from the core rainforest area of Indonesia. Burning appears to have increased around 130 kyr, without any notable change to the vegetation structure, apart from some evidence for increased variability in tree and shrub to herb representation. However, major changes occurred around 46 kyr that included the total disappearance of *Olea* type from the record. This decline in *Olea* may have been associated with a general further increase in burning that has continued to the present.

4.6 THE LAST GLACIAL MAXIMUM (LGM) TO HOLOCENE

A much greater spread of site records as well as more detailed analysis of sites during this period is available than for previous ones (Figure 4.1), allowing more refined investigation of temporal and spatial patterns both within rainforest and between rainforest and more open vegetation communities. Of particular interest are the extent of rainforest and altitudinal shifts in rainforest communities during the LGM that inform debates on contemporary precipitation and temperature levels.

4.6.1 Last Glacial Maximum

The idea that lowland rainforests might have been replaced by grassy savannas at the LGM is clearly not substantiated from the evidence from longer records, but there is some evidence of savanna expansion. The actual degree and areal expression of this expansion is hotly debated. In some more marginal rainforest areas, savanna vegetation did replace rainforest in part—as around Rawa Danau (van der Kaars *et al.*, 2001) and the Bandung Basin (van der Kaars and Dam, 1995)—or totally—as on the Atherton Tableland in northeastern Australia (Kershaw, 1986)—but increased representation of grasses in coastal sites may have been reflecting more open vegetation on exposed continental shelves or an increased aquatic component. Greatest debate has been over the potential existence of a north–south dry corridor extending through Malaysia and between Sumatra and Borneo during the LGM. At one extreme is the view of Morley (2000, 2002) who considered that rainforest massifs, or refugia, were essentially restricted to southwestern Borneo and the adjacent Sunda Shelf, the western part of Sumatra, and very western tip of Java (see Chapter 1). Major migration of rainforest is implied between glacials and interglacials unless high diversity was conserved in river gallery forests, a situation proposed for northeastern Australia during the last glacial period (Hopkins *et al.*, 1993). Kershaw *et al.* (2001), on the other hand, see little evidence for such a dry corridor, at least during the last glacial period. Evidence is sparse and their interpretation is based largely on the almost complete dominance of rainforest pollen in submerged peat cores from the Sunda Shelf off southeastern Sumatra (van der Kaars, unpublished data). Although undated, the peat almost certainly derives from the last glacial period rather than any earlier period as it is unlikely to have survived subsequent low sea level stands.

There is general consensus, however, that—in accordance with the reconstruction of Morley (2000)—much of Borneo retained rainforest and that this forest extended over the continental shelf within the South China Sea region. Confirmation of the maintenance of a rainforest cover within inland West Kalimantan is provided by Anshari *et al.* (2001, 2004), although drier conditions during the later part of the last glacial period are evident, while a marine pollen record from core 17964 in the southern part of the South China Sea (Sun *et al.*, 1999) further substantiates the dominance of a rainforest cover.

In comparison with lowland sites, those from the highlands show the clear maintenance of rainforest through the LGM. Here attention has focused on altitudinal changes in representation of rainforest components. There is good pollen evidence from several sites showing movement of montane tree taxa to lower altitudes. An excellent example of this migration is seen in the pollen diagram from the swamp at the edge of Lake di-Atas in Sumatra (Newsome and Flenley, 1988). The site is at 1,535 m a.s.l., and—where the forest around the lake survives—it is dominated by a variety of tropical oak taxa: *Lithocarpus*, *Castanopsis*, and *Quercus*. It is believed that formerly the tree *Altingia excelsa* (Hamamelidaceae) was abundant also (van Steenis, 1972), but it has been selectively logged. Above 1,800 m the forest changes sharply and becomes dominated by gymnosperms: *Dacrycarpus imbricatus*, *Podocarpus neriifolius* and (in swamps) *Dacrydium* cf. *elatum*. Even *Pinus merkusii* is present, its only natural occurrence in the southern hemisphere. There are also angiosperm trees, the most conspicuous being *Symingtonia populnea*. The diagram (Figure 4.11) shows that in a phase dated to between c. 18 kyr and c. 12 kyr BP (c. 22–14 kcal . yr BP) all those gymnosperms are prominent in the record, only to disappear in the Holocene and be replaced by the distinctive pollen of *Altingia* (previously rare) and peaks of *Quercus* and *Lithocarpus/Castanopsis*. This replacement strongly suggests a climate cooler at the LGM, perhaps by 2°C or more. Interestingly, there is an inversion of radiocarbon dates around ^{14}C kyr BP, which could be explained by lower water tables, permitting erosion of swamp sediments and their redeposition within the core. This depositional event would correlate with the drier lowland Pleistocene climates already mentioned.

Confirmation of these results comes from a site in Java at c. 1,300 m—Situ Bayongbong (Stuijts, 1984)—that is close to the Bandung basin site. Lower altitude sites in Sumatra (Maloney, 1981, 1985, 1998; Morley, 1982; Maloney and McCormac, 1995) also are supportive. These data bring the records of montane gymnosperms down to 1,100 m at c. 22 kcal . yr BP and do not conflict with the Bandung occurrence of *Dacrycarpus* pollen at 650 m, and at Rawa Danau at only c. 100 m (van der Kaars *et al.*, 2001). Collectively, therefore, these results support the Bandung estimate (van der Kaars, 1998) of a climate cooler at the LGM by as much as 4°C or 5°C.

The incursion of montane elements into lowland areas could have been even more marked. The site at Kau Bay in Halmahera (Barmawidjaya *et al.*, 1989) is at present-day sea level. This flooded volcanic crater was a freshwater lake when sea level was

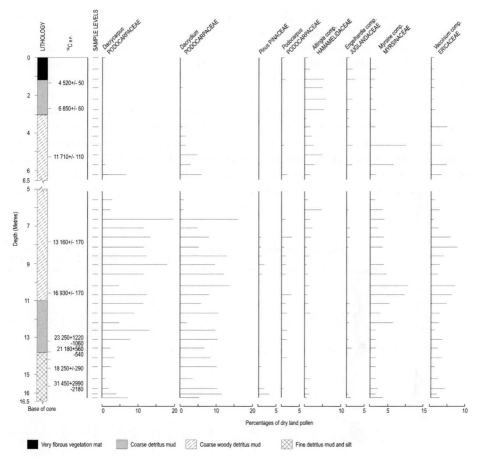

Figure 4.11. Pollen diagram from Danau di Atas Swamp, West Sumatra, altitude 1,535 m. Values are given as percentages of total dry land pollen. Only selected taxa are shown. After Newsome and Flenley (1988) and Stuijts *et al.* (1988).

lower by >100 m at the LGM. Palynology of this site showed occurrences of *Castanopsis*/*Lithocarpus* and *Quercus* at the LGM. While it is not suggested that these taxa necessarily grew at present sea level, they apparently grew close enough for small amounts of their pollen to enter the record. This record would be consistent with a temperature lowering of c. 6°C at the LGM, even in the lowlands. Similar results have been obtained from lowland sites in West Kalimantan (Anshari *et al.*, 2001, 2004).

There is also evidence from the highest mountains of Indonesia for Pleistocene cooling of as much as 6°C. Leaving aside the evidence from New Guinea, we have evidence of Pleistocene glaciation on Mt. Kinabalu in northern Borneo (Koopmans and Stauffer, 1968) and of deglaciation around 8,000 BP (c. 10 kcal . yr BP) at 4,000 m a.s.l. (Flenley and Morley, 1978). Similar evidence (for solifluction at least) is claimed for the slightly lower peak of Gunong Leuser (3,381 m) in northern Sumatra (Beek, 1982).

In New Guinea the best evidence of environmental change comes from upland regions. The site at Sirunki (Walker and Flenley, 1979) appears to cover the last 33 kyr (c. 40 kcal.yr BP), at an altitude of 2,500 m a.s.l. The site currently lies in *Nothofagus* forest (much disturbed), and is some 1,300 m below the altitudinal forest limit at c. 3,800 m a.s.l. Nevertheless the pollen record clearly shows the presence of tropic-alpine herbs (*Astelia, Gentiana, Drapetes*, etc.) in the Late Pleistocene, when forest pollen values decline to a level consistent with unforested conditions. Similar results were obtained from Lake Inim at 2,550 m by Flenley (1972) (Figure 4.12).

There is, of course, geomorphological evidence of lowered snow lines (U-shaped valleys, moraines, etc.) in the New Guinea mountains. On Mt. Wilhelm, a lowering in the Late Pleistocene of c. 1,000 m is indicated (Löffler, 1972), and there are similar findings from Irian Jaya (Hope and Peterson, 1975). One thousand metres translates into perhaps 6°C cooling, using a modern lapse rate.

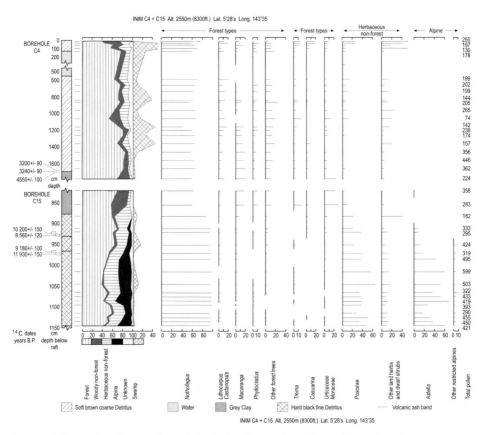

Figure 4.12. Pollen diagram from Lake Inim, boreholes C4 and C15, plotted on the same scales. The results are expressed as percentages of pollen of forest types, except in the summary diagram where the total of dry land pollen and spores forms the pollen sum. Only selected taxa shown. After Flenley (1972).

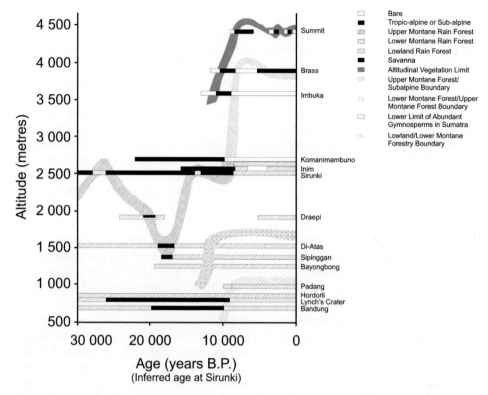

Figure 4.13. Selected vegetation records derived from pollen diagrams from tropical Southeast Asia and the West Pacific. Only the last 30,000 years are shown. Human impact is omitted. References to individual sites are as follows: Summit, Brass, Imbuka, Komanimambuno (Hope, 1976), Inim (Flenley, 1972; Walker and Flenley, 1979), Sirunki (Walker and Flenley, 1979), Draepi (Powell *et al.*, 1975), Di-Atas (Newsome and Flenley, 1988), Sipinggan (Maloney, 1981), Bayongbong (Stuijts, 1984; Stuijts *et al.*, 1988). For other sites see Flenley (1998). After Flenley (1998).

At the LGM, climates cooler than now by as much as 7–11°C can be suggested from the pollen results, but only c. 6°C from the geomorphology. How can this be? Possibly, snow lines were kept artificially high by the reduced precipitation that probably occurred at the LGM. This suggestion has been advanced by Walker and Flenley (1979). But, the precipitation in the mountains cannot have been too reduced, or rainforests would have disappeared there. Late Quaternary vegetational changes are summarized in Figure 4.13.

This whole question has been reviewed by Pickett *et al.* (2004) in a reconstruction of Quaternary biomes for the Southeast Asian region, Australia, and the Pacific. They conclude that the evidence from the Southeast Asian tropics indicates an LGM cooling of 1–2°C at sea level and 6–9°C at high elevation sites. This discrepancy was first noted by Walker and Flenley (1979), who attributed it to a steeper lapse rate, which was itself related to the generally drier conditions at the LGM, leading to a lapse rate closer to the dry adiabatic lapse rate. Such a steeper lapse rate was however criticized as impossible in an environment where the pollen evidence clearly showed the persistence of rainforests (Webster and Streten, 1972; Kutzbach and Guetter, 1986).

The idea has however been revived by Farrera *et al.* (1999) who give a range of plausible mechanisms by which a steeper lapse rate can occur on tropical mountains. One among these was the observation that the moist adiabatic lapse rate steepens anyway as temperature is lowered (Hartmann, 1994). Another relevant point is the possible impact of reduced concentrations of carbon dioxide at the LGM (Street-Perrott, 1994), which would favour C_4 plants (grasses) at the expense of trees.

A steeper lapse rate alone is however unable to explain all aspects of observed vegetation changes. Hope (1976) demonstrated—regarding Mt. Wilhelm—the curious fact that the Upper Montane Rain Forest (UMRF) (cloud forest) did not simply migrate downhill at the LGM: it virtually disappeared. This is not a Late Pleistocene anomaly, for Upper Montane taxa were greatly reduced in each glacial period in the Banda Sea record (see Section 4.5). To understand this phenomenon we must consider the physiognomy and environment of the UMRF at the present time (Flenley, 1992, 1993). The trees are stunted, with short internodes and small thick leaves which possess a hypodermis (an extra layer of cells below the epidermis). Often extra pigments are present as well as chlorophyll: usually flavonoids and/or antho-cyanins. These attributes are typical of plants experiencing stress of various kinds, including high ultraviolet-B and temperature extremes. The soils are unusual in the thickness of their litter layer. The temperature environment shows extreme variations on a diurnal basis: from very cold nights and early mornings, to sunny mornings and misty afternoons with 100% humidity. The morning insolation is high in ultraviolet-B because of the altitude. It is known that high UV-B can produce experimentally in crop plants exactly the same physiognomic peculiarities as the Upper Montane Rain Forest (Teramura, 1983), and can also inhibit insect activity, leading to thick litter layers in soils (Day, 2001). It therefore seems possible that UV-B is involved in the ecology of these forests. It may well be that the extreme diurnal variation of the temperature regime is also involved.

How does all this help to explain the decline of the UMRF in cold phases of the Pleistocene? Presumably, when cooler temperatures forced taxa downhill, they found themselves in an environment where the diurnal extremes of temperature and UV-B no longer existed to the same extent. The tropical lowlands are in fact usually lacking in such extremes. Assuming the UMRF taxa are genetically adapted to their present environment, the disappearance or great restriction of that environment at the LGM would have led to their reduction in the pollen record (Flenley, 1996, 1998). This argument will be elaborated in Chapter 8.

4.6.2   The Pleistocene–Holocene transition

Marine records generally show abrupt pollen shifts from glacial to interglacial con-ditions, suggesting strong Milankovitch forcing of climate and rapid response of the vegetation. However, altered boundary conditions, including those related to coastal landscapes and oceanic and atmospheric circulation, may have played a part in producing this degree of synchroneity. A major exception related to the Coral Sea record of ODP 820 where an increase in rainforest lags the marine isotope change from MIS 2 to 1 by several thousand years. As this lag is also evident in the Lynch's Crater record (Kershaw, 1986), it cannot easily be attributed to global climate forcing. Possible explanations are: the time taken for rainforest patches to expand from glacial "refugia" (unlikely considering the regional extent of rainforest during the last glacial period); the influence of southern hemisphere insolation forcing including ENSO on the record; and the continuing impact of Aboriginal burning; all of which slowed the rainforest advance. More detailed analysis of the earlier part of the ODP 820 record may help resolve this question.

Many terrestrial records are too coarse or insufficiently well-dated to detail local patterns of change during the last termination. However, at an altitude of 3,630 m in Irian Jaya, there is supporting evidence for a rapid replacement of grasslands and scattered shrubs by rainforest at the Pleistocene–Holocene boundary. Similar changes are evident at Lake Inim (Flenley, 1972) and in the elegant suite of sites on Mt. Wilhelm, the highest mountain in Papua New Guinea (Hope, 1976). With four sites at elevations from 2,750 m to 3,910 m, Hope was able to trace the deglaciation of the mountain and the rapid climb of the altitudinal forest limit to about 4,000 m in the early Holocene.

Walker and Flenley (1979) found a hint of a Late Pleistocene oscillation at Sirunki, though its age of c. 17 kcal.yr BP does not correlate well with the Younger Dryas and is more consistent with the Antarctic Reversal. Support for such an oscillation has recently been demonstrated from a detailed analysis of the last termination at Rawa Danau in Java (Turney et al., 2006). Towards the end of the LGM (Turney et al., in press), high values for grass pollen—combined with the presence of the montane trees Dacrycarpus, Podocarpus, and Quercus—indicate much drier and cooler conditions than today. Initial increases in temperature and rainfall are recorded as early as 17 kcal.yr BP with increased representation of lowland rainforest taxa and reduction in Poaceae. There is then a reversal of this trend between 15.4 kcal.yr BP and 14.6 kcal.yr BP, prior to both the Antarctic Reversal and Younger Dryas, suggesting a regional tropical rather than hemispheric control over climate variation. Although rainforest became dominant at 14.6 kcal.yr BP, increased catchment erosion suggests rainfall further increased around 12.9 kcal.yr BP and that the summer monsoon may not have become fully established until the early Holocene.

4.6.3 The Holocene

Rainforest achieved its maximum areal extent in the Early–Middle Holocene under high levels of precipitation and temperature before it opened up again mainly within the last 5,000 years. Reasons for this rainforest reduction include climate factors, although these varied regionally. Hope (1976) attributes a reduction in the altitudinal treeline of about 200 m to a reduction in temperature in New Guinea, while seasonality or increased ENSO influence is considered to have been the major influence on both a change in the composition of rainforest and slight sclerophyll woodland expansion in northeast Queensland (Kershaw and Nix, 1989; McGlone *et al.*, 1992; Haberle, 2005). However, the major impact on Holocene rainforest has been that of people.

Although people have been in the region for around 1.8 million years (Swisher *et al.*, 1994; Huffman, 2001) and have had the ability to manage vegetation through the use of fire within the last 100 kyr, the ability of people to physically clear rainforest for agriculture is essentially a Holocene phenomenon within the region. There are indications of agriculture (for rice-growing) as early as 16 kcal. yr BP in the Yangtze Valley, China (Yasuda, 2002)—an area recognized as one of the cradles of crop domestication (Vavilov, 1951; Diamond, 1998)—and its spread into Southeast Asia, including those parts of the Sunda Platform which were then joined to the Asian Mainland. Several upland sites in Sumatra suggest that swiddening (slash-and-burn) was occurring for cultivation of dry (non-irrigated) rice or root crops as early as c. 10.3 kcal. yr BP at Danau-di-Atas (Newsome and Flenley, 1988), c. 10 kcal. yr BP at Pea Bullok (Maloney and McCormac, 1995), and c. 9 kcal. yr BP at Rawang Sikijang (Flenley and Butler, 2001). The general pattern of evidence for the region was reviewed by Maloney (1998) and Flenley (2000).

A separate center for the origin of agriculture is found in New Guinea, based on rootcrops—such as *Colocasia* (taro)—and palynological evidence of forest destruction presumably for agriculture dates back to c. 9 kcal. yr BP in the Baliem Valley (Haberle *et al.*, 1991), to c. 6 kcal. yr BP or earlier at Draepi Swamp (Powell *et al.*, 1975), and to 5 kcal. yr BP at Sirunki Swamp (Walker and Flenley, 1979). Early human activity in the New Guinea Highlands was confirmed by archeological finds at Kuk Swamp (Golson and Hughes, 1976; Golson, 1977; Denham *et al.*, 2003). These included evidence of swamp drainage, presumably for the growing of taro, back to at least 6.8 kcal. yr BP and possibly c. 10 kcal. yr BP. The destruction of swamp forest on Lynch's Crater about 5 kcal. yr BP suggests that some form of cultivation may have spread into the rainforest areas of northeastern Australia at this time.

The progressive impact of these activities has led to the creation of permanent grasslands in many areas. These include the *cogonales* of the Philippines, the *kunai* of New Guinea, and smaller areas in Sumatra and elsewhere. In general, these areas have been maintained by frequent burning, and they tend to occur in regions where there is a more lengthy dry season (Thomas *et al.*, 1956). Recently, agriculture and logging has of course still further diminished the area of surviving forest, but consideration of that is beyond the scope of this chapter.

4.7   VEGETATION RESPONSES TO CYCLICAL FORCING

Spectral analysis is a powerful tool for examination of cyclical variation within components of well-dated, largely marine, records and has been applied to a number of sequences within the far eastern rainforest region in order to assess responses of various proxies to potential forcing mechanisms on orbital timescales.

Analysis of ODP Site 1144 in the South China Sea was restricted to pine and herb pollen, essentially winter monsoon indicators, and demonstrated clear Milankovitch forcing with prominent 100 kyr (eccentricity), 40 kyr (obliquity), and 20 kyr (precession) periodicities, in phase with northern hemisphere insolation and ice volume and indicating a clear link with monsoon activity (Sun *et al.*, 2003). It is interesting that the herbs demonstrate a closer correspondence with the ice volume signature than pine, reinforcing the suggested higher latitude source of its pollen. Pine displays a higher precessional than obliquity peak, indicating some tropical influence. It also shows a strong semi-precessional frequency that may be the result of an additional southern hemisphere tropical precessional signal resulting from changing mean position of the Intertropical Convergence Zone through time. Unfortunately, as spectral analysis was performed on the whole million-year record, there is no way of determining changes in forcing through time—that may be expected with a change in the global signal—from dominant obliquity to eccentricity forcing around the Brunhes–Matuyama boundary.

A greater range of proxies has been examined from the southern hemisphere sites of Banda Sea, Lombok Ridge, and ODP 820 (Kershaw *et al.*, 2003) and the North Australian Basin (van der Kaars, new data). In general terms, mangroves show similar frequencies to those from associated oxygen isotope records, and indicate strong northern hemisphere forcing. With mangroves, this pattern is not surprising as they are constrained by sea level changes that relate directly to ice volume. Their closer relationship with variation in sea level rather than that of climate has been explained by Grindrod *et al.* (1999, 2002). It is the broad exposure of the continental shelf during marine transgressions that facilitates mangrove colonization and peak mangrove pollen representation.

The major indicators of core rainforest—pteridophytes and rainforest angiosperms—surprisingly exhibit rather different spectral signatures. Pteridophytes exhibit strong glacial–interglacial cyclicity with a prominent obliquity signal that is not evident in the rainforest angiosperms that display a dominant precessional signal, except in the Banda Sea record. The most parsimonious explanation is that the core rainforest area is greatly influenced by the Asian monsoon and this influence extends to other marine sites in the pteridophytes due to the wide dispersal of their spores. The implication that more marginal areas are displaying a more localized tropical precessional influence is well-demonstrated in the eucalypt component of the North Australian Basin, whose variation can be clearly correlated with that of southern hemisphere precession. Although the area expresses a marked monsoon climate, the source of monsoon rainfall is most probably the southern Indian Ocean that is largely divorced from the Asian system.

Rainforest angiosperms also display significant variation in the 30-kyr frequency band, and this frequency is even more clearly expressed in rainforest gymnosperms and some charcoal records. It has been proposed by Kershaw *et al.* (2003) that this non-Milankovitch cycle is related to ENSO, due to the correspondence of peaks in burning and associated declines in fire-sensitive araucarian forests in the ODP record with peaks in ENSO frequency derived from the modelling of Clement *et al.* (1999). From similar spectral signatures in the records of charcoal from cores MD97-2141 and MD97-2140 in similar West Pacific settings, Beaufort *et al.* (2003) and Thevenon *et al.* (2004) propose that the 30-kyr frequency can be attributed to the competing influences of long-term ENSO-like forcing and the glacial–interglacial cycle on the East Asian Summer Monsoon.

4.8   GENERAL DISCUSSION AND CONCLUSIONS

Perhaps the major feature of the Quaternary history of far eastern rainforest is that there has been little overall change in gross distribution over most of the area and through most of the period. Rainforest has shown tremendous resilience in the face of increasingly variable climatic conditions that have seen its partial replacement by more open vegetation types during drier and cooler glacial phases. The major spatial exception to this pattern is in tropical Australia, where the extensive occurrence of small patches of drier rainforest demonstrates survival from a much broader past distribution. This must pay testament to the aggressive nature of fire-promoting eucalypts, as it appears that grasslands alone have had much less of a long-term impact on similar vegetation in the Southeast Asian region, although poor soils and extreme climatic variability over much of tropical Australia may have contributed to this pattern. This patchy distribution is evident through the last 500 kyr in northern Australia, and it can only be surmised that this is a product of the Quaternary rather than a geologically earlier period.

In northeastern Australia, however, the impact of the Quaternary is clear with widespread reduction in drier araucarian forest within the late Quaternary. There has been speculation into the cause of this decline, elaborated by Kershaw et al. (2006). One explanation is human impact. However, without any evidence of the presence of people on the continent before about 50–60 kyr ago, this line of speculation is not particularly constructive, although there seems little doubt that, with the arrival of people, the trend towards more open vegetation was accelerated. A more robust hypothesis is that there has been a long-term trend towards aridity or variability in, at least, the northern part of Australia, and such a pattern is evident in physical as well as biological proxy data. The proposed changes in marine isotope signatures in the Coral Sea may be indicative of the development of the West Pacific Warm Pool, perhaps a threshold response to the movement of Australia into the Southeast Asian region and constriction of the Indonesian Gateway between the Pacific and Indian Oceans. Although an increase rather than decrease in precipitation might be expected by the rise in sea surface temperatures, any accompanying development in the ENSO system would have increased climatic variability and resulted in the frequent drought conditions required for effective biomass burning. A more direct influence of higher sea surface temperatures could be the expansion of mangrove vegetation that seems to have been a general regional feature of the late Quaternary.

Although there are no dramatic changes in overall rainforest distribution within the South China Sea region, the suggestion by Sun et al. (2003) that the climate may have become more seasonal within the last 350 kyr could be significant, if the variability was interannual instead of, or in addition to, being seasonal. This would be consistent with a broad regional ENSO signal.

One important feature of the history of most, if not all, of the region is that conditions during the latter part of the last glacial period to the present day were distinctive, and are limited in the degree they can be applied to an understanding of extreme glacial and interglacial conditions or the nature of abrupt climate changes, in general. Assessment of the extent of savanna expansion during glacial periods is particularly problematic and critical to the understanding of mammal, including hominid, migrations into and establishments within the region. The separation of climatic and human influences is clearly important for patterns of change that relate to moisture variables, but the suggestion that the cooler-adapted communities were extensive during the last glacial period suggests that temperature variation also needs to be taken into consideration when comparison is made of glacial–interglacial cycles.

4.9   REFERENCES

Anshari, G., Kershaw, P., and van der Kaars, S. (2001) A Late Pleistocene and Holocene pollen and charcoal record from peat swamp forest, Lake Sentarum Wildlife Reserve, West Kalimantan, Indonesia. *Palaeogeography, Palaeoclimatology, Palaeoecology* **171**, 213–228.

Anshari, G., Kershaw, A. P., van der Kaars, S., and Jacobsen, G. (2004) Environmental change and peatland forest dynamics in the Lake Sentarum area, West Kalimantan, Indonesia. *Journal of Quaternary Science* **19**, 637–655.

Barmawidjaja, D. M., de Jong, A. F. M., van der Borg, K., van der Kaars, W. A., and Zachariasse, W. J. (1989) Kau Bay, Halmahera, a Late Quaternary palaeoenvironmental record of a poorly ventilated basin. *Netherlands Journal of Sea Research* **24**, 591–605.

Beaufort, L., Garidel, T., Linsley, B., Oppo, D., and Buchet, N. (2003) Biomass burning and oceanic primary production estimates in the Sulu Sea area over the last 380 kyr and the East Asian monsoon dynamics. *Marine Geology* **201**, 53–65.

Beek, C. G. G. van (1982) *A Geomorphological and Pedological Study of the Gunong Leuser National Park, North Sumatra, Indonesia* (187 pp. + plates). Wageningen Agricultural University, Wageningen, The Netherlands.

Chivas, A. R., García, A., van der Kaars, S., Couapel, M. J. J., Holt, S., Reeves, M. R., Wheeler, D. J., Switzer, A. D., Murray-Wallace, C. V., Banerjee, D. *et al.* (2001) Sea-level and environmental changes since the last interglacial in the Gulf of Carpentaria, Australia: An overview. *Quaternary International* **83-85**, 19–46.

Clement, A. C., Seager, R., and Cane, M. A. (1999) Orbital controls on the El Niño/Southern Oscillation and the tropical climate. *Paleoceanography* **14**, 441–456.

Davies, P. J. (1992) Origins of the Great Barrier Reef. *Search* **23**, 193–196.

Day, T. A. (2001) Ultraviolet radiation and plant ecosystems. In: C. S. Cockell and A. R. Blaustein (eds.), *Ecosystems, Evolution and Ultraviolet Radiation*. Springer-Verlag, New York.

Denham, T. P., Haberle, S. G., Lentfer, C., Fullagar, R., Field, J., Therin, M., Porch, N., and Winsborough, B. (2003) Origins of agriculture at Kuk Swamp in the Highlands of New Guinea. *Science* **301**, 189–193.

Diamond, J. (1998) *Guns, Germs and Steel* (480 pp.). Vintage, London.

Farrera, I., Harrison, S. P., Prentice, I. C., Bartlein, P. J., Bonnefille, R., Bush, M., Cramer, W., von Grafenstein, U., Holmgren, K., Hooghiemstra, H. *et al.* (1999) Tropical climates of the

Last Glacial Maximum: A new synthesis of terrestrial palaeoclimate data. 1. Vegetation, lake levels and geochemistry. *Climate Dynamics* **15**, 823–856.

Fedorova, I. T., Volkova, Y. A., and Varlyguin, D. L. (1993) Legend to the World Vegetation Cover Map. Unpublished.

Fedorova, I. T., Volkova, Y. A., and Varlyguin, D. L. (1994) World Vegetation Cover: Digital raster data on a 30-minute Cartesian orthonormal geodetic (lat./long.) 1,080 × 2,160 grid. In: *Global Ecosystems Database Version 2.0*. USDOC/NOAA National Geophysical Data Center, Boulder, CO. Three independent single-attribute spatial layers and one tabular attribute data file (11,806,343 bytes in 19 files).

Flenley, J. R. (1972) Evidence of Quaternary vegetational change in New Guinea. In: P. Ashton and M. Ashton (eds.), *The Quaternary Era in Malesia: Transactions of the Second Aberdeen–Hull Symposium on Malesian Ecology* (Miscellaneous Series 13, pp. 99–109). Department of Geography, University of Hull, U.K.

Flenley, J. R. (1973) The use of modern pollen rain samples in the study of vegetational history of tropical regions. In: H. J. B. Birks and R. G. West (eds.), *Quaternary Plant Ecology: The 14th Symposium of the British Ecological Society* (pp. 131–141). Blackwell, Oxford, U.K.

Flenley, J. R. (1979) *The Equatorial Rain Forest: A Geological History*. Butterworths, London.

Flenley, J. R. (1992) UV-B insolation and the altitudinal forest limit. In: P. A. Furley, J. Proctor, and J. A. Ratter (eds.), *Nature and Dynamics of Forest–Savanna Boundaries* (pp. 273–282). Chapman & Hall, London.

Flenley, J. R. (1993) Cloud forest, the Massenerhebung effect, and ultraviolet insolation. In: L. S. Hamilton, J. O. Juvik, and F. M. Scatena (eds.), *Tropical Montane Cloud Forests* (pp. 94–96). East–West Center, Honolulu, HI.

Flenley, J. R. (1996) Problems of the Quaternary on mountains of the Sunda–Sahul Region. *Quaternary Science Reviews* **15**, 549–555.

Flenley, J. R. (1998) Tropical forests under the climates of the last 30,000 years. *Climatic Change* **39**, 177–197.

Flenley, J. R. (2000) Keynote address: The history of human presence and impact in island S E Asia and the South Pacific. In: M. Roche, M. McKenna, and P. Hesp (eds.), *Proceedings of the 20th New Zealand Geography Conference* (pp. 18–23). New Zealand Geographical Society, Massey University, Palmerston North.

Flenley, J. R. and Butler, K. R. (2001) Evidence for continued disturbance of upland rain forest in Sumatra for the last 7000 years of an 11,000 year record. *Palaeogeography, Palaeoclimatology, Palaeoecology* **171**, 289–305.

Flenley, J. R. and Morley, R. J. (1978) A minimum age for the deglaciation of Mt Kinabalu, East Malaysia. *Modern Quaternary Research in South-East Asia* **4**, 57–61.

Golson, J. (1977) No room at the top: agricultural intensification in the New Guinea Highlands. In: Allen, J., Golson J. and Jones R. (eds.), *Sunda and Sahul: Prehistoric Studies in Southeast Asia, Melanesia and Australia*, pp. 601–638. Academic Press, London.

Golson, J. and Hughes, P. J. (1976) The appearance of plant and animal domestication in New Guinea. *Proceedings of IXth Congress*. Union International des Sciences Préhistoriques et Protohistoriques, Nice, France.

Grindrod, J., Moss, P., and van der Kaars, S. (1999) Late Quaternary cycles of mangrove development and decline on the north Australian continental shelf. *Journal of Quaternary Science* **14**, 465–470.

Grindrod, J., Moss, P. T., and van der Kaars, S. (2002) Late Quaternary mangrove pollen records from Continental Shelf and ocean cores in the North Australian–Indonesian region. In: A. P. Kershaw, B. David, N. Tapper, D. Penny, and J. Brown (eds.),

Bridging Wallace's Line: The Environmental and Cultural History and Dynamics of the Southeast Asian–Australian Region (pp. 119–146). Catena Verlag, Reiskirchen, Germany.

Haberle, S. G. (2005) A 23,000-yr pollen record from Lake Euramoo, Wet Tropics of NE Queensland, Australia. *Quaternary Research* **64**, 343–356.

Haberle, S. G., Hope, G. S., and De Fretes, Y. (1991) Environmental changes in the Baliem Valley, montane Irian Jaya, Republic of Indonesia. *Journal of Biogeography* **18**, 25–40.

Hartmann, D. L. (1994) *Global Physical Climatology.* Academic Press, San Diego.

Hope, G. S. (1976) The vegetational history of Mt Wilhelm, Papua New Guinea. *Journal of Ecology* **64**, 627–663.

Hope, G. S. and Peterson, J. A. (1975) Glaciation and vegetation in the High New Guinea Mountains. *Royal Society of New Zealand Bulletin* **13**, 155–162.

Hopkins, M. S., Ash, J., Graham, A. W., Head, J., and Hewett, R. K. (1993) Charcoal evidence of the spatial extent of the *Eucalyptus* woodland expansion of lowland rainforest in humid, tropical north Queensland. *Journal of Biogeography* **20**, 357–372.

Huffman, O. F. (2001) Geologic context and age of the Perning/Mojokerto *Homo erectus. Journal of Human Evolution* **40**, 353–362.

Isern, A. R., McKenzie, J. A., and Feary, D. A. (1996) The role of sea-surface temperature as a control on carbonate platform development in the western Coral Sea. *Palaeogeography, Palaeoclimatology, Palaeoecology* **124**, 247–272.

Kershaw, A. P. (1970) A pollen diagram from Lake Euramoo, north-east Queensland, Australia. *New Phytologist* **69**, 785–805.

Kershaw, A. P. (1973) The numerical analysts of modern pollen spectra from northeast Queensland ram-forests. *Special Publication Geological Society of Australia* **4**, 191–199.

Kershaw, A. P. (1986) Climate change and Aboriginal burning in north-east Australia during the last two glacial/interglacial cycles. *Nature* **322**, 47–49.

Kershaw, A. P. and Bulman, D. (1994) The relationship between modern pollen samples and the environment in the humid tropics region of northeastern Australia. *Review of Palaeobotany and Palynology* **83**, 83–96.

Kershaw, A. P. and Hyland, B. P. M. (1975) Pollen transport and periodicity in a rainforest situation. *Review of Palaeobotany and Palynology* **19**, 129–138.

Kershaw, A. P. and Nix, H. A. (1989) The use of climatic envelopes for estimation of quantitative palaeoclimatic estimates. In: T. H. Donnelly and R. J. Wasson (eds.), *CLIMANZ 3* (pp. 78–85). CSIRO Division of Water Resources, Canberra.

Kershaw, A. P. and Sluiter. I. R. (1982) Late Cainozoic pollen spectra from the Atherton Tableland, northeastern Australia. *Australian Journal of Botany* **30**, 279–295.

Kershaw, A. P. and Strickland, K. M. (1990) A 10 year pollen trapping record from rainforest in northeastern Queensland, Australia. *Review of Palaeobotany and Palynology* **64**, 281–288.

Kershaw, A. P., McKenzie, G. M., and McMinn, A. (1993) A Quaternary vegetation history of northeastern Queensland from pollen analysis of ODP Site 820. *Proceedings of the Ocean Drilling Program Scientific Results* **133**, 107–114.

Kershaw, A. P., Penny, D., van der Kaars, S., Anshari, G., and Thamotherampillai, A. (2001) Vegetation and climate in lowland southeast Asia at the Last Glacial Maximum. In: I. Metcalfe, M. B. Smith, M. Morwood, and I. Davidson (eds.), *Faunal and Floral Migrations and Evolution in SE Asia–Australasia* (pp. 227–236). A.A. Balkema, Lisse, The Netherlands.

Kershaw, A. P., van der Kaars, S., Moss, P. T., and Wang, S. (2002) Quaternary records of vegetation, biomass burning, climate and possible human impact in the Indonesian–Northern Australian region. In: A. P. Kershaw, B. David, N. Tapper, D. Penny, and J. Brown (eds.), *Bridging Wallace's Line: The Environmental and Cultural History and*

Dynamics of the Southeast Asian–Australian Region (pp. 97–118). Catena Verlag, Reiskirchen, Germany.

Kershaw, A. P., van der Kaars, S., and Moss, P. T. (2003) Late Quaternary Milankovitch-scale climate change and variability and its impact on monsoonal Australia. *Marine Geology* **201**, 81–95.

Kershaw, A. P., Moss, P. T., and Wild, R. (2005) Patterns and causes of vegetation change in the Australian Wet Tropics region over the last 10 million years. In: E. Bermingham, C. Dick, and C. Moritz, C. (eds.), *Tropical Rainforests: Past, Present and Future* (pp. 374–400). Chicago University Press, Chicago.

Kershaw, P., van der Kaars, S., Moss, P., Opdyke, B., Guichard, F., Rule, S., and Turney, C. (2006) Environmental change and the arrival of people in the Australian region. *Before Farming* (online version) 2006/1 article 2.

Koopmans, B. N. and Stauffer P. H. (1968) Glacial phenomena on Mount Kinabalu, Sabah. Borneo Region, Malaysia. *Geological Survey Bulletin* **8**, 25–35.

Kutzbach, J. E. and Guetter P. J. (1986) The influence of changing orbital parameters and surface boundary conditions on climate simulations for the past 18,000 years. *Journal of Atmospheric Science* **43**, 1726–1759.

Löffler, E. (1972) Pleistocene glaciation in Papua and New Guinea. *Zeitschift für Geomorphologie*, Supplement 13, 46–72.

Maloney, B. K. (1981) A pollen diagram from Tao Sipinggan, a lake site in the Batak Highlands of North Sumatra, Indonesia. *Mod. Quat. Res. in South-East Asia* **6**, 57–66.

Maloney, B. K. (1985) Man's impact on the rainforests of West Malesia: The palynological record. *Journal of Biogeography* **12**, 537–558.

Maloney, B. K. (1998) The long-term history of human activity and rainforest development. In: B. K. Maloney (ed.), *Human Activities and the Tropical Rainforest* (pp. 66–85). Kluwer, Dordrecht, The Netherlands.

Maloney, B. K. and McCormac, F. G. (1995) Thirty thousand years of radiocarbon dated vegetation and climatic change in highland Sumatra. *Radiocarbon* **37**, 181–190.

Martin, H. A. and McMinn, A. (1993) Palynology of sites 815 and 823: The Neogene vegetation history of coastal northeastern Australia. *Proceedings of the Ocean Drilling Program Scientific Results* **133**, 115–125.

McGlone, M. S., Kershaw, A. P., and Markgraf, V. (1992) El Niño/Southern Oscillation climatic variability in Australasian and South American palaeoenvironmental records. In: H. F. Diaz and V. Markgraf (eds.), *El Niño: Historical and Palaeoclimatic Aspects of the Southern Oscillation* (pp. 435–462). Cambridge University Press, Cambridge, U.K.

Metcalfe, I. (2002) Tectonic history of the SE Asian–Australian region. In: A. P. Kershaw, B. David, N. Tapper, D. Penny, and J. Brown (eds.), *Bridging Wallace's Line: The Environmental and Cultural History and Dynamics of the Southeast Asian Australian Region* (pp. 29–48). Catena Verlag, Reiskirchen, Germany.

Morley, R. J. (1982) A palaeoecological interpretation of a 10,000 year pollen record from Danau Padang, Central Sumatra, Indonesia. *Journal of Biogeography* **9**, 151–190.

Morley R. J. (2000) *Origin and Evolution of Tropical Rainforests*. John Wiley & Sons, New York.

Morley, R. J. (2002) Tertiary vegetational history of Southeast Asia, with emphasis on the biogeographical relationships with Australia. In: A. P. Kershaw, B. David, N. Tapper, D. Penny, and J. Brown J. (eds.), *Bridging Wallace's Line: The Environmental and Cultural History and Dynamics of the Southeast Asian–Australian Region* (pp. 49–60). Catena Verlag, Reiskirchen, Germany.

Morley, R. J., Morley, H. P., Wonders, A. A., Sukarno, H. W., and van der Kaars, S. (2004) Biostratigraphy of modern (Holocene and Late Pleistocene) Sediment cores from the Makassar Straits. *Proceeding, Deepwater and Frontier Exploration in Asia & Australasia Symposium.* Indonesian Petroleum Association, Jakarta.

Moss, P. T. (1999) Late Quaternary environments of the humid tropics of northeastern Australia. Unpublished PhD thesis, School of Geography and Environmental Science, Monash University, Melbourne.

Moss, P. T. and Kershaw, A. P. (2000) The last glacial cycle from the humid tropics of northeastern Australia: Comparison of a terrestrial and a marine record. *Palaeogeography, Palaeoclimatology, Palaeoecology* **155**, 155–176.

Newsome, J. and Flenley, J. R. (1988) Late Quaternary vegetational history of the Central Highlands of Sumatra, II: Palaeopalynology and vegetational history. *Journal of Biogeography* **15**, 555–578.

Peerdeman, F. (1993) The Pleistocene climatic and sea-level signature of the northeastern Australian continental margin. Unpublished PhD thesis, Australian National University, Canberra.

Peerdeman, F. M., Davies, P. J., and Chivas, A. R. (1993) The stable oxygen isotope signal in shallow-water, upper-slope sediments off the Great Barrier Reef (Hole 820A). *Proceedings of the Ocean Drilling Program Scientific Results* **133**, 163–173.

Pickett, E. J., Harrison, S. P., Hope, G., Harle, K., Dodson, J. R., Kershaw, A. P., Prentice, C. *et al.* (2004) Pollen-based reconstructions of biome distributions for Australia, South East Asia and the Pacific (SEAPAC region) at 0, 6000 and 18,000 ^{14}C yr B.P. *Journal of Biogeography* **31**, 1381–1444.

Powell, J. M., Kulunga, A., Moge, R., Pono, C., Zimike, F., and Golson, J. (1975) *Agricultural Traditions of the Mount Hagen Area* (Department of Geography Occasional Paper No. 12). University of Papua New Guinea, Port Moresby.

Ramage, C. (1968) Role of the tropical "maritime continent" in the atmospheric circulation. *Monthly Weather Review* **96**, 365–369.

Shackleton, N. J., Crowhurst, S., Hagelberg, T., Pisias, N., and Schneider, D. A. (1995) A new late Neogene timescale: Applications to leg 138 sites. *Proceedings of the Ocean Drilling Program Scientific Results* **138**, 73–101.

Soman, M. K. and Slingo, J. (1997) Sensitivity of Asian summer monsoon to aspects of sea surface temperature anomalies in the tropical Pacific Ocean. *Quarterly Journal Royal Meteorological Society* **123**, 309–336.

Street-Perrott, A. (1994) Palaeo-perspectives: Changes in terrestrial ecosystems. *Ambio* **23**, 37–43.

Stuijts, I. (1984) Palynological study of Situ Bayongbong, West Java. *Modern Quaternary Research in S.E. Asia* **8**, 17–27.

Stuijts, I., Newsome, J. C., and Flenley, J. R. (1988) Evidence for Late Quaternary vegetational change in the Sumatra and Javan highlands. *Rev. Palaeobotan. Palynol.* **55**, 207–216.

Sun, X. and Li, X. (1999) A pollen record of the last 37 ka in deep sea core 17940 from the northern slope of the South China Sea. *Marine Geology* **156**, 227–244.

Sun, X., Li, X., and Beug, H-J. (1999) Pollen distribution in hemipelagic surface sediments of the South China Sea and its relation to modern vegetation distribution. *Marine Geology* **156**, 211–226.

Sun, X., Luo, Y., Huang, F., Tian, J., and Wang, P. (2003) Deep-sea pollen from the South China Sea: Pleistocene indicators of East Asian monsoon. *Marine Geology* **201**, 97–118.

Swisher, C. C. III, Curtis, G. O., Jacob, T., Getty, A. G., Suprio, A., and Widiasmoro (1994) Age of the earliest known hominids in Java, Indonesia. *Science* **263**, 1118–1121.

Teramura, H. (1983) Effects of ultraviolet-B radiation on the growth and yield of crop plants. *Physiol. Plant* **58**, 415–427.

Thevenon, F., Bard, E., Williamson, D., and Beaufort, L. (2004) A biomass burning record from the West Equatorial Pacific over the last 360 ky: Methodological, climatic and anthropic implications. *Palaeogeography, Palaeoclimatology, Palaeoecology* **213**, 83–99.

Thomas, W. L., Sauer, C. O., Bates, M., and Mumford, L. (eds.) (1956) *Man's Role in Changing the Face of the Earth*. University of Chicago Press, Chicago.

Turney, C. S. M., Kershaw, A. P., Moss, P., Bird, M. I., Field, L. K., Cresswell, R. G., Santos, G. M., di Tada, M. L., Hausladen, P. A., and Youping, Z. (2001) Redating the onset of burning at Lynch's Crater (North Queensland): Implications for human settlement in Australia. *Journal of Quaternary Science* **16**, 767–771.

Turney, C. S. M., Kershaw, A. P., Lowe, J. J., van der Kaars, S., Rochelle Johnston, R., Rule, S., Moss, P., Radke, L., Tibby J., McGlone, M. S. *et al.* (2006) Climate variability in the southwest Pacific during the Last Termination (20–10 ka BP). *Quaternary Science Reviews* **25**, 886–903.

van der Kaars, S. (1998). Marine and terrestrial pollen records of the last glacial cycle from the Indonesian region: Bandung basin and Banda Sea. *Palaeoclimates—Data and Modelling* **3**, 209–219.

van der Kaars, S. (2001) Pollen distribution in marine sediments from the south-eastern Indonesian waters. *Palaeogeography, Palaeoclimatology, Palaeoecology* **171**, 341–361.

van der Kaars, S. and De Deckker, P. (2003) Pollen distribution in marine surface sediments offshore Western Australia. *Review of Palaeobotany and Palynology* **124**, 113–129.

van der Kaars, S., Wang, X., Kershaw, A. P., Guichard, F., and Setiabudi, D. A. (2000) A Late Quaternary palaeoecological record from the Banda Sea, Indonesia: Patterns of vegetation, climate and biomass burning in Indonesia and northern Australia. *Palaeogeography, Palaeoclimatology, Palaeoecology* **155**, 135–153.

van der Kaars, S., Penny, D., Tibby, J., Fluin, J., Dam, R., and Suparan, P. (2001) Late Quaternary palaeoecology, palynology and palaeolimnology of a tropical lowland swamp: Rawa Danau, West Java, Indonesia. *Palaeogeography, Palaeoclimatology, Palaeoecology* **171**, 185–212.

van der Kaars, W. A. and Dam, M. A. C. (1995) A 135,000-year record of vegetational and climatic change from the Bandung area, West-Java, Indonesia. *Palaeogeography, Palaeoclimatology and Palaeoecology* **117**, 55–72.

van Steenis, C. G. G. J. (1972) *The Mountain Flora of Java* (illustrated by Amir Hamzah and Moehamad Toha, 90 pp.). Brill, Leiden, The Netherlands.

Vavilov, N. I. (1951) The origin, variation, immunity and breeding of cultivated plants. *Chronica Botanica* **13**, 1–364.

Walker, D. and Flenley, J. R. (1979) Late Quaternary vegetational history of the Enga Province of upland Papua New Guinea. *Philosophical Transactions Royal Society B* **286**, 265–344.

Webster, P. J. and Streten, N. A. (1972) Aspects of Late Quaternary climate in tropical Australasia. In: D. Walker (ed.), *Bridge and Barrier: The Natural and Cultural History of Torres Strait* (pp. 39–60). Research School of Pacific Studies, Dept. of Biogeography and Geomorphology, Australian National University, Canberra.

Whiffin, T. (2002) Plant biogeography of the SE Asian–Australian region. In: A. P. Kershaw, B. David, N. Tapper, D. Penny, and J. Brown (eds.), *Bridging Wallace's Line: The Environmental and Cultural History and Dynamics of the Southeast Asian–Australian Region* (pp. 61–82). Catena Verlag, Reiskirchen, Germany.

Yasuda, Y. (ed.) (2002) *The Origins of Pottery and Agriculture* (400 pp.). Roli and Lustre, New Delhi.

Zheng, Z. and Lei, Z-Q. (1999) A 400,000 year record of vegetational and climatic changes from a volcanic basin, Leizhou Peninsula, southern China. *Palaeogeography, Palaeoclimatology, Palaeoecology* **145**, 339–362.

5

Rainforest responses to past climatic changes in tropical Africa

R. Bonnefille

5.1 INTRODUCTION

In Africa the lowland rain forest occurs under significantly drier conditions than in other continents, within an average precipitation of 1,600 to 2,000 mm/yr, although higher rainfall is observed around the Atlantic coast of Cameroon, Gabon, and in the Central Zaire Basin. Seasonal distribution of precipitation is far from being uniform (White, 1983). Variations in the duration of the dry season follow distance from the equator in both hemispheres and also along a west-to-east gradient. The Biafran Gulf is the only region where the minimum monthly precipitation value always exceeds the 50-mm threshold for the driest month, therefore experiencing no dry season. However, great annual rainfall variability is registered at most of the meteorological stations. Mean monthly temperatures remain constant. Inside the area occupied by the African rainforest, topography is not uniform. Low elevations are found in the coastal Atlantic plain, and in the Zaire Basin, that lie below 400 m. The undulating plateaus of Gabon and Cameroon, generally located between 600 and 800 m, can reach up to 1,500 m, whereas the eastern part of the Zaire Basin joins the slopes of high mountains above 2,000 m bordering the Rift in the Kivu region. Mount Cameroon exceeds 4,000 m in elevation.

The geographical distribution of plant species is complex (Richards, 1981). Relationships between geographical plant distribution and ecological variables—such as rainfall, available moisture, and seasonality—within the Guineo-Congolian domain are far from being well-established, although there are significant variations of these factors inside the areas occupied by rainforest. The only comprehensive review relies on the vegetation-mapping done for Africa. However, the classification of the different types (or variants after White, 1983) was difficult. This is partly because variation in floristic composition, physiognomy, and phenology is largely gradual and continuous (Aubreville, 1951). This chapter concerns the oriental part of the Guineo-Congolian domain, where the degree of endemism is high, representing 80% of the total 8,000

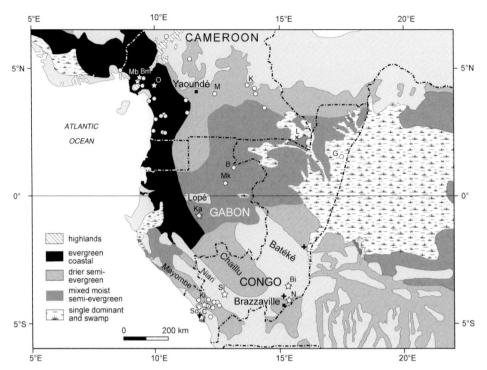

Figure 5.1. Distribution of different vegetation units within the Guineo-Congolian rainforest (6°S and 6°N) (after White 1983; Letouzey, 1965, 1985) with location of modern (○) and fossil pollen (*) and wood sites (+). Bm: Barombi Mbo; Mb: Mboandong, O: Ossa; M: Mengang; K: Kandara; L: Lobéké; B: Belinga; Mk: Makokou; G: Guibourtia; Ka: Kamalete; Bi: Bilanko; N: Ngamakala; Si: Sinnda; Ki: Kitina; C: Coraf; So: Songolo.

species and 25% of the genera, with the greatest number of endemic genera being found among the Leguminosae–Caesalpiniaceae. We present modern and fossil pollen data relating pollen and plant distribution within the present day and for the latest Quaternary (Figure 5.1).

5.2 VEGETATION UNITS WITHIN THE GUINEO-CONGOLIAN RAINFOREST

5.2.1 Hygrophilous coastal evergreen forest

Syn.: forêt biafréenne à Caesalpiniaceae (Letouzey, 1968), forêt dense humide sempervirente à Légumineuses (Aubreville, 1957–58), wet evergreen forest (Hall and Swaine, 1981).

The evergreen Guineo-Congolian wet forest is located between 2°S and 6°N in areas that receive 2,000 to 3,000 mm/yr rainfall, without or with a reduced dry season.

The atmospheric humidity is very high throughout the year. Most individuals of most tree species are evergreen and shed their leaves intermittently. *Lophira alata* (Ochnaceae) is one of the most abundant of the widespread taller trees, although it is not confined to this type. *Lophira alata* is light-demanding with drought-sensitive seedlings, with the potential to be a large tree that can live for several centuries. Historical remains (slave rings, pottery, charcoal, etc.) observed in excavations after deforestation for *Hevea* plantations and road cuttings indicate that such forest now occurs in areas that were cultivated perhaps a few centuries ago (Letouzey, 1968).

At its most typical sites, the hygrophilous coastal evergreen forest rainforest is rich in Caesalpinioideae, many of which are gregarious and include species of various genera such as *Anthonota, Brachystegia, Julbernardia, Berlinia, Monopetalanthus*, associated with *Cynometra hankie* (Caesalpiniaceae) and *Coula edulis* (Ochnaceae).

5.2.2 Mixed moist semi-evergreen

Syn.: forêt congolaise (Letouzey, 1968, 1985), forêt dense humide sempervirente à Légumineuses (Aubreville, 1957–58), forêts semi-caducifoliées sub équatoriales et guinéeennes (Lebrun and Gilbert, 1954).

Most Guineo-Congolian rainforest belongs to the mixed moist semi-evergreen type. It occurs on well-drained soils and covers most of the area at low elevation (600–700 m), throughout the region, comprising northeast Gabon, northern Congo, southeast Cameroon, and most of the Zaire Basin, except for the wettest and driest extremities. Mean annual rainfall is mostly between 1,600 and 2,000 mm and is well-distributed. The prevalent vegetation is moist semi-evergreen of mixed composition, and rich floristically. Some species are evergreen, but many are briefly deciduous. No detailed description of this type of forest exists except for the *Oxystigma oxyphyllum* (Caesalpiniaceae) and *Scorodophloeus zenkeri* (Caesalpiniodeae) association particularly significant in the Zaire Basin (Lebrun and Gilbert, 1954). Some of the most abundant emergent species—for example, *Canarium schweinfurthii* (Burseraceae), *Piptadeniastrum africanum* (Mimosoideae), *Ricinodendron heudelotii* (Euphorbiaceae), *Terminalia superba* (Combretaceae)—are also found in secondary forests of the dry peripheral semi-evergreen type. In Cameroon the forest of this type has few species of Caesalpiniaceae (except *Gilbertiodendron dewevrei* which can form pure stands), but includes large and tall *Baillonella toxisperma* (Sapotaceae) which exhibit no regeneration, suggesting that the over-mature forest is disintegrating, and being locally invaded by dry evergreen (semi-deciduous) forest (Letouzey, 1968).

5.2.3 Single-dominant moist evergreen and semi-evergreen

Several authors have described single-dominant forests which form isolated stands of a few hectares in extent inside the mixed moist semi-evergreen forest or islands which occur in a broad aureole surrounding the Zaire Basin (White, 1983). The upper stratum is formed by tall trees, usually 35–45 m high, that belong to a few or a single species (Evrard, 1968). Among the dominant species, *Gilbertiodendron dewevrei* (Caesalpiniaceae) from southeast Cameroon is normally completely evergreen, whereas the more widespread *Cynometra alexandrii* (Caesalpiniaceae) is irregularly deciduous. In Uganda, *Cynometra* trees, particularly abundant above 700–800 m, shed their leaves simultaneously, and *Julbernardia* (Caesalpiniaceae) inside the Zaire Basin is said to behave in a similar fashion. In the single-dominant moist evergreen forest, heliophitic trees are rare. Lianas and giant monocotyledonous herbs are poorly represented.

5.2.4   Drier peripheral semi-evergreen forest

Syn.: forêt dense humide semi-décidues de moyenne altitude (Letouzey, 1968), forêts semi-caducifoliées, forêts semi-décidues à Malvales et Ulmacées (Aubreville, 1957–58).

This type of forest extends geographically between the moist evergreen forest and occurs in the form of two bands running transversely across Africa to the north and south of the moister forests described above at the limit of areas occupied by savanna. They were called "peripheral" (White, 1983) and have a patchy distribution in the Lake Victoria basin. Rainfall is between 1,200 and 1,600 mm/yr, with a bimodal distribution. The dry season lasts 1 or 2 months during which relative humidity remains high. Mean annual temperatures range from 23.5 to 25°C. Most individuals of the commoner larger tree species are deciduous and lose their leaves during the dry season. But, any individual is deciduous for a few weeks only. This forest type has a distinct floristic composition and includes species virtually absent from the wetter types—such as *Afzelia africana* (Caesalpiniaceae), *Aningeria altissima* (Sapotaceae), *Cola gigantea* (Sterculiaceae). Some other species—such as *Celtis mildbraedi* and *C. zenkeri* (Ulmaceae), *Holoptolea grandis* (Ulmaceae), *Sterculia oblonga* (Sterculiaceae) that are important components of the dry peripheral semi-evergreen forest—also occur in mixed moist semi-evergreen forests. In Cameroon the "drier peripheral semi-evergreen forest" or forêt dense semi-decidue de moyenne altitude (Letouzey, 1968) is characterized by the dominance of *Cola, Sterculia, Celtis,* and *Holoptolea grandis*, together with *Piptadeniastrum africanum* (Mimosaceae), *Funtunia* (Apocynaceae), and *Polyalthia* (Annonaceae). *Terminalia superba* (Combretaceae) and *Triplochiton scleroxylon* (Sterculiaceae) are two rapidly growing, light-demanding, valuable timbers that can regenerate on abandoned farmland.

Besides the main four types of rainforest described above, there are different vegetation types that have also been mapped as separate units. Secondary forests, swamp forests, and a mosaic of forests and grasslands can be found in any of the four main types of forests described before (White, 1983).

5.2.5   Secondary rainforest

Outside the forest reserves, much of the remaining forest occurs on land that has been formerly cultivated and is therefore considered secondary. This regrowth contains abundant heliophytes and pioneers that grow quickly and have a short life; *Tetrorch-idium* (Euphorbiaceae) and *Trema* (Ulmaceae) are characteristic, while *Musanga cecropioides* (Moraceae) is strictly Guineo-Congolian in distribution.

5.2.6 Swamp forest

Swamp forests, including riparian forest, occur throughout the Guineo-Congolian region wherever the conditions are suitable. These forests are floristically distinct, but have a similar appearance to rainforest with the tallest trees reaching 45 m in height. The main canopy, however, is irregular and open. Inside the clearings, climbing palms *Eremospatha* and *Calamus*, shrubs, and lianas fill the gap. Forest types with Mar-antaceae and *Gilbertiodendron* are located between the western side of the inundated wet evergreen forests and the mixed moist semi-evergreen forest (forêt congolaise).

5.2.7 Edaphic and secondary grasslands

On hydromorphic soils, edaphic grasslands surrounded by forest represent a transi-tory stage in the succession from aquatic vegetation to forest. The origin of such grassland patches is still controversial. They may be maintained by frequent fire, but they also occupy superficial soils on rocks that are periodically inundated (Koechlin, 1961). At its northern and southern limits, the Guineo-Congolian rainforest is burned at least once a year. A mosaic of patches of secondary grassland including scattered fire-resistant trees and secondary forest clumps locally replace the forest. In this situation, the mosaic can regenerate into forest. There are also patches of secondary grasslands inside the Guineo-Congolian region (Descoings, 1976), but most of them are distributed at the transition with the Sudanian zone in the north and with the Zambezian zone in the south. They show considerable local variation in floristic composition.

5.2.8 Transitional and Afromontane evergreen forests

In Cameroon, close to the border with Nigeria, Afromontane forests show a great resemblance to those from East Africa. Two distinct altitudinal horizons are distin-guished. The lower, from 800 to 1,800 m, is characterised by *Podocarpus latifolius* (Podocarpaceae) (syn. *P. milanianus*) and is associated with *Olea capensis* (Oleaceae) (syn. *O. africana*). The upper, from 1,800 to 2,800 m, contains *P. latifolius* together with *Prunus africana* (Rosaceae), *Myrsine melanophloeos* (Myrsinaceae), and *Nuxia congesta* (syn. *Lachnopylis*, Loganiaceae). In eastern Zaire a transitional forest ob-served between 1,100 and 1,750 m includes tree species of the Afromontane forest among *Aningeria* (Sapotaceae), *Entandophragma* (Meliaceae), *Mitragyna* (Rubiaceae), and *Ocotea* (Lauraceae), associated with components of the lowland forest—such as *Cynometra alexandrii* (Caesalpiniaceae), *Pycnanthus angolensis* (Myristicaceae)—and some endemics, specific to the transitional forest. The species *Strombosia grandiflora* (Olacaceae), *Symphonia globulifera* (Olacaceae), *Uapaca guineense* (Euphorbiaceae), and *Parinari* sp. (Chrysobalanaceae) are ecological transgressors between lowland and Afromontane forests (White, 1983).

5.3   MODERN POLLEN RAIN STUDIES

No systematic study of modern pollen rain within the African rainforest has addressed the relationship between the distribution of vegetation types and climatic variables. Indeed, in equatorial Africa, this vegetation and climate pattern is very complex. The lack of ecological work and the long-standing, old belief of "equatorial climatic stability" might have discouraged such studies. New modern pollen rain data from tropical evergreen and deciduous forests of southwest India (Barboni *et al.*, 2003) should encourage them now. In India, the western coast supports many different types of forests under a single monsoonal regime, with a south-to-north increased gradient of rainfall, a west-to-east increase in dry season length (seasonality), and temperature decrease along highland slopes reaching up to 2,500 m in elevation. Ecological studies showed that these factors indeed influence the distribution of species and that of vegetation types, following the different bioclimatic regions (Pascal *et al.*, 1984). Modern pollen assemblages reproduce the pattern of bioclimatic regions and associated vegetation-mapped units. The results enable a clear distinction of the different types of forests and their associated pollen markers (Bonnefille *et al.*, 1999; Barboni and Bonnefille, 2001; Anupama *et al.*, 2000). In Africa, recent studies on modern pollen rain of the Guineo-Congolian region provides a first understanding of pollen production and markers of the different types within the rainforest. Many of the main types of vegetation (Figure 5.1) have now been sampled for modern pollen, although the mixed moist semi-evergreen forest of the Zaire Basin remains poorly documented. Detailed investigations were made in connection with floristic "relevés" in the evergreen rainforest of Cameroon and the semi-evergreen from southern Congo. Because different sampling procedures and different plot sizes were used by the various authors, and because original counts from Cameroon have not been included yet in the "African Pollen Data Base", it was not possible to provide a single synthetic diagram for the region's rainforest. Therefore, the presentation of the results on modern pollen studies will be organized according to geographical regions and vegetation-mapping units.

5.3.1   From coastal evergreen to drier semi-evergreen forests in Cameroon

In Cameroon, modern pollen rain studies mainly concern the evergreen and coastal hygrophilous forests ("biafréenne à Caesalpiniacées"), with a few samples collected within the mixed moist semi-evergreen "forêt congolaise" and the drier peripheral semi-evergreen ("forêt semi-caducifoliée") (Letouzey, 1968). Sampling for pollen analysis and counts of tree species over 5 cm in diameter were done within 20 × 20-m plots (Reynaud-Farrera 1995). The pollen counts of several plots have been used by the author to draw a diagram to compare plant (Figure 5.2a) and pollen frequencies of the different taxa (Figure 5.2b).

An interesting result from this work is that in all the samples collected inside forest plots, the ratio of arboreal pollen (AP)—including all trees and shrubs—calculated versus total pollen counts—excluding the spores—always exceeds 75% (Figure 5.2a). The samples showing the lowest AP values were collected in open vegetation areas such as at Kandara near the limit between the peripheral semi-evergreen and the forest/grassland mosaic. Because abundant tree or climber species among common tropical families—such as Annonaceae, Apocynaceae, Violaceae (except *Rinorea*), most Myristicaceae (except *Pycnanthus*), Chrysobalanaceae, Olacaceae (except *Strombosia*), Clusiaceae (except *Symphonia*), etc.—are not represented by their pollen, this was not expected *a priori*. A good correspondence between the total arboreal pollen and the tree coverage of the wet evergreen and semi-evergreen forests in Cameroon confirms results from other tropical regions in Africa (Bonnefille *et al.*, 1993). This interesting result shows that the classic distinction between arboreal pollen (AP) and non-arboreal pollen (NAP), used in the interpretation of pollen analysis from the temperate region, is also valid for palynology in the rainforest region. In the modern pollen study from Cameroon, the high number (279) of pollen taxa identified within the soil samples reflects the great floristic diversity of evergreen forests. Because of the strong heterogeneity in the spatial distribution of the tree species, the occurrence of many intermediate forest types, and the unavoidable bias between over- and under-representation of the pollen of many tropical families, a greater number of pollen samples, more evenly distributed within the different vegetation types, would have been advantageous. Nevertheless, despite evident discrepancies between vegetation cover and pollen representation, there is a certain degree of agreement between the listed plants within ecological plots and the pollen types found in the surface soil samples (Figure 5.2c). The pollen assemblages from the different forest types appear clearly distinct from each other, enabling one to recognize the different forest types. Joint occurrences of *Saccoglottis* and *Lophira*, together with "*Berlinia*-type" pollen including many Caesalpiniaceae, are good markers of the coastal Biafran forest, in agreement with the dominant trees. Pollen of *Irvingia* (Irvingiaceae), *Lophira*, *Diospyros* (Ebenaceae), and Sapotaceae (which might include *Baillonella toxisperma*, the pollen of which cannot be identified at the generic level), show highest representation in the mixed semi-evergreen "Congolese" forest. The *Pycnanthus* (Myristicaceae) and *Piptadeniastrum* (Mimosoideae) pollen association characterizes the dry evergreen (semi-deciduous) forest, although *Pycnanthus* alone can be abundant within the coastal Biafran forests. *Macaranga*, *Alchornea*, *Celtis*, and sometimes *Uapaca* tend to be over-represented in the pollen rain. The two pollen assemblages corresponding to the mountain forest are well-characterized by the association of *Podocarpus*, *Olea*, *Nuxia*, and Ericaceae pollen (Figure 5.2c).

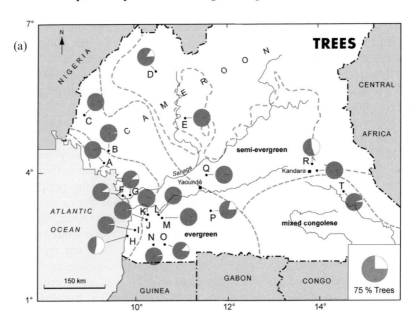

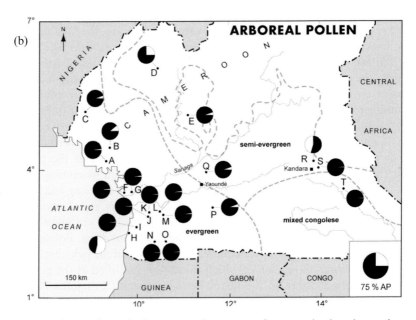

Figure 5.2. Modern pollen rain from coastal evergreen forests, mixed moist semi-evergreen, and drier semi-evergreen forests from Cameroon: (a) % of trees and shrubs among plants in the plots; (b) % of arboreal pollen in soil samples from the same plots. Some trees, identified in floristic counts, but not found as pollen are not illustrated here, such as the Marantaceae among the herbs, fern spores are not represented (% calculated versus pollen sum after excluding spores and Cyperaceae).

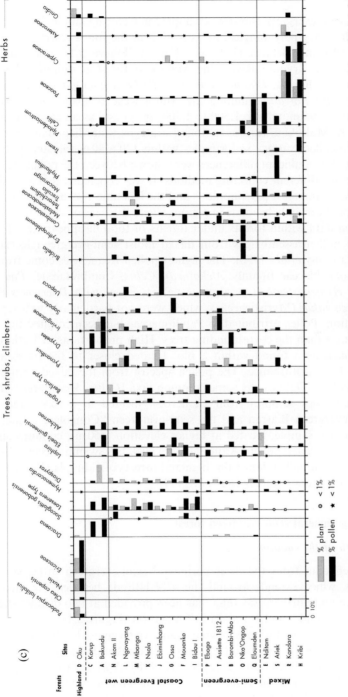

Figure 5.2 (*cont.*). (c) Distribution of relative frequencies of plant and pollen for the main taxa at pollen sampling sites as in (a) and (b) (after Reynaud-Farrera, 1995).

5.3.2   Mixed moist semi-evergreen forest

Mixed moist semi-evergreen forest was sampled in Central Gabon (Jolly *et al.*, 1996) during a preliminary palynological investigation in the forest reserves at Makokou (0°30′N, 12°50′) and Belinga (1°06′N, 13°10′E) (Figure 5.1). Located at 470 m elevation, near the Irvindo River, the Makokou forest is dominated by *Scorodophloeus zenkeri, Baphia, Dialium, Pancovia, Dichostemma glaucescens* among the Papilionaceae and *Polyalthia* among the Annonaceae (Caballe, 1986; Aubreville, 1967). Precipitation amounts to 1,500 to 1,700 mm/yr and mean annual temperature is over 24°C. At Makokou, various sampling strategies (line, diagonals, random, etc.) were tested to collect surface soil samples within 120 × 40 m forestry plots along a 1 km transect. No significant differences were shown between the pollen assemblages obtained with the different collecting methods. It was concluded that a random procedure represents the most parsimonious and the less time-consuming one. Random collection should certainly be made in regions where there is an urgent need to obtain modern pollen data within forests under threats of total disappearance. The pollen counts of 16 pollen assemblages from mixed moist semi-evergreen forest include 82 pollen taxa (Jolly *et al.*, 1996). The highest pollen frequencies come from Moraceae and Euphorbiaceae (mainly *Alchornea*), *Celtis*, Combretaceae, *Pausinystalia* (Rubiaceae), *Hymenostegia* (Caesalpiniaceae)—together with *Dacryodes* (Burseraceae) and *Pycnanthus* (Myristicaceae), which were not particularly well-represented in the vegetation. Pollen attributed to Papilionaceae has been counted but in much lesser abundance than the corresponding trees. High proportions of Urticaceae and *Macaranga* have been found within a plot located closer to the river, and were attributed to local disturbance. The pollen spectra from the mixed semi-evergreen forest at Makokou (Gabon) has some common taxa (*Celtis, Macaranga, Alchornea,* Combretaceae), with samples from Cameroon located in a forest intermediate between the evergreen Biafran and the semi-evergreen "Congolese". Interestingly, the Belinga pollen sample, located at higher elevation (950 m), contains significant percentages of *Syzygium* pollen. Although *Syzygium* has swamp species, this finding agrees with modern pollen data from highland forests of Ethiopia (Bonnefille *et al.*, 1993) and South India (Bonnefille *et al.*, 1999).

5.3.3   Drier peripheral semi-evergreen forest

5.3.3.1   North of the equator

Modern pollen data from the drier peripheral semi-evergreen forest (forêt semicaducifoliées), which extends around the 4°N latitude between the mixed moist semi-evergreen and northern savanna, has been provided at two distinct sites: Mengang (east of Yaounde) and Kandara in Cameroon (Figure 5.1).

The Mengang forest station is located in an area receiving 1,600-mm/yr precipitation interrupted by two dry months (<50 mm)—December, January—and a minimum in July. Atmospheric pollen rain was captured in 1987 (April to October) by an aboveground framed trap, and the results presented at monthly intervals (Fredoux and Maley, 2000). They indicate that the greatest amount of pollen is produced by the

forest trees (c. 90% of total counts), greatest monthly flux averaging 1,000 grains/m^3.
Altogether, 118 tree taxa were identified distributed among 58 families. The
Euphorbiaceae show the greatest diversity (21 taxa) and total 27% of the pollen
counts dominated by Ulmaceae (33%) and Moraceae (26%). They are associated with
Urticaceae (3%)—and Papilionaceae, Caesalpiniaceae, Mimosaceae, Anacardiaceae,
Sterculiaceae, Rubiaceae, and Sapindaceae, each accounting for less than 1%. The
greatest pollen producers indicated by flux calculations are *Macaranga* (1 to 1,000/m^3
maximum in May), *Celtis* (1 to 500/m^3 in April), *Musanga* (or *Myrianthus*) and
Combretaceae (10 to 100/m^3). The amounts of *Pycnanthus*, *Trema*, Urticaceae, and
Poaceae pollen are produced in much lower proportion (<1 to 10/m^3). Although
established for 1 year, such discrepancies have to be remembered while interpreting
fossil pollen results. It would have been most interesting to compare this distribution
with parallel analysis of surface soil samples within the forest that provide an average
of modern pollen rain over several years, minimizing monthly variations.

In southeastern Cameroon—east of Mengang (Figure 5.1), near the Kandara
village (4°20′N, 13°43′E, 640 m)—the dry semi-evergreen forest encloses a few square
kilometer area of shrub tall-grass savanna. The wet tropical climate is characterized by
the same precipitation (1,600 mm/yr) and dry season, with mean annual temperature
averaging 24°C. In this area the forest had expanded at a rate of 1 m/yr between 1951
and 1993 (Youta Happi, 1998; Achoundong *et al.* 2000), in agreement with observations made on aerial photographs in Cameroon (Letouzey, 1968). The forest succession begins with *Raphia monbuttorum* (Palmae) swamp along the Soukato River,
followed by a semi-evergreen *Rinorea* (Violaceae) forest, a young *Albizia* forest, and a
savanna–forest transition (ecotone) surrounding the savanna dominated by Panicoi-
deae grasses with some *Albizia* clusters (Figure 5.3a). Associated with *Rinorea dentata*
and *Rinorea batesii* (Achoundong *et al.*, 1996), the most abundant trees of the mature
forest are *Triplochiton scleroxylon* (Sterculiaceae) and *Piptadeniastrum africanum*
(Mimosaceae), good indicators of dry peripheral semi-evergreen forest (Letouzey,
1968). The young *Albizia* forest includes *Albizia adiantifolia* (Mimosaceae), *Funtumia
elastica* (Apocynaceae), *Canthium* (Rubiaceae), *Tabernaemontana crassa* (Apocynaceae), *Sterculia rhinopetala* (Sterculiaceae), and *Myrianthus arboreus* (Moraceae). At
Kandara, 26 surface soil samples collected in contiguous 20 × 30-m plots, along a 750-
m transect across the succession (Figure 5.3a), provided modern pollen rain data.
Each pollen sample consists of about 20 sub-samples randomly distributed (Vincens *et
al.*, 2000). The relative frequencies of the most abundant ones among the 101 pollen
taxa altogether identified are illustrated here (Figure 5.3c). All samples from the dry
semi-evergreen forests register percentages of arboreal pollen (AP) ranging from 60 to
80%, in good correspondence with the canopy coverage estimated by field measurements of the leaf area index (Cournac *et al.*, 2002) and abundance of tree phytoliths
counted in the same samples (Bremond *et al.*, 2005). The total AP drops abruptly to
less than 10% in the nearby savanna, at less than 100 m from the forest limit,
indicating very little transport from the forest into the savanna (Figure 5.3b). The
pollen assemblages from this transect include some pollen of the trees characterizing
dry peripheral semi-evergreen forest. These are *Celtis*, *Chaetachme*, *Holoptolea*
among the Ulmaceae, *Triplochiton* (Sterculiaceae) associated with *Piptadeniastrum*

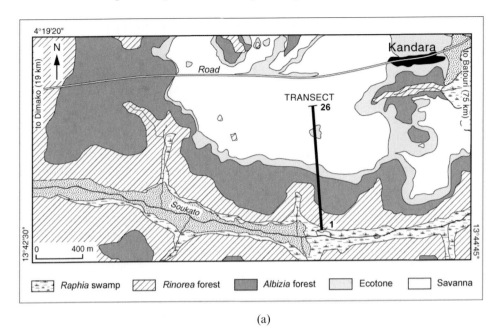

(a)

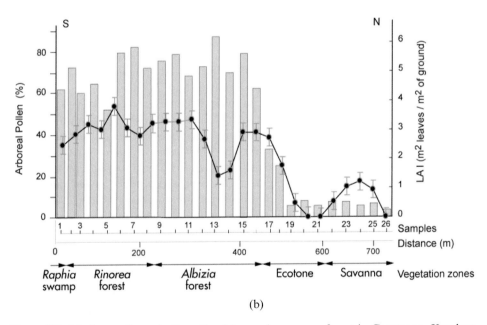

(b)

Figure 5.3. Modern pollen rain from the drier semi-evergreen forest in Cameroon, Kandara site: (a) location map of the studied transect (samples 1 to 26); (b) comparison between arboreal pollen and leaf area index (Cournac *et al.*, 2002). After Bremond *et al.*, 2005 and Vincens *et al.*, 2000.

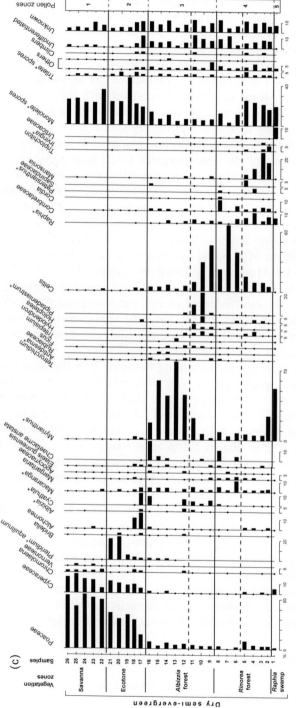

Figure 5.3. (c) Simplified pollen diagram (after Vincens *et al.*, 2000) (% calculated versus pollen sum including all identified taxa, Cyperaceae, and spores). * Denotes names representing the greatest probability for identification of pollen type, but for some pollen types other genera or species having the same pollen morphology cannot be excluded (pollen types <1%). Dots = pollen % < 1.

(Mimosaceae), *Margaritaria discoidea* (Apocynaceae) and Combretaceae. These taxa can be considered as good indicators of semi-evergreen forest under a seasonal climate, with a short dry season. However, and as expected, their pollen abundances do not exactly correspond to plant abundances in the vegetation cover. This is unavoidable due to differential pollen production. The soil samples from the *Albizia* forest contain a few grains of *Albizia* but are dominated by 50 to 70% *Myrianthus arboreus* (Moraceae). Near the contact with the savanna, *Chaetachme aristata* (Ulmaceae) dispersed more pollen (20%) than *Albizia* (5 to 10%) despite the latter being more strongly represented in the vegetation cover. *Tetrorchidium*, *Ficus*, *Trilepisium* (all Moraceae), *Antidesma* (Euphorbiaceae), Rubiaceae, and Sapindaceae are normal components of the semi-evergreen forest pollen rain. At contact with the *Rinorea* forest, *Piptadeniastrum* pollen was relatively abundant whereas none of the different species of *Rinorea* were recorded as pollen. *Raphia* pollen dominates near the swamp, which is also marked by an abundance of fern spores (*Pteridophytes monoletes*). At the northern end of the transect, close to the savanna, the abundance of *Pteridium aquilinum* spores indicates frequent burning. The pollen diagram illustrates the spatial colonization of a burnt savanna by a dry semi-evergreen forest. The succession starts with Moraceae, followed by *Celtis*, and then by Sterculiaceae/ *Raphia*/Pteridophyta, following dry to wetter local conditions under climatic conditions favorable to the establishment of dry semi-evergreen forest. Although this succession might be valid only for the northern semi-evergreen forest region, the results improve our understanding of tropical rainforest dynamics in the past and should stimulate similar studies for other regions.

While there is some agreement among botanists that the forest dominated by shade-intolerant tree species established from relatively open conditions several hundred years ago, the origin of savanna "islands" in the semi-evergreen forest of southeast Cameroon still remains controversial. Analyses of opal phytolith assemblages from topsoil samples of the forests and the included savanna patches at Lobéke—2°17′N, 15°42′E, 300 to 700 m in elevation, south of Kandara (Figure 5.1) in a region receiving 1,600–1,700-mm/yr precipitation—suggest stable conditions with no evidence of recent disturbances, such as fire or logging (Runge and Fimbel, 1999). Elsewhere, evidence for recent invasion of forest into the savanna is also provided from soil organic carbon isotopic studies (Guillet *et al.*, 2001) whereas $\delta^{13}C$ of organic matter in soil profiles provides information on previous forest development in areas now occupied by savanna (Schwartz *et al.*, 1996). Constant re-organization of the distribution of forest and savanna patches at the limit of the drier semi- evergreen forest may be forced by the seasonal distribution of the rain through the year and variability in dry-season length. But, recent human deforestation has modified the pattern. Sorting out the respective effect of climate from that of human impact on modern vegetation deserves further thorough investigation.

5.3.3.2 *South of the equator*

The coastal semi-evergreen forest does not extend to southern Congo beyond 4°S latitude (White, 1983). Instead, a mosaic of forest–grassland occupies the Atlantic

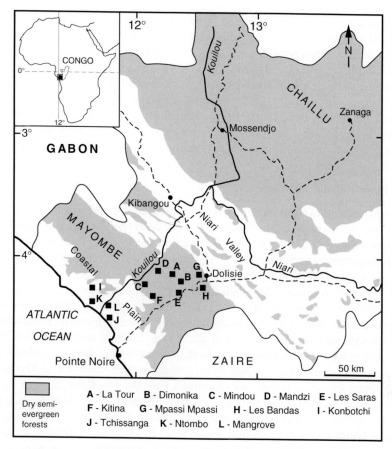

Figure 5.4. Location map of sites for floristic inventories and modern pollen surface soil samples collected within the semi-evergreen forests from Congo (after Elenga *et al.*, 2000a).

coast, receiving less than 1,100 mm/yr of rainfall and with temperatures averaging 18°C during the dry season. Several types of semi-evergreen forests are present further inland, in the Mayombe massif, which reaches 730 m maximum elevation and receives 1,400 to 1,600 mm/yr of precipitation. Botanical investigation was made at 12 geographical sites (A to L), distributed between 4°05′ to 4°50′S latitude and 11°45′ to 12°35′E longitude (Figure 5.1). Among the investigated sites, 8 are distributed in Mayombe and 4 correspond to the coastal plain (including the mangrove) (Figure 5.4). At each site, trees with a diameter at breast height >5 cm were counted within 20 × 100-m plots (Elenga *et al.*, 2000a). All the floristic inventories highlight the great number of tree species and the great spatial heterogeneity of their distribution. Out of a total of 620 individual trees, 352 species were identified, distributed among 47 botanical families. Of the represented tree species, the most important ones are Annonaceae (15 species), Euphorbiaceae (8 species), Caesalpiniaceae (7 species), Rubiaceae (6 species), followed by Burseraceae, Olacaceae, Sapotaceae, and Moraceae. Within all the studied plots, 110 species have tree frequencies >1%, but only 41 species were present in at least two sites where the average density of trees reached 120 individuals per 100 m^2 (Elenga *et al.*, 2000a). The forests of the Mayombe are remarkable for high floristic diversity and little overlap between tree composition at different localities. Correlation between floristic richness and climatic variables— such as insolation, water availability, or primary production—are now being explained by a new hypothesis. Palynological studies explore how such floristic diversity is reproduced in pollen assemblages extracted from 50 surface soil samples collected at the 12 sites where tree counts had been made (Figure 5.5).

5.3.4   Forests of the Mayombe

Within all the samples from the semi-evergreen forests of the Mayombe massif, c. 19 to 23 plant species make up more than 75% of the total tree counts. But, the dominant tree varies greatly within the 8 studied plots and the results of pollen analysis show that this dominance is not necessarily reproduced in the corresponding pollen assemblages. At La Tour (site A), *Treculia obovoidea* (Moraceae) represents 8.5% of the trees, but its pollen frequencies reach much higher percentages (30 to 60%) in all of the 10 pollen assemblages (Figure 5.5). Other trees—such as *Plagiostyles africana* (Euphorbiaceae), *Maranthes* (Chrysobalanaceae), *Irvingia*, *Uapaca* (Euphorbiaceae), common to the floristic list (Elenga *et al.*, 2000a, table 2) and the pollen diagram (Figure 5.5)—show similar percentages both as plants and pollen. *Strombosia* (Olacaceae) are under-represented by their pollen, whereas Annonaceae and Myristicaceae have not been found as pollen. At Dimonika (site B), 11 of the 22 most abundant trees have been recognized as pollen taxa. Among them, *Treculia obovoidea* (Moraceae), *Anisophyllea myriostricta* (Anisophyllaceae), *Trichoscypha* (Anacardiaceae), *Dacryodes* (Burseraceae), Caesalpiniaceae, and *Allanblachia* (Guttifereae) have significant percentages, both as plant and pollen. However, other genera among Burseraceae, Apocynaceae, Clusiaceae, and Annonaceae were found weakly or not represented by their pollen. At Mindou (site C), the most abundant trees—*Anthostema* (Euphorbiaceae, 13%), *Dialium* and *Guibourtia* (all Caesalpiniaceae totaling 22% trees)—are represented by significant pollen percentage values. *Syzygium*, not in the tree list, provided the greatest amount of pollen (20 to 50%) in samples 16 to 22 (Figure 5.5). The Mindou site, located on the humid western slopes of Mayombe, is close to the coastal forest. Proximity of *Syzygium* clumps in local swamps perhaps explains such high pollen percentages. At Mandzi (site D), of the most abundant trees—such as *Microdesmis* sp. (Pandanaceae 10%), *Grewia* (Tiliaceae 8%), and *Tessmania* sp. (Caesalpiniaceae 6%)—only *Grewia* is recorded by its pollen (2 to 5%). Other taxa—such as *Irvingia*, *Pancovia* (Sapindaceae), and *Aidia micrantha* (Rubiaceae)—have about the same representation both in trees and in pollen. However, associated pollen from *Piptadeniastrum*, *Calpocalyx* (Mimosaceae), *Dacryodes/Santiria*, *Ganophyllum* (Sapindaceae), *Fagara* (Rutaceae), *Macaranga* and *Elaeis* do not correspond to the listed trees at this site (Elenga *et al.*, 2000a, table 5.2), a discrepancy possibly explained by the fact that pollen origin may be found outside the sampled plots used for tree counts. At les Saras (site E), *Treculia*—an abundant tree

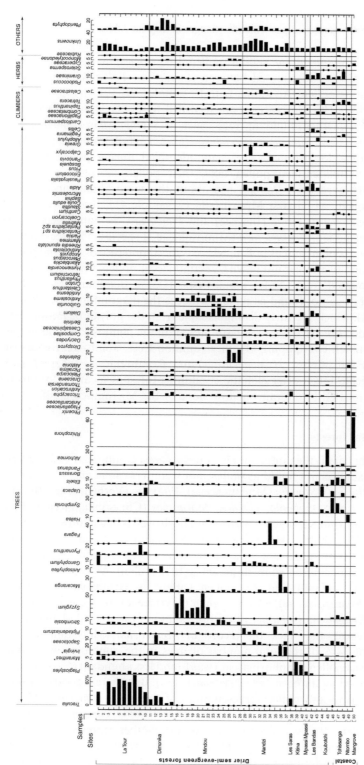

Figure 5.5. Frequencies of the main pollen taxa identified within modern soil surface samples from Southern Congo (after Elenga *et al.*, 2000a) (% calculated versus pollen sum including all identified taxa; Cyperaceae and Pteridophyta spores being unimportant, except at site B).

(12.6%)—is represented by similar pollen frequencies, whereas three species of Annonaceae (13.5% distributed between *Anonidium*, *Polyalthia*, and *Enantia*) are totally absent as pollen. At Kitina (site F, samples 39, 40), the dominant (12.4%) *Anthostema* trees (Euphorbiaceae) are recorded by pollen, but at much lower percentages (<5%). Other abundant trees—such as *Scytopetalum klaineanum* (8% Scytopetalaceae), *Ctenolophon englerianus* (6% Ctenolophonaceae), *Spathandra blackeoides* (6% Melastomataceae)—not identified during pollen analysis might correspond to unidentified pollen taxa, their pollen morphology not currently well-known (Figure 5.5). *Dialium* has about the same abundance (5%) both as plant and pollen. At Mpassi Mpassi (site G, sample 41), the two most abundant trees—*Hua gabonensis* (16% Huaceae) and *Pancovia* (7% Sapindaceae)—were not represented as pollen. *Pentaclethra* (Mimosoideae, 2 species, 5% trees) has the same representation as plant and as pollen, whereas the abundance of *Berlinia* pollen (20%) may correspond to another unknown Caesalpiniaceae pollen, since *Berlinia* was not in the list of counted trees. At les Bandas (site H), the dominant tree (56%) *Parkia* (Mimosaceae) was not found as pollen, whereas the diagram includes *Pentachlethra* pollen (10 %), which is clearly distinguishable from the polyad of *Parkia*. Except for *Dacryodes* and *Aidia* (Rubiaceae), both present in the plant record and in the pollen, there is not much overlapping of other common trees. Joint occurrences of pollen from *Allophyllus*, *Celtis*, *Hymenocardia*, and Combretaceae together with more Poaceae indicate much drier climatic conditions for sites located close to the drier Niari valley located in the rain shadow slope of the Mayombe massif.

5.3.5   Coastal forests

At Koubotchi (site I) the dominant (30% trees) forest component *Celocaryon preussii* (Myristicaceae) was not found as pollen, nor were *Xylopia aethiopica* (Annonaceae), *Carapa* (Meliaceae), *Staudia* (Myristicaceae), *Vitex* sp. (Verbenaceae). But the pollen representation of *Symphonia*, *Uapaca*, *Maranthes* (Chrysobalanaceae), *Pycnanthus* (Myristicaceae), and Sapotaceae correspond fairly well to the number of trees counted in the plot. Remarkably, *Macaranga* and *Alchornea* (<1.4% in the tree counts) are over-represented by significant pollen percentages (>20%), and *Tetracera* (Dilleniaceae) is also over-represented by its pollen (>10% pollen). At Tchissanga (site J), two pollen samples (47, 48) indicate pollen values for *Symphonia* and Sapotaceae, *Fegimanra* (Anacardiaceae), and *Syzygium* (Myrtaceae), that are also characteristic trees of the *Symphonia globulifera* forest in valleys of the coastal plain (Elenga *et al.*, 2000b). However, they lack the record of *Memecylon* (Melastomataceae) which accounted for 34% of the total number of trees in the same plot. At Ntombo (site K), *Anthostema* pollen (Euphorbiaceae) was found less abundant (5%) than in the tree cover dominated by *Anthostema aubryanum* (51%). But, there is good correspondence between the plant and pollen representation of *Syzygium guineensis*, *Hallea ciliata* (Rubiaceae), *Elaeis guineensis* (Palmae), and *Alstonia congensis* (Apocynaceae). *Tetracera* pollen (5%) was found, whereas this climber represents less than 1% in botanical inventories. This may indicate an over-representation of pollen from climbers, or an under-estimation of plant specimens, those with diameter lower than 5 cm not being counted. The mangrove (site L) is dominated by the abundance of *Rhizophora*, associated with *Phoenix* and *Pandanus*, also abundant plants in the plant cover.

In order to clarify the distribution of plants and pollen versus environmental factors, correspondence analysis done on all the samples compared the composition of the floristic inventories from lowland coastal forests with those of Mayombe mid-elevation forests. Indeed, there is a west/east increasing rainfall gradient from 1,100 to 1,600 mm/yr and an increased elevation between the coastal plain (sea level) and the Mayombe (700 m). But, the authors favoured an explanation involving different soil composition. They distinguished the pollen association of *Syzygium*, *Symphonia globulifera*, *Phoenix*, *Tetracera*, *Sclerosperma* (Arecaceae) as characterizing swamp forests (Elenga *et al.*, 2000b). However, differences in elevation and also strong variations in precipitation could partly explain the differences in floristic and pollen composition of the coastal and Mayombe forests. Variations in the amount of rainfall are not negligible. Moreover, during the dry season, the effect of clouds on evapo-transpiration (Maley and Elenga, 1993) and that of the Benguela Current induce cooler temperature in southern Congo (Maley, 1997).

In conclusion, the study on modern pollen rain from Congo shows that arboreal pollen percentages from 70 to 90% characterize samples collected under closed forest, lower values being found within disturbed forest. These high values are obtained despite the fact that important families—such as Annonaceae (all species), most Myristicaceae (except *Pycnanthus*), Chrysobalanaceae, Olacaceae (except *Strombosia*), Clusiaceae (except *Symphonia*), Apocynaceae, Meliaceae, Melastomataceae, etc.—were poorly documented in modern pollen rain. Well-diversified pollen assemblages from southern Congo document the floristic diversity of semi-evergreen forests south of the equator. Although there is no direct overlap between pollen and tree composition, pollen assemblages clearly distinguish the different types of forest that have produced them. Associated *Symphonia globulifera*, *Uapaca*, *Hallea*, *Dacryodes*, *Anthostema*, *Dialium*, *Plagiostyles*, and Sapotaceae, both in the vegetation and in the corresponding modern pollen rain, characterize the dry semi-evergreen forest of Mayombe. This is enough to appreciate that the semi-evergreen forests south of the equator appear palynologically distinct from the same vegetation unit mapped north of the equator in Cameroon (sites Q and E, Figure 5.2), including Kandara. Possible explanation for such differences may be searched for in the long-term geological history and (or) in the differential ecological requirements and threshold climatic limits of the various forest trees. More investigation on this line is needed.

5.3.6 Swamp forest

Preliminary information about modern pollen rain from the inundated evergreen swamp forests of Central Congo was provided by three samples collected within the *Guibourtia demeusii* (Caesalpiniaceae) dominated association. Located in the central Congo Basin below 400 m in elevation (1°34′N, 17°30′E), the area receives more than 1,600 mm/yr of precipitation with a very short dry season. The results show that arboreal pollen (AP) again ranges from 75 to 90%. Pollen assemblages are dominated by *Lophira* (up to 60% in one sample), followed by *Guibourtia, Alchornea, Macaranga, Uapaca,* Combretaceae, and *Myrianthus,* and a few pteridophyte spores. These pollen assemblages are different from those collected from the hygrophilous evergreen forest of Cameroon and from any types of the semi-evergreen drier peripheral forest (Elenga, 1992).

The pioneer studies summarized in this chapter represent significant progress in tropical modern pollen rain of Africa. First, they clearly demonstrate that a high proportion of tree pollen can identify forest cover. Second, they indicate that the main mapping vegetation units (evergreen, semi-evergreen, and mixed) as well as secondary subdivisions in the vegetation (from Cameroon and Congo) are characterized by different pollen assemblages. Third, there is partial overlap between pollen and plant representation. In conclusion, differences in pollen (taxa composition and abundance) can be used to recognize the vegetation units and sub-units within the Guineo-Congolian rainforest, despite the lack of pollen representation of some dominant trees. Although not covering continuous climatic gradients, the results discussed bear critical information for interpreting fossil pollen data from the region. Extracting individual or associated pollen markers for all the vegetation units within the rainforest, however, requires additional and more homogenously distributed samples before being statistically valid. A complete inventory of forest types is essential because of the high diversity of rainforests. Collecting along two distinct transects—one from south to north crossing the equator to address the climatic influence of the ITCZ (Haug *et al.*, 2001), the other from west to east, to address inland monsoon penetration—would be most valuable. This section has shown that modern pollen from the rainforest can be studied in the same way as other forests in the world. It is a long, but feasible task.

5.4   QUATERNARY HISTORY

Despite recent progress and new fossil pollen sequences obtained by coring swamps and lakes in Africa, it is not possible to reconstruct the Quaternary history of all the different forest units described in Section 5.2. Not all of them being documented yet, the presentation of fossil pollen records follows a chronological order, starting with the last glacial period, and then discussing the last 10,000 years of the Holocene. The list of sites where fossil pollen are available is given in Table 5.1 and Figure 5.1. Evidence for climate changes based upon other sources of information—such as lake sediments, paleosols, stable isotopic, phytoliths, diatoms—has been summarized in Battarbee *et al.* (2004).

5.4.1   Ice-age record

Marine sediments have provided pollen data related to vegetation change on the continent. These sequences have the great advantage of providing land–sea linkage on a straightforward isotopic chronology (Bengo and Maley, 1991; Dupont and Wienelt,

Table 5.1. List of fossil pollen sites located within the African lowland rainforest.

Sites	Coordinates	Elevation (m)	Rainfall (mm/yr)	Authors
Barombi Mbo	4°67'N, 09°40'E	300	2,400	Maley and Brenac (1998),
Mboandong	4°30'N, 09°20'E	120	2,400	Richards (1986)
Ossa (OW4)	3°40'N, 10°05'E	8	2,950	Reynaud-Farrera et al. (1996)
Kamalete	0°43'S, 11°46'E	350	1,500	Ngomanda et al. (2005)
Bilanko	3°31'S, 15°21'E	600	1,500	Elenga et al. (1991)
Sinnda	3°50'S, 12°48'E	130	1,100	Vincens et al. (1994, 1998)
Coraf	4°00'S, 11°00'E	0	1,260	Elenga et al. (1992)
Kitina	4°15'S, 11°59'E	150	1,500	Elenga et al. (1996)
Ngamakala	4°04'S, 15°23'E	400	1,300	Elenga et al. (1994)
Songolo	4°46'S, 11°52'E	5	1,260	Elenga et al. (2001)

1996; Dupont and Behling, 2006). Offshore transported pollen include taxa from different vegetation zones that are mixed together and can hardly be interpreted in terms of past expansion of each of the different forest types. Indeed, marine records are the only source of information for older geological time periods (see Chapter 1). Here, we will discuss the oldest continental evidence provided by lacustrine pollen sequences. Lake records span the last 30,000 years including part of the last glacial period (ice age) and its maximum in the Last Glacial Maximum (LGM).

The ice-age record for African tropical lowland forests is known from two sites: Barombi Mbo within a forested region of Cameroon (Maley and Brenac, 1998) and Ngamakala within secondary grassland nearby the Congo River (Elenga et al.,1994). A third record, from lake Bosumtwi, West Africa, remains poorly documented through a preliminary pollen diagram (Maley, 1991). Except for *Celtis, Olea* it does not contain any detail about the forest composition. Although the Bosumtwi record is informative about the lowland rainforest along the Guinean gulf, this review does not include it. A long core spanning the last 1.1 Myr was raised in 2003 and is still being analyzed. This record will be of immense significance to the reconstruction of West African paleoclimates.

5.4.1.1 *Barombi Mbo, evergreen and semi-evergreen forests (Cameroon)*

North of the equator, the small crater lake Barombi Mbo ("Mbo" means lake in the local language) (4°40'N, 9°24'E) is located 15 km north of Mount Cameroon and 50 km inland from the Atlantic coast. Core MB-6 recovered from the deepest part of the lake (Maley et al., 1990) yields a remarkably complete record for the last 32,000 years (27 ^{14}C kyr BP), including the LGM. The lake is situated at low elevation (300 m a.s.l.) and presently surrounded by forest (Figure 5.6, see color section). The crater lies within the wide belt of lowland evergreen Biafran forest dominated by Caesalpinia-ceae bordered by two large bands of semi-evergreen forest. Patches of semi-evergreen forest occupy areas under the rain shadow of Mount Cameroon, which causes a decrease in precipitation and a reduction in the length of the rainy season. Lying

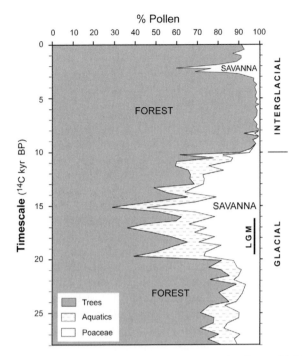

Figure 5.7. Synthetic pollen diagram from core BM-6, Lake Barombi Mbo, Cameroon, presented according to interpolated ^{14}C ages (after Maley and Brenac, 1998) (% calculated versus pollen sum including all identified taxa, excluding spores).

within Mount Cameroon's rainshadow, Barombi Mbo receives 2,350 mm/yr with a 3-month dry season from December to February. This relatively low rainfall contrasts with 9,000 mm/yr of windward (coastal) precipitation on the side of Mount Cameroon. The laminated sediments of the 23.5-m core were regularly deposited and present no hiatus, an exceptional situation for African lakes. Twelve AMS radiocarbon dates provide a reliable depth/age curve (Giresse *et al.*, 1991, 1994). In the pollen diagram each sample corresponds to a 1-cm thickness of sediment averaging c. 10 to 15 years of deposition. Pollen analyses were made at c. 200-yr intervals in the Holocene, and c. 300 yr in the glacial period (Maley and Brenac, 1998; Elenga *et al.*, 2004). All the results discussed here follow the ^{14}C chronology provided by the authors.

At Barombi Mbo, the curve of total arboreal pollen (Figure 5.7) provides a good average estimate of the forest cover surrounding the lake, although pollen deposition into the lake integrates a much larger basin. It clearly shows that, during the last glacial period—from 27 to 10 ^{14}C kyr BP (c. 32–11.5 kcal.yr BP)—the area around the lake remained forested. However, between 20 to 10 ^{14}C kyr BP (c. 24–11.5 kcal.yr BP) the tree cover was significantly reduced. Some fluctuations are depicted by the curve of total arboreal pollen, which would have been less marked if aquatics (sedges) had been eliminated from the pollen sum on which relative frequencies are calculated. High abundances of sedges (aquatics illustrated in Figure 5.7) coincide with the Last Glacial Maximum (LGM). Such peaks attest to enlarged herbaceous wetlands (including grass) that occupied emerged land on the shore line. A probable explanation for wetland expansion is falling lake levels in response to drier climatic conditions.

Glacial period and refuge hypothesis
Regarding the detailed pollen composition of the tree component, the last glacial period can be subdivided into two distinct phases (Figure 5.8). During the first phase from c. 32–24 kcal . yr BP, the total AP pollen (c. 80%) indicates a dense canopy cover (Figure 5.7) which remained fairly stable throughout and includes the highest frequencies (7 to 10%) of the Caesalpiniaceae evergreen component. Out of the 150 identified pollen taxa, 20 different genera are included in the Caesalpiniaceae curve (Maley and Brenac, 1998). As their pollen is normally under-represented in modern pollen rain (Reynaud-Farrera, 1995, and Section 5.3), the Caesalpiniaceae may have been a very important and diverse component within the forest at that time. Sapotaceae, a component of mature semi-evergreen forest (Elenga *et al.*, 2004), were also present (1 to 3%), together with other components of this subflora. Among markers of the mountain forest, *Podocarpus* pollen was found at such low frequencies (<1.5%) that it is unlikely that the trees occurred close to the lake. Today, *Podocarpus* is present on Mount Koupé (2,050 m) and its fossil occurrence could well be attributed to long-distance transport from this mountain. In marine cores *Podocarpus* pollen is quite abundant in sediment dating from the glacial period (Marret *et al.*, 1999). In contrast, *Olea* was recorded at higher pollen percentages (>10%) indicating that their trees may have been present near the lake. The *Olea* pollen curve shows a remarkable pattern through time. An increasing trend started at 24^{14}C kyr BP (28 kcal . yr BP), reaching a maximum (30%) at 20^{14}C kyr BP (24 kcal . yr BP), and then decreasing again to 5% at 17^{14}C kyr BP (c. 20 kcal . yr BP). From 24 kcal . yr BP to 20 kcal . yr BP, the decreasing trend of *Olea* pollen is in good correspondence with the 4,000-yr duration of the LGM chronozone placed between 23 kcal . yr BP and 19 kcal . yr BP on marine records (Mix *et al.*, 2001). Today, *Olea capensis* grows on Mount Cameroon at an elevation of 1,600 m and much higher, such as in cloud forest. Its abundance in the fossil pollen record has been explained by the impact of stratiform clouds and associated fogs produced by sea surface temperature cooling of the Atlantic (Maley, 1989; Maley and Elenga, 1993). That such processes may have played a role cannot be ruled out. Interestingly, the maximum of *Olea* percentages occurred slightly before the LGM and corresponds to the timing of the Dansgaard–Oeschger event 2 (DO2) and the last δ^{18}O maximum of the Antarctica Byrd ice core (Mix *et al.*, 2001). The pollen/climate transfer function in East Africa indicates a glacial continental cooling of $3 \pm 2°$C in the tropical region (Bonnefille *et al.*, 1990, 1992; Vincens *et al.*, 1993)—a maximum value—since the effect of lower carbon dioxide content of the atmosphere could not be taken into account. Using the present day lapse rate, such an estimate corresponds to a 600-m shift in elevation, much less than the 1,300-m necessary lowering for *Olea* to reach the Barombi Mbo lowlands. Originally, tropical cooling was inferred from a significant shift in altitudinal distribution of vegetation zones on East African mountains (Flenley, 1979). The descent of vegetation on tropical mountains results from the

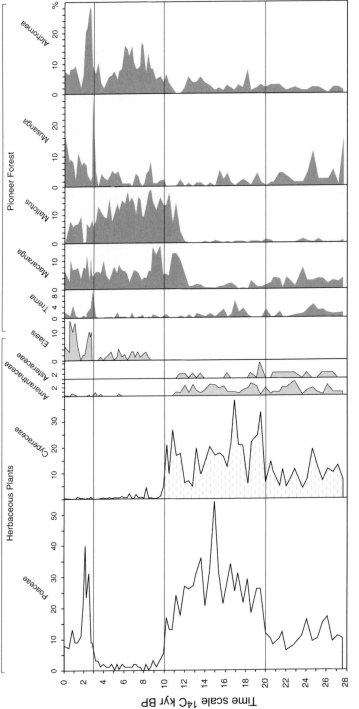

Figure 5.8(a). Detailed pollen diagram from core BM-6, Lake Barombi Mbo, Cameroon, presented according to interpolated ^{14}C ages: trees and shrubs (after Maley and Brenac, 1998) (% calculated versus pollen sum including all identified taxa, excluding spores).

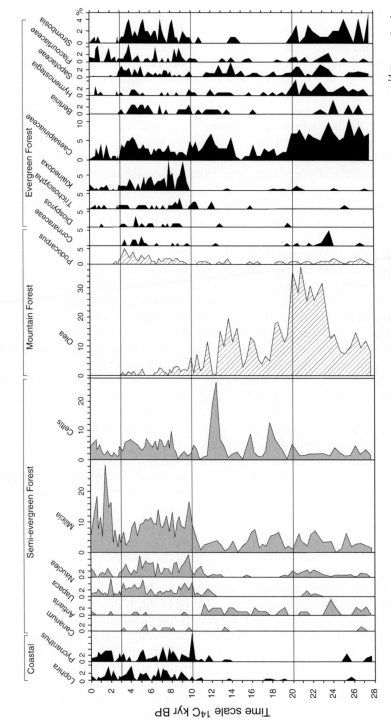

Figure 5.8(b). Detailed pollen diagram from core BM-6, Lake Barombi Mbo, Cameroon, presented according to interpolated ^{14}C ages: herbs (after Maley and Brenac, 1998) (% calculated versus pollen sum including all identified taxa, excluding spores).

associated effects of both decreasing temperature and rainfall. The Barombi Mbo record clearly demonstrates an individualistic movement of *Olea* into the lowland vegetation during glacial time. This cannot be forced by lower CO_2 concentration as it affected a tree which is a C_3 plant. Applying 3°C-cooling at the Barombi Mbo will lead to a value of 21°C (24 − 3 = 21°C) for mean annual temperature, a value above the 18°C threshold for tropical highland forests in India (Bonnefille *et al.*, 1999; Barboni and Bonnefille, 2001), and above the 15°C threshold used to define the tropical biome (Prentice *et al.*, 1992). Under such conditions, *Olea* could reach the lowland rainforest where other tropical trees remained. The pattern shown by the *Olea* curve in the Barombi Mbo fossil record provides a good example of how plants individualistically responded to climatic changes. Significant rainfall decrease, during glacial time at the equatorial latitude, was estimated around 20 to 30% of the present value (Bonnefille *et al.*, 1990; Bonnefille and Chalie, 2000). Applying this estimate at Barombi Mbo leaves enough precipitation (1,500 mm/yr) to maintain a forest cover, during the glacial period, prior to the LGM.

During the second phase of the glacial period (24–11.5 kcal.yr BP), total AP pollen dropped with decreasing abundance of typical Biafran evergreen forest taxa, whereas semi-evergreen components—such as *Celtis* and *Antiaris* (Moraceae)—become more abundant, although their pollen frequencies show large fluctuations. Lowland species of *Strombosia*, Flacourtiaceae, Sapotaceae, *Antiaris*, *Hymenostegia* (Caesalpiniaceae), *Berlinia*, and other Caesalpiniaceae are still present, but decreased significantly (Maley and Brenac, 1998). Isotopic studies from the same core point to an increased proportion of C_4 grasses, likely favored by low CO_2 concentration of the global atmosphere at that time (Giresse *et al.*, 1994). The increase in grass pollen does not overlap the *Olea* phase (29–22 kcal.yr BP), but follows it, becoming more abundant between 24 kcal.yr BP and 11.5 kcal.yr BP, synchronously with the increase in Cyperaceae (Figure 5.8). The different patterns of the Poaceae and the Cyperaceae curves may indicate that the peak of Poaceae is not related to subaquatic grasses, but rather come from open grassland inside the forest. Pollen/biome reconstruction at 22 kcal.yr BP emphasized the replacement of rainforest by a tropical seasonal forest (Elenga *et al.*, 2000c). However, during the minimum extent of forest which lasted 5,000 years (24–19 kcal.yr BP), two sharp increases in tree cover are observed. They attest that forest expanded significantly during glacial time, although fluctuations in tree pollen percentages would have been minimized by excluding Cyperaceae from the pollen sum in the calculation of relative percentages. The maximum of grass pollen associated with the greatest opening of the forest is dated at 18 kcal.yr BP, a radiocarbon date that fits Heinrich Event H1 (Mix *et al.*, 2001), and therefore occurred a long time after the LGM. If the peak of *Olea* registers the maximum cooling and the peak of grasses the maximum aridity, these were delayed by at least 5,000 years. Aridity and cooling were decoupled and a complex pattern of forest dynamics is evidenced during the glacial period when the climatic impact of the two Heinrich Events H1 and H2 affected the lowland rainforest at Barombi Mbo. Nevertheless, rainforest appears very sensitive to global climatic changes. While considering the high topography of Mount Cameroon and the high precipitation gradient, a great variety of climatic conditions must have prevailed in the region in the past, just as it does today. During glacial time, enough precipitation could have existed on the western slopes, allowing the persistence of evergreen forests there at the same time as semi-evergreen forests at Barombi Mbo, at the eastern base of Mount Cameroon. During glacial time, the coastal area expanded as sea level fell and offered new opportunities for new land occupation. Various forest refuges could have existed during glacial time and could be located on direct evidence by means of new palynological studies, rather than postulated on various hypotheses (Maley, 1996).

5.4.1.2   *Ngamakala, savanna contact with semi-evergreen rainforest (Congo)*

On the right bank of the Congo River, the small Ngamakala lake (4°04'S, 15°23'E, 400 m), is located at the southern end of the Bateke plateau (Figure 5.1) where mesophilous, hygrophytic forests are related to humid edaphic conditions (Descoing, 1960). The lake—1 km wide—is now covered by *Sphagnum* (Sphagnaceae) and clumps of trees of *Alstonia boonei* (Apocynaceae). It is surrounded by a wooded *Loudetia demeusei* (Poaceae) savanna, with *Pentaclethra* (Mimosoideae) new growth (Makany, 1976). The results of pollen analysis of a 160-cm core show that, from c. 30–17 kcal. yr BP, the fossil pollen sequence was dominated by Sapotaceae and *Syzygium* (Figure 5.9). With Fabaceae (Leguminosae) and *Canthium*, later identified among the unknown (Elenga *et al.*, 2004), percentages of tree pollen exceed 80%. This record clearly indicates a forested environment during the glacial period. This forest was developed on a swamp attested by the occurrence of aquatic plants—such as *Xyris, Laurembergia* and the floating *Nymphea*. The forest existing there during the last glacial period included significant Fabaceae with Combretaceae, *Alchornea, Campylospermum* (Ochnaceae), *Cleistanthus* (Euphorbiaceae), *Canthium*, and *Celtis*. Rare pollen of other trees—such as *Crudia gabonensis* (Caesalpinioideae), *Guibourtia*, and *Tetracera* (Dilleniaceae)—provide a link with modern surface samples of the central Congo Basin (Elenga *et al.*, 1994). But, except for *Celtis* and Sapotaceae, the glacial forest in southern Congo had no floristic resemblance to that documented at Barombi Mbo in Cameroon at the same time. More specifically, the Ngamakala record does not show any of the highland taxa pointing to cooler temperatures, such as observed in the Cameroon record. The swamp environment may not have been favorable to the growth of *Olea*, but the lack of *Podocarpus* is more surprising. If *Podocarpus* had been present at mid-elevation on the Bateke plateau, its abundant pollen would have been blown away, and at least a few grains found in the Ngamakala sediment. However, no valid conclusion can be drawn until another glacial sequence from southern Congo confirms it. As in the Cameroon record, variations within the relative abundance of the different trees are observed during the glacial period, although masked by abundant *Syzygium* and Sapotaceae. The Ngamakala core had a very low sedimentation rate with only an 80-cm thickness of sediment deposited during the glacial interval (from 29 kcal. yr BP to 17 kcal. yr BP), and conventional dating is not accurate enough. The time resolution interval between adjacent samples (350 years) is greater than that of Barombi Mbo and insufficient to address short-term climatic variability during the glacial period. At Ngamakala, no

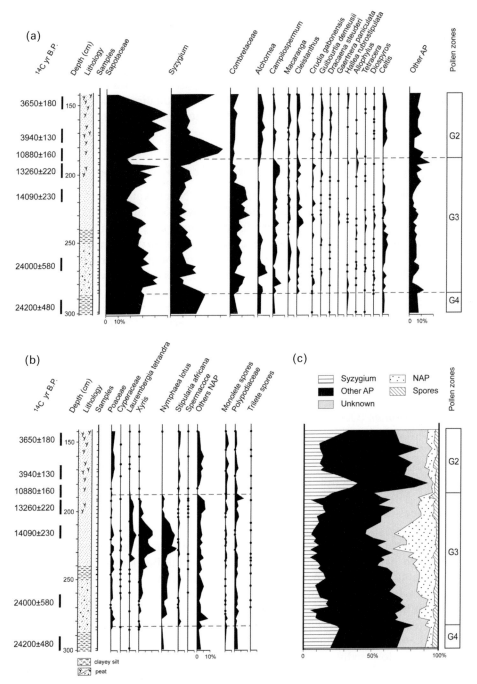

Figure 5.9. Simplified pollen diagram from Ngamakala, presented according to depth (Congo, after Elenga *et al.*, 1994): (a) arboreal and climber taxa; (b) herbaceous taxa; (c) synthetic diagram (% calculated versus pollen sum including all identified taxa, aquatics, and spores).

significant change at 24 kcal.yr BP is clearly identified in the pollen diagram, although postulated within a change of sediment. The minimum forest cover (60% AP) occurred at c. 17 kcal.yr BP when Sapotaceae and *Syzygium* decreased and Combretaceae reached their maximum relative abundance. Although the conventional dates of the Ngamakala core have a great experimental error, the minimum tree cover at Ngamakala appears synchronous with grass maximum frequencies at Barombi Mbo from which maximum aridity has been inferred. Simultaneous abundance of aquatic herbs, among which *Xyris* and *Nymphaea* dominate, may be explained by a water depth lower than 1 m. The Ngamakala record lacks evidence for cooler indicators, and therefore the decoupled effect of coolness and aridity during glacial time is not established yet for the southern tropics until further high-resolution pollen sequences are provided.

Past vegetation history of the rainforest during glacial time is only documented at two sites so far. It clearly indicates the persistence of two well-diversified rainforests of different composition on both sides of the equator.

5.4.1.3   *Bateke plateau, savanna (Congo)*

Bilanko

Located within the Bateke plateau (600–800 m), the Bilanko site (3°31′S, 15°21′E, 700 m) is a 10-km closed depression occupied by floating sedges and grass mats with

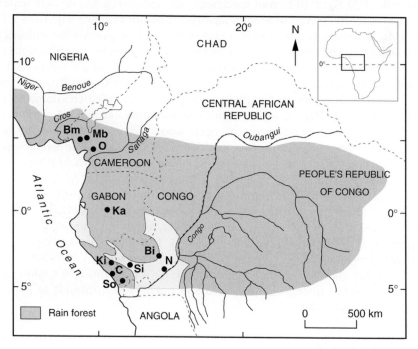

Figure 5.10. Location map of fossil pollen sites. Bm: Barombi Mbo; Mb: Mboandong; O: Ossa; Ka: Kamalete; N: Ngamaka; Si: Sinnda; Bi: Bilanko; Ki: Kitina; C: Coraf; So: Songolo.

abundant *Syzygium* (Figure 5.10). Shrub savanna vegetation occupies sandy soil in this region, characterized by 1,600-mm/yr rainfall, 4-month dry season and 4 to 6°C seasonal temperature range. The 60-cm short core, recovered from the 4 m thick peat deposit inside the depression, contained a wood fragment at the bottom that was dated at $10,850 \pm 200\ ^{14}$C yr BP, the top part of the core not being dated. Although the pollen flora were fairly diverse (103 pollen taxa), the pollen sequence was dominated by *Syzygium* (90% of the total count) associated with a few dispersed forest trees of semi-deciduous forest. This indicated the development of a *Syzygium* swamp at Bilanko. The sediment of the core—yielding dated fossil wood—contained up to 30% *Podocarpus* pollen (Elenga *et al.*, 1991). *Podocarpus milanjianus* does not occur in the Congo Basin today, but it has been collected as isolated trees in the Chaillu Mountains, in Gabon, at the same elevation as Bilanko. Although its pollen can be transported over long distances, such high percentages imply the proximity of abundant trees nearby Bilanko, suggesting cooler temperatures during a time period corresponding to the Younger Dryas. Development of stratified low clouds has been proposed to explain occurrences of mountain plants in the lowlands (Maley and Elenga, 1993).

Bateke plateau
Macrobotanical remains have been identified from two glacial-aged sites (Figure 5.1) on the Bateke plateau (Dechamps *et al.*, 1988a). At Gaganlingolo, 17 km north of Brazzaville (3°55′S, 15°10′E) root specimens older than $30\ ^{14}$C kyr BP were tentatively attributed to several species of *Monopetalanthus* (Caesalpiniaceae), with *Grewia* (Tiliaceae) and *Pterocarpus* (Fabaceae). So many species of *Monopetalanthus* within a small area is surprising and brings into question the validity of specific identifications of this fossil wood. Nevertheless, the wood documents the presence of a forest in the Congo Basin during the last glacial period on the right bank of the Congo River, in a region now occupied by savanna. At Gambona (2°S, 16°E), a few hundred kilometers north, closer to the Congo Basin, wood remains of the same age yielded a more mixed assemblage including components of open woodland—such as *Nauclea latifolia* (Rubiaceae)—and evergreen forest components—such as *Detarium senegalense* (Caesalpiniaceae), *Connarus griffonianus* (Connaraceae), and *Brachystegia*.

The recovery of fossil wood at sites many kilometers apart attests to a certain geographical extension of the rainforest on land along the Congo River during the Last Glacial Maximum. Fossil wood of *Podocarpus* was not found in these terrestrial deposits dated prior to the Last Glacial Maximum, although its pollen was abundant in marine cores off the Congo coast (Jahns, 1996). Transported pollen from rainforest was indeed less important during the LGM than during stages 3 and 4 of the glacial period, but they showed their minimum percentages during stages 5b (90 kyr) and 5d (110 kyr) of the last interglacial, which cannot possibly be attributed to lower CO_2 content of the atmosphere.

5.4.2   Holocene record

In west Central Africa, the past history of lowland forests for the last 10,000 years is documented by a total of ten fossil pollen sequences located between 5°N and 5°S latitude (Figure 5.10). However, except for Barombi Mbo, most pollen records remain incomplete, discontinuous, and insufficiently well-dated. The persistence of rich, diversified forests until the last few thousand years, evidence for several drastic forest declines, use of oil palm accompanying (or not accompanying) Iron Age civilizations and the Bantu expansion in Africa will be discussed in this section.

5.4.2.1 Sites located north of the equator

Evergreen wet and semi-evergreen (Barombi Mbo, Cameroon)

The pollen sequence from the Barombi Mbo core, yielding a record for the last glacial period discussed above, also contains one of the most complete records for the last 10,000 years (Figure 5.8). In this sequence the maximum of tree density started around 11.5 kcal . yr BP and remained constant until 2.5 kcal . yr BP, with more than 90% AP indicating a dense tree canopy cover. *Olea* frequencies lower than 2% indicate that the tree had progressively disappeared from the surroundings of the lake since the beginning of the Holocene. Dominated by pioneer trees (*Mallotus, Macaranga, Musanga*, and *Alchornea*), and decreasing occurrences of *Antiaris*, the Holocene forest contained more Euphorbiaceae and fewer Caesalpiniaceae than the forest existing during the glacial period. The three types of forests occurring in the area today were already in place. The Biafran wet evergreen forest is attested by *Klaine-doxa, Trichoscypha*, Connaraceae, the coastal by *Lophira/Saccoglottis* and *Pyc-nanthus*, the dry evergreen semi-evergreen by *Milicia, Celtis, Nauclea*, and *Uapaca*. The Biafran evergreen forest already existing during the glacial period appears more diversified during the Holocene with greater abundance of *Klainedoxa* (Irvingiaceae), *Trichoscypha* (Anacardiaceae), *Diospyros*, and Connaraceae. But, there were fewer Caesalpiniaceae altogether than between 31 kcal . yr BP and 24 kcal . yr BP. The forest succession started at 14.7 kcal . yr BP with abrupt increases of *Macaranga*, then *Mal-lotus*, followed by that of *Milicia* (Moraceae) which culminate around 11.5 kcal . yr BP. *Alchornea* reached its first peak at c. 9 kcal . yr BP. Such a progressive trend of forest took more than 2.5 kcal . yr BP to be fully established. Although it may have been triggered by increased monsoon rains, it cannot be qualified as an abrupt onset. Higher resolution pollen analysis for this interesting transition period would have allowed a valuable comparison with the fine-resolution isotopic record obtained on the same core (Giresse *et al.*, 1994). During the whole Holocene, several fluctuations in the pollen percentages of various components are depicted (Figure 5.8)—such as decreasing trends of *Klainedoxa, Mallotus*, and an increasing trend for *Musanga*. But, the c. 200-yr sampling interval of the pollen record remains too large to interpret these variations in terms of the vegetation dynamic under climatic variability attested by global changes. The strongest change depicted by the available data occurred around 3 kcal . yr BP when most components of the Biafran evergreen decreased or totally disappeared, while pioneers—such as *Musanga*—increased significantly. The sharp 40% peak of Poaceae attests to a short phase of open savanna around 2.5 kcal . yr BP. At that time, the increase in grass pollen was not accompanied by that of Cyperaceae, having almost disappeared during the Holocene when the lake was high and its shore immediately surrounded by forest. This is an indication that grasses may have developed inside forest openings rather than along the shoreline itself. To what extent this savanna phase is the result of human deforestation or of significant climatic aridity will be discussed later.

Ossa, hygrophilous evergreen forest (Cameroon)
The shallow Lake Ossa (7-m depth during wet season) is located a few kilometers to the west of the Sanaga River within the *Lophira/Saccoglottis* hygrophilous evergreen forest (Figure 5.10). *Elaeis guineense* and *Hevea brasiliensis* plantations have been recently established on the western shore line of the lake. The region is characterized by high rainfall—c. 3,000 mm/yr—with a long rainy season from March to November, followed by a short dry season from December to February (Nguestop *et al.*, 2004). Of the three cores recovered from Lake Ossa, one with a basal age of c. 10 kcal.yr BP shows reversed dates that indicate perturbed deposition, although the pollen attests to a well-diversified Caesalpiniaceae forest well-established around the lake since that date (Reynaud-Farrera, 1995). Another 5-m core OW 4 (3°48′N, 10°01′E, at 8 m a.s.l.), collected from the western side, provided an accurately dated pollen sequence for the last 5.5 kcal.yr BP. The pollen data include high percentages of sedges (Cyperaceae) reaching up to 50% of the total count, these perhaps due to the proximity of the river drainage, the shallowness of the lake, and associated land partially emerged during the dry season. The exclusion of Cyperaceae as well as that of fern spores from the total pollen sum for calculating percentages of other forest components is justified by the local context (Renaud-Farrera *et al.*, 1996). The Poaceae were kept in the pollen sum, although some may belong to local edaphic wetlands. Poaceae reached 15% at most and, except around 2.8 kyr BP, their stable percentages would not have strongly influenced variations in other elements (Figure 5.11).

The Ossa pollen diagram clearly indicates the persistence of an evergreen forest (AP percentages >75%) in the region throughout the last 5.5 kcal.yr BP. Forest composition around Lake Ossa differs significantly from that around Barombi during the same period (Renaud-Farrera *et al.*, 1996). Two distinct periods of forest development could be individualized, separated by an abrupt change at c. 2.85 kcal.yr BP, in surprisingly good correspondence with the limit between subboreal and subatlantic palynological subdivisions in Europe (Van Geel *et al.*, 1998). From 5.5 kcal.yr BP to 2.8 kcal.yr BP, forest composition is marked by the predominance of Caesalpiniaceae, Sapotaceae, and *Martretia* (Euphorbiaceae), associated with few *Lophira* and *Saccoglottis*, good markers of wet evergreen forest. Pollen input from highland forest is attested by minor percentages of *Podocarpus* and *Olea* (<5%), whereas the deciduous *Celtis*, *Holoptolea*, and *Piptadeniastrum* indicate a mixed semi-evergreen influence. From 2.8 kcal.yr BP to the last century, *Alchornea*, *Macaranga*, and then *Elaeis* increased significantly, while the evergreen components decreased. Although these taxa have been considered as pioneer heliophytics, it is unlikely that such strong changes in tree composition can be attributed to human impact. Earlier settlements in the Lake Ossa catchment have been dated at a much younger period, between 700 and 500 cal yr BP (Wirrmann and Elouga, 1998). From 5.5 kcal.yr BP to 2.8 kcal.yr BP, vegetation changes reflect climatic changes that may be looked for within the distribution pattern of the rainfall, rather than within its total

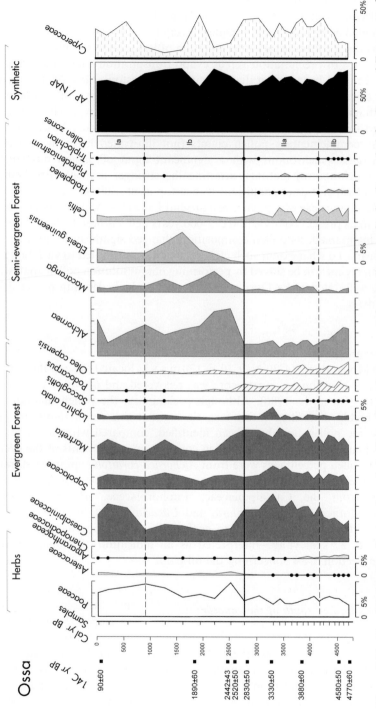

Figure 5.11. Simplified pollen diagram from Lake Ossa, Cameroon, presented according to calendar timescale (after Reynaud-Farrera *et al.*, 1996) (% calculated versus pollen sum including all the identified taxa, but excluding Cyperaceae plotted at the right-hand side of the diagram).

amount. Vegetation change appears in fairly good agreement with interpretation of the mineral composition of the sediment. High fluxes of orthoclase feldspar have been interpreted as an indication of higher precipitation than today (Wirmann *et al.*, 2001). However, a contradictory interpretation of less rainfall but high water level— indicating excess of precipitation versus evapo-transpiration—was supported by an hydrological model based upon diatom studies (Nguestop *et al.*, 2004). Interesting variability, documented by the 50-yr resolution of diatom analysis, is not depicted in the pollen diagram provided at c. 200-yr resolution only. Discrepancies in resolution prevents further comparison between the two indicators. Between 2.2 kcal. yr BP and 2 kcal. yr BP, complementary pollen counts (not illustrated here) showed another Poaceae increase synchronous with a significant input of Sahara dust that also contained allochthonous diatoms, attesting the stronger influence of north trade winds and subsequent drier conditions (Nguestop *et al.*, 2004).

The Caesalpiniaceae evergreen forest around Lake Ossa appears to have been re-established in its present composition c. 1,000 years ago. Although the wet evergreen forest was maintained, its pollen composition changed significantly throughout the last 5,000 years. To what extent such changes could be attributed to variability in the precipitation regime can be solved by performing finer resolution pollen analysis. This interesting, well-dated Lake Ossa sequence with its high deposition rate deserves further analysis.

Mboandong, evergreen wet and semi-evergreen (Cameroon)

From Lake Mboandong, (4°30′N, 9°20′E, 120 m a.s.l.) located close to Barombi Mbo (Figure 5.1), a 13 m long sediment core was recovered and provides a basal age of 6.8 kcal. yr BP (Richards, 1986). Although preliminary, the pollen diagram shows interesting highly diversified forest taxa throughout this period until the present day. Many tree pollen taxa have been identified, compared with just a few from herbaceous plants. The majority of pollen comes from components of the evergreen forest. High pollen percentages were from *Alchornea cordifolia*, an understorey tree, not specific to evergreen forest. Although there are indisputable variations in relative percentages from the Caesalpiniaceae, Euphorbiaceae, *Macaranga, Uapaca, Pycnanthus,* Moraceae, *Celtis, Lophira,* and *Uncaria* during the past 5,000 years, the author did not attribute them to ecological or climatic changes. The event recorded at c. 2.5 kcal. yr BP—interpreted as an indication of human impact—is discussed in Section 5.4.4 concerning the oil palm.

5.4.2.2   Sites located south of the equator

Coastal Congo, evergreen forest

The sandy coastal plain, north of Pointe Noire, is now occupied by a *Loudetia* (grass) savanna with *Manilkara lacera* (Sapotaceae). *In situ* fossil wood trunks with roots have been discovered included in a humic podzol bed (4°S, 11°45′E, Figure 5.1) that spans 7.4 kcal. yr BP to 4 kcal. yr BP (Schwartz *et al.*, 1990). The fossil wood specimens themselves have provided dates within the same time interval. Among the 117 collected specimens, a total of 20 species have been identified. These include *Saccoglottis gabonenesis*, *Agelaea* sp. (Connaraceae), and *Jundea* cf. *pinnata* (Connaraceae), *Uvariopsis angolana* (Annonaceae), *Cassipourea barteri*, *C.* sp. (Rhizophoraceae), *Dicranolepis* sp. (Thymeleaceae), *Dictyandra arborescens* (Rubiaceae), *Grewia* sp., *Neuropeltis acuminata* (Convolvulaceae), *Rinorea* cf. *gracilipes* (Violaceae), *Combretum* sp. (Combretaceae), *Rheedia* (Clusiaceae), and an unknown legume (possibly *Anthanota* according to Dechamps *et al.*, 1988b). Many species of *Monopetalanthus* (*M. microphyllus*, *M. pellegrinii*, *M. letestui*, and *M. durandii*), among the Caesalpiniaceae, indicate that a moist evergreen Caesalpiniaceae forest existed along the coastal plain for 3,000 years c. 6.8 kcal . yr BP to 3 kcal . yr BP. The forest was well-diversified and, according to the authors, resembles the evergreen forest described in Mount Cristal in Gabon. Fossil wood remains indicate that the moist evergreen coastal forest, now observed in Gabon, occurred 500 km south of its present southern limit between 7.4 kcal . yr BP and 3 kcal . yr BP, and therefore had a greater southern extension. The taphonomic conditions of the preserved fossil—such as standing trunks in the living position—tend to indicate that the Caesalpiniaceae forest had disappeared suddenly from the coastal region of Congo and that such disappearance was not due to human deforestation. The coastal Congo region receives 1,200 mm/yr of rainfall and experiences a 5-month dry season. There is a great contrast in seasonal temperature due to proximity of the cold oceanic upwelling offshore. The existence of a moist evergreen forest implies a greater rainfall (at least 2,000 mm/yr) or a shorter dry season. Its disappearance post-4 kcal . yr BP suggests strong changes in climatic or edaphic coastline conditions (Schwartz, 1992).

Two pollen sequences extracted from nearby depressions complete the Holocene vegetation history of southern Congo. The Coraf pollen sequence (4°15′S, 11°59′E, 150 m a.s.l.) located nearby the fossil wood site (Figure 5.10) was extracted from peat deposited inside a white sand horizon, and dated from 3 kyr to 950 yr BP. It documents a swamp forest progressively decreasing toward a modern savanna. The forest decrease was interrupted by a minor *Syzygium*/Combretaceae/*Tetracera* forest phase, dated around 1.4 kcal . yr BP. No sign of an abrupt forest decline around 2.5 kcal . yr BP was observed at Coraf, such as that in Cameroon, but the core was poorly sampled within that interval (Elenga *et al.*, 1992). On the Atlantic coast, north of Pointe Noire, the Songolo pollen sequence S2 (4°45′, 11°51′E, 5 m a.s.l., 1,260-mm/yr rainfall)—dated to 7.5 kcal . yr BP at its base—indicates a well-developed *Rhizophora* mangrove, following the marine transgression at 10 kcal . yr BP. The mangrove was associated with a swamp forest dominated by *Symphonia globulifera*, *Hallea*, and *Uapaca*. Humid conditions required for the development of swamp forest are confirmed by $\delta^{13}C$ values on total organic matter. There are significant and well-dated changes depicted in the mineralogical analysis of this core. These have no equivalent in the pollen data illustrating the coastal swamp forest, but the resolution of the pollen study is too low. The swamp forest persisted until c. 3.6 kcal . yr BP, when sedges, palms, and ferns became dominant (Elenga *et al.*, 2001).

In summary, the coastal plain of southern Congo was occupied by two different types of forests: a highly diversified Caesalpiniaceae forest documented by fossil woods, and more localized swamp forests and mangrove documented by pollen evidence since at least c. 7 kcal. yr BP. The forests persisted until 3 kcal. yr BP, clearly attesting to higher humidity or precipitation than now between 7 kcal. yr BP and 3 kcal. yr BP. The pattern of forest disappearance from the area remains unsolved and variations in sea levels should be considered. The discrepancies in dating really preclude any synchronism in various changes as proposed for an environmental event occurring at 4 kyr (Marchant and Hooghiemstra, 2004). A rather more complicated pattern seems to emerge, which needs comparison of fossil data at the same resolution interval with firmly established short-term Holocene climatic variability. Nevertheless, coastal grassland in the Congo seems to have been established only after 3 kcal. yr BP, and was interrupted by a minor forest increase again at c. 1.4 kcal. yr BP.

Mayombe, evergreen transitional forests
Evergreen transitional forests dominated by Meliaceae, Fabaceae, and Irvingiaceae with a great diversity of representatives from other families occupy the Mayombe massif today (Descoings, 1976; Cusset, 1987). On the western slope of the Mayombe, Lake Kitina (4°15'S, 11°59'E, 150 m a.s.l.) is located in a valley which receives about 1,500 mm/yr of rainfall and experiences a 4-month dry season with heavy cloud cover. The lake is surrounded by swamp vegetation with *Cyperus papyrus*, *Anchomanes* (Araceae), ferns, and a few trees—such as *Alstonia* and *Alchornea*—followed by a swamp forest including *Uapaca*, *Santiria*, and *Memycelon*. A transitional semi-evergreen dry forest—with *Dacryodes*, *Klainedoxa gabonensis* (Ixonanthaceae), *Piptadenia* (Leguminosae), *Plagiostyles africana* (Euphorbiaceae), *Anthostema* (Euphorbiaceae), and Sapotaceae—is developed on slopes (Elenga *et al.*, 1996). The Holocene pollen record is dated $5,460 \pm 70$ yr BP at a 620-cm depth (Figure 5.12). From 5.5 to 2.7 ^{14}C kyr BP, the synthetic pollen diagram indicates two well-developed forest associations: one from dry land (*Dacryodes*, *Martretia*, and *Anopyxis* (Rhizophoraceae)), the other from swamp (*Syzygium*, *Hallea*, *Anthostema*). Between 2.7 and 1.3 ^{14}C kyr BP, four AMS dates within a 40-cm depth interval demonstrate a very low sedimentation rate (or perhaps a discontinuity). During that interval—which lasted almost 1.4 kyr—significant changes occurred. First, a sharp and important increase in dry land forest (40%) accompanied by a significant increase in amorphous silica (Bertaux *et al.*, 1996) shows that Lake Kitina partially dried out. The successive increase in pioneers (*Macaranga* and *Alchornea*), associated ferns, and Poaceae (c. 10%) indicates a greater extent of swampy conditions contemporaneous with the appearance of *Elaeis guineense*. Although these pollen changes appear depicted as an abrupt 2.7 ^{14}C kyr BP event, several episodes of reduced lake level are shown by the lithology (Figure 5.12). No detailed modifications of pollen taxa composition have been provided for documenting the effect of aridity on the composition of the forest. There is no evidence for human impact either (Elenga *et al.*, 1996). The return to humid conditions after 1.3 ^{14}C kyr BP is attested by swamp forest pollen rather than a decrease in dry land forest. But, pollen from dry land forest increased again at 0.5 ^{14}C kyr BP. The Kitina diagram clearly indicates that the composition of the Mayombe forest is recent—no older than 500 years at most.

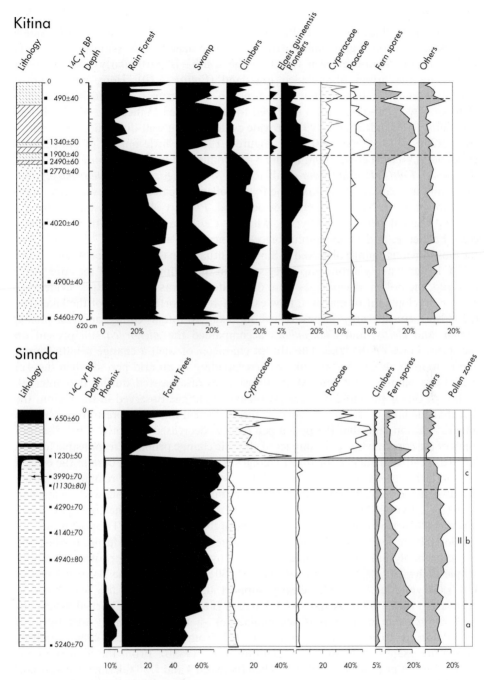

Figure 5.12. Simplified pollen diagrams from Lake Kitina (from Elenga *et al.*, 1996) and Sinnda, presented according to depth (after Vincens *et al.*, 1998) (% calculated versus pollen sum including all identified taxa, Cyperaceae, and spores).

Niari Valley, grasslands

The geographical location of Lake Sinnda (3°50'S, 12°48'E, 128 m a.s.l.)—inside a large band of wooded grassland separating the two massifs of Mayombe and Chaillu within the dry semi-evergreen forest to the south—is particularly interesting for addressing the origin of included grassland (Figure 5.10). Here, mean annual rainfall is about 1,100 mm/yr with a 5-month dry season from mid-May to mid-October. Two well-dated cores analyzed for both pollen (Vincens *et al.*, 1994) and phytoliths (Alexandre *et al.*, 1997) provide evidence for a well-diversified dense forest from $5,240 \pm 70$ to $3,990 \pm 70$ yr BP (Figure 5.12) with little impact of pollen from grasslands (Vincens *et al.*, 1998, Figure 5.4). Dominated by *Chlorophora* (Moraceae), *Alchornea, Celtis, Pausinystalia* (Rubiaceae), *Macaranga*, and *Lannea* (Anacardiaceae), and a great variety of taxa, the fossil pollen indicates a semi-evergreen (semi-deciduous forest). Noticeable changes occurred. *Phoenix, Myrianthus*, and ferns are well-represented before 5^{14}C kyr BP, whereas *Celtis*—associated with Caesalpiniaceae, Fabaceae, and many Euphorbiaceae—increased at about 4.5^{14}C kyr BP. But, the most obvious change observed at Sinnda is situated in the interval between c. 4 and 1.2^{14}C kyr BP when a strong gap is registered in the sedimentation of the core. After that gap, the pollen assemblages were dominated by grasses and sedges, evidencing greater development of marshes, now existing at the northeast end of the lake. The forest was considerably impoverished in taxa, although Ulmaceae (*Holoptolea, Trema*, and *Celtis*) and *Chlorophora* accompanied the *Alchornea* still present on the steep shores of the lake. The abrupt character of such a change results from a gap of more than 2,000 years, likely corresponding to an arid period when the 4 m deep Lake Sinnda dried out. Many forest trees disappeared during that interval. Although this is the most obvious post-4 kyr BP change observed in the region, the fact that the lake is shallow, located at the ecological limit for forest growth with a long dry season, and that the pollen percentages decreased abruptly at the discontinuity of the core maximize the impact of climatic change probably responsible for this pattern. Positive evidence for forest changes is missing. A second core in the central part of the lake contains black organic layers with abundant plant remains indicating that two subsequent drying phases occurred: one at 4.5 kyr BP followed by another post-1.7^{14}C kyr BP. An increase in tree forest taxa (AP > 40%) at the upper part of the core dated 650 ± 60 yr BP points to a recovery of the forest nearby. The pollen curve of the possibly cultivated *Elaeis guineense* became significant only during the last 600 years, probably linked to possible settlement by Bantu-speaking proto-agricultural people (Schwartz, 1992). Charcoal dated $2,130 \pm 70$ yr BP, on the Bateke plateau, and 1,600 yr BP in Mayombe clearly support the long use of fire in the region (Schwartz *et al.*, 1990). A date of $1,350 \pm 70$ yr BP obtained on burned specimens of *Erythrophleum suaveloens* (Caesalpiniaceae)—a ubiquitous tree, used by local people for its toxic alkaloids and extracted from archeological sites—confirms human fire usage in Mayombe (Dechamps *et al.*, 1988b). From the Sinnda record itself, it cannot be exactly established when (between 4 and 1.2^{14}C kyr BP) the savanna spread in the Niari Valley, but grasslands are present once fire is introduced at c. 1.2 kcal.yr BP.

Close to the Congo River, grassland/semi-evergreen forests
Near the right bank of the Congo River three cores were recovered from the
Ngamakala pond (4°04′S, 15°23′E, 400 m a.s.l.) which yielded records from the
glacial period (see above). None of them yielded a complete Holocene sequence.
In one core, there is only 20-cm sediment deposited between the two conventional
radiocarbon dates of $10,880 \pm 160$ and $3,940 \pm 130$ yr BP (Elenga *et al.*, 1994). Their
pollen content corresponds to a *Syzygium*/Sapotaceae swamp forest, but accurate age
determination is not available. Another core provided a 1 m thick accumulation of
peat deposited between c. 3 kyr and 1 kyr BP, attesting to the retreat of swamp forest at
the expense of grasslands during that interval. A significant expansion of swamp
Syzygium forest re-occupied the pond again at 930 ± 140 yr BP. Although far from
being complete, the succession of forest associations nearby the Congo River emphasizes many successive steps of changes during the Holocene. Among those the opening
of the forest at Ngamakala is sharply registered at c. $3.3\,^{14}$C kyr BP. But, this cannot be
considered synchronous with any of the events at 4 or $2.5\,^{14}$C kyr BP previously
discussed.

5.4.3 History of the rainforest during the last 5,000 years

During the last 5,000 years the evolution and dynamics of the rainforest forced by
climatic changes appear rather complicated. Apparently, they do not depict the same
pattern north and south of the equator, but the lower Holocene record is missing (or
discontinuous) at all the sites except at Barombi Mbo in Cameroon. There, the
depicted changes indicate significant climatic variability during the Holocene, but
there is no clear identification of an "African humid period" prior to 6 kcal. yr BP. At
most sites, negative evidence and lack of sedimentation during the early Holocene may
attest to drier climatic conditions. Hence the status of rainforest existence at that time
and composition remains unknown. The results summarized here firmly establish that
evergreen and semi-evergreen rainforests persisted between 6 kcal. yr BP and c.
3 kcal. yr BP at all of the nine investigated sites. Slightly before or after 3 kcal. yr BP,
major vegetation changes are documented (Vincens *et al.*, 1999). Because of discrepancies in the resolution of the analysis it is not clear, however, whether these changes
reflect several short-term episodes of climatic variability at a millennial scale concerning seasonal distribution of rainfall, or a longer single event. One major vegetation change documented in Cameroonian sites is well-bracketed between c.
2.8 kcal. yr BP and 2.2 kcal. yr BP. Elsewhere, more attention needs to be paid to
their chronological control (Russel *et al.*, 2003). Discrepancies in the stratigraphy,
gaps in the sedimentation, different time resolutions of pollen data, as well as local
climatic conditions relevant to length of the dry season, prevent the correlation of
changes at all the sites within a single event. The proximity of a lake or rivers could
maintain edaphic forests. Although reduction in the hydrological budget and lowering
lake levels have been documented simultaneously by geological studies (Servant and
Servant-Vildary, 2000), Upper Holocene aridity was not an irreversible climatic event.

Many sites indicate the return of wetter conditions during the last millennium, creating favorable conditions for new forest expansion again between 0.9 kcal . yr BP and 0.6 kcal . yr BP (Vincens *et al.*, 1996a, b; Elenga *et al.*, 2004).

On the other hand, the synchronous expansion of grasslands at c. 2.2 kcal . yr BP in lowlands, on the eastern side of Mount Cameroon, nearby lakes Barombi Mbo (Maley, 1992), Mboandong (Richards, 1986), and Ossa (Reynaud-Farrera, *et al.*, 1996) is in good correspondence with the Iron Age. But, whether such expansion was climatically controlled or human-induced does not reach a consensus among different specialists and remains a matter of speculation (Maley, 1992; Schwartz, 1992). High-resolution diatom analysis of the Ossa core provides new, interesting data for the climatic interpretation of the complexity of that event. In striking correspondence with the 2.7 kcal . yr BP North Atlantic cold event (Bond *et al.*, 1997), in Africa an increase in lake level and rainfall is bracketed by two strong amplitude shifts to drier conditions. The first was interpreted as a decrease in rainfall. The second lasted a few hundred years when decrease in water level and stronger input of allochthonous diatoms transported from Lake Chad indicate reinforced northern winds responsible for stronger aridity between 2.3 kcal . yr BP and 2 kcal . yr BP. The high-amplitude oscillations registered at Ossa suggest that a further high-resolution study of the nearby high-elevation Bambili lake (Stager and Anfang-Stutter, 1999) would be worthwhile. Synchronous vegetation change between a Caesalpiniaceae/*Lophira*-dominated forest before 2.5 kcal . yr BP replaced by an *Alchornea/Macaranga* open forest after 2.3 kcal . yr BP could well be explained by an increased length in the dry season and possible effect of low subsidence and storms following movement of the ITCZ (Nguestop *et al.*, 2004). Such an interesting hypothesis might receive more support when higher resolution pollen data have been obtained. An increased proportion of *Elaeis guineense* (oil palm) in the Ossa pollen record occurred simultaneously, which brings to the fore the fact that several dry episodes—also expressed at Kitina and Sinnda—created consecutive openings through the equatorial forest. These gaps offered a possible direct route for the Bantu migrations from Cameroon to the south.

5.4.4   The oil palm, evidence for human impact?

Evidence for human impact on past vegetation can come from the findings of abundant pollen from used or cultivated plants—such as *Dioscorea* (the yam), *Elaeis guineensis* (oil palm), and *Canarium schweinfurthii*, the last two producing fruits and seeds extensively used as a source of edible oil (Shaw, 1976). *Dioscorea* pollen has been identified in modern surface samples at the contact between the forest and the savanna at Kandara (Vincens, pers. commun.), but no evidence for its fossil pollen has been found, whereas fossil *Elaeis guineensis* has been reported at many sites. At Mboandong (Cameroon), a forest decrease associated with a significant grass increase, and followed by a well-marked increase in oil palm pollen (up to 20%) at 2.4 kcal . yr BP, has been interpreted as evidence for human impact on the basis that this date was close enough to that of the Nok archeological Iron Age site. It was suggested that forest clearance could be attributed to iron technology (Richards, 1986). But, in the pollen diagram opening of the forest occurred first. Pollen of

Canarium schweinfurthii, clearly distinguished from other Burseraceae—notably *Aucoumea* (Harley and Clarkson, 1999; Sowunmi, 1995)—was recorded at c. 5 kcal. yr BP, suggesting an earlier presence of humans nearby the fossil pollen site. Early evidence of oil palm pollen is known since the Tertiary in the Niger delta (Zeven, 1964), and was reported in Upper Pleistocene marine sediments offshore the Niger River (Dupont and Weinelt, 1996). However, it is only during the last 3,000 years that the proportion of *Elaeis guineensis* pollen reached significant values (up to 24%) in the Niger delta (Sowunmi, 1999). As a "palm belt", present day natural distribution of the oil palm follows the gulf of Guinea and largely penetrates inside the Congo Basin. *Elaeis guineensis* is an heliophytic pioneer species, also fire-resistant (Swaine, 1992), which occurs naturally in a great variety of habitats inside the rainforest including swamp and at its periphery (Letouzey, 1978). All these characteristics largely contribute to the difficulty of interpreting their fossil record. The pattern of occurrences and abundance of *Elaeis guineensis* pollen through time is not similar at all the investigated sites, and this led to controversial interpretation. In Ghana, archeological remains indicate an earliest use at 5.8 kcal. yr BP (Shaw, 1976), but its fossil pollen became abundant only after 3 kcal. yr BP in the Niger delta. The common pattern in the three Cameroon sites, starting with scattered *Elaeis guineensis* occurrences from c. 4 kcal. yr BP to 3 kcal. yr BP, followed by a significant increase at c. 2 kcal. yr BP (Richards, 1986; Reynaud Farrera *et al.*, 1996; Maley and Brenac, 1998) is remarkable. At Barombi Mbo the peak of oil palm pollen (2.9 kcal. yr BP to 2.4 kcal. yr BP) corresponds to the 2.75 kcal. yr BP crisis (Van Geel *et al.*,1998) or "dramatic forest decline". It occurs a few hundred years earlier than the peak of grasses (2.7 kcal. yr BP to 2.2 kcal. yr BP) and much earlier than the Bantu invasion (2.2 kcal. yr BP to 2 kcal. yr BP) (Schwartz, 1992; Maley, 2001). Therefore, *Elaeis guineensis* increased in response to increased aridity happening in the rainforest a few centuries before the spreading of oil palm by Bantu speakers. In contrast, several papers discussed the past record of *Elaeis guineensis* in West Africa (see Sowunmi, 1995 for an exhaustive literature), possibly enhanced by Bantu speakers (Schwartz, 1992; Maley, 2001). In Gabon, Neolithic populations had occupied the savanna of the Ogoué Valley since c. 3.7 kcal. yr BP and the main expansion of iron-smelting dates at c. 2.2 kcal. yr BP (Oslisly and Fontugne, 1993; Oslisly, 2001). Abundance of oil palm pollen happened at a coastal site around 3 kcal. yr BP, following forest reduction 1,000 years later. At Kamalete, inland, it is registered much later, during the warm medieval period from 1.2 kcal. yr BP to 0.3 kcal. yr BP. There, the oil palm phase is registered simultaneously within the pioneer forest phase, showing a significant peak at 0.9 kcal. yr BP, during a gap in human occupation (Ngomanda, in press). At the Songolo mangrove site (Congo coast line), scattered occurrences of *Elaeis guineensis* pollen have been found from 3.6 kcal. yr BP onward (Elenga *et al.*, 2001), but no significant increase has been observed until the last centuries, neither at Ngamakala and Bilanko nor Kitina and Sinnda (Vincens *et al.*, 1998). At inland sites, climatic conditions may have been too dry for the development of oil palm (Elenga *et al.*, 1996, 2001). Occurrences of oil palm are apparently associated with increased grass pollen or a more deciduous character of the forests (Vincens *et al.*, 1996a, b). But, their peaks in abundance are not synchronous events and the geographical distribution pattern through time is far from being spatially and geographically consistent. Our available knowledge can be summarized by the following conclusions. First, there was a greater evidence of oil palm pollen during the last 3,000 years over the Guineo-Congolian rainforest. Second, simultaneity of forest openings and use of oil palm by human populations is not demonstrated at all the sites with sufficiently accurate time control or fine-resolution interval. Third, geographical progression of the use of oil palm, following Bantu migration, remains to be demonstrated. The oil palm has a very short reproductive cycle, and it can indicate simultaneous use by humans following—by a few years—anthropogenic or natural forest perturbation. Fourth, proof of anthro-pologically enhanced modifications of the rainforest, prior to the last few centuries, has yet to be provided.

5.4.5   The last historical period (Gabon)

Important past vegetation changes inside the moist evergreen Guineo-Congolian rainforest have been evidenced in the historical period within core sediments obtained from Lake Kamalete in central Gabon (Ngomanda *et al.*, 2005). Within the forest of the Lopé National Park, Lake Kamalete is located inside a 60 km long strip of savanna interrupting the Maranthaceae forest, whereas the closed canopy forest stands at c. 20 km west of the lake (Figure 5.1). Situated at the end of a valley, not far from the Ogooué River, the shallow Lake Kamalete is surrounded by a mosaic of savanna and isolated fragments of Marantaceae forest. Sedges (Cyperaceae) and ferns colonize its shoreline (Figure 5.13, see color section). In this area, precipitation oscillates around 1,500 mm/yr, a low value for Gabon due to the rain shadow effect of the Cristal Mounts on the western side. But, present day meteorological values show great inter-annual variability related to variable timing and duration of the dry summer season (June to September) when the Inter Tropical Convergence Zone (ITCZ) moved to the northern hemisphere. During the dry season, dense cloud cover maintains high relative humidity and slightly lower temperature when sea surface temperature drops out.

A c. 4-m core recovered from Lake Kamalete (0°43′S, 11°46′E, 350 m a.s.l.) was dated by four conventional radiocarbon dates, providing a basal age of c. 1.3 kcal. yr BP. Fine-resolution pollen analysis of the order of one or two decades documents important past vegetation changes and detailed composition of the moist evergreen Guineo-Congolian rainforest (Ngomanda *et al.*, 2005). The original fossil pollen sequence includes 80 samples and counts of 124 identified pollen taxa. A simplified version is presented here (Figure 5.14).

In this context, fern spores, grasses, and sedges produced the highest quantities of pollen. Grass pollen percentages average 70% of the total pollen sum (after spores, aquatics, and sedges had been excluded) indicating persistence of a forest–savanna mosaic at the site. Because significant pollen counts had been achieved at each level, a detailed diagram could be drawn for forest arboreal pollen (Figure 5.15).

Among palynologists it is assumed that pollen abundance is related to abundance of trees, rather than to direct yearly pollen production, although this might deserve consideration when 1-cm sampling represents 2 or 3 years of pollen deposition.

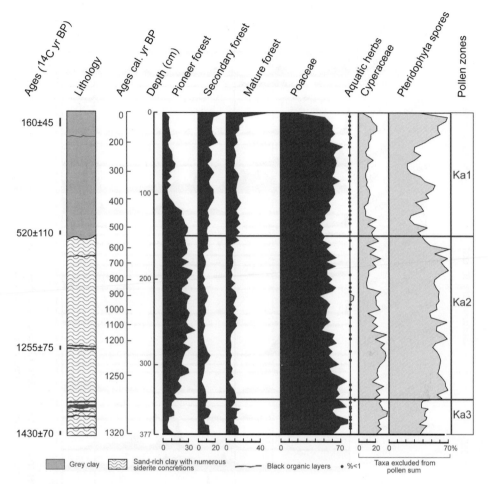

Figure 5.14. Synthetic pollen diagram from Kamalete, Gabon (after Ngomanda *et al.*, 2005), presented according to depth with corresponding time scale (% calculated versus pollen sum including all identified taxa, Cyperaceae, and spores).

Significant changes in abundance and floristic composition of the regional forest occurred at Kamalete during the last c. 1.5 kcal . yr BP. Successive fluctuations in dominant taxa are documented by trends that lasted a few hundred years or more. Before c. 1325 to 1240 cal yr BP, the first forest phase includes high representation of *Cnestis* (Connaraceae), *Celtis*, *Bosqueia* (Moraceae), Caesalpiniaceae, and *Lophira alata* pollen. It ends rather abruptly, being replaced by a disturbed forest dominated by pollen from trees—such as *Tetrorchidium, Macaranga, Musanga, Alchornea*—that dominate in opening gaps within forests today, whereas mature forest trees simultaneously decrease. The disturbance phase lasted about 700 years (c. 1,250 to 550 cal yr BP) and may have been caused by stronger winds, changes in rainfall regime, or human impact. Indeed, Iron Age settlements occupied the Lopé region since

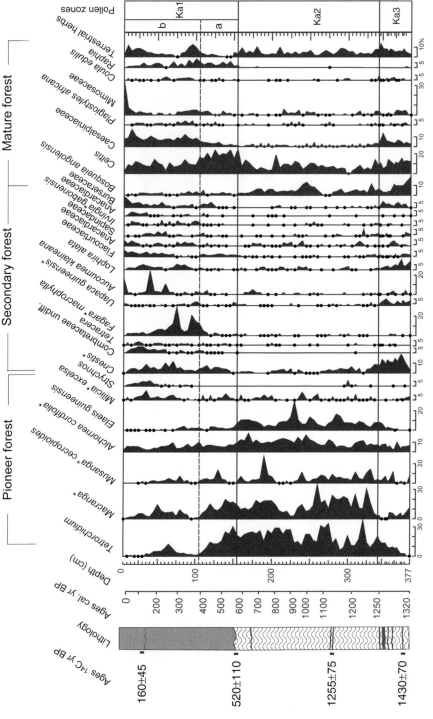

Figure 5.15. Detailed pollen diagram of the forest trees from Kamalete, Gabon (after Ngomanda *et al.*, 2005), presented as in Figure 5.12 (% calculated versus pollen sum after excluding Cyperaceae, spores, and Poaceae). Symbols for lithology as in Figure 5.12. * = pollen types < 1%.

2.6 kcal. yr BP—that is, long before the disturbed forest phase documented at Kamalete. But, from 1,400 to 800 yr BP the number of archeological sites in Gabon were
fewer than before, a reduction attributed to a "population crash" (Oslisly, 2001). If
there was any significant human impact on vegetation due to Iron Age population,
one would expect it to be less important when the population was smaller than before
or after. The opposite pattern being observed, Iron Age impact was considered not
responsible for the observed disturbed forest (Ngomanda *et al.*, 2005). The period of
maximum forest disturbance includes the time span of the "Medieval Warm Period"
(MWP, 900–600 cal. yr BP), although the disturbed forest phase in Gabon (c.
1.25 kcal. yr BP to 550 cal. yr BP) started before the beginning of the MWP climatic
anomaly. Well-known in the northern hemisphere (Bradley, 2000), the "Little Ice
Age" (c. 600–200 cal. yr BP) that followed the MWP had a major effect on the
evergreen forest of Gabon (Ngomanda *et al.*, in press), and is probably present in
records from Cameroon (Nguetsop *et al.*, 2004), Congo (Elenga *et al.*, 1996), East
Africa (Verschuren *et al.*, 2000), and Lake Victoria (Stager *et al.* 2005). In Gabon, a re-
expansion of the forest took place at c. 550 cal. yr BP, after the disturbance phase. At
this time *Celtis* and *Raphia* dominate, indicating a colonizing stage resembling the
modern one described for Kandara (Vincens *et al.*, 2000). Two hundred years later,
the site was both a "secondary forest" and a well-diversified mature forest occupied by
Lophira, Fagara (Rutaceae), *Pycnanthus, Irvingia, Plagiostyles*, Caesalpiniaceae, and
Burseraceae—bearing floristic affinities with the present day forest in the region (Jolly
et al., 1996). At Kamalete the "Little Ice Age" period was favorable for forest
development. In Gabon the re-establishment of a well-diversified mature forest
dates from 400 years at most. Since 250 years ago, *Aucoumea* pollen frequencies
exhibit three sharp peaks, indicating significant recoveries after forest cutting for
timber. We have already discussed that the origin of mature forest seems no older than
a few hundred years, in strong contradiction with the long-standing belief of rainforest
stability over time. Past hydrological changes across the equatorial tropics are partly
explained by amplified shifts in mean latitude of the ITCZ (Haug *et al.* 2001). The
sensitivity of the rainforest to short-term past climatic forcing is clearly demonstrated
and suggests the great impact that greenhouse warming can have in the future.

5.5   CONCLUSIONS

For many decades, palynological studies in the tropics were limited by the lack of
knowledge in tropical pollen morphology. Important reference collections of African
tropical plants have been made in Durham (NC), Montpellier and Marseille (France),
Germany, Nigeria (Africa), the United Kingdom, etc. Description of the pollen
morphology for most common trees is available in several publications and pollen
atlases (Maley, 1970; Bonnefille 1971a, b; Sowunmi, 1973, 1995; Caratini *et al.*, 1974;
Bonnefille and Riollet, 1980; Salard-Cheboldaeff, 1980, 1981, 1982), from which some
pollen photographs are now being included in the African Pollen Data Base. Therefore, pollen analysis in the African tropics is feasible, although it still requires lengthy
training. The bias between pollen assemblages and plant associations will remain. But,

interpretation of fossil data will be facilitated by modern pollen rain studies concerning many types of rainforest, encountered under different ecological conditions. Taxonomic composition and taxa frequencies of pollen assemblages can be used to characterize habitat types, despite the absence of pollen from important tropical families. Modern pollen assemblages help us to interpret tropical forest dynamics in the past, as forced by climatic changes. They have also been successfully used in elaborating a biome approach to modeling past global vegetation (Jolly *et al.*, 1998), the quality of which depends upon homogeneity in sampling and completeness along continuous climatic gradients. With the major threat of rainforest disappearance in mind, completing modern pollen data from all the African forests constitutes an urgent task.

Fossil pollen data from different vegetation types inside the Guineo-Congolian rainforest of the African continent—now available—document strong modifications in the floristic composition of both evergreen, semi-evergreen forests, and associated mixed types through time, the most important occurring during the last glacial maximum. Proof exists that rainforest persisted during glacial time, at two investigated sites, although it included more semi-evergreen taxa, a few highland taxa (*Olea* and *Podocarpus*) and was more open. Site-based maps of global vegetation at the LGM including Africa (Elenga *et al.*, 2000c) were used for a comparison between different vegetation models (Kohfeld and Harrison, 2000) addressing land surface feedbacks (Clausen, 1997) or the effect of lower CO_2 in global modelling. Although it has been generally considered that "tropical forests in Africa, Australia and Asia were partly replaced by more open vegetation" (Harrison and Prentice, 2003), the persistence of rainforest during the last glacial period, even when lower CO_2 content in the atmosphere was not favorable to tree growth, is clearly attested at two tropical sites located within the rainforest region. A greater extension of forest can even be postulated from abundant fossil wood remains.

Lowlands on the eastern side of Mount Cameroon and near the Congo River have been proved to be rainforest refuges during the LGM. Clearly, forest refuges were not geographical areas or spots bearing forests of similar composition to those of today, and there may have been many other refuges of different groups of species that remain to be found. Throughout the inter-tropical region ecological constraints related to hydrological constraints may have varied. Different species may have found refuge in different localities. Direct evidence can be provided by further research on fossil pollen or macrobotanical remains.

During the Holocene, significant variations in tree pollen assemblage percentages have also been observed at all the ten investigated sites presented in this chapter. These are not induced by variations in the CO_2 content of the atmosphere, which stayed fairly stable throughout this period. Replacement of dominant taxa or fluctuations in pollen abundance continuously occurred, suggesting the dynamic behavior and successive replacement and dominance within the rainforest forced by variations in the hydrological system. Some of these modifications reflect changes in the amount and distribution of rainfall in the tropics, a pattern determined by several causes not unanimously explained yet. Simultaneity of forest changes at different localities cannot be firmly established until homogeneity in resolution is reached and the lower Holocene is more fully documented at more than one site. A strong-amplitude century-scale arid/humid/arid oscillation in the interval 2.5 kcal . yr BP to 2 kcal . yr BP is only found north of the equator. This change corresponds fairly well to the Iron Age, but its reversal character seems to be reflecting natural climatic change rather than being attributed to human impact. Evidence for disturbed vegetation in Central Gabon during the warm medieval period, when the influence of the Iron Age decreased, is followed by a different forest composition from 600 to 200 cal yr BP when wetter conditions seem to have been a response to the "Little Ice Age". That the tropical world was not a stable environment in the past is now amply demonstrated. Permanent changes in ecological conditions may have triggered speciation leading to high floristic diversity. The great sensitivity of the rain forest to any short-term global climatic changes draws attention to the consequences of future climatic changes induced by the greenhouse effect on the rainforest. The modern composition of the different types of rainforests only dates from a few centuries or 1,000 years at most at all the investigated sites. This bears some consideration to rainforest conservation and biodiversity issues.

Acknowledgements

My warmest thanks are due to A. Ngomanda, J. Maley, D. Jolly, A. Vincens, D. Wirrmann for providing valuable information and stimulating discussion. I also thank H. Hooghiemstra and an anonymous reviewer whose comments contributed to the improvement of this manuscript, J. J. Motte for precious help in preparing the drawings, and G. Buchet for taxomical precision on pollen identification.

5.6 REFERENCES

Achoundong, G., Youta Happi, J., Guillet, B., Bonvallot, J., and Kamganag Beyla, V. (2000) Formation et evolution des recrus sur savanes. In: S. Servant-Vildary and M. Servant (eds.), *Dynamique à long terme des écosystèmes forestiers intertropicaux* (pp. 31–41). UNESCO, Paris [in French].

Alexandre, A., Meunier, J. D., Lezine, A. M., Vincens, A., and Schwartz, D. (1997) Phytoliths: Indicators of grassland dynamics during the late Holocene in intertropical Africa. *Palaeogeography, Palaeoclimatology, Palaeoecology* **136**, 213–229.

Anupama, K., Ramesh, B. R., and Bonnefille, R. (2000) The modern pollen rain from the Biligirirangan–Melagiri hills of Southern Eastern Ghats, India. *Review of Palaeobotany and Palynology* **108**, 175–196.

Aubreville, A. (1951) Le concept d'association dans la forêt dense équatoriale de la basse Côte d'Ivoire. *Mémoire de la Société botanique de France* **1950–51**, 145–158 [in French].

Aubreville, A. (1957–58) A la recherche de la forêt en Côte d'Ivoire. *Bois Forêts Trop.* **56**, 17–32 (1957), **57**, 12–27 (1958) [in French].

Aubreville, A. (1967) La forêt primaire des montagnes de Belinga. *Biologica Gabonica* **3**, 95–112 [in French].

Barboni, D. and Bonnefille, R. (2001) Precipitation signal in modern pollen rain from tropical forests, South India. *Review of Palaeobotany and Palynology* **114**, 239–258.

Barboni, D., Bonnefille, R., Prasad S., and Ramesh, B. R. (2003) Variation in modern pollen rain from tropical evergreen forests and the monsoon seasonality gradient in S W India. *Journal of Vegetation Science* **14**, 551–562.

Battarbee, R., Gasse, F., and Stickley, C. E. (2004) *Past Climatic Variability through Europe and Africa* (Developments in Paleoenvironmental Research Series 6, 638 pp.). Springer-Verlag, Berlin.

Bengo, M. D. and Maley, J. (1991) Analyses des flux polliniques sur la marge sud du Golfe de Guinée depuis 135.000 ans. *Comptes Rendus de l'Académie des Sciences, Paris, série II* **313**, 843–849 [in French].

Bertaux, J., Sifeddine, A., Schwartz, D., Vincens, A., and Elenga, H. (1996) Enregistrement sédimentologique de la phase sèche d'Afrique équatoriale c. 3000 BP par la spectrométrie IR dans les lacs Sinnda et Kitina (Sud-Congo). *Symposium "Dynamique à long terme des écosystèmes forestiers intertropicaux"* (pp. 213–215). CNRS-ORSTOM, Bondy, France [in French].

Bond, G., Showers, W., Cheseby, M., Lotti, R., Almasi, P., de Menxal, P., Prime, P., Cullen, H. H., Ajolas, I., Bonani, G. *et al.* (1997) A persuasive millennia-scale cycle in North Atlantic Holocene and Glacial climates. *Science* **278**, 1257–1266.

Bonnefille, R. (1971a) Atlas des pollens d'Ethiopie. Principales espèces des forêts de montagne. *Pollen et spores* **13/1**, 15–72 [in French].

Bonnefille, R. (1971b) Atlas des pollens d'Ethiopie. Pollens actuels de la basse vallée de l'Omo, récoltes botaniques 1968. *Adansonia* (2) **11/3**, 463–518 [in French].

Bonnefille, R. and Chalié, F. (2000) Pollen-inferred precipitation time-series from equatorial mountains, Africa, the last 40 kyr BP. *Global and Planetary Change* **26**, 25–50.

Bonnefille, R. and Riollet, G. (1980) *Pollens des savanes d'Afrique orientale* (140 pp.). CNRS, Paris [in French].

Bonnefille, R., Roeland, J-C., and Guiot, J. (1990) Temperature and rainfall estimates for the last 40,000 years in equatorial Africa. *Nature* **346**, 347–349.

Bonnefille, R., Chalié, F., Guiot, J., and Vincens, A. (1992) Quantitative estimates of full glacial temperatures in equatorial Africa from palynological data. *Climate Dynamics* **6**, 251–257.

Bonnefille, R., Buchet, G., Friis, I., Kelbessa E., and Mohammed, M. U. (1993) Modern pollen rain on an altitudinal range of forests and woodlands in South West Ethiopia. *Opera Botanica* **121**, 71–84.

Bonnefille, R., Anupama, K., Barboni, D., Pascal, J. P., Prasad, S., and Sutra, J. P. (1999) Modern pollen spectra from tropical South India and Sri Lanka, altitudinal distribution. *Journal of Biogeography* **26**, 1255–1280.

Bradley, R. (2000) 1000 years of climate change. *Science* **288**, 1353–1355.

Bremond, L., Alexandre, A., Hély, C., and Guiot, J. (2005) A phytolith index as a proxy of tree cover density in tropical areas: Calibration with Leaf Area Index along a forest–savanna transect in southeastern Cameroon. *Global and Planetary Change* **45**, 277–293.

Caballe, G. (1986). Sur la biologie des lianes ligneuses en forêt gabonaise, Thèse Fac. Sci. Montpellier (341 pp.) [in French].

Caratini, C., Guinet, Ph., and Maley, J. (eds.) (1974) *Pollens et spores d'Afrique tropicale* (Travaux Documentaires Geographie tropicale, 282 pp.). CNRS/CEGET, Bordeaux, France [in French].

Claussen, M. (1997) Modeling bio-geophysical feedback in the African and Indian monsoon region. *Climate Dynamics* **13**, 247–257.

Cournac, L., Dubois, M-A., Chave, J., and Riéra, B. (2002) Fast determination of light availability and leaf area index in tropical forests. *Journal of Tropical Ecology* **18**, 295–302.

Cusset, G. (1987) *La flore et la végétation du Mayombe congolais. Etat des connaissances* (Rapport, 46 pp.). UNESCO, Paris [in French].

Dechamps, R., Lanfranchi, R., Le Cocq, A., and Schwartz, D. (1988a) Reconstitution d'environnements quaternaires par l'étude de macrorestes végétaux (Pays Bateke, R.P. du Congo). *Palaeogeography, Palaeoclimatology, Palaeoecology* **66**, 33–44 [in French].

Dechamps, R., Lanfranchi, R., and Schwartz, D. (1988b). Découverte d'une flore forestière miholocène (5800–3100 yr B.P.) conservée in situ sur le littoral ponténégrin (R.P. du Congo). *Comptes Rendus de l'Académie des Sciences (Paris), série II* **306**, 615–618 [in French].

Descoings, B. (1960) *Les steppes loussékés de la zone de Gabouka (Plateau Batéké, République Congo-Brazzaville)* (34 pp.). ORSTOM, Brazzaville, Congo [in French].

Descoings, B. (1976) Notes de phyto-écologie équatoriale, 3: Les formations herbeuses de la vallée de la Nyanga (Gabon). *Adamsonia* (2), **15**, 307–329 [in French].

Dupont, L. and Behling, H. (in press, on line) Land–sea linkages during deglaciation: High-resolution records from the eastern Atlantic off the coast of Namibia and Angola (ODP Site 1078). *Quaternary International*.

Dupont, L. and Weinelt, M. (1996) Vegetation history of the savanna corridor between the Guinean and the Congolian rain forest during the last 150,000 years. *Vegetation History and Archaeobotany* **5**, 273–292.

Elenga, H. (1992) Végétation et climat du Congo depuis 24 000 ans B.P. Analyse palynologique de séquences sédimentaires du Pays Bateke et du littoral. Unpublished thesis, University Aix-Marseille III (238 pp.) [in French].

Elenga, H., Vincens, A., and Schwartz, D. (1991) Présence d'éléments forestiers montagnards sur les plateaux Batéké (Congo) au cours du Pléistocène supérieur. Nouvelles données palynologiques. *Palaeoecology of Africa* **22**, 239–252 [in French].

Elenga, H., Schwartz, D., and Vincens, A. (1992) Changements climatiques et action anthropique sur le littoral congolais au cours de l'Holocène. *Bulletin de la Société géologique de France* **163**, 83–90 [in French].

Elenga, H., Schwartz, D., and Vincens, A. (1994) Pollen evidence of Late Quaternary vegetation and inferred climate changes in Congo. *Palaeogeography, Palaeoclimatology, Palaeoecology* **109**, 345–356.

Elenga, H., Schwartz, D., Vincens, A., Bertaux J., de Namur, C., Martin, L., Wirrmann, D., and Servant, M. (1996) Diagramme pollinique holocène du lac Kitina (Congo): mise en évidence de changements paléobotaniques et paléoclimatiques dans le massif forestier du Mayombe. *Comptes Rendus de l'Académie des Sciences, Paris*, série IIa, **323**, 403–410 [in French].

Elenga, H., de Namur, C., Vincens A., and Roux, M. (2000a) Use of plots to define pollen–vegetation relationships in densely forested ecosystems of Tropical Africa. *Review of Palaeobotany and Palynology* **112**, 79–96.

Elenga, H., de Namur, C., and Roux, M. (2000b) Etudes des relations pollen-végétation dans les formations forestières du Sud Congo (Massif de Mayombe et forêts littorales): apport de la statistique. *Symposium "Dynamique à long terme des écosystèmes forestiers intertropicaux", Paris, 20–22 mars 1996* (pp. 121–132) [in French].

Elenga, H., Peyron, O., Bonnefille, R., Jolly, D., Cheddadi, R., Guiot, J., Andrieu, V., Bottema, S., Buchet, G., and de Beaulieu, J. L. (2000c) Pollen-based biome reconstruction for Southern Europe and Africa 18,000 years ago. *Journal of Biogeography* **27**, 621–634.

Elenga, H., Vincens, A., Schwartz, D., Fabing, A., Bertaux J., Wirrmann, D., Martin, L., and Servant, M. (2001) Le marais estuarien de la Songolo (Sud Congo) à l'Holocène moyen et récent. *Bulletin de la Société géologique de France* **172**, 359–366 [in French].

Elenga, H., Maley J., Vincens, A., and Farrera, I. (2004) Palaeoenvironments, palaeoclimates and landscape development in Atlantic Equatorial Africa: A review of key sites covering the last 25 kyrs. In: R. W. Battarbee, F. Gasse, and C. E. Stickley (eds.), *Past Climate Variability through Europe and Africa* (pp. 181–198). Springer-Verlag, Dordrecht, The Netherlands.

Evrard, C. (1968) Recherches écologiques sur le peuplement forestier des sols hydromorphes de la Cuvette centrale congolaise. *Publications INEAC, sér. sci.* **110**, 1–285 [in French].

Flenley, J. (1979) *The Equatorial Rain Forest: A Geological History* (162 pp.). Butterworths, London.

Fredoux, A. and Maley, J. (2000) Le contenu pollinique de l'atmosphère dans les forêts du Sud Cameroun près de Yaoundé. Résultats préliminaires. *Symposium "Dynamique à long terme des écosystèmes forestiers intertropicaux", Paris, 20–22 mars 1996* (pp. 139–148) [in French].

Giresse, P., Maley, J., and Kelts, K. (1991) Sedimentation and palaeoenvironment in crater lake Barombi Mbo, Cameroon during the last 25,000 years. *Sedimentary Geology* **71**, 151–175.

Giresse, P., Maley, J., and Brénac, P. (1994) Late Quaternary palaeoenvironments in the lake Barombi Mbo (West Cameroon) deduced from pollen and carbon isotopes of organic matter. *Palaeogeography, Palaeoclimatology, Palaeoecology* **107**, 65–78.

Guillet, B., Achoundong, G., Youta Happi, J., Kamgang Kabeyene Beyala, V., Bonvallot, J., Riera, B., Mariotti, A., and Schwartz, D. (2001) Agreement between floristic and soil organic carbon isotope ($^{13}C/^{12}C$, ^{14}C) indicators of forest invasion of savannas during the last century in Cameroon. *Journal of Tropical Ecology* **17**, 809–832.

Hall, J. B. and Swaine, M. D. (1981) *Distribution and Ecology of Vascular Plants in a Tropical Rain Forest: Forest Vegetation in Ghana* (383 pp.). Junk, The Hague.

Harley, M. M. and Clarkson, J. J. (1999) Pollen morphology of the African Burseraceae and related genera. *Palaeoecology of Africa*, 225–241.

Harrison, S. and Prentice, C. I. (2003) Climate and CO_2 controls on global vegetation distribution at the last glacial maximum: Analysis based on palaeovegetation data, biome modelling and palaeoclimate simulations. *Global Change Biology* **9**, 983–1004.

Haug, G. H., Hughen, K. A., Sigman, D. M., Peterson, L. C., and Rôhl, U., (2001) Southward migration of the intertropical convergence through the Holocene. *Science* **293**, 1304–1308.

Jahns, S. (1996). Vegetation history and climate changes in West Equatorial Africa during the the late Pleistocene and Holocene, based on a marine pollen diagram from the Congo fan. *Vegetation History and Archaeobotany* **5**, 207–213.

Jolly, D., Bonnefille, R., Burcq, S., and Roux, M. (1996) Représentation pollinique de la forêt dense humide du Gabon, tests statistiques. *Comptes Rendus de l'Académie des Sciences Paris, série IIa*, **322**, 63–70 [in French].

Jolly, D., Prentice, C., Bonnefille, R., Ballouche, A., Bongo, M., Brenac, P., Buchet, G., Burney, D., Casey, J. P., and Cheddadi, R. (1998) Biome reconstruction from pollen and plant macrofossil data for Africa and the Arabian peninsula at 0 and 6000 years. *Journal of Biogeography* **25**, 1007–1027.

Koechlin, J. (1961) *La végétation des savanes dans le Sud de la République du Congo* (Mémoire 1, 305 pp.). ORSTOM, Paris [in French].

Kohfeld, K. E. and Harrison, S. P. (2000) How well can we simulate past climates? Evaluating the models using global environmental data sets. *Quaternary Science Reviews* **19**, 321–346.

Lebrun, J. P. and Gilbert, G. (1954) Une classification écologique des forêts du Congo. *Publications INEAC, sér. sci.*, **63**, 1–89 [in French].

Letouzey, R. (1968). *Etude phytogéographique du Cameroun* (508 pp.). Lechevalier, Paris [in French].

Letouzey, R. (1978). Notes phytogeographiques sur les palmiers du Cameroun. *Adansonia* **18**, 293–325 [in French].

Letouzey, R. (1985) *Notice de la carte phytogéographique du Cameroun au 1:500 000*. IRA Yaoundé et Institut Carte internationale de la Végétation, Toulouse, France [in French].

Makany, L. (1976) Végétation des plateaux Téké. *Travaux Université Brazzaville* **1**, 301 [in French].

Maley, J. (1970) Contribution à l'étude du Bassin du Tchad. Atlas des pollens du Tchad. *Bulletin du jardin Botanique national belge* **40**(25), 29–48 [in French].

Maley, J. (1989) Late Quaternary climatic changes in the African rain forest: Forest refugia and the major role of sea surface temperature variations. In: M. Leinen and M. Sarnthein (eds.), *Paleoclimatology and Paleometeorology: Modern and Past Patterns of Global Atmospheric Transport* (NATO Advanced Sc. Inst. Series C, 282, pp. 585–616). Kluwer Academic, Dordrecht, the Netherlands.

Maley, J. (1991) The African rainforest vegetation and palaeoenvironments during late Quaternary. *Climatic Change* **19**, 79–98.

Maley, J. (1992) Commentaire à la note de D. Schwartz: Mise en évidence d'une péjoration climatique entre ca. 2500 et 2000 ans B.P. en Afrique tropicale humide. *Bulletin de la Société géologique de France* **163**, 363–365 [in French].

Maley, J. (1996) The African rainforest, main characteristics of changes in vegetation and climate from the Upper Cretaceous to the Quaternary. In: I. J. Alexander, M. D. Swaine, and R. Watling (eds.), *Essays on the Ecology of the Guinea-Congo Rainforest. Proceedings of the Royal Society of Edinburgh B Biological Sciences* **104**, 31–73.

Maley, J. (1997) Middle to late Holocene changes in Topical Africa and other continents: Palaeomousson and sea surface temperature variations. In: H. N. Dalfes, G. Kukla, and H. Weiss (eds.), *The Third Millennium BC Climate Change and Old World Collapse* (NATO ASI Series 149, pp. 611–640). Springer-Verlag, Berlin.

Maley, J. (2001) *Elaeis guineensis* Jacq. (oil palm) fluctuations in central Africa during the late Holocene: Climate or human driving forces for this pioneering species? *Vegetation History and Archaeobotany* **10**, 117–120.

Maley, J. and Brénac, P. (1998) Vegetation dynamics, palaeoenvironments and climatic changes in the forests of western Cameroon during the last 28,000 years B.P. *Review of Palaeobotany and Palynology* **99**, 157–187.

Maley, J. and Elenga, H. (1993). Le rôle des nuages dans l'évolution des paléoenvironnements montagnards de l'Afrique Tropicale. *Veille Climatique Satellitaire* **46**, 51–63 [in French].

Maley, J., Livingstone, D. A., Giresse, P., Thouveny, N., Brenac, P., Kelts, K., Kling, G., Stager, C., Haag, M., Fournier, M. *et al.* (1990). Lithostratigraphy, volcanism, palaeomagnetism and palynology of Quaternary lacustrine deposit from Barombi Mbo (West Cameroon): Preliminary results. *J. Volcanol. Geotherm. Res.* **42**, 319–335.

Marchant, R. and Hooghiemstra, H. (2004) Rapid environmental change in African and South American tropics around 4000 years before present: A review. *Earth Science Reviews* **66**, 217–260.

Marret, F., Scourse, J., Jansen, J. F., and Shneider, R. (1999) Changements climatiques et paléoocéanographiques en Afrique central Atlantique au cours de la dernière déglaciation: contribution palynologique. *Comptes Rendus de l'Académie des Sciences, Paris, série IIa* **329**, 721–726 [in French].

Mix, A., Bard, E., and Schneider, R. (2001) Environmental processes of the ice age: Land, oceans, glaciers (EPILOG). *Quaternary Science Reviews* **20**, 627–657.

Ngomanda, A., Chepstow-Lusty, A., Makaya, M., Schevin, P., Maley, J., Fontugne, M., Oslisly, R., Rabenkogo, N., and Jolly, D. (2005) Vegetation changes during the past

1300 years in Western Equatorial Africa: A high-resolution pollen record from Lake Kamalete, Lopê Reserve, Central Gabon. *The Holocene* **15**, 1021–1031.

Ngomanda, A., Jolly, D., Bentaleb, I., Chepstow-Lusty, A., Makaya, M., Maley, J., Fontugne, M., Oslisly, R., and Rabenkogo, N. (in press). Response of African lowland rainforest to hydrological balance changes during the past 15 centuries years in Gabon, Western Equatorial Africa. *Quaternary Research.*

Nguetsop, V. F., Servant-Vildary, S., and Servant M. (2004) Late Holocene climatic changes in West Africa: A high resolution diatom record from equatorial Cameroon. *Quaternary Science Reviews* **23**, 591–609.

Oslisly, R. (2001) The history of human settlement in the Middle Ogooué Valley (Gabon). In: W. Weber, L. J. T. White, A. Vedder, and L. Naughton-Treves (eds.), *African Rain Forest Ecology and Conservation* (pp. 101–118). Yale University Press, New Haven, CT.

Oslisly, R. and Fontugne M. (1993) La fin du stade neolithique et le début de l'Age du Fer dans la moyenne vallée de l'Ogoué au Gabon. Problèmes chronologiques et changements culturels. *Comptes Rendus de l'Académie des Sciences, Paris, série IIa* **31**, 997–1003 [in French].

Pascal, J.-P., Shyam Sunder, V., and Meher-Homji, V. M. (1984) *Forest Maps of South India, Sheet: Belgaum-Dharwar-Panaji* (Travaux Scientifiques et Techniques). Karnataka Forest Department and the French Institute, Pondicherry, India.

Prentice, I. C., Cramer, W., Harrison, S. P., Leemans, R., Monserup, R A., and Soloman, A. M. (1992) A global biome model based on plant physiology and dominance, soil properties and climate. *Journal of Biogeography* **19**, 117–134.

Reynaud-Farrera, I. (1995) Histoire des paléoenvironnements forestiers du Sud-Cameroun à partir d'analyses palynologiques et statistiques de dépôts holocènes et actuels. Unpublished thesis, University Montpellier II (230 pp.) [in French].

Reynaud-Farrera, I., Maley, J., and Wirrmann, D. (1996) Végétation et climat dans les forêts du Sud-Ouest Cameroun depuis 4770 ans B.P.: analyse pollinique des sédiments du lac Ossa. *Comptes Rendus de l'Académie des Sciences, Paris, série IIa* **322**, 749–755.

Richards, K. (1986) Preliminary results of pollen analysis of a 6000 year core from Mboandong: A crater lake in Cameroon. *Hull Univ. Geogr. Dep. Misc. Ser.* **32**, 14–28.

Richards, P. W. (1981) *The Tropical Rain Forest: An ecological study* (450 pp.). Cambridge University Press, London.

Runge, F. and Fimbel, R. A. (1999) Opal phytoliths as evidence of the formation of savanna islands in the rain forest of Southeast Cameroon. *Palaeoecology of Africa* **27**, 171–187.

Russell, J., Talbot, M., and Haskell, B. J. (2003) Mid-holocene climate change in Lake Bosumtwi, Ghana. *Quaternary Research* **60**, 133–141.

Salard-Cheboldaeff, M. (1980). Palynologie camerounaise. I Pollens de la mangrove et des fourrés arbustifs côtiers. *Compte rendus Congres National des Sociétés savantes* **106**, 125–136 [in French].

Salard-Cheboldaeff, M. (1981) Palynologie camerounaise. II Grains de pollen de la forêt littorale de basse altitude. Mangrove et des fourrés arbustifs côtiers. *Compte rendus Congres National des Sociétés savantes* **107**, 127–141 [in French].

Salard-Cheboldaeff, M. (1982) Palynologie camerounaise. III Grains de pollen de la forêt dense humide de basse et moyenne altitude. *Compte rendus Congres National des Sociétés savantes* **108**, 117–129 [in French].

Schwartz, D. (1992) Assèchement climatique vers 3000 B.P. et expansion Bantu en Afrique centrale atlantique: quelques réflexions. *Bulletin de la Société géologique de France* **163**, 353–361 [in French].

Schwartz, D., Guillet, B., and Dechamps, R. (1990) Etude de deux flores forestières mi-holocène (6000–3000 BP) et subactuelle (500 BP) conservées in situ sur le littoral pontenegrin (Congo). In: R. Lanfranchi and D. Schwartz (eds.), *Paysages quaternaires de l'Afrique centrale atlantique* (pp. 283–297). ORSTOM, Bondy, France [in French].

Schwartz, D., de Foresta, H., Mariotti, A., Balesdent, J., Massimba, J. P., and Girardin, C. (1996a) Present dynamics of the savanna–forest boundary in the Congolese Mayombe: A pedological, botanical and isotopic (^{13}C and ^{14}C) study. *Oecologia* **106**, 516–524.

Schwartz, D., Dechamps, R., Elenga, H., Mariotti, A., and Vincens, A. (1996b) Les savanes d'Afrique Centrale: des écosystèmes à l'origine complexe, spécifiques de l'Holocène supérieur. *Symposium "Dynamique à long terme des écosystèmes forestiers intertropicaux"* (pp. 179–182). CNRS-ORSTOM, Bondy, France [in French].

Shaw, T. (1976) Early crops in Africa: A review of the evidence. In: J. R. Harlan, J. M. J. de Wet, and A. B. L. Stemler (eds.), *Origin of African Plant Domestication* (pp. 107–153). Mouton, The Hague.

Sowunmi, M. A. (1973) Pollen of Nigerian Plants, I: Woody species. *Grana* **13**, 145–186.

Sowunmi, M. A. (1995) Pollen of Nigerian Plants, II: Woody species. *Grana* **34**, 120–141.

Sowunmi, M. A. (1999) The significance of the oil palm (*Elaeis guineensis* Jacq.) in the late Holocene environments of west and west central Africa: A further consideration. *Vegetation History and Archaeobotany* **8**, 199–210.

Stager, J. C. and Anfang-Stutter, R. (1999) Preliminary evidence of environmental changes at Lake Bambili (Cameroon, West Africa) since 24,000 BP. *Journal of Palaeolimnology* **22**, 319–330.

Stager, J. C., Ryves, D., Cumming, B. F., Meeker, D., and Beer J. (2005) Solar variability and the levels of Lake Victoria, East Africa, during the last millennium. *Journal of Palaeolimnology* **33**, 243–251.

Swaine, M. (1992) Charateristics of dry forests in West Africa and the influence of fire. *Journal of Vegetation Science* **3**, 365–374.

Van Geel, B., Van der Plicht, J., Kilian, M. R., Klaver, E. R., Kouvenberg, J. H. M., Renssen, H., Reynaud-Farrera I., and Waterbolk, H. T. (1998) The sharp rise of δ^{14}C ca 800 cal BC: Possible cause, related climatic tele connections and the impact on human environments. *Radiocarbon* **40**, 535–550.

Verschuren, D., Laird, K. R., and Cumming, B. F. (2000) Rainfall and drought in equatorial east Africa during the past 1,100 years. *Nature* **403**, 410–414.

Vincens, A., Chalié, F., Bonnefille, R., Guiot, J., and Tiercelin, J. J. (1993) Pollen-derived rainfall and temperature estimates from Lake Tanganyika and their implication for late Pleistocene water levels. *Quaternary Research* **40**, 343–350.

Vincens, A., Buchet, G., Elenga, H., Fournier, M., Martin, L., de Namur, C., Schwartz, D., Servant, M., and Wirrmann, D. (1994) Changement majeur de la végétation du lac Sinnda (vallée du Niari, Sud-Congo) consécutif à l'assèchement climatique holocène supérieur: apport de la palynologie. *Comptes Rendus de l'Académie des Sciences, Paris, série IIa* **318**, 1521–1526 [in French].

Vincens, A., Elenga, H., Schwartz, D., de Namur, C., Bertaux, J., Fournier, M., and Dechamps, R. (1996a). Histoire des écosystèmes forestiers du Sud-Congo depuis 6000 ans. *Symposium "Dynamique à long terme des écosystèmes forestiers intertropicaux"* (pp. 291–294), CNRS-ORSTOM, Bondy, France [in French].

Vincens, A., Alexandre, A., Bertaux, J., Dechamps, R., Elenga, H., Maley, J., Mariotti, A., Meunier, J. D., Nguetsop, F., Reynaud-Farrera, I. *et al.* (1996b) Evolution de la forêt tropicale en Afrique équatoriale atlantique durant les 4000 dernières années et héritage sur les paysages végétaux actuels. *Symposium "Dynamique à long terme des écosystèmes forestiers intertropicaux"* (pp. 287–289). CNRS-ORSTOM, Bondy, France [in French].

Vincens, A., Schwartz, D., Bertaux, J., Elenga, H., and de Namur, C. (1998) Late Holocene climatic changes in Western equatorial Africa inferred from pollen from lake Sinnda, Southern Congo. *Quaternary Research* **50**, 34–45.

Vincens, A., Schwartz, D., Elenga, H., Reynaud-Farrera, I., Alexandre, A., Bertaux, J., Mariotti, A., Martin, L., Meunier, J. D., and Nguetsop, F. (1999) Forest response to climatic changes in Atlantic Equatorial Africa during the last 4000 years BP and inheritance on the modern landscapes. *Journal of Biogeography* **26**, 879–885.

Vincens, A., Dubois, M. A., Guillet, B., Achoundong, G., Buchet, G., Kamgang Kabeyene Beyala, V., de Namur, C., and Riera, B. (2000) Pollen–rain–vegetation relationships along a forest–savanna transect in southeastern Cameroon. *Review of Palaeobotany and Palynology* **110**, 191–208.

White, F. (1983). *The Vegetation of Africa* (a descriptive memoir to accompany the UNESCO/AETFAT/UNSO Vegetation Map of Africa, 356 pp.). UNESCO, Paris.

Wirrmann, D. and Elouga, M. (1998) Discovery of an Iron Age site in Lake Ossa, Cameroonian Littoral Province. *Comptes Rendus de l'Académie des Sciences, Paris, série IIa* **323**, 139–146.

Wirrmann, D., Bertaux, J., and Kossoni, A. (2001) Late Holocene palaeoclimatic changes in Western Central Africa inferred from mineral abundance in dated sediments from Lake Ossa (Southwest Cameroon). *Quaternary Research* **56**, 275–287.

Youta Happi (1998) Arbres contre graminées; la lente invasion de la savane par la forêt au Centre-Cameroun. Unpublished thesis, Université Paris IV (237 pp.) [in French].

Zeven, A. C. (1964) On the origin of the oil palm (*Elaeis guineensis* Jacq.). *Grana Palynologica* **5**, 121–123.

6

Tropical environmental dynamics: A modeling perspective

R. Marchant and J. Lovett

6.1   INTRODUCTION

Records of fluctuating biotic and chemical characteristics from numerous sedimentary archives—including oceans (Guilderson *et al.*, 1994), continental lakes, bogs, swamps (Hooghiemstra and van der Hammen, 1998; Behling and Hoogiemstra, 2000; Haberle and Maslin, 1999), and ice caps (Thompson *et al.*, 1995)—demonstrate that tropical environments are dynamic at a range of timescales. One of the most recent extremes of these variations in geological time concerns the 100,000-yr glacial–interglacial cycles associated with the Quaternary geological period over the past 2.2 million years. Climate change involves massive reorganization of global climate systems with major impacts on ecosystem form and function that give rise to complex interactions between the atmosphere, geosphere, hydrosphere, and biosphere (Kohfeld and Harrison, 2000). Such changes can be documented by accessing sedimentary archives, whereas the nature and implications of such changes can be investigated through modeling. Numerous modeling approaches address different components of the Earth system. The majority of these initiatives have been developed within the temperate regions, particularly North America and Europe, with relatively few studies focused on the tropics. Indeed, as we will see, the modeled environmental history of the tropics remains poorly resolved despite its increasing importance in understanding global climates (Marchant and Hooghiemstra, 2004), biogeochemical cycles (Prentice *et al.*, 1996), developing biogeographical theory (Tuomisto and Ruokolainen, 1997), and understanding issues concerned with biodiversity and human–environment interactions (Marchant *et al.*, 2004b). Modeling climate change impacts in tropical environments represents a special challenge for computer climate models, in part due to the sharp relief of the mountain areas, where a single 1° grid-cell may encompass a wide range of environmental and climate gradients. These difficulties are compounded by the poor availability of marine and terrestrial data important for accurate simulations (Valdes, 2000). Nevertheless, as the complexity of the tropical

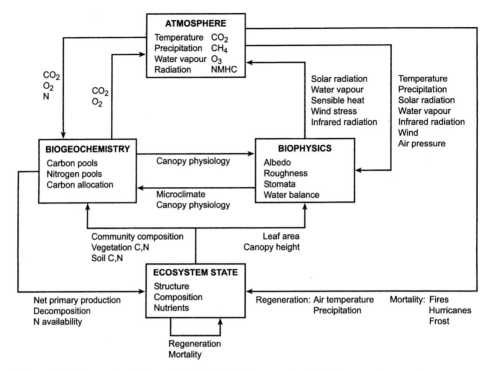

Figure 6.1. Biogeochemical and physical links between the Earth's atmosphere and ecosystems that need to be accounted for and parameterized within an Earth system model.

environmental system, and its importance to contributing to global climate models (GCMs), is being realized, models take an ever increasing number of parameters into account (Figure 6.1). However, such complex models require significant computing power and can be relatively static in time (non-dynamic). A compromise is available in models of intermediate complexity that can simulate climate evolution over thousands of years but use a relatively large grid-cell or a reduced number of inputs to attain a concomitant reduction in computational time.

This chapter will provide an overview of these different approaches, review the present understanding, and identify future areas for research development by investigating two main areas of modeling: first, biosphere models focusing on vegetation change, and the links from these to biogeochemical fluxes; second, climate modeling. First, we need to investigate the variety of modeling approaches—understanding model limitation is critical for future development and useful application (Peng, 2000). Such application may concern differences in the timing, intensity, and duration of the seasons that can have huge impacts on human prosperity, health, and surrounding environment. For example, the winter of 1982/1983 was exceptional: the dry seasons in Peru and Chile were very wet and the rainy season in Indonesia was extremely dry. Modeling can be focused to predict when a rainy season might fail, or when flooding or temperature extremes might be likely. However, if we cannot predict the weather next week, why should we trust climate predictions for next season, or in 25 years' time, or indeed 21,000 years ago! We can however say something useful about the climate trends, how certain components of Earth's system respond to such trends, and how these trends can influence our future climate associated environment and resources.

There are two basic modeling approaches: inverse and forward modeling techniques. In the former, comparisons rely on establishing empirical relations between modern and past environment observations through a transfer function (Kohfeld and Harrison, 2000)—that is, geological data are translated into climatic (e.g., mean monthly temperature, precipitation) or bioclimatic (plant distribution) parameters that can then be compared with simulated results. In the forward modeling approach, models are used to produce a predicted response that can be compared with current observation. These two approaches should not be seen as independent but as complementary (Kageyama, 2001); such a hierarchy of models is useful to allow the move from large to regional-scale investigation. This chapter will focus on modeling biosphere changes, although it should be noted that other components of Earth's system are required to understand these fully. For example, geophysiological modeling can assess the feedback mechanisms acting between geosphere and biosphere, the nature of the land surface, and impacts of changes in this—such as altering climate by albedo change that influences the reflectivity and absorption of incoming solar radiation, water vapor, convective precipitation, soil moisture, transpiration, hydrology, and latent heat flux (Figure 6.2).

6.2   BIOSPHERE MODELING

Vegetation is a major factor affecting land surface processes (Figure 6.3), and a good measure of ambient climate and change in this. Therefore, a central part of understanding climate change impact is to understand the correspondence between climate and the distribution of vegetation, both in terms of major vegetation formations, or biomes, and individual plants or species. Modeling the biosphere is not a new field of research. Initial attempts were carried out by Kostitzin (1935). Working on ideas about the interdependence between vegetation and climate he created the first mathematical model for the co-evolution of atmosphere and biota. Individual-based models describe vegetation dynamics in terms of interactions between individual plants with little emphasis on ecophysiology and climate (Farquhar, 1997). A compromise between these schools is the popular forest gap models (Tongeren and Prentice, 1986) that describe species-specific establishment, growth, death of trees, and interactions between them resulting in successional patterns. Species-based approaches create bioclimatic envelopes delimited by a combination of climate variables based on the correlation between spatial distribution of individual species and climatic parameters. The bioclimatic envelopes can then be used to model potential species responses to climate change. Genetic algorithms are used to create the envelopes because of their ability to take into account a large number of climate variables (McClean et al., 2005).

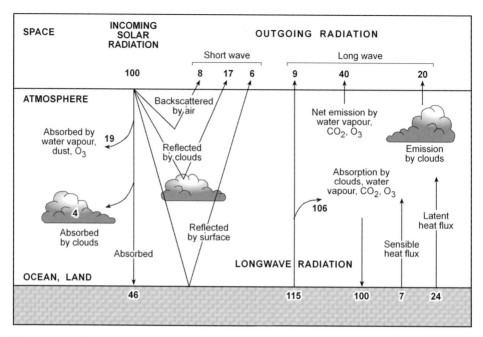

Figure 6.2. Indication of the amount of incoming and outgoing radiation, and the percentage absorbed and reflected by the various atmospheric components.

The best-known approach for predicting the equilibrium response of broad-scale potential vegetation types to climate change is the climate–vegetation classification approach (Holdridge, 1947)—using this approach Holdridge recognizes some 37 "life zones". The disadvantage of such an approach is that the climatic variables may not be the factor to which vegetation responds (Peng, 2000). Vegetation responds to a range of climatic influences, geomorphic substrates, ecological disturbances (natural and human-induced) with an incredible array of different species, growth and competition habits, and basic life forms (Figure 6.4). For example, simulations of vegetation patterns with and without fire show that large areas of C_4 grasslands in Africa and South America have the climate potential to support forests (Bond *et al.*, 2004). Functional groups work well—for example, the FORMIND model was applied to lowland forest data from Indonesia. The analysis used 22 functional groups, comprising 436 species, based on diameter growth and light demand, with an additional criterion being based on height (Köhler and Huth, 1998). Plant-functional types that group plant species by their physiognomic and morphological traits and responses to climate (e.g., tropical evergreen broad-leaf rainforest tree) are a very useful classification tool, particularly when dealing with the complexity of a tropical flora (Figure 6.4).

By including climate dependencies, biosphere models can account for climatic, immigration, and competitional influences, thus providing a forecast of ecosystem impacts under various climate scenarios (Figure 6.5). In earlier studies, climate input

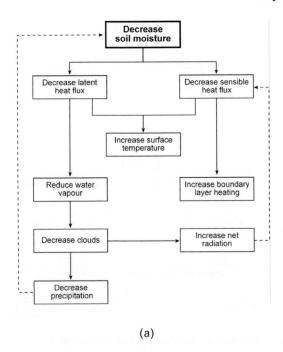

(a)

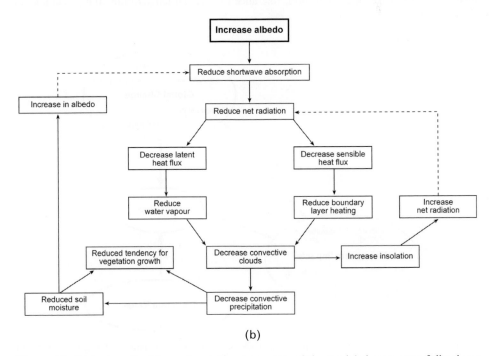

(b)

Figure 6.3. Impact on the biogeochemical components of the modeled ecosystem following a change in below-ground (a) and above-ground (b) biomass.

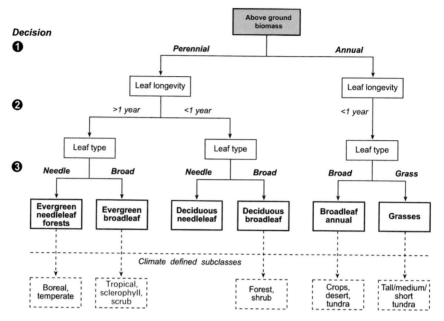

Figure 6.4. Schematic division to determine plant-functional types based on a series of divisions, in this case on growth form, tolerance to seasonal temperature, and physiology of the parent plant.

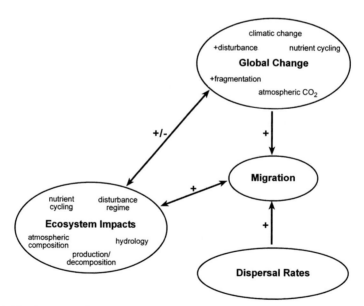

Figure 6.5. The impacts of migration and dispersal that are increasingly being incorporated within dynamic vegetation models as feedbacks, in conjunction with other ecosystem impacts— such as disturbance type.

data and timing of species appearance were inferred from the pollen records to which the simulation results were compared. Such circular reasoning between cause and effects limits insights gleaned about factors controlling vegetation response, and has been easily broken by using climatic input from independent data sources. Additionally, there has been an increasing interest in parameterizing the role of dispersal (Figure 6.5), in part driven by the development of more appropriate (realistic) models and partly due to the increasing understanding of the dynamic nature of populations. These approaches are currently confined to the temperate realm, but have interesting applications to tropical areas (Paradis *et al.*, 2002). This complexity has resulted in a number of approaches for studying biosphere responses to climate change that incorporate migration and adaptation (Kirilenko *et al.*, 2000), dispersal factors (seed production, fecundity, dispersal vector interaction), additional environmental factors (such as climate seasonality, edaphic factors, aspect), and ecological factors (such as association, parasite/disease, inertia) (Figure 6.5). In addition to changing areal extent and composition of the vegetation, changes in vertical structure are an important part of vegetation change (Cowling, 2004).

6.2.1   Biome modeling

A continental-scale study—the VEMAP project (VEMAP members, 1995)—demonstrated the interdependence of biogeographical and biogeochemical aspects of the ecological response to climate change; assessing how global change will affect ecosystems and must therefore include both aspects. Such a combined approach is exemplified by the process-based equilibrium terrestrial biosphere model BIOME-3 that simulates vegetation distribution and biogeochemistry (Haxeltine and Prentice, 1996). BIOME-3 predicts plant functional type (PFT) dominance based on environmental conditions, ecophysiological constraints, and resource limitations. The model uses the inputs of temperature, precipitation, cloudiness (Crammer and Leemans, 1991), soil texture, atmospheric pressure, and $[CO_2]_{atm}$ (Figure 6.6). The level of $[CO_2]_{atm}$ prescribed to BIOME-3 has a direct influence on gross primary productivity via a photosynthetic algorithm and competitive balance between C_3 and C_4 plants (Haxeltine and Prentice, 1996). Combining these inputs, a coupled carbon and water flux model calculates leaf area index (LAI) and net primary productivity (NPP) for each PFT. The NPP is translated to a series of prescribed PFTs, which then combine to form biomes (Figure 6.6). Although developed as a global vegetation model, BIOME-3 allows simulating changing environmental conditions on vegetation at regional and local scale (Jolly and Haxeltine, 1997; Marchant *et al.*, 2002, 2004a). BIOME-3 can be modified to represent vegetation change within a single pixel that can be used to isolate and manipulate environmental variables—such as temperature, precipitation, seasonal variations of these, and changes in CO_2 concentration (Foley *et al.*, 1996; Haxeltine and Prentice, 1996). BIOME-3 model output has been tested and compared against maps of potential vegetation at a global (Prentice *et al.*, 1992; 1993), continental (Jolly *et al.*, 1998; Williams *et al.*, 1998), and regional scale (Marchant *et al.*, 2001a, 2004b). Often, there are discrepancies between model-based

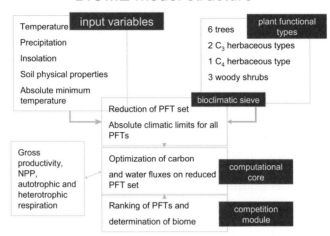

Figure 6.6. A basic summary of the steps taken to calculate biomes on the basis of climatic input variables.

reconstructions and potential vegetation; these discrepancies result from numerous reasons, such as over-simplified soil hydrology (Marchant *et al.*, 2001b).

Of particular interest for understanding vegetation dynamics is how climate system–biosphere interactions have developed since the last glacial maximum (LGM) (Cowling and Sykes, 1999). With this LGM focus in mind, BIOME-3 is run to demonstrate the impact of reducing $[CO_2]_{atm}$ to levels (200 p.p.m.V) ambient at the LGM (Petit *et al.*, 1999). Reduced concentrations of $[CO_2]_{atm}$ have a very significant impact on pan-tropical vegetation (Jolly and Haxeltine, 1997; Boom *et al.*, 2002), as we can see from the modeled output of vegetation (Figure 6.7, see color section). Under low $[CO_2]_{atm}$ (Figure 6.7, *bottom*) the amount of grassland (short and tall), xeric woodland, and scrub increase dramatically, particularly at the expense of Tropical Seasonal Forest (Figure 6.7). Interestingly, the amount of Tropical Rain Forest remains relatively constant as a consequence of largely being under control of changes in temperature and precipitation; the later component needing to change significantly to effect notable changes in Tropical Rain Forest distribution. There are likely to be major within-biome composition dynamics, both in terms of the importance of individual taxa and structure of the vegetation (not portrayed in the biome output). As gaseous exchange at the leaves involves both H_2O and CO_2 (Figure 6.8), changes in $[CO_2]_{atm}$ not only impact on the processes of photosynthesis and photo-respiration but also water-use efficiency (WUE) (Cowling and Sykes, 1999). WUE is linearly related to the level of $[CO_2]_{atm}$; under low $[CO_2]_{atm}$ plants have to transpire more to achieve the same level of photosynthesis and hence NPP—in other words, halving the $[CO_2]_{atm}$ is comparable to halving the rainfall (Farquhar, 1997). Thus, it is likely to be the hydrological impact rather than physiological impacts of lowered CO_2 that causes the vegetation to change. Although it has been shown that some C_3 plants

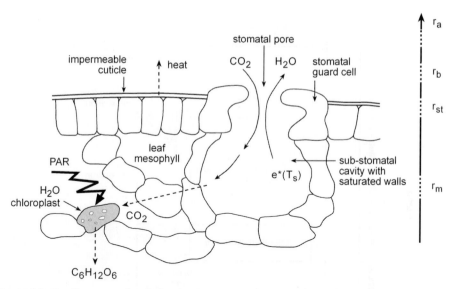

Figure 6.8. Leaf cross-section indicating gaseous exchanges taking place—in particular, those of CO_2 and H_2O.

can respond to decreased $[CO_2]_{atm}$ by increasing the amount of stomatal area on the leaf lamella (Wagner *et al.*, 1999), this is difficult to apply to the late glacial as the main impact—rather, the physiological response to low $[CO_2]_{atm}$—appears to be reduced WUE (Cowling and Sykes, 1999). Thus, if the stomata have a wider aperture, or are more frequent, this will result in more water being evaporated. Therefore, no matter how the stomata compensate for the variation in $[CO_2]_{atm}$, C_4 or CAM plants will always have a competitive advantage over C_3 plants in warm environments subject to water stress (Ehleringer *et al.*, 1997; Boom *et al.*, 2002).

An interesting application of vegetation models is to use them as a vehicle to display output from climate (Claussen, 1994, 1997) and biogeochemical models (Prentice *et al.*, 1993; Peng *et al.*, 1998). This use allows model output to be translated into maps of potential natural vegetation (Claussen and Esh, 1994; Foley *et al.*, 1996; Prentice *et al.*, 1996; Williams *et al.*, 1998) used for the coupling of biosphere, atmosphere, and oceanic components (Claussen, 1994; Texier *et al.*, 1997), and testing of biogeochemical dynamics (Peng *et al.*, 1998). Climate shifts are displayed as shifts in biome boundaries and areal extent, which can be used to investigate the feedback between atmosphere, biosphere, and oceanic systems under changing boundary conditions (Figure 6.9). However, it must be stressed that, as there are a number of different scenarios available to drive the vegetation model, the results will vary depending on the model output used and feedbacks to the climate system (Figure 6.1). Such differences, particularly when outputs are compared with independent data, can be used to assess model performance and determine the importance of model components—such as land–ocean feedback, or dynamic rather than fixed ocean temperatures.

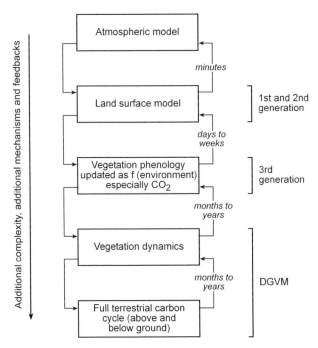

Figure 6.9. Schematic of increasing levels of details being added to surface modeling approaches. In this case the requirements are additive—that is, a fully coupled DGVM requires a traditional land surface model.

6.2.2   Dynamic global vegetation models

Extreme environmental events—such as droughts and fires—are important factors in global vegetation processes, yet data on them are sparse, unreliable, or completely lacking. Furthermore, how these are parameterized within models is even worse! Techniques to estimate the rates, extent, and magnitudes of extreme events, and the ability to quantify uncertainty in the results need to be developed. The role of biogeophysical vegetation feedback is not considered, but can be an important aspect of the tropical climate by altering the wet season and thus amplifying the response to orbital forcing (Doherty *et al.*, 2000). These changes affect local, regional, and global climates, which feed back to the biogeography and physiology of the vegetation (Cao and Woodward, 1998). The influence of such dynamics and impact on vegetation can be assessed within a vegetation model developed to run over time (Thonicke *et al.*, 2001). Similarly, internal feedbacks (Figures 6.3 and 6.9) can be incorporated by developing dynamic global vegetation models (DGVMs); of course, this results in a more complex model with associated computing limitations. The potential applications of DGVMs can be summarized within three main areas: simulating the transient changes in global vegetation patterns under future climate change, investigating human disturbance scenarios and estimating the transient behaviors of carbon pools and fluxes to provide fully interactive representation of biosphere within

GCMs to investigate potential vegetation–climate feedback mechanisms (Peng, 2000). To express these changes, the vegetation model needs to run to equilibrium—taking into account vegetation development and impacts of factors, such as fire. Such an approach will have advantages over equilibrium models in driving our understanding of ecosystem dynamics and the impact of a suite of forcing factors.

6.2.3   Models of biogeochemical cycles

With the premise that terrestrial ecosystems play a central role in regulating global biogeochemical and climate systems, a suite of models concerned with biogeochemical cycles, particularly carbon and nitrogen (Melillo *et al.*, 1993), have been developed. Such models commonly use a series of mass balance equations to obtain values for given biogeochemical components and rates of fluxes. The relationship between biosphere carbon stocks and $[CO_2]_{atm}$ during glacial to interglacial cycles may help us understand the mechanisms that drive change during the 21st century and beyond. How tropical vegetation contributes to the global carbon budget is an area of continued discrepancy and hence research activity (Lewis *et al.*, 2004). With rising atmospheric carbon derived from human activities a focus on investigating changes in the global carbon cycle has resulted in dozens of models to simulate atmosphere–biosphere CO_2 exchange processes. For example, the terrestrial biosphere model "Carbon Assimilation in the Biosphere" (CARAIB), now in its fourth incarnation, records a slight increase in the extent of tropical rainforest linked to the colonization of exposed continental shelf, such as Sunda. Carbon stocks are commonly calculated from biosphere models that use standing stock and NPP. Efforts can be divided into models orientated towards global simulations and those orientated towards plot-scale simulations that generally have a complex structure and require a large number of parameters. Indeed, there remain many challenges to scale up the single leaf process, and to parameterize respiration, allocation, deposition, and sequestration (Ito and Oikawa, 2002).

6.3   CLIMATE MODELING

A detailed and comprehensive discussion on climate modeling is outside the scope of this chapter, but can be found in the excellent review by McGuffie and Henderson-Sellers (1997). Climate models are used in a range of applications from investigating atmospheric processes to fluctuations of deep-ocean circulation and, increasingly, the impact of perturbations caused by the world's human population (Pitman, 2003). Models of the global climate system have several uses: interpolating and explaining paleodata, studying climate processes, separating natural variability from anthropogenic impact in the climate record, and investigating the nature of the climatic response and influence of various components—such as oceanic feedback. The ultimate aim of such approaches is to be able to forecast future climate where two factors are paramount: first, investigating impacts of higher $[CO_2]_{atm}$ levels (Houghton *et al.*, 2001); and, second, the impacts and feedback of land-cover change on the climate system (Cao and Woodward, 1998).

Atmospheric general circulation models (AGCMs) are continually improving in their ability to simulate the major features of today's climate, and over time have proved useful for investigating mechanisms of past climate changes (Joussaume, 1999). There is broad similarity in the reconstruction of different AGCMs as they are ultimately based on laws of physics, constants and particle motion, and only a small number of external boundary conditions such as the solar constant and atmospheric concentrations of radioactively active gases and aerosols. Models often have prescribed lower boundary conditions of sea surface temperatures (SSTs) and sea ice amounts with a three-dimensional representation of atmosphere motions. Dynamic climate models exhibit a problem not evident in those with prescribed SSTs and sea ice: in particular, fluxes of heat, momentum, and water across the ocean–atmosphere interface lead to *ad hoc* flux adjustments regarded as necessary to correct problems. These so-called "flux adjustments" can be relatively large in models that use them, raising questions about the realism of derived climate reconstructions and associated feedback processes (Gates *et al.*, 1998).

Despite their similar foundations, different models give somewhat contradictory results mainly due to subtle differences in assumptions about clouds and their role in absorption, reflectance, and moisture transport (Figure 6.2). To try to deal with such discrepancies, a series of working groups were established—such as the Atmospheric Model Inter-comparison Program (AMIP) (Gates *et al.* 1998) and the Paleoclimate Model Inter-comparison Program (PMIP) (Joussaume and Taylor, 1995)—to evaluate, test, and develop climate models, coordinate the systematic study of AGCMs, and to assess their ability to design experiments to simulate changes in climate such as those that occurred at the LGM. For example, in the AMIP simulations, sea ice and SST are prescribed to match recent observations, and the atmospheric response to boundary conditions; in CMIP the complete physical climate system including the oceans and sea ice adjust to prescribed $[CO_2]_{atm}$. These projects also provide climate scientists with a database of coupled GCM simulations under standardized boundary conditions, and thus have permitted an assessment as to why different models give different output in response to the same input.

Thus, there are a variety of climate models. Just two are mentioned here: these latest Earth system models incorporate fully coupled ocean–atmosphere–biosphere components that are thought to be the way forward to assess the impact of climate change on ecosystems. One such model, HadCM3LC, when run with a LGM climate demonstrates structural, as well as purely distributional change, in tropical forest with LAI reducing significantly as the tropical canopy became more open (Cowling, 2004). Another model, GENESIS-IBIS, incorporates an assessment of the global ecosystem through the integrated biosphere simulator (IBIS) and terrestrial hydrology through another model (HYDRA), as well as a fully coupled climate–vegetation model that incorporates new data sets of global climate systems. IBIS is one of the few computer models to incorporate a range of ecosystem processes within a single framework that is then integrated within a climate model. IBIS simulates:

- energy, water, and carbon dioxide exchange between plants, the atmosphere, and the soil;
- physiological processes of plants and soil organisms, including photosynthesis and respiration;
- seasonal changes of vegetation, including spring budburst, fall senescence, and winter dormancy;
- plant growth and plant competition;
- nutrient cycling and soil processes.

6.4   MODELING THE LAST GLACIAL PERIOD

Part of the PMIP effort was focused on the LGM to understand the impacts of extreme cold conditions and to study the feedbacks associated with a decrease in $[CO_2]_{atm}$ and ice sheet elevation of 2 to 3 km above North America and northern Europe. The LGM was simulated by 17 models through the PMIP program that prescribe a series of set boundary conditions (Pinot et al., 1999). An update of the ice sheet extent and height was provided by (Peltier, 1994). As a result of the Laurentide and Fennoscandian ice sheets being 1,000 meters lower than the previous reconstruction (CLIMAP, 1981), $[CO_2]_{atm}$ was prescribed to be 200 p.p.m.V, as inferred from Antarctic ice cores (Raynaud, et al., 1993), and Earth's orbital parameters were changed according to their values at 21,000 yr BP. Over the oceans, two sets of experiments were defined: one with SST changes prescribed from estimates (CLIMAP, 1981), the other with SSTs computed using coupled atmosphere–mixed layer ocean models and assuming no change in ocean heat transport. Each approach has its advantages and problems:

1. Computing SSTs permits an evaluation of models used for future climate prediction, but does not account for any changes in oceanic heat transport, despite evidence for thermohaline circulation changes.
2. Prescribing SSTs should yield better results over land; however, SST estimates are subject to substantial uncertainty (Duplessy et al., 1988).

Indeed, a limitation of model simulations has been the role of ocean dynamics, which is crucial for understanding changes in tropical upwelling (Bush and Philander, 1998; Ruter et al., 2004) and resultant rainfall regimes. Model simulations still underrepresent tropical cooling at the LGM, apart from over Eastern Africa where most of the models indicate a cooling (relatively minor) similar to the observational data. Following the debate on the degree of tropical SST cooling raised by Rind and Peteet (1985), Farrera et al. (1999), and Pinot et al. (1999) it can be shown that: (1) all PMIP simulations using the relatively warm tropical SSTs given by CLIMAP (1981) tend to be too warm over land, except over equatorial Africa; (2) computed SSTs are colder than CLIMAP, especially over the tropical Pacific where the erroneous warm pools of CLIMAP are not reproduced (Pinot et al., 1999); (3) models with computed SSTs show a range of terrestrial cooling strongly related to the intensity of tropical SST cooling. Some models produce a strong terrestrial cooling consistent with the paleodata, but this is associated with SST cooling that is too large when compared against recent alkenone data on SST. However, one model gives reasonable results over both land and oceans: "CLIMBER", a model of intermediate complexity, reconstructs a tropical land cooling of 4.6°C with SST cooling of 3.3°C in the Atlantic, 2.4°C in the Pacific and 1.3°C in the Indian Ocean (Ganopolski *et al.*, 1998). This is in broad agreement with the data that show tropical SSTs were 5°C colder in Barbados corals (the coldest throughout the tropics), although 2–3°C is a more common value (Guilderson *et al.*, 1994). Terrestrial temperatures simulated by 17 models within PMIP are relatively similar showing temperature was reduced by 5–6°C about the LGM. For some models run for both fixed and computed SSTs (UGAMP and GEN2), the annual mean change in temperature remains very similar on global average, although regional differences can be substantial (Dong and Valdes, 1998) but are comparable with data-based reconstructions that also document considerable variation.

According to PMIP simulations, vegetation change in the tropical realm is primarily driven by precipitation changes. According to these models, the LGM was relatively dry apart from East Africa and throughout high elevations of South America and Papua New Guinea (Pinot *et al.*, 1999). Reduced precipitation, particularly in mid-latitude western South America, is likely to result from a reduction in the intensity of westerly climate systems. In a comparison of two models within the PMIP suite (CCM3 and CSM) differences are quite small in most measurements of atmospheric circulation, with one exception that involves tropical precipitation (Joussaume and Taylor, 1995). Moisture changes tend to be associated with changes at the regional scale when model simulations are characterized by a number of common features, including a reduction in the strength of the Afro-Asian monsoon and increased inter-tropical aridity, corroborated by various paleoindicators. Climates of all the continents have monsoonal climates; change in insolation, such as that occurring about the LGM due to changes in Earth's orbital parameters, would cause changes in the monsoonal climate (Joussaume *et al.*, 1999) and the associated feedbacks. For example, application of a coupled ocean–atmosphere model (FOAM) indicate that SST feedbacks produce a much larger enhancement of precipitation in Central America than direct radiative forcing alone (Harrison *et al.*, 2003).

Discrepancies between LGM model runs and comparison with paleoenvironmental data sets are likely to result from missing feedbacks—for example, all simulations omit possible influences of vegetation change due to climate-induced shifts, and CO_2-induced changes in vegetation and leaf conductance. Indeed, numerous factors are not included—for example, mineral aerosol (dust) concentrations were many times higher than today at the LGM, especially in the polar stratosphere, and this could have provided an extra cooling effect. Taking into account climate and CO_2-induced vegetation changes to infer the dust distribution (which was in fairly good agreement with proxy data), computations indicate a small positive change in radiative forcing in high latitudes, but a larger negative change in the tropics. Ocean dynamic changes are also likely to be important at the LGM, as demonstrated by models of intermediate complexity (Ganopolski, *et al.*, 1998; Weaver, *et al.*, 1998); indeed, dramatic changes in ocean circulation are likely to be responsible for abrupt climate change during the last ice age and transition to the Holocene (Stocker and Marchal., 2000).

6.5 TESTING MODELS WITH DATA

The accuracies of climate models are often compared with the only means of validation available: climate estimates derived from proxy data. Data–model comparisons are useful for assessing the relative influence of past climate change over various spatial and temporal scales. To enable comparisons to be carried out in a systematic fashion, compiled data sets are required to have two fundamental contributions to make to Earth system modelling: as input boundary conditions and as datasets for model evaluation (Kohfeld and Harrison, 2000). For example, the LGM Tropical Data Synthesis (Farrera *et al.*, 1999) contains quantitative reconstructions of the mean temperature of the coldest month and mean annual temperature (MTCO and MAT) and qualitative reconstructions of plant-available moisture and run-off. The data set combines pollen, plant macrofossil records, noble gas, and speleothem data sets between 32°N and 33°S. Such data sets need to contain sufficient information and documentation to evaluate the assumptions involved, interpretation of the data, exercise quality control, and select data appropriate for specific goals (Kohfeld and Harrison, 2000). These need to:

1. Classify and describe errors and uncertainties in Earth observation data.
2. Represent dependencies and the associated uncertainty.
3. Develop computational techniques for synthesis of multiple factors, allowing for model and data uncertainty.

Numerous databases of proxy indicators on past climate are amassed within the World Centre for Paleoclimatology (*http://www.noaa.gov*)—such as the BIOME 6000 dataset, the Global Lake Status Database and the Tropical Terrestrial Data Synthesis. Central to the integration of Earth observation data with environmental models is the development of techniques for quantifying and manipulating uncertainties in both data and models. To establish links between the climatic modeling and paleoecological communities, the BIOME-6000 project has developed techniques to portray changes in pollen data as changes in biome distribution (Prentice and Webb, 1998). This link has allowed pollen data based environmental reconstructions to provide a validation tool for climate model based reconstructions (Prentice *et al.*, 1996; Jolly *et al.*, 1998; Weaver *et al.*, 1998). However, displaying pollen data as biomes has a greater utility than just providing a benchmark for climate model output validation. Transformed pollen data can be used in conjunction with other data to understand the causal factors driving vegetation change over the recent geological past. Additionally, methods need to be developed to encompass errors and uncertainties: errors in raw Earth observation data, errors in input data (e.g., meteorological data), and inherent limitations of the models themselves (e.g., scale effects) all need to be assessed.

Farrera *et al.* (1999) and Pinot *et al.* (1999) provide a detailed discussion on the comparison between PMIP LGM simulations and the available data from the tropics. Model comparisons with this data set show that models continue to under-estimate the pattern of cooling, which is significantly spatially different: varying from 5–6°C in South America to 2–3°C in East Africa and Indonesia. Most models in the PMIP suite provide a uniform cooling; however, the United Kingdom Meteorological Office (UKMO) model provides a meridional patterning in tropical temperature reduction that is comparable with observations. Paleodata–model mismatches arise from numerous reasons because of (a) inappropriate comparisons, (b) errors or uncertainties in data, and (c) model failures (Kohfeld and Harrison, 2000). Certainly within the tropics—despite valiant efforts of a relatively few, and thankfully growing, number of paleoecologisits—data coverage remains too low to make meaningful comparisons and increased data coverage will lead to a re-evaluation of the patterns observed.

6.6 PRACTICAL APPLICATION OF MODEL OUTPUT AND FUTURE DEVELOPMENTS

The drivers of land-use and land-cover change are a complex mix of environmental and climatic factors that make a fully integrated modeling perspective highly challenging, and outside our reach at present. Although modeling the full complexity of landscape processes and associated feedbacks is making remarkable progress, and structures are in place to combine research communities and develop new integrated approaches that will produce more valid outputs and hence develop their use in management, there is much work to be done. To engender such a process, the collection and generation of new environmental data sets on past variability is crucial. In addition to developing new data sets, there are a number of methodological developments—for example, neural networks (Grieger, 2002). Bayes theory applied to vegetation dynamics that can provide good results where the floristic diversity is high and the mechanistic understanding of the ecosystem relatively low (Hilbert and Ostendorf, 2001). Improved comparisons between model output and proxy data can be made by upscaling from the resolution appropriate from proxy data (catchment scale) and/or through the downscaling of regional climate models (Sailor and Li, 1999). Unfortunately, the application of scale changes and regional modeling remains in its infancy (Sewalle *et al.*, 2000), particularly concerning the tropics, with most regional applications remaining a subset of larger investigations.

Fully coupled land–ocean–atmosphere models—currently under active development with increasingly detailed (realistic) models—combined with improved climate data sets will be required to arrive at a robust quantitative understanding of climate changes and associated environmental impact. A number of developments underway fall into three main areas. First, increase the number of components taken into account: complex models that allow the whole Earth–atmosphere–ocean–climate system to be interconnected and permit the analysis of feedback between the elements. Second, increase the duration of model runs to study climatic transitions rather than equilibrium: this approach offsets the development of the more complex models suggested above with some elements of the climate system needing to be simplified, so models are affordable to run given present (and developing) computer limitations. Third, increased spatial scale of the simulations that will aid in the understanding of data in their local context. This approach, still in its infancy, will help us to improve our understating of small-scale phenomena that are important for the behavior of the proxy records of paleoclimate (Kageyama, 2001).

6.7   CONCLUSIONS

Climate models have evolved from a very simple, implicit approach representing surface energy balance and hydrology to complex models that represent many of the key processes through which land surface and climate interact (Pitman, 2003). Model development has tended to focus on GCMs that often perform relatively poorly in tropical regions, or have been developed in temperate areas where there is better input data in terms of climate and environmental parameters, the ecology of the taxa is generally better known, and—as the total number of taxa are relatively few—the system is relatively simple. The time is right to combine expertise from chemical, biological, geoscience, and remote-sensing communities to work with climate modelers to build an integrated framework that is mutually beneficial; indeed, the breadth of knowledge required to develop such a fully integrated approach is intimidating, but can be developed through umbrella organizations such as IGBP-PAGES. Although there are numerous improvements, it is apparent that the current generation of coupled ocean–atmosphere models still have relatively poor simulations of tropical climate change, possibly due to the spatially and temporally variable nature of tropical environments (Ruter *et al.*, 2004). Environmental change is rarely spatially uniform and, therefore, necessitates an even greater wealth of data on present and past environmental states to determine the complexity and patterns behind this. New sites located in key areas, combined with the application of a range of proxies of environmental change, are required to refine our understanding of tropical ecosystem responses to Late Quaternary climatic variations. There also needs to be a realization that the tropics must not be treated with the same assumptions applied to Europe and North America. Accurate representation of tropical areas in Earth system models is a continuing challenge, particularly due to the incredible environmental and ecological diversity.

6.8   REFERENCES

Behling, H. and Hooghiemstra, H. (2000) Holocene Amazon rain forest–savanna dynamics and climatic implications: High-resolution pollen record from Laguna Loma Linda in eastern Colombia. *Journal of Quaternary Science* **15**, 687–695.

Bond, W. J. G., Woodward, F. I., and Midgley, G. F. (2004) The global distribution of ecosystems in a world without fire. *New Phytologist* **165**, 525–538.

Boom, A., Marchant, R. A., and Hooghiemstra, H. (2002) Pollen-based biome and $\delta^{13}C$ reconstructions for the past 450,000, Colombia-ecosystem relationships to late Quaternary CO_2. *Palaeogeography, Palaeoclimatology, Palaeoecology* **177**, 151–168.

Bush, A. B. G. and Philander, G. H. (1998) The role of ocean–atmosphere interactions in tropical cooling during the Last Glacial Maximum. *Science* **279**, 1341–1344.

Cao, M. and Woodward, I. (1998) Dynamic responses of terrestrial ecosystem carbon cycling to global climate change. *Nature* **393**, 249–252.

Claussen, M. (1994) On coupling global climate models with vegetation models. *Climate Research* **4**, 203–221.

Claussen, M. (1997) Modelling biogeophysical feedback in the Africa and Indian monsoon region. *Climate Dynamics* **13**, 247–257.

Claussen, M. and Esch, M. (1994) Biomes computed from simulated climatologies. *Climate Dynamics* **9**, 235–243.

CLIMAP (1981) *Seasonal Reconstructions of the Earth's Surface at the Last Glacial Maximum* (Map Series, Technical Report MC-36). Geological Society of America, Boulder, CO.

Cowling, S. A. (2004) Tropical forest structure: A missing dimension to Pleistocene landscapes. *Journal of Quaternary Science* **19**, 733–743.

Cowling, S. A. and Sykes, M.T. (1999) Physiological significance of low atmospheric CO_2 for plant–climate interactions. *Quaternary Research* **52**, 237–242.

Doherty, R., Kutzbach, J., Foley, J., and Pollard, D. (2000) Fully coupled climate/dynamical vegetation model simulations over northern Africa during the mid-Holocene. *Climate Dynamics* **16**, 561–573.

Dong, B. and Valdes, P. J. (1998) Simulations of the last glacial maximum climate using a general circulation model: Prescribed versus computed sea surface temperatures. *Climate Dynamic* **14**, 571–591.

Duplessy, J. C., Shackleton, N. J., Fairbanks, R. G., Labeyrie, L., Oppo D., and Kallel, N. (1988) Deep water source variations during the last climatic cycle and their impact on the global deep water circulation. *Paleoceanography* **3**, 343–360.

Ehleringer, J. R., Cerling, T. E., and Helliker, B. R. (1997) C_4 photosynthesis, atmospheric CO_2 and climate. *Oecologica* **112**, 285–299.

Farquhar, G. D. (1997) Carbon dioxide and vegetation. *Science* **278**, 1411.

Farrera, I., Harrison, S. P., Prentice, I. C., Ramstein, G., Guiot, J., Bartlein, P. J., Bonnefille, R., Bush, M., Cramer, W., von Grafenstein, U. *et al.* (1999) Tropical climates at the last glacial maximum: A new synthesis of terrestrial palaeoclimate data. I. Vegetation, lake-levels and geochemistry. *Climate Dynamics* **11**, 823–856.

Foley, J. A., Prentice I. C., Ramankutty, N., Levis, S., Pollard, D., Sitch, S., and A. Haxeltine (1996) An integrated biosphere model of land surface processes, terrestrial carbon balance, and vegetation dynamics. *Global Biogeochemical Cycles* **104**, 603–628.

Ganopolski, A., Rahmstorf, S., Petoukhov, V., and Claussen, M. (1998) Simulation of modern and glacial climates with a coupled global model of intermediate complexity. *Nature* **391**, 351–356.

Gates, W. L., Boyle, J., Covey, C., Dease, C., Doutriaux, C., Drach, R., Fiorino, M., Gleckler, P., Hnilo, J., Marlais, S. *et al.* (1998) An overview of the results of the Atmospheric Model Intercomparison Project (AMIP I). *Bulletin of the American Meteorological Society* **73**, 1962–1970.

Grieger, B. (2002) Interpolating paleovegetation data with an artificial neural network approach. *Global and Planetary Change* **34**, 199–208.

Guilderson, T. P., Fairbanks, R. G., and Rubenstone, R. G. (1994) Tropical temperature variations since 20,000 years ago: Modulating inter-hemispheric climate change. *Science* **263**, 663–665.

Haberle, S. and Maslin, M. (1999) Late Quaternary vegetation and climate change in the Amazon Basin based on a 50,000 year pollen record from the Amazon fan, ODP site 932. *Quaternary Research* **51**, 27–38.

Harrison, S. P., Kohfeld, K. E., Roelandt, C., and Claquin, T. (2003) The role of dust in climate change today, at the last glacial maximum and in the future. *Earth Science Reviews* **54**, 43–80.

Haxeltine, A. and Prentice, I. C. (1996) BIOME3: An equilibrium terrestrial biosphere model based on ecophysiological constraints, resource availability, and competition among plant functional types. *Global Biogeochemical Cycles* **10**, 693–709.

Hilbert, D. W. and Ostendorf, B. (2001) The utility of artificial neural networks for modelling the distribution of vegetation in past, present and future climates. *Ecological Modelling* **146**, 311–327.

Holdridge, L. R. (1947) Determination of world plant formations from a simple climate data set. *Science* **105**, 367–368.

Hooghiemstra, H. and van der Hammen, T. (1998) Neogene and Quaternary development of the neotropical rain forest: The forest refugia hypothesis, and a literature overview. *Earth Science Reviews* **44**, 147–183.

Houghton, J. T., Ding, Y., Grigg, D. J., Noyer, M., van der Linden, P. J., Dai, X., Maskell, K., and Johnson, C. A. (eds.) (2001) *Climate Change, 2001: The Scientific Basis*. Cambridge University Press, Cambridge, U.K.

Ito, A. and Oikawa, T. (2002) A simulated model of the carbon cycle in land ecosystems (Sim-CYLE): A description based on dry-matter production theory and plot-scale validation. *Ecological Modelling* **151**, 143–176.

Jolly, D. and Haxeltine A. (1997) Effect of low glacial atmospheric CO_2 on tropical African montane vegetation. *Science* **276**, 786–788.

Jolly, D., Prentice, I. C., Bonnefille, R., Ballouche, A., Bengo, M., Brenac, P., Buchet, G., Burney, D., Cazet, J-P., Cheddadi, R. *et al.* (1998) Biome reconstruction from pollen and plant macrofossil data for Africa and the Arabian peninsula at 0 and 6 ka. *Journal of Biogeography* **25**, 997–1005.

Joussaume, S. (1999) Modelling extreme climates of the past 20,000 years with general circulation models. In: W. R.Holland, S. Joussaume, and F. David (eds.), *Modelling the Earth's Climate and Its Variability, Les Houches, Session LXVII* (pp. 527–565). Elsevier, North-Holland, The Netherlands.

Joussaume, S. and Taylor, K. E. 1995. Status of the Paleoclimate Modelling Intercomparison Project (PMIP). *Proceedings of the First International AMIP Scientific Conference* (WCRP Report, pp. 425–430). World Climate Research Programme, World Meteorological Organization, Geneva.

Joussaume, S., Taylor, K. E., Braconnot, P., Mitchell, J. F. B., Kutzbach, J. E., Harrison, S. P., Prentice, I. C., Broccoli, A.J., Abe-Ouchi, A., Bartlein, P. J. *et al.* (1999) Monsoon changes for 6000 years ago: Results of 18 simulations from the Paleoclimate Modelling Inter-comparison Project (PMIP). *Geophysical Research Letters* **26**, 859–862.

Kageyama, M. (2001) Using model hierarchies to better understand past climate change. In: T. Matsuno and H. Kida (eds.), *Present and Future of Modelling Global Environmental Change: Toward Integrated Modelling* (pp. 243–252). Terra Scientific, Tokyo.

Kirilenko, A., Belotelov, N. V., and Bogatyrev, B. G. (2000) Global model of vegetation migration: Incorporation of climatic variability. *Ecological Modelling* **132**, 125–133.

Kohfeld, K. and Harrison, S. P. (2000) How well can we simulate past climate? Evaluating the models using global palaeoenvironmental datasets. *Quaternary Science Reviews* **19**, 321–346.

Köhler, P. and Huth, A. (1998) The effects of tree species grouping in tropical rainforest modelling: Simulations with the individual-based model FORMIND. *Ecological Modelling* **109**, 301–321.

Kostitzin, V. A. (1935) *L'évolution de l'atmosphère: Circulation organique, epoques glaciaries.* Hermann, Paris [in French].

Lewis, S. L., Malhi, Y., and Philips, O. L. (2004) Fingerprinting the impacts of global change on tropical forests. *Philosophical Transactions of the Royal Society of London, Series B: Biological Sciences* **359**, 437–462.

Marchant, R. A. and Hooghiemstra, H. (2004) Rapid environmental change in tropical Africa and Latin America about 4000 years before present: A review. *Earth Science Reviews* **66**, 217–260.

Marchant, R. A., Behling, H., Berrio, J. C., Cleef, A., Duivenvoorden, J., Van Geel, B., Van der Hammen, T., Hooghiemstra, H., Kuhry, P., Melief, B. M. *et al.* (2001a) Mid- to Late-Holocene pollen-based biome reconstructions for Colombia. *Quaternary Science Reviews* **20**, 1289–1308.

Marchant, R. A., Behling, H., Berrio, J. C., Cleef, A., Duivenvoorden, J., van Geel, B, van der Hammen, T., Hooghiemstra, H., Kuhry, P., Melief, B. M. *et al.* (2001b) A reconstruction of Colombian biomes derived from modern pollen data along an altitude gradient. *Review Palaeobotany and Palynology* **117**, 79–92.

Marchant, R. A., Behling, H., Berrio, J. C., Cleef, A., Duivenvoorden, J., van Geel, B, van der Hammen, T., Hooghiemstra, H., Kuhry, P., Melief, B. M. *et al.* (2002) Colombian vegetation derived from pollen data at 0, 3000, 6000, 9000, 12,000, 15,000 and 18,000 radiocarbon years before present. *Journal of Quaternary Science* **17**, 113–129.

Marchant, R. A., Boom, A., Behling, H., Berrío, J. C., van Geel, B, van der Hammen, T., Hooghiemstra, H., Kuhry, P., and Wille, M. (2004a) Colombian vegetation at the Late Glacial Maximum: A comparison of model and pollen-based biome reconstructions. *Journal of Quaternary Science* **19**, 721–732.

Marchant, R. A., Behling, H., Berrío, J. C., Cleef, A., Duivenvoorden, J., van Geel, B, van der Hammen, T., Hooghiemstra, H., Kuhry, P., Melief, B. M. *et al.* (2004b) Mid to Late Holocene vegetation disturbance in Colombia: A regional reconstruction. *Antiquity* **78**, 828–838.

McClean, C. J., Lovett, J. C., Küper, W., Hannah, L., Sommer, J. H., Barthlott, W., Termansen, M., Smith, G. F., Tokumine, S., and Taplin, J. R. D. (2005) African plant diversity and climate change. *Annals of the Missouri Botanical Garden* **92**, 139–152.

McGuffie, K. and Henderson-Sellers, A. (1997) *A Climate Modelling Primer* (253 pp.). John Wiley & Sons, New York.

Melillo, J. M., McGuire, A. D., Kicklighter, D. W., Moore III, B., Voromurty, C. J., and Schloss, A. L. (1993) Global climate change and terrestrial net primary production. *Nature* **363**, 234–240.

Paradis, E., Baillie, S. R., and Sutherland, W. J. (2002) Modelling large-scale dispersal distances. *Ecological Modelling* **151**, 279–292.

Peltier, R. W. (1994) Ice age paleotopography. *Science* **265**, 195–201.

Peng, C. (2000) From static biogeographical model to dynamic global vegetation model: A global perspective on modelling vegetation dynamics. *Ecological Modelling* **135**, 33–54.

Peng, C. H., Guiot, J., and Van Campo, E. (1998) Estimating changes in terrestrial vegetation and carbon storage: Using palaeoecological data and models. *Quaternary Science Reviews* **17**, 719–735.

Petit, J. R., Jouzel, J., Raynaud, D., Barkov, N. I., Barnola, J-M., Basile, I., Bender, M., Chappellaz, J., and Davis, M. (1999) Climate and atmospheric history of the past 420,000 years from the Vostok ice core, Antarctica. *Nature* **399**, 429–436.

Pinot, S., Ramstein, G., Harrison, S. P, Prentice, I. C., Guiot, J., Stute, M., Joussaume, S., and PMIP-participating groups (1999) Tropical paleoclimates at the Last Glacial Maximum: comparison of Paleoclimate Modelling Intercomparison Project (PMIP) simulations and paleodata. *Climate Dynamics* **15**, 857–874.

Pitman, A. J. (2003) The evolution of, and revolution in, land–surface schemes designed for climate models. *International Journal of Climatology* **23**, 479–510.

Prentice, I. C. and Webb, T. (1998) BIOME 6000: Reconstructing global Mid-Holocene vegetation patterns from palaeoecological records. *Journal of Biogeography* **25**, 997–1005.

Prentice, I. C., Cramer, W., Harrison, S. P., Leemans, R., Monserud, R. A., and Solomon, A. M. (1992) A global biome model based on plant physiology and dominance, soil properties and climate. *Journal of Biogeography* **19**, 117–134.

Prentice, I. C., Sykes, M. T., Lautenschlager, M., Harrison, S. P., Denissenko, O., and Bartlein, P. (1993) Modelling global vegetation patterns and terrestrial carbon storage at the last glacial maximum. *Global Ecology and Biogeography Letters* **3**, 67–76.

Prentice, I. C., Guiot, J., Huntley, B., Jolly, D., and Cheddadi, R. (1996) Reconstructing biomes from palaeoecological data: A general method and its application to European pollen data at 0 and 6 ka. *Climate Dynamics* **12**, 185–194.

Raynaud, D., Jouzel, J., Barnola, J.-M., Chappelaz, J., Delmas R., and Lorius, C. (1993) The ice record of greenhouse gases. *Science* **259**, 926–934.

Rind, D. and Peteet, D. (1985) Terrestrial conditions at the Last Glacial Maximum and CLIMAP sea-surface temperature estimates: Are they consistent? *Quaternary Research* **24**, 1–22.

Ruter, A., Arzt, J., Vavrus, S., Bryson, R. E., and Kutzbach, J. E. (2004) Climate and environment of the subtropical and tropical Americas (NH) in the mid Holocene: Comparisons of observations with climate model simulations. *Quaternary Science Reviews* **23**, 663–679.

Sailor, D. J. and Li, X. (1999) A semiempirical downscaling approach for predicting regional temperature impacts associated with climatic change. *Journal of Climate* **12**, 103–114.

Sewall, J. O., Sloan, L. C., Huber, M., and Wing, S. (2000) Climate sensitivity to changes in land surface characteristics. *Global and Planetary Change* **26**, 445–465.

Stocker, T. and Marchal, O. (2000) Abrupt climate change in the computer: Is it real? *Proceedings of the National Academy of Sciences U.S.A.* **97**, 1362–1365.

Texier, D., de Noblet, N., Harrison, S. P., Haxeltine, A., Jolly, D., Joussaume, S., Laarif, F., Prentice, I. C., and Tarasov, P. (1997) Quantifying the role of biosphere–atmosphere feedbacks in climate change: Coupled model simulations for 6000 years BP and comparison with paleodata for northern Eurasia and northern Africa. *Climate Dynamics* **13**, 865–882.

Thompson, L. G., Mosely-Thompson, E., Davis, E. M., Lin, P. E., Henderson, K. A., Cole-Dai, B., and Liu, K. (1995) Late glacial stage and Holocene tropical ice core records from Huascarán, Peru. *Science* **269**, 46–50.

Thonicke, K., Venevsky, S., Sitich, S., and Crammer, W. (2001) The role of fire disturbance for global vegetation dynamics: Qoupling fire into a dynamic global vegetation model. *Global Ecology and Biogeography* **10**, 661–677.

Tongeren, O. and Prentice, I. C. (1986) A spatial simulation model for vegetation dynamics. *Plant Ecology* **65**, 163–173.

Tuomisto, H. and Ruokolainen, K. (1997) The role of ecological knowledge in explaining biogeography, and biodiversity in Amazonia. *Biodiversity and Conservation* **6**, 347–357.

Valdes, P. (2000) South American palaeoclimate model simulations: How reliable are the models. *Journal of Quaternary Science* **15**, 357–368.

VEMAP members (1995) Vegetation/ecosystem modelling and analysis project: Comparing biogeography and biogeochemistry models in a continental-scale study of terrestrial ecosystem responses to climate change and CO_2 doubling. *Global Biogeochemical Cycles* **9**, 407–437.

Wagner, F., Bohncke, S. J. P., Dilcher, D. L., Kürschner, W. M., Van Geel, B., and Visscher, F. H. (1999) Century-scale shifts in early Holocene atmospheric CO_2 concentration. *Science* **284**, 1971–1973.

Weaver, A. J., Eby, M., Fanning, A. F., and Wiebe, E. C. (1998) Simulated influence of carbon dioxide, orbital forcing and ice sheets on the climate of the Last Glacial Maximum. *Nature* **394**, 847–853.

Williams, J. W., Summer, R. S., and Webb III, T. (1998) Applying plant functional types to construct biome maps from eastern North American pollen data: Comparisons with model results. *Quaternary Science Reviews* **17**, 607–627.

Figure 5.6. Lake Barombi Mbo in Cameroon, surrounded by rainforest (photo Jean Maley).

Fig. 11.1. Kork..... in the Loré National Park, Gabon (photo Alfred Ngomanda).

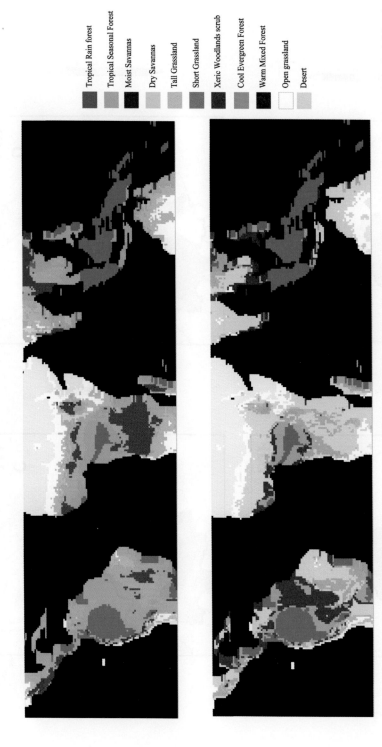

Figure 6.7. BIOME-3 run for the tropics with inputs of modern climate; in the bottom map output CO_2 has been reduced to glacial levels (200 p.p.m.V) with no change in other environmental inputs.

Tropical Rain forest
Tropical Seasonal Forest
Moist Savannas
Dry Savannas
Tall Grassland
Short Grassland
Xeric Woodlands scrub
Cool Evergreen Forest
Warm Mixed Forest
Open grassland
Desert

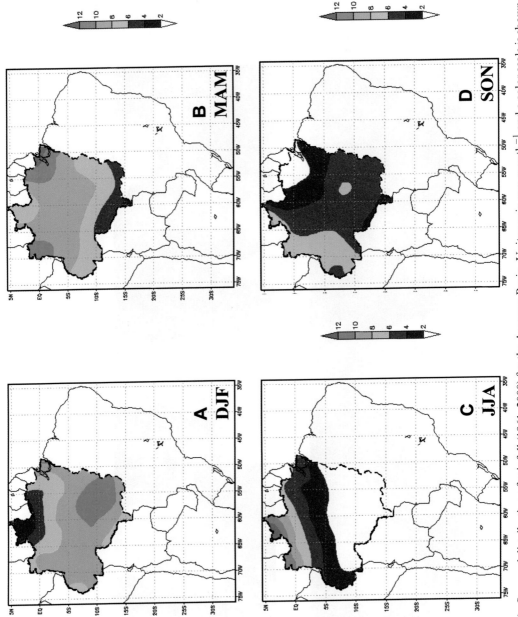

Figure 9.1. Seasonal distribution of rainfall (1961–2000) for the Amazon Basin. Units are in mm month^{-1} and a color scale is shown at the right

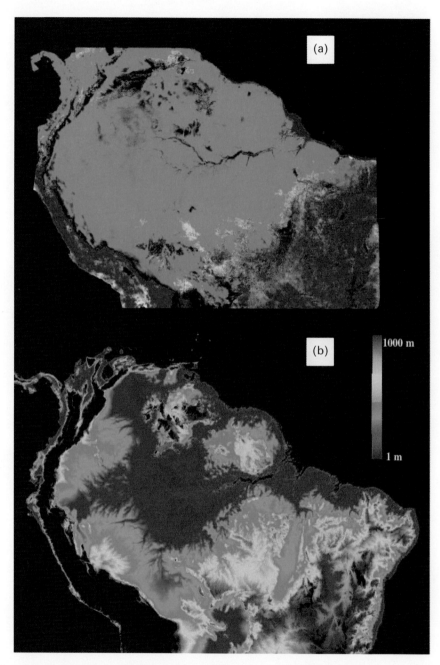

Figure 10.1. Overview of Amazonian geography: (a) Vegetation cover shown as false color composite of MODIS reflectance data. Areas from light to dark green represent open to densely covered forest, grasslands and woodland savannas appear pink to red, and areas of no vegetation including water and wetlands appear black. (b) Topographical variation across the Amazon Basin represented by SRTM (Shuttle Radar Topography Mission) elevation data.

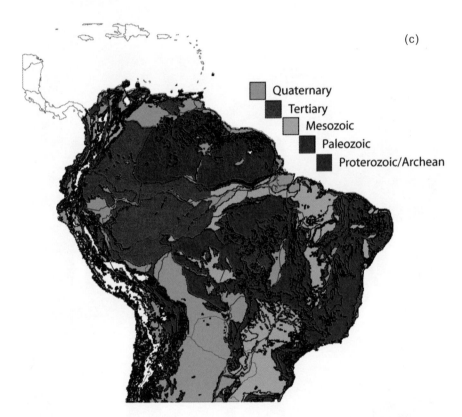

(c)

Quaternary
Tertiary
Mesozoic
Paleozoic
Proterozoic/Archean

Figure 10.1 (*cont*.). Overview of Amazonian geography: (c) Ages of surface geology.

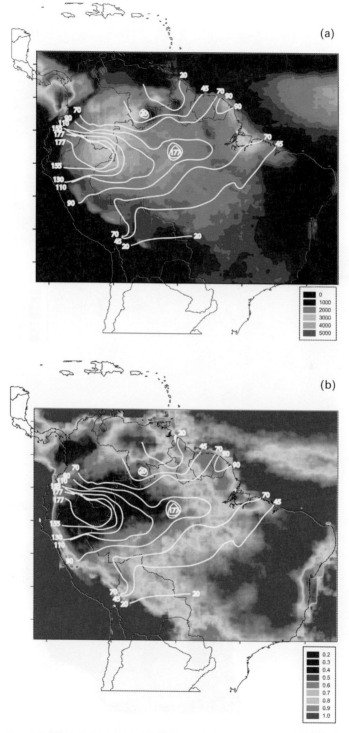

Figure 10.5. Local (alpha) diversity versus rainfall patterns and geologic age. (a) Accumulated precipitation. (b) Variability in precipitation.

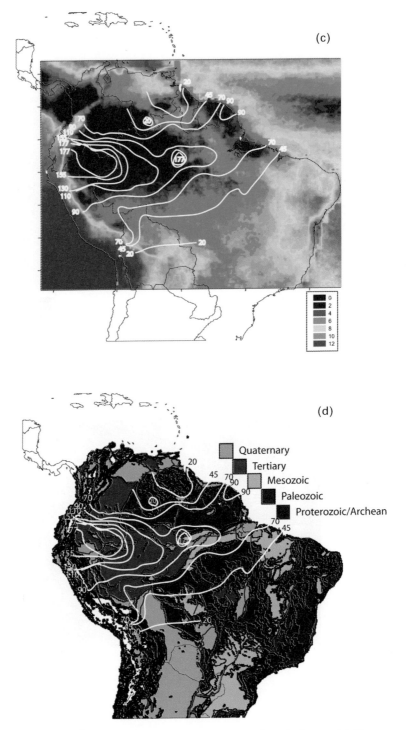

Figure 10.5. Local (alpha) diversity versus rainfall patterns and geologic age. (c) Dry-season length. (d) Geologic age.

7

Prehistoric human occupation and impacts on Neotropical forest landscapes during the Late Pleistocene and Early/Middle Holocene

Dolores R. Piperno

7.1 INTRODUCTION

This chapter presents a review of the evidence for human occupation and modification of the lowland Neotropical forest during the pre-Columbian era. Late Pleistocene through Early and Middle Holocene temporal frames (c. 16 kcal. yr BP to 4.8 kcal. yr BP) will be covered most thoroughly. These are the periods during which human colonization of low latitudes first took place in the Americas and agricultural societies emerged and spread throughout the lowland forest. Prehistoric agricultural practices, which often included types of swidden or slash and burn cultivation, sometimes resulted in profound landscape alteration and forest clearing. It is beyond the scope of this chapter to adequately summarize the information relevant to the Old World. I provide brief comments on some particularly salient work carried out recently that bears on early habitation and modification of tropical forest in Africa, mainland southern and southeast Asia, and New Guinea.

7.2 SOME BRIEF COMMENTS ON THE OLD WORLD

Even this brief review of some of the pre-agrarian Old World evidence should make it clear that occupation of tropical forest by archaic (non-*Homo sapiens*) human relatives and subsequently by our own species has deep antiquity. When insular Southeast Asia was colonized by *Homo erectus* more than 1.2 million years ago, tropical forest was probably widespread, for example (Sémah *et al.*, 2003). It seems all but certain now that our own species evolved once in Africa little more than 150,000 years ago (e.g., Forster and Mausumura, 2005). The timing, routes, and particular habitats associated with the earliest dissemination of *Homo sapiens* out of Africa are under active investigation and discussion (Macaulay *et al.*, 2005; Thangaraj *et al.*, 2005). It appears from genetic evidence obtained from modern African and Eurasian human populations that some Pygmy groups of the African rainforest represent an ancient human lineage (Forster and Matsumura, 2005), implying possible rainforest occupation at or shortly after the dawn of human emergence. Moreover, the tropical coastal areas of southern and southeast Asia were rapidly settled at some time between 85 kyr BP and 65 kyr BP by descendants of the first groups to leave Africa, indicating successful adaptations to tropical forest by some of the first human colonizers of Asia (Forster and Matsumura, 2005; Thangaraj et al., 2005).

Furthermore, available archeological and associated paleobotanical evidence from Africa robustly indicates a succession of human occupations of species-diverse tropical forest in Equatorial Guinea, Cameroon, and Zaire between c. 40 kcal . yr BP and 23 kcal . yr BP (Mercader et al., 2000; Mercader and Martí, 2003). Thus, by about 16 kcal . yr BP, when hunters and gatherers of the New World were first approaching tropical environments in what is now southwestern Mexico, tropical forest on a number of Old World continental land masses had already been well-settled.

Despite the fact that the Neotropics were colonized far later than Old World tropical regions, the transition from hunting and gathering to agriculture occurred around the same time, shortly after the Pleistocene ended, in both places. Recent multidisciplinary investigations at Kuk Swamp, Papua New Guinea, for example, have convincingly identified the onset of banana (*Musa* spp.) and probably taro (*Colocasia* spp.) cultivation by 10 kcal . yr BP using phytolith and starch grain evidence (Denham et al., 2003). Associated with these data is pollen and phytolith evidence for significant human modification of the forest near the Swamp along with alterations of the Swamp terrain itself for plant cultivation in the form of drainage ditches and stakeholes for tethering plants. Indications for an ancient and substantial human impact on tropical forest elsewhere in the Old World tropical forest is reviewed by Willis et al. (2004) (evidence for the Neotropics is summarized later).

7.3　HUMAN COLONIZATION OF NEOTROPICAL FORESTS: AN ICE-AGE ENTRY

7.3.1　The early evidence for human occupation

As elsewhere, there are formidable holes in the archeological records relating to initial human entrance into the now-humid Neotropics, and new evidence tends to accumulate slowly because early human groups were small and often highly mobile, leaving behind few tangible remains. Nonetheless, we know considerably more about the subject now than we did 10 to 15 years ago. There is convincing evidence of human occupation dating to the Late and terminal Pleistocene periods—between c. 15.2 kcal . yr BP and 11.8 kcal . yr BP (13k ^{14}C yr BP and 10k ^{14}C yr BP)—from a number of localities (Figure 7.1). The best-studied regions are often those of seasonally dry areas of southern Central America and northern South America, where focused archeological research has been of longer duration and broader scale (e.g., Cooke, 1998; López Castaño, 1995; Mora and Gnecco, 2003; Ranere and Cooke, 2003;

Stothert *et al.*, 2003). A related and significant factor is that forests in many of these areas have unfortunately long been cleared, making it much easier for archeologists to find and excavate ancient human occupations.

In much of tropical lowland Mesoamerica, including Mexico (but see Pope *et al.*, 2001), research has been more spotty, owing to the relative ease of finding sites in the sparsely vegetated areas of the region's arid and semi-arid zones. The dry highlands of Mexico, which have seen seminal work (MacNeish, 1967; Flannery, 1986), have trad-itionally been more politically stable as well. Research in potentially important seasonally dry tropical areas in the states of Guerrero, Michoacán, and Chiapas (Mexico) directed toward documenting Late Pleistocene and Early through Middle Holocene human adaptations is just beginning (Voorhies *et al.*, 2002; Piperno *et al.*, 2004). Similarly, in the seasonal forests of Bolivia and southwestern Brazil, where it appears that famous staple crops—such as manioc (*Manihot esculenta* Crantz) (Olsen and Schaal, 1999)—were domesticated, archeological research is presently under-developed and very little information is available on early human settlement and economic systems.

Significant information has been generated over the last 15 years concerning earliest human presence in the Amazon Basin, which at the site of Caverna de Pedra Pintada, located near Santarém just 10 km north of the main Amazon river channel, is convincingly dated to c. 12.9 kcal. yr BP (Roosevelt *et al.*, 1996). Human settlements were present in the wet forests of the western Amazon Basin (middle Caquetá river area of Colombia) before c. 10.1 kcal. yr BP (Cavelier *et al.*, 1995; Mora and Gnecco, 2003). However, the Amazon Basin is so vast (about the size of con-tinental U.S.A.) and often well-forested that it will be many years before any consensus is reached concerning the timing, passage routes, and possible ecological foci of early human settlement. These questions are particularly important with regard to the cultural and environmental history of the *terra firme* forests (those not under the influence of watercourses), which occupy 98% of the land area of Amazonia, and contain some of the poorest soils and lowest concentrations of plant and animal resources found anywhere in the tropics (Piperno and Pearsall, 1998). There has been a long-term and vocal dichotomy of views over whether the *terra firme* forests were well-occupied and farmed in prehistory, which only will be resolved with much more archeological research (see Neves, 1999 for an excellent review of these and other issues in Amazonian archeology).

Figure 7.1 is a map showing the Late and terminal Pleistocene archeological sites that have been located and studied. Also extremely relevant to the issue of early human settlement in the tropics is that a long and classic debate has just been con-cluded concerning the initial peopling of North and South America as a whole. Tom Dillehay's excavations at the site of Monte Verde, located in a wet temperate forest in southern Chile (Dillehay, 1997), overturned the "Clovis First" paradigm that had previously dominated archeology, especially in North America, and that held that humans were present in North America no earlier than c.11,200 BP (Meltzer, 1997). The oldest radiocarbon dates at Monte Verde go back to nearly 13,000 BP. The site is now accepted as convincing proof for an early human presence in South America by a strong consensus of archeologists. Other archeological sites located to the north of

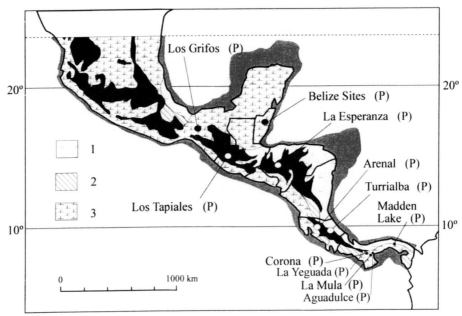

Figure 7.1. Locations of archeological sites in the Neotropics (*opposite page*) that date to between c. 15.4 kcal. yr BP and 11.4 kcal. yr BP placed against a reconstruction of Central and South American Pleistocene vegetation (*above*). Modified from Piperno and Pearsall (1998, figs. 4.1a and b). PC = Pre-Clovis site. P/P = Pre-Clovis and Paleoindian site. P = Paleoindian site. More detailed information on the sites can be found in Dillehay *et al.* (1992), López Castaño (1995), Roosevelt *et al.* (1996), Cooke (1998), Piperno and Pearsall (1998), and Ranere and Cooke (2003). Black areas are mountain zones of 1,500 m a.s.l. and greater. Grey area along coastlines is land exposed by sea level drop; in most cases, exposed land probably contains vegetation similar to adjacent terrestrial zones. The vegetation reconstruction is based on available paleoecological sequences and, for regions where such information is not available, reasonable extrapolations of data. The reconstruction is intended to provide broad guides to glacial-age vegetation. For more information on the archeological sites see Piperno and Pearsall, 1998, pp. 168–175. Explanation for the vegetational reconstruction in the figure:

1. Largely unbroken moist forest, often with a mixture of presently high-elevation and lowland forest elements. In some areas, montane forest elements (e.g., *Podocarpus, Quercus, Alnus, Ilex*) are conspicuous. Annual precipitation is lower than today, but sufficient precipitation exists to support a forest.
2. Forest containing drier elements than characteristic today. High-elevation forest elements occur, especially in moister areas of the zone. Areas near the 2,000-mm precipitation isohyet and areas with sandy soils may contain savanna woodland. The vegetation may be patchy.
3. Mostly undifferentiated thorn woodland, low scrub, and wooded savanna vegetation. Some regions (e.g., Guatemala) have temperate elements (e.g., *Juniperus*). Areas receiving greater than 2,000 mm of rainfall today may still support a drier forest, as in 2. River- and stream-side locations support a forest.
4. Quite possibly, a drier vegetation formation than 5 (below), with fewer trees and more open-land taxa. Paleoecological data are lacking for the zone.
5. Fairly open and humid forest containing many presently high-elevation taxa (e.g., *Ilex, Podocarpus, Rapanea, Symplocos*) combined with elements of the modern semi-evergreen forest and cerrado. Precipitation is lower than today but northward shifts in the southern polar fronts and other factors ameliorate precipitation reduction. The modern, seasonal forest–cerrado vegetational formations of the region are not present until about 10,000 BP
6. Desert/cactus scrub.

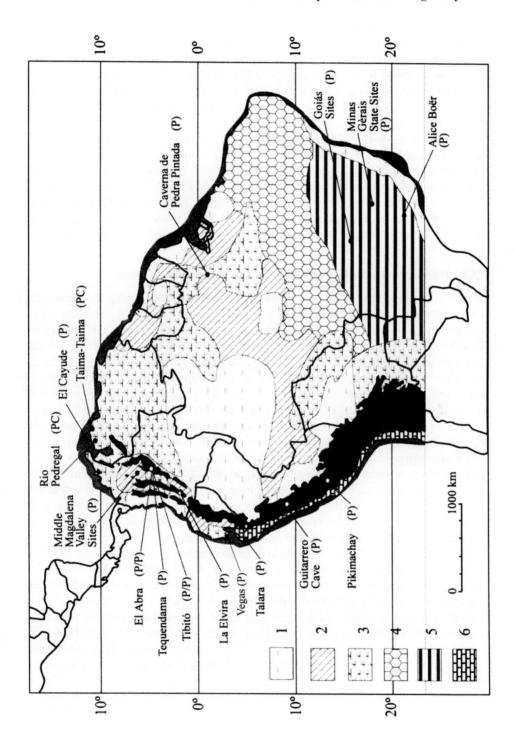

Monte Verde—such as El Abra and Tibitó, Colombia, and Taima-Taima, Venezuela, that were controversial before Monte Verde was excavated—also contain convincing proof of human occupation at c. 15.4–13.8 kcal.yr BP (Figure 7.1) (Dillehay et al., 1992; Cooke, 1998; Ranere and Cooke, 2003). Therefore, human populations must have first moved through Central America at an earlier time, although no incontrovertible sites dating to before c. 12.9 kcal.yr BP have been found yet (Cooke, 1998; Ranere and Cooke, 2003).

Claims for a pre-13,000 BP human presence in South America, including a cave in northern Brazil known as Pedra Furada, are not well-supported because artifacts were not recovered from securely dated contexts, or pieces of stone attributed to human manufacture do not display convincing signs that they had been altered and used by people (e.g., Meltzer et al., 1994; Piperno and Pearsall, 1998, p. 169). Thus, there is no convincing evidence at the present time that people occupied the Neotropics during the Last Glacial Maximum and the 3,000 or so years immediately following it.

7.3.2   Pleistocene landscapes and early human modification of them

Of course, showing that people were living at tropical latitudes during the Pleistocene is not the same as proving they were living in tropical forest and surviving off its resources. In fact, early scholars assumed that the earliest hunters and gatherers of the New World preferentially exploited the numerous, now-extinct large game animals that were available to them. These investigators proposed that, because megafauna and other sizeable animals would have been rare in tropical forest, humans migrating from north to south would have largely avoided densely wooded areas, living instead in the more open landscapes they believed characterized the Pacific watershed of Central America, northern South America, and the intermontane valleys of the Andes (e.g., Sauer, 1944; Lothrop, 1961; see Ranere, 1980 for one of the first robust counterarguments based on archeological excavation and analysis). Some scholars writing later would agree with this assessment, going so far as to argue that Holocene hunters and gatherers could not have survived for long in tropical forest without access to a cultivated food supply because wild food resources, especially carbohydrates, were supposedly scarce (e.g., Bailey et al., 1989; see Colinvaux and Bush, 1991 and Piperno and Pearsall, 1998 for responses and further discussion).

The large corpus of paleoecological data accumulated during the past 20 years, discussed in detail in other chapters of this book, shows that Late Pleistocene environmental conditions were indeed significantly different from those of the Holocene in ways that could have influenced early human colonization and the specific kinds of habitats that early hunters and gatherers exploited. Reconstructions from lacustrine pollen and phytolith data demonstrate the presence of a variety of vegetation communities. These range from dense, species-diverse forest to open, shrub- and grass-dominated formations. There is strong evidence indicating some regions were considerably drier than today during the Late Pleistocene—for example, Petén, Guatemala, much of Pacific-side Central America, and parts of northern South America. At the present time, these regions receive between 1.2 and 2.6 meters of precipitation and their potential vegetation is deciduous or drier forms of semi-evergreen forest. During

the Late Pleistocene, their vegetational formations were dominated by low woodlands, thorn scrublands, and savannas (Leyden, 1984, 1985; Leyden *et al.*, 1993;
Piperno and Pearsall, 1998; Piperno and Jones, 2003). It was during the first 2,000
years of the Holocene that tropical forest developed on these landscapes. Recently
accumulated data from lakes and large swamps located in the Río Balsas watershed
in tropical southwest Mexico (Guerrero state), where the potential vegetation is a
tropical deciduous forest, indicates a drier Late Pleistocene climate and more
open vegetation there as well (Piperno *et al.*, 2004). Thus, when humans first
penetrated tropical latitudes, forests did not cover landscapes to the extent they
do today.

   Where, however, annual rainfall is above about 2.6 meters today and the actual or
potential vegetation is evergreen and semi-evergreen forest, the evidence is strong that
Pleistocene landscapes were mostly forested. This is empirically demonstrated in
Caribbean-side Panama at the Gatun Basin (Bartlett and Barghoorn, 1973), and
probably was the case throughout most of the Caribbean watershed of Panama,
Costa Rica, Nicaragua, and Honduras. It is also demonstrated at Pacific watershed
sites in Panama at elevations of between 500 and 700 m above sea level, such as La
Yeguada and El Valle (Bush *et al.*, 1992; Bush and Colinvaux 1990), and at various
locations in South America (see Colinvaux *et al.*, 1996a, b and Chapter 3 in this book).
Therefore, considerable portions of the Pleistocene Neotropical landscape were
forested.

   How can we relate this corpus of data on environmental history to questions
concerning early cultural adaptations to tropical latitudes? Arguably, one of the best
ways is to correlate reconstructed Pleistocene vegetation with archeological sites of the
same age located nearby. If, for example, human settlement before c. 11.4 kcal . yr BP is
largely confined to open areas, the implication would be strong that forests were not
persistently lived in and that human populations were surviving for the most part off
resources typical of non-wooded environments (e.g., large animal game and plants
like cacti and tree legumes found in drier types of vegetation). Another way to assess
the issue is to examine actual dietary evidence from archeological sites to directly
determine what kinds of resources people were exploiting. There is presently more
evidence to consider from the first than from the second option. Fortunately, some of
the best-documented archeological sites are located near lakes from which detailed
paleoenvironmental information has been generated. Furthermore, in the cases where
early archeological sites are not in the vicinity of old lakes, major characteristics of the
Pleistocene environment can still be reasonably inferred for them by using paleoecological information recovered from zones with a similar modern potential vegetation. When the relevant data are evaluated, the following patterns emerge (see Piperno
and Pearsall, 1998, pp. 169–175, for more details) (Figure 7.1).

   The few available archeological sites where a pre-Clovis (pre-12.9 kcal . yr BP)
occupation is indicated are located in deserts/grasslands/open woodlands at low
elevations in northern South America (two sites in Venezuela) and open environments
(páramo) at two northern Andean locations. No incontrovertible human occupation
is located so far in an area reconstructed as having supported tropical forest vegetation. However, the number of sites is still far too few to draw firm conclusions as to

whether people of this time period were preferentially selecting one type of habitat and its plant and animal resources over another.

Archeological sites of later, Paleoindian age—c. 12.9–11.4 kcal.yr BP (11–10k ^{14}C yr BP)—are greater in number and more likely to be representative of habitat choices that people made, and the sites were located in a diverse array of environments. (For purposes of simplicity, I call all human occupations dated to between c. 12.9 kcal.yr BP and 11.4 kcal.yr BP Paleoindian, even though all of them do not contain characteristic Clovis culture types of tools.) They included alpine meadow (in Guatemala), low- and higher-elevation forest (e.g., in Panama, Costa Rica, Colombia, and Brazil), and open, thorny, and/or temperate scrub/savanna types of vegetation (e.g., in Mexico, Belize, Panama, and Venezuela) (see Cooke, 1998; Piperno and Pearsall, 1998, pp. 169–175; and Ranere and Cooke, 2003 for further descriptions of these sites).

Out of the 24 Paleoindian localities included in this survey, 10 were located in some kind of tropical forest. Moreover, moving through southern Central America and entering South America without encountering and living in forest some of the time may not have been possible. Data on human dietary patterns recovered from the sites that can buttress arguments of tropical forest occupation and resource exploitation are often scant because people typically did not stay in one spot long enough for a sizeable midden of food and other remains to accumulate. However, Caverna de Pedra Pintada in Brazil yielded abundant carbonized nut and seed fragments from a variety of trees—such as palms and Brazil nuts—as well as faunal remains of large and small mammals that were clearly derived from the forest and were dietary items (Roosevelt et al., 1996). The archeological phytolith and carbonized seed and nut record from central Panama also contains indications that tropical forest plants were being exploited and eaten between c. 12.9 kcal.yr BP and 11.4 kcal.yr BP (Piperno and Pearsall, 1998; Dickau, 2005).

We should remember that Pleistocene forests were often considerably different in their floristic compositions when compared with forests that grow in the same areas today; few of them appear to have modern analogs. Thus, we cannot expect to be able to directly compare potential plant and animal resources of modern forests with those that existed during the Late and terminal Pleistocene periods. In many areas, forests probably contained more trees tolerant of lower rainfall than extant examples, and they perhaps had more open canopies due to reduced precipitation and also lowered light use efficiencies during photosynthesis that resulted from reduced atmospheric CO_2 concentrations (Sage, 1995; Cowling and Sykes, 1999). Phytolith studies of the Gatun Basin core sequence from Panama, which is located very near the Madden Lake Paleoindian archeological sites, add some empirical weight to these inferences (Piperno et al., 1992). They indicate that prior to c. 11.4 kcal.yr BP arboreal associations contained more trees characteristic of modern deciduous forests—such as the Chrysobalanaceae. Pre-11.4 kcal.yr BP phytolith records are, in fact, a very good match with those constructed from directly underneath modern deciduous forest in Guanacaste, Costa Rica. In contrast, plants that are significant components of modern semi-evergreen forests of the area (e.g., the Annonaceae [*Guatteria*], bamboos [*Chusquea*], and palms) don't enter the Gatun Basin record until after 11.4 kcal.yr BP (Piperno et al., 1992).

Thus, these records appear to reveal a significant change in forest composition over the Pleistocene/Holocene boundary characterized by increases of trees and understorey plants that were likely responses to rising moisture levels. In Panama and probably elsewhere, the Pleistocene forests that humans occupied were likely to have been drier and more open—and possibly contained a higher animal biomass—than those that grow in the same regions today.

We should also remember that early human populations appear to have been actively modifying on their own the new landscapes they were encountering—largely, it seems, by fire. For example, at Lake La Yeguada, Panama, where detailed, multi-proxy assessments of fire and other vegetational disturbances were carried out by the author, Mark Bush, and Paul Colinvaux, frequencies of charcoal and burnt phytoliths indicate that human firing of the vegetation around the 16,000-year-old lake began at c. 12.9 kcal. yr BP and continued without pause through the Early and Middle Holocene periods (Piperno *et al.*, 1990, 1991a, b). Archeological data document initial human colonization of the lake's watershed when vegetational disturbances are first apparent. In fact, a projectile point that dates to some time in the 12.9–11.4 kcal. yr BP period was found on the La Yeguada shoreline (Ranere and Cooke, 2003). The lacustrine evidence, which includes detailed comparisons of the phytolith, pollen, and charcoal information with modern analog data from different types of mature forests and culturally modified vegetation, has been discussed in detail and illustrated elsewhere and will not be repeated here (e.g., Piperno 1993, in press c; Piperno *et al.*, 1990, 1991a, b; Bush *et al.*, 1992). A summary of some of the data is provided in Figure 7.2, where it can be seen that records of burnt phytoliths at La Yeguada show close resemblances to phytolith profiles constructed from modern vegetation undergoing active human disturbance, but not from modern forests experiencing little to no human activity.

7.3.3   After the Pleistocene: the origins and spread of tropical forest agriculture during the Early Holocene

As in several other areas of the world, the lowland Neotropical forest witnessed an independent emergence of plant food production and domestication not long after the Pleistocene ended (Piperno and Pearsall, 1998; Diamond, 2002). Combined information from archeological, molecular, and ecological research tells us that—out of the more than 100 species of plants that were taken under cultivation and domesticated by native Americans before Europeans arrived—more than half probably came from the lowland tropical forest (Piperno and Pearsall, 1998; Piperno, in press a). Both hemispheres of tropical America were involved. In lower Central and northwestern South America, where the greatest amount of work has been carried out to date, the domestication and spread of important native crops like maize (*Zea mays*), manioc (*Manihot esculenta*), at least two species of squash (*Cucurbita moschata* and *C. ecuadorensis*), arrowroot (*Maranta arundinacea*), yams (*Dioscorea trifida*), and líren (*Calathea allouia*) between 10,000 and 5,000 years ago has been empirically documented through phytolith, pollen, and starch grain research (Pearsall, 1978; Monsalve, 1985; Mora *et al.*, 1991; Piperno and Pearsall, 1998; Bray, 2000;

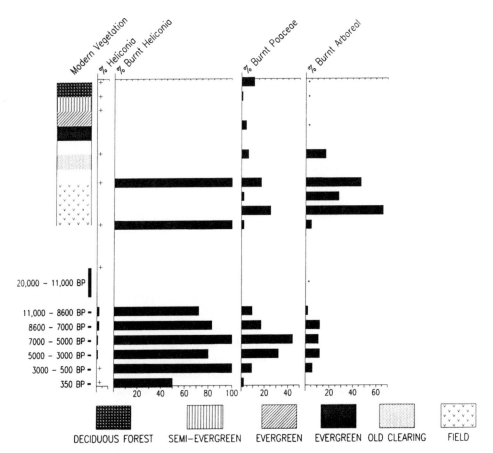

Figure 7.2. The frequencies of early successional phytoliths and burnt successional and arboreal phytoliths in modern tropical forests and through time at Lake La Yeguada. The data profiles from the Late Pleistocene period at El Valle, where human disturbance was not detected, are also displayed for comparison. At c. 12.9 kcal. yr BP at La Yeguada, charcoal levels also increase by several orders of magnitude, and pollen and phytoliths from grasses and other invasive taxa increase substantially. Reprinted from Piperno and Pearsall (1998, fig. 4.4). Description of modern vegetation in the figure:

Evergreen forests are from El Cope, Panama (hatched symbol) and north of Manaus, Brazil (black symbol). Semi-evergreen forest is Barro Colorado, Panama. Deciduous forest is Guanacaste Province, Costa Rica. Old clearing (cleared from forest and planted with banana 50 years ago) is from Guancaste, Costa Rica. Fields are present day slash and burn agricultural plots from Panama planted in manioc and maize. Phytolith frequencies for each modern site are averages from a series of soil transects or "pinch samples" taken from the upper soil surface at the sites (see Piperno, 1988 for details). Circa 23.9–12.9 kcal. yr BP records are from El Valle and La Yeguada. Circa 12.9–0.35 kcal. yr BP records are from La Yeguada. + = Observed at a frequency of less than 1%.

Piperno *et al.*, 2000 a, b; Pope *et al.*, 2001; Mora and Gnecco, 2003; Pearsall *et al.*, 2003, 2004; Piperno and Stothert, 2003; Dickau, 2005; Piperno, in press a and b).

Interestingly, joining the evidence from archeology, molecular biology, and botany also tells us that most important lowland crops in both Central and South

America were originally brought under cultivation and domesticated in the seasonal tropical forest (e.g., Piperno and Pearsall, 1998; Olsen and Schaal, 1999; Matsuoka *et al.*, 2002; Sanjur *et al.*, 2002; Westengen *et al.*, 2005; Piperno, in press a). Figure 7.3 provides a guide to the geography of origins for various crops and shows the locations of archeological sites with early—c. 11.4–5.7 kcal. yr BP (10–5k ^{14}C yr BP)—remains of domesticated plants. Particularly important were regions such as the Balsas River Valley, southwestern Mexico (domesticated there were maize and quite possibly the lowland Mesoamerican squash *Cucurbita argyrosperma*, the cushaw and silver-seeded squashes); the Cauca and Magdalena Valleys of Colombia and adjacent mid-eleva-tional areas (for sweet potato, *líren*, arrowroot, and possibly the South American lowland squash, *Cucurbita moschata*); southwestern Brazil/eastern Bolivia (the probable birth place of manioc and probably other crops), and southwestern Ecuador and possibly northwestern Peru (for a species of *Cucurbita* [*C. ecuadorensis*], South American cotton [*Gossypium barbadense*], and probably the South American jackbean [*Canavalia plagiosperma*]).

The Amazon Basin has long been an area of interest for crop plant origins. However, although some crops like manioc were domesticated on the fringes of the Basin, few to no others that would become staple foods with the exception of the peach palm (*Bactris gasipaes*) appear to have been domesticated within its core area (Piperno and Pearsall, 1998). And as Harlan (1971) predicted, there appears to be no single, major center of agricultural origins in South America at all. Even after plants were domesticated and dispersed out of their geographic cradles of origin, peoples in other regions continued to experiment with, modify them, and significantly change them phenotypically. One prominent example of this is maize. There are hundreds of different varieties adapted to a wide range of ecological conditions. Paleoecological and archeological evidence indicates that the crop had been well-dispersed and established in South America by c. 6.3 kcal. yr BP (Monsalve, 1985; Bush *et al.*, 1989; Mora *et al.*, 1991; Piperno and Pearsall, 1998; Pearsall *et al.*, 2003, 2004; Iriarte *et al.*, 2004).

A significant number of investigators interested in the origins of agriculture, including this one, believe that changing ecological circumstances at the end of the Pleistocene combined with a consideration of how efficiently (in calories obtained per person per hour) full-time hunters and gatherers could exploit their post-glacial landscapes, may provide satisfactory answers for why and when agriculture arose. These end-Pleistocene transitions have often been depicted as a kind of environmental amelioration for human populations in the literature on cultural adaptations during this period. In all likelihood, however, subsistence options for low-latitude hunters and gatherers, and perhaps those of other areas of the world, became a great deal *poorer* when the ice age ended.

For example, during the Pleistocene more than 30 genera of now-extinct, large- and medium-sized grazers and browsers—including horses, mammoths, and giant ground sloths—roamed the tropical landscape, and it is clear that humans routinely hunted some of them (Cooke, 1998; Piperno and Pearsall, 1998; Ranere and Cooke, 2003). The animals were gone by c. 11.4 kcal. yr BP, at which point hunting and gathering became a far different enterprise. When compared with the Pleistocene

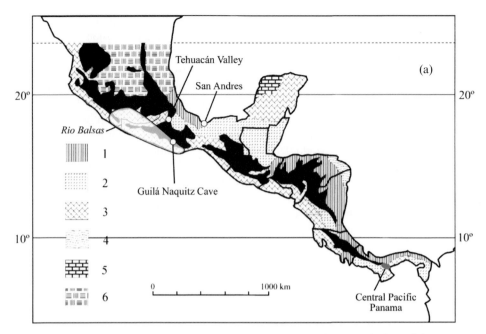

Figure 7.3. Domestication areas for various lowland crop plants as indicated by present molecular, archeological, and ecological evidence. Also shown are the locations of archeological and paleoecological sites in Central America (a) and South America (b) with early (11.4–5.7 kcal . yr BP) domesticated seed and root crop remains. Detailed information on the sites can be found in MacNeish (1967), Pearsall (1978), Monsalve (1985), Flannery (1986), Bush *et al.* (1989), Mora *et al.* (1991), Cavelier *et al.* (1995), Smith (1997), Piperno and Pearsall (1998), Piperno *et al.* (2000a, b), Pope *et al.* (2001), Pearsall *et al.* (2003, 2004), Piperno and Stothert (2003), and Piperno (in press a). Domestication areas for Mesoamerica:

Mexico: maize (*Zea mays*) and squash (*Cucurbita argyrosperma*); also possibly *jícama* (*Pachyrhizus* spp).

Domestication areas for South America:

D1. Sweet potato (*Ipomoea batatas*), squash (*Cucurbita moschata*), arrowroot (*Maranta arundinacea*), achira (*Canna edulis*—lower, mid-elevational in origin); also possibly sieva beans (*Phaseolus lunatus*), *yautia*, or cocoyam (*Xanthosoma saggitifolium*), and *lirén* (*Calathea allouia*).
D2. Yam (*Dioscorea trifida*); also possibly *yautia* (*Xanthosoma saggitifolium*), *lirén* (*Calathea allouia*), and chile peppers (*Capsicum baccatum*).
D3. Manioc or yuca (*Manihot esculenta*), peanut (*Arachis hypogaea*), chili pepper (*Capsicum baccatum*), and possibly squash (*C. maxima*).
D4: Cotton (*Gossypium barbadense*), *Cucurbita ecuadorensis*, possibly jack bean (*Canavalia plagiosperma*).

Notes: Probable areas of origin for other lowland, pre-Columbian cultivars include the wet forests of the northwestern Amazon Basin (*Bactris gasipaes* [the peach palm] and possibly *Sicana odorifera* [cassabanana]), eastern Mexico (chili peppers), and the Yucatan Peninsula (*G. hirsutum* [cotton]).

Explanation of modern vegetation:

(a)   1. Tropical evergreen forest (TEF). 2. Tropical semi-evergreen forest (TSEF). 3. Tropical deciduous forest (TDF). 4. Savanna. 5. Low scrub/grass/desert. 6. Mostly cactus scrub and desert.
(b)   1. TEF. 2. TSEF. 3. TDF. 4. Mixtures of TEF, TSEF, and TDF (TSEF and TDF grow over substantial areas of the southern parts of the Guianas and south of the Orinoco River). 5. Mainly semi-evergreen forest and drier types of evergreen forest. Floristic variability can be high in this zone. 6. Savanna. 7. Thorn scrub. 8. Caatinga. 9. Cerrado. 10. Desert.

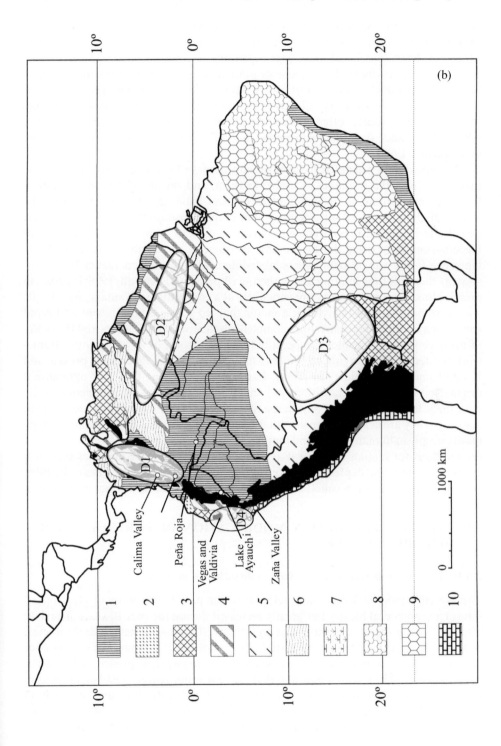

Calima Valley
Peña Roja
Vegas and Valdivia
Lake Ayauchi
Zaña Valley

1
2
3
4
5
6
7
8
9
10

D1
D2
D3
D4

10°
0°
10°
20°

1000 km

0

(b)

fauna, animals that were available to human hunters between c. 12.5 kcal.yr BP and 11.4 kcal.yr BP occurred at much lower biomass and were also typically small-sized. Moreover, because tropical forest was expanding into the considerable areas where tree cover had previously been sparse or more discontinuous, hunters and gatherers had to more routinely exploit forest plants, but would find them to be a generally poor source of calories and widely dispersed in space. The most starch-dense examples (roots, rhizomes, tubers) often contained high amounts of toxic chemicals and other defenses that made them time-consuming and difficult to convert into food (Piperno and Pearsall, 1998).

Empirical data generated recently on how modern hunters and gatherers choose their diets from the resources available to them, and on the relative efficiencies of foraging and farming in various modern tropical habitats, have also proved to be significant illuminators of subsistence change at the transition to agriculture (Kennett and Winterhalder, in press). These data can be used to predict that, in contrast to the situation that existed during the Late Pleistocene, plant cultivation in the Early Holocene forest was probably a less labor-intensive and more energetically-efficient strategy than full-time hunting and gathering (Piperno and Pearsall, 1998; Piperno, in press). Thus, nascent farmers were very likely at a competitive advantage over people who were not growing their food, a factor that led to the establishment and rapid spread of agricultural systems (for a complete discussion of these issues and the utility of using evolutionary ecology, especially foraging theory, as an explanatory framework for agricultural origins, see Piperno and Pearsall, 1998, Piperno, in press a, and Kennett and Winterhalder, in press). Explanations such as these for agricultural origins and other major transitions in human lifeways are not environmental determinism, at least not the form of it that a fair number of anthropologists are prone to deriding. They are acknowledgments that ecological factors and evolutionary biology matter deeply in human affairs, and that scientists need not shy away from nomothetic explanations for human behavior if available empirical evidence indicates that such kinds of generalizing explanations are supportable (see Piperno and Pearsall, 1998, Piperno, in press a, and Kennett and Winterhalder, in press, for further discussions).

7.4 EVIDENCE FOR HUMAN MODIFICATION OF FORESTS BY PRE-HISTORIC FARMERS

7.4.1 Lake records and detecting human disturbance in them

The available paleoecological data relevant to the past 11,000 years of Neotropical forest environmental history has greatly increased during the past 10 years. The data often go hand-in-hand with the discovery and investigations of nearby archeological sites, which provide allied information largely unobtainable from lake studies concerning which forest plants and animals were most frequently exploited and manipulated by people. The research indicates that the development and spread of agriculture in the American tropics exerted profound influences on the structure and composition of the vegetation. Many regions far removed from ancient centers of civilizations experienced systematic interference with, and sometimes removal of, tropical forest thousands of years ago. The records show that fire was an important instrument of vegetational modification for people practicing agriculture in Central and South America. Figures 7.4 and 7.5 contain summaries of this information from a representative sample of the paleoecological sites that have been examined.

A caveat is that the records were not all examined by the same methods. Detailed pollen information is available for all of them, but a minority contain both pollen and phytolith information, meaning that the resolution of the data is not the same for every one. Phytoliths, for example, are sometimes of greater utility than pollen for documenting shifts in taxa of primary and older types of forest, while pollen data can usually more accurately document increases and decreases of secondary forest growth. This all has to do with differences in production characteristics and taxonomic specificity (see Piperno, 1993, in press b). Charcoal data were accumulated for most of the sequences considered here. Charcoal frequencies were not all computed in exactly the same manner, but most studies employed counts of charcoal made from the pollen or phytolith slides which then were converted into concentration and/or influx values in the same way as was done for associated pollen and phytolith data. For all of the sequences discussed, the sampling resolution was no greater than 300 years and was often less than 100 years.

A commonality of views has emerged among paleoecologists working in the tropical forest about how to interpret charcoal, pollen, phytolith, and other lacustrine data sets, making it possible for trends recognized from sequence to sequence to be meaningfully compared and evaluated. For example, in a long depositional sequence sampled at an appropriate resolution, most investigators interpret the continual presence of high frequencies of charcoal over thousands of years of a lake's history as being indicative of human-set fires. This is because the high humidity, moisture, and shaded understorey of forests make the likelihood of natural ignitions occurring that frequently very low. Similarly, when large proportions of phytoliths from early invasive herbaceous plants—such as the Poaceae and *Heliconia*—are continually burned over long periods of time human interference is indicated. It is hard to imagine a natural process that would create and then ignite large areas of early successional plant growth that often. (Burnt phytoliths are easily recognizable because they obtain charred surfaces when exposed to fire while retaining their diagnostic morphological features—see Piperno, in press b.)

Relying on pollen and phytolith profiles constructed from modern, old-growth forests where vegetational censuses and other detailed plant inventories are available, as well as vegetation currently experiencing varying types of anthropogenic pressure, various investigators have also developed pollen and phytolith population markers of human disturbance and agricultural activity for different types of Neotropical forest (e.g., Piperno, 1988, 1993, 1994, in press c; Bush, 1991; Rodgers and Horn, 1996; Bush and Rivera, 1998; Clement and Horn, 2001; Piperno and Jones, 2003). These efforts went hand-in-hand with those dedicated to constructing large modern reference collections of Neotropical pollen grains and phytoliths (references above; see also Colinvaux *et al.*, 1999). They have resulted in the identification of many tree and shrub taxa that were unknown microfossils in older paleoecological work, and paleoenvironmental reconstructions that are more finely resolved.

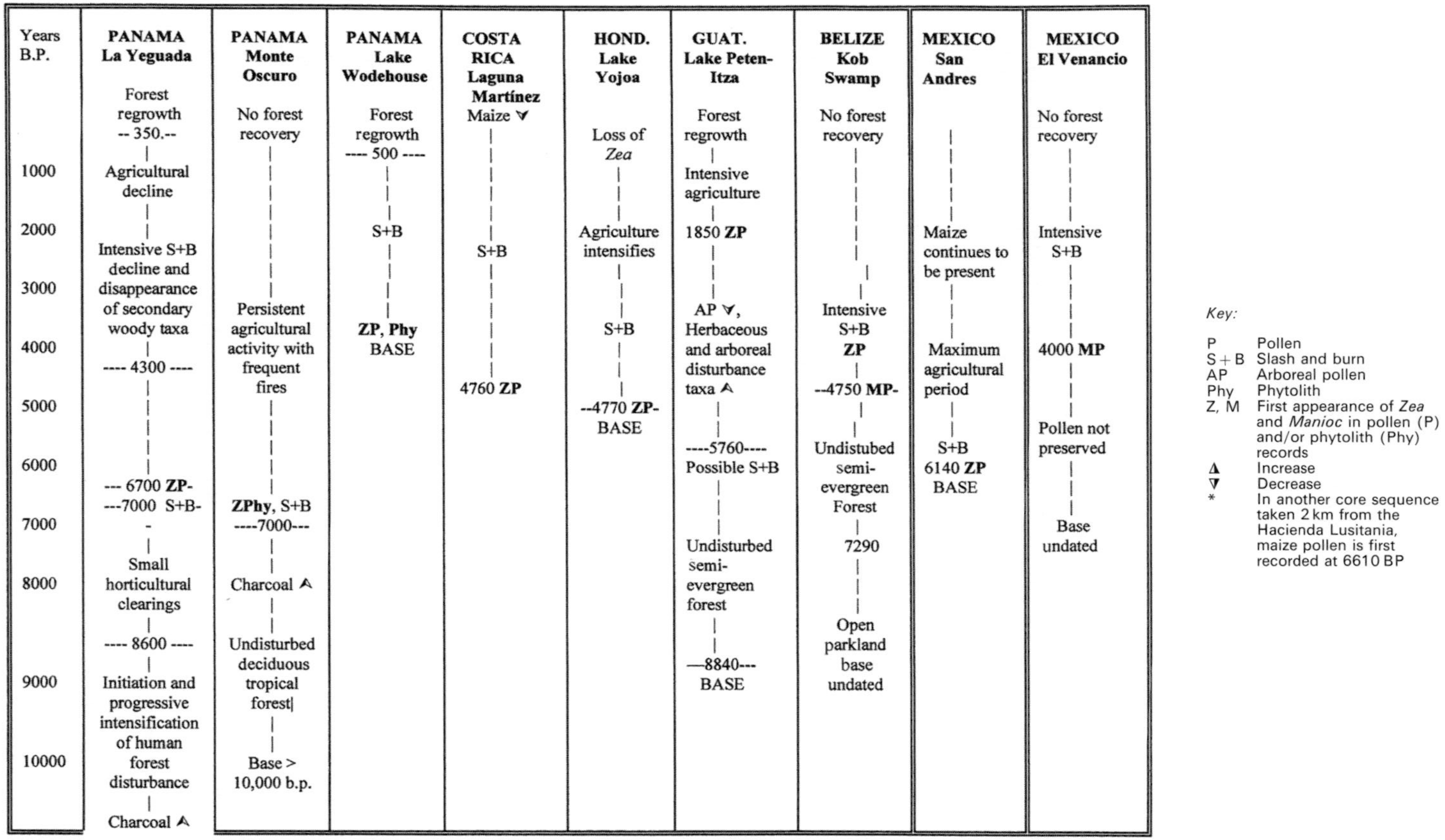

Figure 7.4. Summary of charcoal, pollen, and phytolith data for vegetational history and human impacts on tropical forests in Central America from various paleoecological sites. *Sources:* La Yeguada (Piperno *et al.*, 1991a, b; Bush *et al.*, 1992); Monte Oscuro (Piperno and Jones, 2003); Laguna Martínez (Arford and Horn, 2004); Lake Wodehouse (Bush and Colinvaux, 1994; Piperno, 1994); Lake Yohoa (Rue, 1987); Lake Peten-Itza (Islebe *et al.*, 1996); Kob Swamp (Pohl *et al.*, 1996); San Andrés (Pope *et al.*, 2001); El Vinancio (Piperno, in press c). Dates in radiocarbon years.

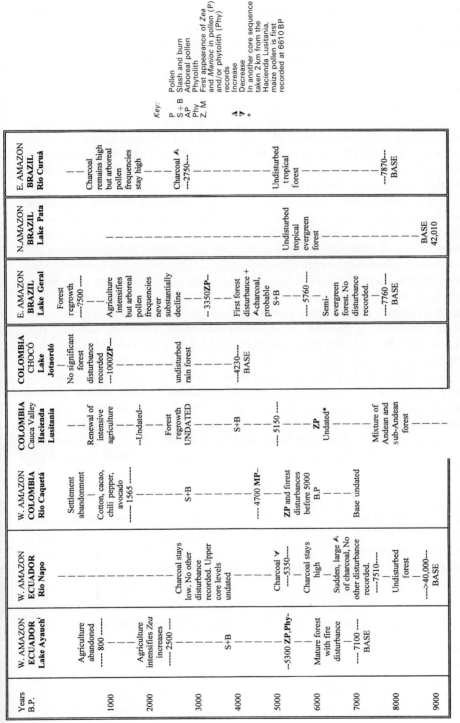

Figure 7.5. Summary of charcoal, pollen, and phytolith data for vegetational history and human impacts on tropical forests in South America from various paleoecological sites. *Sources:* Lake Ayauchi (Bush *et al.*, 1989; Piperno, 1990); Rio Napo (Athens and Ward, 1999); Rio Caquetá (Mora *et al.*, 1991); Hacienda Lusitania (Monsallve, 1985); Lake Jotaordó (Berrio *et al.*, 2000); Lake Geral (Bush *et al.*, 2000); Lake Pata (Bush *et al.*, 2004; Colinvaux *et al.*, 1996a); Rio Curuá (Behling and Lima de Costa, 2000). Dates in radiocarbon years.

For example, in wetter forests, increases of weedy herbaceous taxa along with pollen from secondary forest trees—such as *Pilea*, *Trema*, and *Alchornea*—occurring at the same time that pollen or phytoliths from a variety of more mature forest taxa are declining usually can be confidently interpreted as human disturbance (Bush *et al.*, 1989, 1992, 2000; Bush and Colinvaux, 1994; Clement and Horn, 2001). In regions with greater seasonality of rainfall, where tree diversity is lower and polleniferous families like the Moraceae are important components of the mature forest, rapid and steep declines of Moraceae and other tree pollen grains combined with simultaneous increases in weedy herbaceous plants and trees—such as *Trema* and *Cecropia*—mark human forest clearance (Piperno *et al.*, 1991b; Jones, 1994; Pohl *et al.*, 1996). It is extremely unlikely that ecological changes such as these can have taken place during the last 10,000 years in the absence of substantial human interference. After the ice age ended, climate change and its effects were never of a sufficient magnitude to cause these kinds and degrees of vegetational deflection and replacement in tropical forest.

The identification of pollen and phytoliths from cultivars provides another important source of information for documenting past human influences and agricultural practices. It should be remembered, however, that while maize micro-remains can be straightforwardly recovered and identified, few other important crop plants routinely leave recognizable pollen or phytolith signatures in lacustrine records. This is especially the case for the many root crops that probably were significant components of prehistoric agricultural systems from ancient times in South America. It is further noted that maize pollen grains are heavy and don't travel very far, so that when recovered they may be reflecting mainly local cultivation at distances of no more than 50 to 60 m from the lake edge (Clement and Horn, 2001). The result is that there may be strong indications of forest disturbances and clearing in lake records without evidence of the crops that caused the clearing. Phytolith and starch grain records from archeological sites located in or near the lakes' watersheds may help to fill in these lacunae, as they have already done in areas of Central America and northern South America (e.g., Piperno and Pearsall, 1998; Piperno *et al.*, 2000b; Perry, 2002; Pearsall *et al.*, 2004).

Often, all of the indicators just noted were registered in sequences from Central and South America—to be discussed shortly—where forests appeared to have been significantly modified by humans. Sometimes, agricultural intensification from pre-existing, and presumably longer fallow, slash and burn systems could be detected because forest-clearing began to involve trees of early secondary forest thousands of years after the initial clearing of older forest was documented.

7.4.2   Major trends and patterns of prehistoric tropical forest modification

Very significant to more moderate human impacts can be identified on the full spectrum of tropical forest vegetation—evergreen, semi-evergreen, and deciduous formations—from Mexico to the Amazon Basin (Figures 7.4 and 7.5). Often, especially in the highly seasonal forests of Central America, the impacts were intense, and a significant number of arboreal taxa were reduced to low levels by agricultural activity. From a large swamp called El Venancio, located in Guerrero, Mexico—very close to the presumed cradle of maize domestication—we have our first indications of ancient agriculture and associated deforestation in the deciduous forests of the Central Balsas River Valley (Piperno, in press, c). Here, maize pollen, high amounts of charcoal, and pollen indications of severe forest-clearing are present from the base of the sequence dated to c. 4.5 kcal. yr BP. An older Mexican sequence from San Andres, located on the Caribbean coast of Tabasco in what would become the Olmec heartland 3,000 years later, records slash and burn cultivation with maize starting at 6,200 BP (Pope et al., 2001). This record provides additional confirmation that maize had been domesticated and dispersed out of its area of origin by the seventh millennium BP.

Two pollen and charcoal records studied by John Jones from swamps called Kob and Cobweb, located about 55 km apart in northeastern Belize, show an early phase of intensive deforestation resulting from slash and burn agriculture starting at c. 4.5 kcal. yr BP (Jones, 1994; Pohl et al., 1996). Moving farther south to Lake La Yeguada, Panama, the initiation of slash and burn agriculture is indicated by 7,000 BP, when primary forest trees decrease greatly in the phytolith record and pollen from early secondary woody growth (Cecropia, Ficus) increases significantly. At c. 4.7 kcal. yr BP even secondary growth taxa and charcoal decline greatly, suggesting that a significant portion of the woody growth was being cut (and could no longer contribute charcoal to the record) because fallow periods were being shortened. This region of Panama has also seen long-term archeological research, and its pre-Columbian cultural records are among the best in the lowland Neotropics (e.g., Piperno and Pearsall, 1998; Cooke, 2005). Archeological foot surveys and excavations document types of settlements, small hamlets, or hamlet clusters—during the c. 7.8–2.7 kcal. yr BP period—near La Yeguada that are very similar in size and other aspects to those of modern shifting cultivators.

Another aspect of paleoecological sequences that draws attention is the variability in human land-use patterns that seems to be evident on a macro-regional scale, remembering that the records cannot tell us what human populations may have been doing much beyond the immediate watersheds of the lakes. For example, forest disturbance resulting from human agricultural pressure starts in the western and eastern Amazon Basin (Lakes Ayauch[1] and Geral, respectively) during the sixth millennium BP, and intensifies during the next few millennia (Piperno, 1990; Piperno and Pearsall, 1998, pp. 280–281; Bush et al., 1989, 2000). However, during the periods of maximal agricultural intensification recorded in the Amazonian pollen and phytolith records, which occurred between c. 3.6 kcal. yr BP and the Conquest period, arboreal pollen and phytolith frequencies never decline and frequencies of pollen from grasses and other weedy herbs never increase to the point indicating a large-scale destruction of the forest (compare Figures 7.4 and 7.5). A record from the Chocó, Colombia (Lake Jotaordó) is an example of even more minimal human disturbance on a wet northern South American landscape. In this case, maize is not recorded until very late in the sequence at c. 1 kcal. yr BP, but this signal of agriculture is still not accompanied by signs of significant forest-clearing (Berrío et al., 2000). In a long sequence from the Río Napo region of the Ecuadorian Amazon, charcoal is common during certain periods, but neither maize nor signs of forest-clearing are recorded at all (Athens and Ward, 1999).

On the whole, the Central American sequences contain evidence for significant to severe forest-clearing more often than do the South American records. These differences may relate to the following factors: presence of smaller populations of shifting cultivators, smaller-scale agricultural systems with longer fallow periods, and the greater importance of root crops like manioc and sweet potato than of soil-demanding crops such as maize in the Amazonian and other South American lake regions. One or all of these could have resulted in less expansive and destructive agriculture. It may also be highly significant that many of the Central American lakes are in areas whose real or potential vegetation is deciduous or otherwise highly seasonal tropical forest. The seasonal tropical forest offers less heavily leached and more highly fertile soils for agriculture, and the vegetation can be effectively cleared and prepared for planting using simple, slash and burn techniques that do not require the use of stone axes. It is no accident that population densities today in the lowland tropics are higher and agriculture is more developed and sustained in areas of former drier forest, and that far fewer of these forests remain than of semi-evergreen and evergreen formations. Although they do not carry the distinction enjoyed by their rainforest relatives, highly seasonal forests likely have been of far more use to humans for a longer period of time.

The lake and other paleoecological sequences in which a human modification of the regional forest—through burning, cultivar presence, or tree-felling cannot be detected at all using any of the available markers developed for this purpose—should also be highlighted. They include long, continuous records covering the entire Holocene period or substantial portions of it from three different lakes and watersheds in northern Brazil, including Lake Pata (Colinvaux *et al.*, 1996a; Bush *et al.*, 2004). Similarly, phytolith-rich records from soils sampled from directly underneath a forest preserve just north of Manaus could not detect a human influence on the vegetation during the past 7,000 years (Piperno and Becker, 1996). Although prehistoric human impacts on the tropical forest can be identified in many regions of Central and South America, the chronologies and trajectories of these impacts varied considerably. The spatial extent of significant human settlement and modification of the Neotropical forest before Europeans arrived is a largely answerable, empirical question on which information will be steadily, sometimes incrementally, accumulated through continued archeological and paleoecological explorations.

7.5   SUMMARY

Neotropical forests were first settled during the final phases of the last ice age by hunters and gatherers who in short order began to modify some of them, especially with fire. Not long after the Pleistocene %nded, humans created systems of plant cultivation that, during the following 5,000 years, would result in the widespread development of slash and burn agriculture. The changes wrought to the lowland forests by early Neotropical farmers were often more severe and of more widespread extent than had been inflicted by the extreme physical elements of the Last Glacial Maximum and later Pleistocene, when intense drying, cooling, and reduced levels of atmospheric CO_2 impacted the vegetation forcefully. In many areas, these ice-age conditions had not caused forests to disappear, just to change some of their floristic affinities, but humans then burned, cut, and eventually removed large tracts of some of these wooded landscapes not long after they settled them. Beliefs that pre-Columbian human populations used tropical landscapes in ways that fostered conserving or protecting most of the natural flora (see Piperno, in press c for a discussion of this issue) must be considered against the evidence presented here.

7.6 REFERENCES

Arford, M. R. and Horn, S. P. (2004) Pollen evidence of the earliest maize agriculture in Costa Rica. *Journal of Latin American Geography* **3**, 108–115.

Athens, J. S. and Ward, J. V. (1999) The Late Quaternary of the western Amazon: Climate, vegetation, and humans. *Antiquity* **73**, 287–302.

Bailey, R. C., Head, G., Jenike, M., Owen, B., Rechtman, R., and Zechenter, E. (1989) Hunting and gathering in tropical rain forest: Is it possible? *American Anthropologist* **91**, 59–82.

Bartlett, A. S. and Barghoorn, E. S. (1973) Phytogeographic history of the Isthmus of Panama during the past 12,000 years (a history of vegetation, climate and sealevel change). In: A. Graham (ed.), *Vegetation and Vegetational History of Northern Latin America* (pp. 203–209). Elsevier, New York.

Behling, H. and Lima de Costa, L. (2000) Holocene environmental changes from the Rio Curuá record in the Caxiuaná region, eastern Amazon Basin. *Quaternary Research* **53**, 369–377.

Berrío, J. C., Behling, H., and Hooghiemstra, H. (2000) Tropical rain-forest history from the Colombian Pacific area: A 4200-year pollen record from Laguna Jotaordó. *The Holocene* **10**, 749–756.

Bray, W. (2000) Ancient food for thought. *Nature* **408**, 145–146.

Bush, M. B. (1991) Modern pollen-rain data from South and Central America: A test of the feasibility of fine-resolution lowland tropical palynology. *The Holocene* **1**, 162–167.

Bush, M. B. and Colinvaux, P. A. (1990) A pollen record of a complete glacial cycle from lowland Panama. *Journal of Vegetation Science* **1**, 105–118.

Bush, M. B. and Colinvaux, P. A. (1994) A paleoecological perspective of tropical forest disturbance: records from Darien, Panama. *Ecology* **75**, 1761–1768.

Bush, M. B. and Rivera, R. (1998) Pollen dispersal and representation in a Neotropical forest. *Global Ecology and Biogeography Letters* **7**, 379–392.

Bush, M. B., Piperno, D. R., and Colinvaux, P. A. (1989) A 6,000 year history of Amazonian maize cultivation. *Nature* **340**, 303–305.

Bush, M. B., Piperno, D. R., Colinvaux, P. A., De Oliveira, P. E., Krissek, L., Miller, M., and Rowe, W. (1992) A 14,300 year paleoecological profile of a lowland tropical lake in Panama. *Ecological Monographs* **62**, 251–275.

Bush, M. B., Miller, M. C., De Oliveira, P. E., and Colinvaux, P. A. (2000) Two histories of environmental change and human disturbance in eastern lowland Amazonia. *The Holocene* **10**, 543–553.

Bush, M. B., De Oliveira, P. E., Colinvaux, P. A., Miller, M. C., and Morenov, J. E. (2004) Amazonian paleoecological histories: One hill, three watersheds. *Palaeogeography, Palaeoclimatology, Palaeoecology* **214**, 347–358.

Cavelier, I., Rodríguez, C., Herrera, L. F., Morcote, G., and Mora, S. (1995) No sólo de caza vive el hombre: Ocupación del bosque Amazónico, Holoceno temprano. In: I. Cavalier and S. Mora (eds.), *Ambito y Ocupaciones Tempranas de la América Tropical* (pp. 27–44). Instituto Colombiano de Antropología, Fundación Erigaie, Sante Fé de Bogotá [in Spanish].

Clement, R. M. and Horn, S. P. (2001) Pre-Columbian land use history in Costa Rica: A 3000-year record of forest clearance, agriculture and fires from Laguna Zoncho. *The Holocene* **11**, 419–426.

Colinvaux, P. A. and Bush, M. B. (1991) The rain-forest ecosystem as a resource for hunting and gathering. *American Anthropologist* **93**, 153–160.

Colinvaux, P. A., De Oliveira, P. E., Moreno, J. E., Miller, M. C., and Bush, M. B. (1996a) A long pollen record from lowland Amazonia: Forest and cooling in glacial times. *Science* **274**, 85–88.

Colinvaux, P. A., Liu, K-B., De Oliveira, P. E., Bush, M. B., Miller, M. C., and Steinitz-Kannan, M. (1996b) Temperature depression in the lowland tropics in glacial times. *Climatic Change* **32**, 19–33.

Colinvaux, P., De Oliveira, P. E., and Moreno, J. E. (1999) *Amazon Pollen Manual and Atlas*. Harwood Academic, Amsterdam.

Cooke, R. G. (1998) Human settlement of Central America and northern South America. *Quaternary International* **49/50**, 177–190.

Cooke, R. G. (2005) Prehistory of native Americans on the Central American landbridge: Colonization, dispersal, and divergence. *Journal of Archeological Research* **13**, 129–187.

Cowling, S. A. and Sykes, M. T. (1999) Physiological significance of low atmospheric CO_2 for plant–climate interactions. *Quaternary Research* **52**, 237–242.

Denham, T., Haberle, S. G., Lentfer, C., Fullagar, R., Field, J., Therin, M., Porch, N., and Winsborough, B. (2003). Origins of agriculture at Kuk Swamp in the highlands of New Guinea. *Science* **301**, 189–193.

Diamond, J. (2002) Evolution, consequences, and future of plant and animal domestication. *Nature* **418**, 700–707.

Dickau, R. (2005) Resource use, crop dispersals, and the transition to agriculture in prehistoric Panama: Evidence from starch grains and macroremains. Unpublished PhD dissertation, Department of Anthropology, Temple University, Philadelphia, PA.

Dillehay, T. D. (1997) *Monte Verde: A Late Pleistocene Settlement in Chile, Vol. 2: The Archaeological Context and Interpretation*. Smithsonian Institution Press, Washington, D.C.

Dillehay, T. D., Calderón, G. A., Politis, G., and da C. Coutinho Beltrão, M. (1992) Earliest hunters and gatherers of South America. *Journal of World Prehistory* **6**, 145–204.

Flannery, K. V. (1986) *Guilá Naquitz: Archaic Foraging and Early Agriculture in Oaxaca, Mexico*. Academic Press, Orlando, FL.

Forster, P. and Matsumura, S. (2005) Did early humans go north or South? *Science* **308**, 965–966.

Harlan, J. R. (1971) Agricultural origins: Centers and noncenters. *Science* **174**, 468–474.

Iriarte, J., Holst, I., Marozzi, O., Listopad, C., Alonso, E., Rinderknecht, A., and Montaña, J. (2004). Evidence for cultivar adoption and emerging complexity during the Mid-Holocene in the La Plata Basin, Uruguay. *Nature* **432**, 614–617.

Islebe, G. A., Hoohiemstra, H., Brenner, M., Curtis, J. H., and Hodell, D. A. (1996) A Holocene vegetation history from lowland Guatemala. *The Holocene* **6**, 265–271.

Jones, J. G. (1994) Pollen evidence for early settlement and agriculture in Northern Belize. *Palynology* **18**, 205–211.

Kennett, D. and Winterhalder, B. (eds.) (in press). *Foraging Theory and the Transition to Agriculture*. University of California Press, Berkeley, CA.

Leyden, B. (1984) Guatemalan forest synthesis after Pleistocene aridity. *Proceedings of the National Academy of Sciences U.S.A.* **81**, 4856–4859.

Leyden, B. (1985) Late Quaternary aridity and Holocene moisture fluctuations in the Lake Valencia Basin, Venezuela. *Ecology* **66**, 1279–1295.

Leyden, B., Brenner, M., Hodell, D. A., and Curtis, J. H. (1993) Late Pleistocene climate in the Central American lowlands. In: P. K. Swart, K. C. Lohmann, J. McKenzie, and S. Savin (eds.), *Climate Change in Continental Isotopic Records* (Geophysical Monograph 78, pp. 165–178). American Geophysical Union, Washington, D.C.

López Castaño, C. E. (1995) Dispersión de puntas de proyectil bifaciales en la cuenca media del Río Magdalena. In: I. Cavalier and S. Mora (eds.), *Ambito y Ocupaciones Tempranas de la América Tropical* (pp. 73–82). Instituto Colombiano de Antropología, Fundación Erigaie, Santa Fe de Bogotá [in Spanish].

Lothrop, S. K. (1961) Early migrations to Central and South America: An anthropological problem in light of other sciences. *Journal of the Royal Anthropological Institute* **91**, 97–123.

Macaulay, M., Hill, C., Achilli, A., Rengo, C., Clarke, D., Meehan, M., Blackburn, J., Semino, O., Scozzari, R., Cruciani, F. *et al.* (2005) Single, rapid coastal settlement of Asia revealed by analysis of complete mitochondrial genomes. *Science* **308**, 1034–1036.

MacNeish, R. S. (1967). A summary of subsistence. In: D. S. Byers (ed.), *The Prehistory of the Tehuacan Valley, Vol. 1: Environment and Subsistence* (pp. 290–309). University of Texas Press, Austin, TX.

Matsuoka, Y., Vigouroux, Y., Goodman, M. M., Sanchez, J., Buckler, E., and Doebley, J. (2002) A single domestication for maize shown by multilocus microsatellite genotyping. *Proceedings of the National Academy of Sciences U.S.A.* **99**, 6080–6084.

Meltzer, D. J. (1997) Monte Verde and the Pleistocene peopling of the Americas. *Science* **276**, 754–755.

Meltzer, D. J., Adovasio, J. M., and Dillehay, T. D. (1994) On a Pleistocene human occupation at Pedra Furada, Brazil. *Antiquity* **68**, 695–714.

Mercader, J. and Martí, R. (2003) The Middle Stone Age occupation of Atlantic central Africa: New evidence from equatorial Guinea. In: J. Mercader (ed.), *Under the Canopy: The Archaeology of Tropical Rain Forests* (pp. 64–92). Rutgers University Press, New Brunswick, NJ.

Mercader, J., Runge, F., Vrydaghs, L., Doutrelepont, H., Ewango, C. E. N, and Juan-Tresseras, J. (2000) Phytoliths from archaeological sites in the tropical forest of Ituri, Democratic Republic of Congo. *Quaternary Research* **54**, 102–112.

Monsalve, J. G. (1985) A pollen core from the Hacienda Lusitania. *Pro Calima* **4**, 40–44.

Mora, S. and Gnecco, C. (2003) Archaeological hunter-gatherers in tropical forests: A view from Colombia. In: J. Mercader (ed.), *Under the Canopy: The Archaeology of Tropical Rain Forests* (pp. 271–290). Rutgers University Press, New Brunswick, NJ.

Mora, S. C., Herrera, L. F., Cavelier, I., and Rodríguez, C. (1991) *Cultivars, Anthropic Soils and Stability* (University of Pittsburgh Latin American Archaeology Report No. 2). University of Pittsburgh Department of Anthropology, Pittsburgh, PA.

Neves, E. (1999). Changing perspectives in Amazonian archaeology. In: G. G. Politis and B. Alberti (eds.), *Archaeology in Latin America* (pp. 216–243). Routledge, London.

Olsen, K. M. and Schaal, B. A. (1999) Evidence on the origin of cassava: Phylogeography of *Manihot esculenta*. *Proc. Nat. Acad. Sci. U.S.A.* **96**, 5586–5591.

Pearsall, D. M. (1978) Phytolith analysis of archaeological soils: Evidence for maize cultivation in formative Ecuador. *Science* **199**, 177–178.

Pearsall, D. M., Chandler-Ezell, K., and Chandler-Ezell, A. (2003) Identifying maize in Neotropical sediments and soils using cob phytoliths. *Journal of Archaeological Science* **30**, 611–627.

Pearsall, D. M., Chandler-Ezell, K., and Zeidler, J. A. (2004) Maize in ancient Ecuador: Results of residue analysis of stone tools from the Real Alto site. *Journal of Archaeological Science* **31**, 423–442.

Perry, L. (2002) Starch analyses reveal multiple functions of quartz "manioc" grater flakes from the Orinoco Basin, Venezuela. *Interciencia* **27**, 635–639.

Piperno, D. R. (1988) *Phytolith Analysis: An Archaeological and Geological Perspective*. Academic Press, San Diego, CA.

Piperno, D. R. (1990) Aboriginal agriculture and land usage in the Amazon Basin, Ecuador. *Journal of Archaeological Science* **17**, 665–677.

Piperno, D. R. (1993) Phytolith and charcoal records from deep lake cores in the American tropics. In: D. M. Pearsall and D. R. Piperno (eds.), *Current Research in Phytolith Analysis: Applications in Archaeology and Paleoecology* (MASCA Research Papers in Science and Archaeology Vol. 10, pp. 58–71). MASCA, University Museum of Archaeology and Anthropology, University of Pennsylvania, PA.

Piperno, D. R. (1994) Phytolith and charcoal evidence for prehistoric slash and burn agriculture in the Darien rainforest of Panama. *The Holocene* **4**, 321–325.

Piperno, D. R. (in press a) The origins of plant cultivation and domestication in the Neotropics: A behavioral ecological perspective. In: D. Kennett and B. Winterhalder (eds.), *Foraging Theory and the Transition to Agriculture*. University of California Press, Berkeley, CA.

Piperno, D. R. (in press b) *Phytoliths: A Comprehensive Guide for Archaeologists and Paleoecologists*. AltaMira Press, Lanham, MD.

Piperno, D. R. (in press c) Agricultural impact on vegetation and Quaternary vegetational history in Central America. In: A. Graham and P. Raven (eds.), *Latin American Biogeography: Causes and Effects*. Missouri Botanical Garden Press, St. Louis.

Piperno, D. R. and Becker, P. (1996) Vegetational history of a site in the central Amazon Basin derived from phytolith and charcoal records from natural soils. *Quaternary Research* **45**, 202–209.

Piperno, D. R. and Jones, J. (2003) Paleoecological and archaeological implications of a late Pleistocene/early Holocene record of vegetation and climate from the Pacific coastal plain of Panama. *Quaternary Research* **59**, 79–87.

Piperno, D. R. and Pearsall, D. M. (1998) *The Origins of Agriculture in the Lowland Neotropics*. Academic Press, San Diego, CA.

Piperno, D. R. and Stothert, K. E. (2003) Phytolith evidence for early Holocene *Cucurbita* domestication in southwest Ecuador. *Science* **299**, 1054–1057.

Piperno, D. R., Bush, M. B., and Colinvaux, P. A. (1990) Paleoenvironments and human occupation in Late-Glacial Panama. *Quaternary Research* **33**, 108–116.

Piperno, D. R., Bush, M. B., and Colinvaux, P. A. (1991a) Paleoecological perspectives on human adaptation in Central Panama, I: The Pleistocene. *Geoarchaeology* **6**, 210–226.

Piperno, D. R., Bush, M. B., and Colinvaux, P. A. (1991b) Paleoecological perspectives on human adaptation in Central Panama, II: The Holocene. *Geoarchaeology* **6**, 227–250.

Piperno, D. R., Bush, M. B., and Colinvaux, P. A. (1992) Patterns of articulation of culture and the plant world in prehistoric Panama: 11,500 BP–3000 BP In: O. R. Ortiz-Troncoso and

T. Van der Hammen (eds.), Archaeology and Environment in Latin America (pp. 109–127). Universiteit van Amsterdam, Amsterdam.

Piperno, D. R., Holst, I., Andres, T. C., and Stothert, K. E. (2000a) Phytoliths in *Cucurbita* and other Neotropical Cucurbitaceae and their occurrence in early archaeological sites from the lowland American tropics. *Journal of Archaeological Science* **27**, 193–208.

Piperno, D. R., Ranere, A. J., Holst, I., and Hansell, P. (2000b) Starch grains reveal early root crop horticulture in the Panamanian tropical forest. *Nature* **407**, 894–897.

Piperno, D. R., Ranere, A. J., Moreno, J. E.., Iriarte, J., Lachniet, M., Holst, I., and Dickau, R. (2004) Environmental and agricultural history in the Central Balsas Watershed, Mexico: Results of preliminary research. *Annual Meeting of the Society for American Archaeology, Montreal, Canada.*

Pohl, M. D., Pope, K. O., Jones, J. G., Jacob, J. S., Piperno, D. R., de France, S., Lentz, D. L., Gifford, J. A., Valdez, F. Jr., Danforth, M. E. *et al.* (1996) Early agriculture in the Maya lowlands. *Latin American Antiquity* **7**, 355–372.

Pope, K. O., Pohl, M. E. D., Jones, J. G., Lentz, D. L., von Nagy, C., Vega, F. J., and Quitmyer, I. R. (2001) Origin and environmental setting of ancient agriculture in the lowlands of Mesoamerica. *Science* **292**, 1370–1373.

Ranere, A. J. (1980) Human movement into tropical America at the end of the Pleistocene. In: L. B. Harten, C. N. Warren, and D. R.Tuohy (eds.), *Anthropological Papers in Memory of Earl H. Swanson* (pp. 41–47). Idaho Museum of Natural History, Pocatello, ID.

Ranere, A. J. and Cooke, R. G. (2003) Late glacial and early Holocene occupation of Central American tropical forests. In: J. Mercader (ed.), *Under the Canopy: The Archaeology of Tropical Rain Forests* (pp. 219–248). Rutgers University Press, New Brunswick, NJ.

Rodgers III, J. C. and Horn, S. P. (1996) Modern pollen spectra from Costa Rica. *Palaeogeography, Palaeoclimatology, Palaeoecology* **124**, 53–71.

Roosevelt, A. C., da Costa, M. L., Machado, C. L., Michab, M., Mercier, N., Valladas, H., Feathers, J., Barnett, W., da Silveira, M. I., Henderson, A. *et al.* (1996) Paleoindian cave dwellers in the Amazon: The peopling of the Americas. *Science* **272**, 373–384.

Rue, D. J. (1987) Early agriculture and early postclassic Maya occupation in western Honduras. *Nature* **326**, 285–286.

Sage, R. F. (1995) Was low atmospheric CO_2 during the Pleistocene a limiting factor for the origin of agriculture? *Global Change Biology* **1**, 93–106.

Sanjur, O., Piperno, D. R., Andres, T. C., and Wessell-Beaver, L. (2002) Phylogenetic relationships among domesticated and wild species of *Cucurbita* (Cucurbitaceae) inferred from a mitochondrial gene: Implications for crop plant evolution and areas of origin. *Proceedings of the National Academy of Sciences U.S.A.* **99**, 535–540.

Sauer, C. O. (1944) A geographic sketch of early man in America. *Geographic Review* **34**, 529–573.

Sémah, F., Sémah, A-M., and Simanjuntak, T. (2003) More than a million years of human occupation in insular southeast Asia: The early archaeology of eastern and central Java. In: J. Mercader (ed.), *Under the Canopy: The Archaeology of Tropical Rain Forests* (pp. 161–190). Rutgers University Press, New Brunswick, NJ.

Smith, B. D. (1997) The initial domestication of *Cucurbita pepo* in the Americas 10,000 years ago. *Science* **276**, 932–934.

Stothert, K., Piperno, D. R., and Andres, T. C. (2003) Terminal Pleistocene/early Holocene human adaptation in coastal Ecuador: The Las Vegas evidence. *Quaternary International* **109–110**, 23–43.

Thangaraj, K., Chaubey, G., Kivisild, T., Reddy, R., Vijay, A. G., Singh, K., Rasalkar, A. A., and Singh, L. (2005) Reconstructing the origin of Andaman Islanders. *Science* **308**, 995–996.

Voorhies, B., Kennett, D. J., Jones, J., and Wake, T. A. (2002) A Middle Archaic site on the west coast of Mexico. *Latin American Antiquity* **13**, 179–185.

Westengen, O. T., Huamán, Z., and Heun, M. (2005) Genetic diversity and geographic pattern in early South American cotton domestication. *Theoretical and Applied Genetics* **110**, 392–402.

Willis, K. J., Gillson, L., and Brncic, T .M. (2004). How "virgin" is virgin rain forest? *Science* **304**, 402–403.

8

Ultraviolet insolation and the tropical rainforest: altitudinal variations, Quaternary and recent change, extinctions, and biodiversity

J. Flenley

8.1 INTRODUCTION

Ultra-violet (UV) light occurs in three wavebands. UV-A is the longest waveband (>315 nm) which is close to visible light and is of limited biological significance. UV-B (280–315 nm) is damaging and mutagenic to living organisms. UV-C (<280 nm) is lethal to all life, but is fortunately absorbed in the stratosphere, so does not reach the surface of the Earth in sunlight. It is therefore to UV-B that we must turn our chief attention. This, like UV-C, is also partly absorbed by ozone in the stratosphere, but some reaches the Earth's surface. Recent concerns about the "Ozone Hole" have focused attention on polar regions, but in fact tropical regions have fairly low ozone concentrations in the stratosphere above them (Smith and Warr, 1991). The result is that, given their high overall insolation resulting from the low latitude, tropical regions have rather high UV-B levels.

UV-B insolation is greater at high altitudes because of less atmosphere being traversed (Caldwell, 1971). It therefore follows that UV-B is particularly significant to vegetation on tropical mountains, which are the highest altitude vegetated surfaces in the world. UV-B is known to be particularly high in the Alpine Zone of tropical mountains (Caldwell *et al.*, 1980; Sullivan *et al.*, 1992).

8.2 ALTITUDINAL VARIATIONS

On tropical mountains, there are marked altitudinal changes of vegetation (Troll, 1959). The concept of an actual zonation is rather too strict to be useful, but there is no doubt that as one ascends a mountain there are gradual changes. Most tree species have an ecological amplitude in terms of mean annual temperature (MAT), of about 6°C (van Steenis, 1934–36). There are some conspicuous exceptions to this, but it applies to many species. There is thus a continuum of change seen as the climber ascends to cooler altitudes. In New Guinea, for example, lowland trees start to be replaced around 1,000 m by tropical oaks (*Lithocarpus* spp. and *Castanopsis* spp.), and at a higher altitude by tropical beeches (*Nothofagus* sect. *brassospora* spp.) which continue up to about 2,800 m on the main ranges. Above the "beech" forest we find the Dwarf Forest. This is also known as the "Mossy Forest", "Cloud Forest", or "Upper Montane Forest", but the term "Dwarf Forest" is adopted in this chapter to avoid assumptions about causation. It is characterized not only by abundance of epiphytic bryophytes, but also by morphological peculiarities: stunted tree growth; small, thick leaves with a hypodermis; and presence of extra pigments (Grubb, 1977). Above this we approach the altitudinal forest limit, usually marked by subalpine forest or shrubbery, over a vertical extent of c. 200 m, before the "Alpine Grassland" begins (Corlett, 1984). These changes all tend to be lower on isolated peaks near the sea or on islands (the Massenerhebung effect, Figure 8.1).

What are the controlling factors of these variations? It has always been assumed that temperature must exert overall control. Mean annual values decline at about 0.6°C per 100 m, with increasing altitude, and the Alpine Grassland begins at about 6°C mean annual temperature (Walker and Flenley, 1979). Diurnal variation is extreme, so that the grassland experiences nightly frost and high temperatures in the day. Even the Massenerhebung effect may partly be explicable by temperature, for temperature lapse rates are steeper on isolated mountains near the sea than on large ranges which provide their own geothermal heat (Hastenrath, 1968; Flenley and Richards, 1982). How temperature operates other than by frost action is not well-understood. Various workers (e.g., Brass, 1941, 1964; Grubb and Whitmore, 1966) have suggested that the Dwarf Forest is associated with cloudiness. It has also been suggested that temperature could operate via the soil (Grubb, 1977), lower temperatures leading to greater accumulation of organic matter and to changes in nutrient status.

It may well be that the extreme diurnal temperature variation is partly responsible. Many tropical mountains, although cloud-covered in the afternoons, lose their cloud cover during the night and early morning. Given the low atmospheric pressure

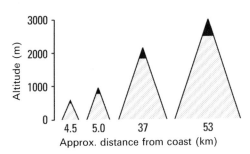

Figure 8.1. The Massenerhebung (mass elevation) effect illustrated by the occurrence of Dwarf Forest on mountains in Indonesia. *From left to right*: Mount Tinggi (Bawean), Mount Ranai (Natura Island), Mount Salak (W. Java) and Mount Pangerango (W. Java) (after van Steenis, 1972).

resulting from the elevation, out-radiation is very high, resulting in extreme day/night temperature differentials (DIF). This thermo-periodicity has been studied in a number of cultivated plant species (Atwell *et al.*, 1999). In general, a small positive DIF (i.e., day temperature a few degrees higher than night temperature) encourages growth. For instance, in *Chrysanthemum* there is a strong positive correlation between stem elongation and positive DIF (Carvalho *et al.*, 2002). In *Begonia*, however, growing the plants with a day temperature of 22°C and a night temperature of 16°C, and adding in a 2-hour temperature drop to 12°C after sunset led to inhibition of total plant height and width (Son *et al.*, 2002). In general, a DIF of 20°C in several species can discourage growth, leading to stunting (Atwell *et al.*, 1999), which is a feature of the trees of the Dwarf Forest. As it is likely that few, if any, species from this forest have been investigated in this regard so far, further consideration of the role of DIF must await research. It remains, however, a distinct possibility that it is a factor of importance on tropical high mountains, where the "summer every day and winter every night" environment is normal (Troll, 1959). It seems surprising that modelers of tropical climate (e.g., Farrera *et al.*, 1999) have chosen to use mean annual temperature and mean temperature of the coldest month, indicating seasonality, as parameters in their models while apparently ignoring DIF.

Most of the explanations of altitudinal variation so far offered are based mainly on association. Experimental evidence of causation is usually scanty, or rather inconclusive. Alternative possible causative factors—such as the decline in atmospheric pressure with altitude, or the increase in ultraviolet light with altitude—have rarely been considered.

This section investigates the hypothesis that the variation of vegetation on the upper parts of tropical high mountains (above c. 2,000 m) is related to insolation by ultraviolet light. This hypothesis was advanced by the late Francis Merton (pers. commun.) in 1973, but there was no serious consideration given to it at the time because of lack of evidence. Since then, however, discoveries have justified a revival of the hypothesis. These data relate to the effects of UV-B on plant growth.

Controlled experiments which simulate natural conditions are difficult with UV-B, for artificial light sources do not correctly reproduce the solar spectrum, and species differ widely in their response to individual wavelengths. Nevertheless, it has been possible to experiment on a range of plants (Lindoo and Caldwell, 1978; Teramura, 1983; Murali and Teramura, 1986a, b; Caldwell *et al.*, 1995). In general, the plants became stunted and developed small, thick leaves with a hypodermis: precisely the characteristics of upper montane and subalpine forests. They also developed extra flavonoid pigments, which is also a common characteristic of the Dwarf Forest and shrubbery. In fact, the puña of Peru (a subalpine scrub) has a distinct yellowish appearance possibly caused by such pigments.

The correspondence between the features induced in crop plants, and the features present in the Dwarf Forest and shrubbery, is rather striking. It must be remembered, however, that the former is phenotypic and the latter (presumably) genotypic. This need not be an insuperable difficulty. Probably, genetic fixation of an initially induced feature would happen by natural selection. Is it possible, therefore, that the upper woody vegetation of tropical mountains is genetically adapted for resistance to UV-B?

It would be good to test this idea by considering the Massenerhebung effect, since this involves the occurrence of Dwarf Forest at anomalously low altitudes.

8.3   THE MASSENERHEBUNG EFFECT

As usually defined, the Massenerhebung or "mountain mass elevation" effect means the occurrence of physiognomically and sometimes floristically similar vegetation types at higher altitudes on large mountain masses than on small isolated peaks, especially those in or near the sea. Although the effect was first reported in the European Alps (Schroeter, 1908) and in North America (where it is known as the "Merriam effect"; Martin, 1963), it is best known in the tropics. Perhaps its clearest expression is the occurrence of Dwarf Forest at lower altitudes on isolated peaks than on the main mountain masses, which are taken as the norm (Figure 8.1).

Some explanations of this phenomenon have involved mean temperatures and cloud formation. Cloud formation is often observed on isolated peaks at quite low altitudes, and it was shown by Hastenrath (1968)—using radiosonde balloons in Mexico—that lapse rates were somewhat steeper over lowlands than over large mountain masses. Similarly, a steep lapse rate (0.74°C per 100 m) was recorded on the 735 m high island of Krakatau in Indonesia (Forster, 1982); the regional average is 0.61°C per 100 m (Walker and Flenley, 1979). Presumably, the afternoon clouding seen on isolated peaks is related to this temperature regime, and also, in the case of islands, to the greater evaporation from the sea (M. Bush, pers. commun.). The clouding does have pronounced ecological effects. For instance, it increases humidity to 100% and reduces total insolation received by about 30% compared with unclouded sites in Sabah, Malaysia (Bruijnzeel *et al.*, 1993). The difficulty comes in relating these changes to the morphological peculiarities of the vegetation. Usually, if plants are grown in high humidity and low insolation, they become etiolated (i.e., they are tall, have long internodes, and large, thin, pale green leaves). This is the exact opposite of the attributes of tropical mountain Dwarf Forest, which are stunted growth, short internodes, small, thick leaves (with a hypodermis), and often extra pigments (anthocyanins or flavonoids).

Now, let us consider the UV-B hypothesis as an explanation for the Massenerhebung effect. UV-B light, like visible light, may experience total reflection at water or cloud surfaces. Total insolation may thus be increased by up to 70% through reflection from clouds (Figure 8.2). In the early morning, low peaks are surrounded by a sea of clouds (Figure 8.3), which will reflect sunlight strongly up onto vegetation. Islands receive similar reflection from the sea surface. Rayleigh scattering of light from air molecules in the sky will also be particularly effective in clear mornings and favours UV above all other wavelengths (Dave and Halpern, 1976).

Later in the day the clouds move uphill and envelop the upper forests, reducing them almost to darkness (Hope, 1986). This later reduction of insolation may serve to exacerbate the effects of any UV-B damage caused earlier in the day. This is because of photo-reactivation, a process by which plants repair themselves from UV damage. Photo-reactivation is strongly dependent on visible light insolation (Caldwell, 1971).

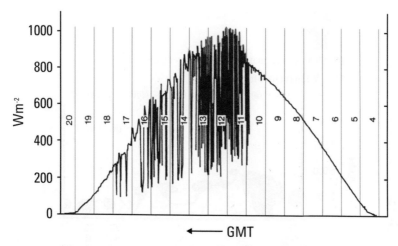

Figure 8.2. Solar radiation on a day of broken cloud (11 June 1969) at Rothamsted (52°N, 0°W) taken directly from recorder charts. Note very high values of irradiance immediately before and after occlusion of the Sun by cloud (after Monteith, 1973).

The daily tropical regime of a heavy dose of UV-B in the morning, followed by semi-darkness, may therefore be particularly harmful to plants that are not adapted to it. There have been several attempts to relate the Massenerhebung effect to soil attributes. Among the more successful of these was that by Bruijnzeel *et al.* (1993), who found a correlation between the occurrence of stunted forest and phenolic compounds in leaf litter. The latter were thought to cause stunting by harmful effects on plant physiology. This theory is complementary to the UV theory, however. The usual response of plants to excess UV-B is to produce protective compounds that absorb UV-B. These are usually flavonoids or alkaloids (Caldwell, 1981), or anthocyanins (Lee and Lowry, 1980b). Many of these compounds are phenolic or are likely to break down into phenolic compounds in litter. They could provide a reinforcement mechanism, exaggerating the original stunting caused by the UV-B (Bruijnzeel and Proctor, 1993).

I conclude, therefore, with the tentative hypothesis that the Massenerhebung effect could be partly the result of a high dose of UV-B due to reflection from clouds or the sea in the mornings. Obviously, further research is needed to test this idea, although there is little doubt that the full explanation will be a multivariate one. Some factors may be more important in one location, while other factors may dominate elsewhere.

8.4   QUATERNARY VARIATIONS OF VEGETATION

Palynologically-based paleoecology has given strong evidence in recent years about altitudinal variations in the period c. 25,000–15,000 years ago, the coldest period of

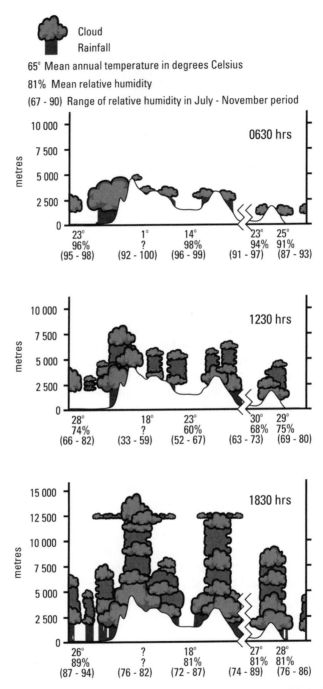

Figure 8.3. The daily weather regime in the New Guinea Highlands (modified after Brookfield, 1964). Note especially the cloud below the peaks at 6 : 30 a.m. followed by heavy clouding at 12 : 30 p.m. and later.

the last glaciation. The evidence from the New Guinea Highlands consists of pollen diagrams from a variety of altitudes between 1,900 and >4,000 m (Flenley, 1979). These are consistent with a lowering of the altitudinal forest limit to c. 2,000 m in the Late Pleistocene (Figure 4.13). This could be explained by a lowering of mean annual temperatures.

There is, however, an anomaly which cannot be explained by mean annual temperature change alone: the fact that in the Late Pleistocene the Dwarf Forest apparently disappeared almost completely (Walker and Flenley, 1979). Its constituent taxa must have survived somewhere, presumably as rare individuals near the altitudinal forest limit. In its place, the alpine grassland was greatly expanded, and was apparently rich in tree ferns. This phenomenon has now been reported to be repeated during each glacial phase (see Chapter 4).

If this absence of the Dwarf Forest in New Guinea in the Late Pleistocene had been an isolated instance, one could perhaps have ignored it. A similar phenomenon has, however, been reported from the Colombian Andes (Salomons, 1986). In this case it was the subpáramo, the subalpine shrubbery of the Andes, which was discontinuous in the Pleistocene. Both cases amount, however, to a great reduction in the Late Pleistocene of the upper woody formations which are characterized by stunted growth, small thick leaves, and a hypodermis.

The usual explanation advanced for this is that the climate of the Last Glacial Maximum was somewhat drier than the present one, as well as cooler. Thus, the Dwarf Forest disappears. However, since desiccation is more likely to favor stunting than to discourage it, this explanation does not appear completely satisfactory. If, however, the UV-B hypothesis advanced in the last section has any credibility, the occurrence of Dwarf Forest is related—at least, partially—to high UV-B. The disappearance of the Dwarf Forest during the glaciation can then be explained in the following way (Figure 8.4). In the Holocene, at lower altitudes—below c. 3,000 m— the genetically stunted trees of the Dwarf Forest would be at a selective disadvantage compared with the larger trees of the Lower Altitude Forest. Above c. 3,000 m they would be at an advantage, because of the greater UV-B insolation there. In the Late Pleistocene, lower temperatures brought the forest limit below 3,000 m, therefore the stunted species became rare. In the Holocene, warmer temperatures allowed forests to expand uphill, but only those species genetically adapted to high UV-B insolation could take advantage of this. The present forest limit thus could be controlled by temperature or UV-B, or a combination of both.

8.5   PRESENT AND POSSIBLE FUTURE TRENDS IN UV INSOLATION AND THEIR EFFECTS

Since the discovery of the Ozone Hole, there has been much interest in the elevated levels of UV-B experienced by natural ecosystems (Caldwell et al., 1995). Each southern spring, as the Ozone Hole that develops over Antarctica at the end of winter is filled in, this results in a depletion of stratospheric ozone in middle southern latitudes, and to a rise in UV levels there. It might have been expected that a knock-on effect

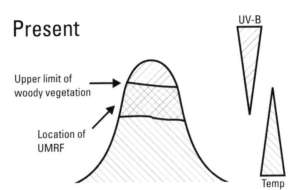

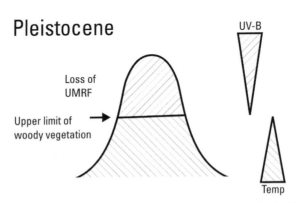

Figure 8.4. The hypothesis that lack of Upper Montane Rain Forest in the Pleistocene may be explained by absence of a habitat with a suitable combination of mean annual temperature and UV-B insolation (diagram by G. Rapson, unpublished).

would then lead to a rise of UV in the tropics, but measurements show little or no change there (Stolarski *et al.*, 1992; Madronich *et al.*, 1995; Gleason, 2001).

The possible effect of global warming on UV insolation has also been studied. More greenhouse gases lead to a warmer atmosphere, but this leads to a cooler stratosphere (Austin *et al.*, 1992; Shindell *et al.*, 1998a). This might be expected to lead to more widespread conditions favoring the destruction of ozone. If this were to happen, it could result in a rise in UV levels in temperate regions, and possibly in the tropics also. Mathematical modeling shows, however, that this is unlikely. A model simulation did indeed suggest a decrease of ozone in middle latitudes, but a small increase (2–6%) in the tropics (Shindell *et al.*, 1998b).

It thus appears that the tropical UV climate is rather stable. One possible exception to this could be the effect of volcanic eruptions. For example, the major eruption of Mount Pinatubo in the Philippines in 1991 resulted in the release of sulphur dioxide aerosols which destroyed stratospheric ozone and led to an observed small increase in ground-level UV for several years on a global scale (Gleason *et al.*, 1993).

Possible increases in tropical UV have been blamed for two major phenomena: the bleaching of tropical corals, and the decline and extinction of many Amphibia. Corals are indeed very sensitive to UV-A and UV-B, and many cases of coral bleaching and death have been reported. Corals have been shown to possess powerful UV-screening compounds, some of which are of commercial importance. It now appears, however, that a more likely cause of coral bleaching and death is a rise of water temperature. Even a rise of 1°C can be effective in this way (Gleason, 2001).

The Amphibia story is more complex, since the reported decline of Amphibia is worldwide. Many Amphibia are indeed sensitive to UV-B, which may significantly reduce the hatching of eggs, or may induce abnormalities (Blaustein *et al.*, 2001, 2003). 427 species of Amphibia (7.4% of known species) are critically endangered and 1,856 species (32.5%) are globally threatened. A high proportion of these (76.5%) are from the Neotropics, Afrotropics, or Australasia (Stuart *et al.*, 2004). Nevertheless, the latest evidence suggests that increasing UV-B is unlikely to be the major single factor, and a fungal disease is now being investigated as the likely principal cause (Daszak *et al.*, 2003). The complete answer may involve many interacting factors (Carey and Alexander, 2003; Collins and Storfer, 2003; Kats and Ferrer, 2003; Storter, 2003).

The possibility of investigating past ultraviolet radiation environments by the record of fossil pigments in lakes (Leavitt *et al.*, 1997) is an interesting development which may prove to be of considerable value.

8.6 BIODIVERSITY

Over the past 30 years, the principal research problem in ecology has been the explanation of biodiversity. More specifically, the aim has been to explain the gradient of diversity from its high values near the equator to the very low values in subpolar regions. Most hypotheses about this assume that the present situation is in equilibrium with the environment, and thus try to explain matters in terms of currently operating processes. Although the complete explanation may well be multivariate— that is, involving the interaction of many separate factors—there has still been a tendency to search for an underlying mechanism.

One such mechanism that has been popular in the last 20 years is the refuge theory (Haffer, 1997; Haffer and Prance, 2001), which was one of the first ideas involving paleoecology. If the tropical rainforest was divided into many separate blocks by dry Pleistocene climates, this could have isolated populations for long enough to allow allopatric speciation. In fact, repetition of this by multiple Milankovitch cycles could have produced a "species pump" leading to many closely related species, which is what we find in the tropical rainforest. Despite the fact that the refuge theory had no explanation as to why it operated only in the tropics, and not also in temperate regions (Flenley, 1993), it remained popular even after it was found that there was little palynological support for the existence of the supposed dry phases (e.g., Bush, 1994; Bush *et al.*, 1992). The theory only foundered completely when it was demonstrated that the Quaternary had not been a time of great speciation in the tropics at all. In fact, using palynological richness as a proxy for diversity (Flenley, 2005), it could be suggested that diversity had actually declined during the Quaternary (Morley, 2000; Bennett, 2004; Flenley, 2005). It seems that mutation rates were too low to produce the required genetic isolation in the time available (Willis and Niklas, 2004).

The opportunity therefore exists for promulgation of a new hypothesis, which might retain some of the attractive features of the refuge hypothesis—such as the species pump concept, related to cyclical climatic change—while avoiding the pitfalls of the earlier hypothesis. This is the aim of the present section.

To be at all plausible, the new hypothesis must account for the origin and survival of large numbers of closely related species, now living in close proximity with each other. Assuming normal evolutionary processes, the most likely procedure for achieving this is allopatric speciation—that is, the original population of a species must be split up and the sub-populations isolated. There must then be mutation at a sufficient rate within each sub-population such that when the sub-populations are recombined, they remain reproductively isolated.

Referring again to the geological record, and using palynological richness as a proxy for diversity (Morley, 2000; Flenley, 2005) we find that the times of rapidly increasing diversity were the Eocene and the Early Miocene, especially the former (Wilf et al., 2003). There is good evidence that both these periods were exceptionally warm. In the Eocene, megathermal forests spread well beyond their present limits (Morley, 2000). For instance, the tropical palm Nypa occurred in Britain. Even allowing for continental movement, world climates must have been much warmer than now, and this could have been the result of either greater insolation, or higher concentration of greenhouse gases, or both (Morley, 2000; Willis and Niklas, 2004). Since the altitudinal temperature lapse rate is dependent on the amount of atmosphere above the surface, it is difficult to see how the lapse rate could have differed very much from the present rate found in the wet tropics, of c. 6°C per 1,000 m. At the peak of Eocene warmth, the sea surface temperature in the tropics rose as high as 32°C (Pearson et al., 2001). This is about 5°C higher than now. The Late Paleocene thermal maximum may have been even warmer (Zachos et al., 2001; Willis and Niklas, 2004).

Thus, using the above lapse rate, we might expect a MAT of 27°C to 21°C (the present range of lowland forest species) to occur at c. 800-m to 1,800-m altitude, and thus what are now lowland species could have grown at that altitude, assuming that their climatic tolerances have not changed. Since the usual MAT tolerance of a species has a range of about 6°C MAT (van Steenis, 1934–36), other species would have occurred at lower altitudes, although they might well have been congeners of the present day lowland species.

At 800 m to 1,800 m we might expect UV-B insolation to be slightly above that at sea level. This altitude effect is usually quite small, with values around 15% per 1,000 m (Blumthaler et al., 1997). Since, however, there were major volcanic eruptions in Greenland during the Eocene (Aubrey et al., 1998) there might have been a considerably greater temporary rise in UV radiation (see previous section).

UV-B is a well-known mutagen. Its effect in producing human skin cancer has been well-researched (Lodish et al., 2000) and it is clear that skin cancer is more

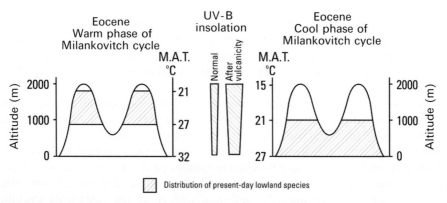

Distribution of present-day lowland species

Figure 8.5. Diagram to show how a combination of appropriate topography, Eocene warmth, and enhanced UV-B after vulcanicity could lead to isolation, mutation, allopatric speciation, a species pump, and increased biodiversity.

common in the tropics (Smith and Warr, 1991). The latitudinal gradient of UV-B is well-established (Caldwell *et al.*, 1980). UV-B is widely used by geneticists and plant breeders to promote mutation (Jansen *et al.*, 1998; Atwell *et al.*, 1999). Since plant reproductive apparatus is necessarily exposed to the atmosphere (for pollination, whether by wind, insects, or other means) it seems possible that UV-B induced mutations in reproductive DNA would occur at an enhanced rate. The effect of elevated UV-B on plant reproductive apparatus has already been reported (van der Staaij *et al.*, 1997). Extreme elevation of UV-B in the geological past is thought to have caused worldwide dieback of woody plants (Visscher *et al.*, 2004).

There is evidence that Milankovitch cycles have affected world climate through-out geological time (Bennett, 1990; Pietras *et al.*, 2003; Willis and Niklas, 2004). The effect of such cycles on Eocene megathermal forest taxa would have been to drive them up and down hills. In warm phases, what are now lowland taxa (already adapted to MAT of 27–21°C) would have migrated to the hills. The possibility of isolation on individual peaks would occur and this would be repeated with each cycle (Figure 8.5). Thus, we have all the requirements for allopatric speciation and a species pump: mutation, geographic isolation, and cyclic change. Furthermore, this process would work preferentially in the area then covered by megathermal vegetation, which included mid-latitudes as well (Morley, Chapter 1 in this book). Outside that area, strong seasonality would have restricted tree growth to low altitudes.

It would support this hypothesis if present day tropical montane taxa were experiencing similar high speciation rates, since they experience high UV-B and the possibility of isolation on mountain peaks during interglacials. One does not have to look far for examples. The case of *Espeletia* in the Colombian Andes is well-known: while some species are widespread, others are restricted to single peaks (van der Hammen, 1974). It is likely that *Rhododendron* may provide another example, in New Guinea and the Himalayas (Sleumer, 1966; Leach, 1962). Interestingly, many *Rhododendron* species are epiphytes, and thus exposed to strong insolation in the canopy. Possibly, many epiphytic orchid genera would be another set of examples (e.g., *Cymbidium*) (Du Puy and Cribb, 1988). The possible origin of varieties of *Leptospermum flavescens* on Mt. Kinabalu (Borneo) by the action of UV-B has already been proposed (Lee and Lowry, 1980a). The abundance of endemic species in several genera on individual peaks in the Andes has already been noted by Gentry (1989).

In summary, the hypothesis presented here is that major speciation in the tropics may have occurred especially in warm periods in the past (Eocene and Early Miocene), because in those times lowland taxa were able to live at higher altitudes where they experienced enhanced UV-B induced mutation rates (increased further by intermittent vulcanicity), and isolation on individual mountain peaks. Cyclical climate changes could have led to a species pump. Occasional major volcanic eruptive phases could have provided the "punctuated equilibrium" which is the pattern of evolution accepted by many (Eldredge and Gould, 1972). It is not suggested that this is a complete explanation for biodiversity. While it could apply to some insects and some other animals, it is unlikely to apply to marine life or to nocturnal or soil animals. Probably, the full explanation will turn out to be multivariate, but it is hoped that this hypothesis may make a contribution.

8.7 CONCLUSIONS

Generalizing, it may be said that the role of ultraviolet insolation (especially UV-B) in the tropics is still largely a matter of speculation and hypothesis. It may well be much more significant in sub-tropical regions, where total solar radiation is so much greater than in the tropics (Landsberg *et al.*, 1966). Nevertheless, there do seem to be some grounds for considering UV-B to be significant. This could be especially so after large volcanic eruptions, when sulphurous aerosols may damage the ozone layer. Even in normal conditions, UV-B appears to be a significant factor in tropical high mountains, especially where cloud conditions are intermittent on a daily basis. It seems meaningful that the unusual features of the Dwarf Forest (stunting, small thick leaves with a hypodermis, presence of extra pigments) can be produced in a range of cultivated plants by applying additional UV-B radiation. Nevertheless, it must be borne in mind that some of these symptoms are typical of stress in plants from various causes. For instance, stunting can be induced by water shortage, nutrient deficiency, or by very cold night temperatures. The last of these is a known factor on tropical mountains.

There is little evidence of a change in UV-B insolation during the Quaternary. This stability of the UV environment, while temperature change was occurring, could account for the near-disappearance of the Dwarf Forest in successive cold phases of the Pleistocene, as the required combination of high UV-B and cool temperatures was much reduced.

The possible role of UV-B as an agent of extinction or endangerment of amphibian species since recent ozone depletion is discussed, but other factors seem to be involved.

Finally, the possible significance of UV-B as a mutagen is discussed. It would be easy to over-estimate the importance of this as an evolutionary mechanism, but it seems possible that it could have contributed to the appearance of high tropical biodiversity during the Tertiary. This would only have been likely during periods of enhanced UV-B resulting from volcanic eruptions, but could have led to a "punctuated equilibrium" type of speciation. It could have been aided by a species pump mechanism involving allopatric speciation in hilly topography. This hypothesis has the advantage of being more effective in tropical than in temperate regions, which is in accordance with biogeographic reality.

8.8   REFERENCES

Atwell, B. J., Kriedemann, P. E., and Turnbull, C. G. N. (eds.) (1999) *Plants in Action: Adaptation in Nature, Performance in Cultivation.* Macmillan, South Yarra, Australia.

Aubry, M.-P., Lucas, S. G., and Berggren, W. A. (eds.) (1998) *Late Paleocene–Early Eocene Climatic and Biotic Events in the Marine and Terrestrial Records.* Columbia University Press, New York.

Austin, J., Butchart, N., and Shine, K. P. (1992) Possibility of an Arctic ozone hole in a doubled-CO_2 climate. *Nature* **360**, 221–225.

Bennett, K. D. (1990) Milankovitch cycles and their effects on species in ecological and evolutionary time. *Paleobiology* **16**, 11–21.

Bennett, K. D. (2004) Continuing the debate on the role of Quaternary environmental change for macroevolution. *Phil. Trans. R. Soc. Lond. B* **359**, 295–303.

Blaustein, A. R., Belden, L. K., Olson, D. H., Green, D. M., Root, T. L., and Kiesecker, J. M. (2001) Amphibian breeding and climate change. *Conservation Biology* **15**, 1804–1809.

Blaustein, A. R., Romansic, J. M., Kiesecker, J. M., and Hatch, A. C. (2003) Ultraviolet radiation, toxic chemicals and amphibian population declines. *Diversity and Distributions* **9**, 123–140.

Blumthaler, M., Ambach, W., and Ellinger, R. (1997) Increase in solar UV radiation with altitude. *Journal of Photochemistry and Photobiology B, Biology* **39**, 130–134.

Brass, L. J. (1941) The 1938–39 expedition to the Snow Mountains, Netherlands New Guinea. *J. Arnold Arbor.* **22**, 271–342.

Brass, L. J. (1964) Results of the Archbold Expeditions No. 86: Summary of the Sixth Archbold Expedition to New Guinea. *Bulletin of the American Museum of Natural History* **127**, 145–215.

Brookfield, H. C. (1964) The ecology of highland settlement: Some suggestions. *American Anthropologist* **66**, 20–38.

Bruijnzeel, L. A. and Proctor, J. (1993) Hydrology and biogeochemistry of tropical montane cloud forests: What do we really know? In: L. S. Hamilton, J. O. Juvik, and F. N. Scatena (eds.), *Tropical Montane Cloud Forests* (pp. 25–46). East-West Center, Honolulu, HI.

Bruijnzeel, L. A., Waterloo, M. J., Proctor, J., Kuiters, A. T., and Kotterink, B. (1993) Hydrological observations in montane forests on Gunung Silam, Sabah, Malaysia, with special reference to the "Massenerhebung" effect. *Journal of Ecology* **81**, 145–167.

Bush, M. B. (1994) Amazonian speciation: A necessarily complex model. *Journal of Biogeography* **21**, 5–17.

Bush, M. B., Piperno, D. R., Colinvaux, C. A., De Oliveira, P. E., Krissek, L. A., Miller, M. C., and Rowe, W. L. (1992) A 14,300-year palaeoecological profile of a lowland tropical lake in Panama. *Ecological Monographs* **62**, 251–275.

Caldwell, M. M. (1971) Solar UV irradiation and the growth and development of higher plants. In: A. C. Giese (ed.), *Photophysiology* (pp. 131–177). Academic Press, New York.

Caldwell, M. M. (1981) Plant response to solar ultraviolet radiation. In: O. L. Lange, P. S. Nobel, C. B. Osmond, and H. Ziegler (eds.), *Physiological Plant Ecology, I: Encyclopedia of Plant Physiology* (New Series, Vol.12A, pp. 169–197). Springer-Verlag, Berlin.

Caldwell, M. M., Robberecht, R., and Billings, W. D. (1980) A steep latitudinal gradient of solar ultraviolet-B radiation in the arctic–alpine life zone. *Ecology* **61**, 600–611.

Caldwell, M., Teramura, A. H., Tevini, M., Bornman, J. F., Bjorn, L. O., and Kulandaivelu, G. (1995) Effects of increased ultraviolet-radiation on terrestrial plants. *Ambio* **24**, 166–173.

Caldwell, M. M., Bjorn, L. O., Bornman, J. F., Flint, S. D., Kulandaivelu, G., Teramura, A. H., and Tevini, M. (1998) Effects of increased solar ultraviolet radiation on terrestrial ecosystems. *Journal of Photochemistry and Photobiology B, Biology* **46**, 40–52.

Carey, C. and Alexander, M. A. (2003) Climatic change and amphibian declines: Is there a link? *Diversity and Distributions* **9**.

Carvalho, S. M. P., Heuvelink, E., Cascais, R., and van Korten, O. (2002) Effect of day and night temperature on internode and stem length in *Chrysanthemum*: Is everything explained by DIF? *Annals of Botany* **90**, 111–118.

Collins, J. P. and Storfer, A. (2003) Global amphibian declines: Sorting the hypotheses. *Diversity and Distributions* **9**, 89–98.

Corlett, R. T. (1984) Human impact on the subalpine vegetation of Mt. Wilhelm, Papua New Guinea. *Journal of Ecology* **72**, 841–854.

Daszak, P., Cunningham, A. A., and Hyatt, A. D. (2003) Infectious disease and amphibian population declines. *Diversity and Distributions* **9**, 141–150.

Dave, J. V. and Halpern, P. (1976) Effect of changes in ozone amount on the ultraviolet radiation received at sea level of a model atmosphere. *Atmospheric Environment* **10**, 547–555.

Du Puy, D. and Cribb, P. (1988) *The Genus Cymbidium*. Helm, Bromley, U.K.

Eldredge, N. and Gould, S. J. (1972) Punctuated equilibria: An alternative to phyletic gradualism. In: T. M. Schopf (ed.), *Models in Paleobiology*. Freeman, Cooper & Co., San Francisco.

Farrera, I., Harrison, S. P., Prentice, I. C., Bartlein, P. J., Bonnefille, R., Bush, M., Cramer, W., von Grafenstein, U., Holmgren, K., Hooghiiiemstra, H. *et al.* (1999) Tropical climates of the Last Glacial Maximum: A new synthesis of terrestrial palaeoclimate data. 1. Vegetation, lake levels and geochemistry. *Climate Dynamics* **15**, 823–856.

Flenley, J. R. (1979) *The Equatorial Rain Forest: A Geological History*. Butterworths, London.

Flenley, J. R. (1993) The origins of diversity in tropical rain forests. *Trends in Ecology and Evolution* **8**, 119–120.

Flenley, J. R. (2005) Palynological richness and the tropical rain forest. In: E. Bermingham, E. C. Dick, and C. Moritz, C. (eds.), *Tropical Rainforests: Past, Present, and Future*. Chicago University Press, Chicago.

Flenley, J. R. and Richards, K. (eds.) (1982) *The Krakatoa Centenary Expedition: Final Report* (Miscellaneous Series No. 25). Geography Department, University of Hull, Hull, U.K.

Forster, R. M. (1982) A study of the spatial distribution of bryophytes on Rakata. In: J. R. Flenley, and K. Richards (eds.), *The Krakatoa Centenary Expedition: Final Report* (Miscellaneous Series No. 25, pp. 103–126). Geography Department, University of Hull, Hull, U.K.

Gentry, A. H. (1989) Speciation in tropical forests. In: L. B. Holm-Nielsen, I. C. Nielsen, and H. Balslev (eds.), *Tropical Forests: Botanical Dynamics, Speciation and Diversity* (pp. 113–134). Academic Press, London.

Gleason, D. F. (2001) Ultraviolet radiation and coral communities. In: C. S. Cockell and A. R. Blaustein, (eds.), *Ecosystems, Evolution and Ultraviolet Radiation* (pp. 118–149). Springer-Verlag, New York.

Gleason, J. F., Bhartia, P. K., Herman, J. R., McPeters, R., Newman, P., Stolarski, S., Flynn, I., Labow, G., Larko, D., Seftor, C. *et al.* (1993) Record low global ozone in 1992. *Science* **260**, 523–526.

Grubb, P. J. (1977) Control of forest growth and distribution on wet tropical mountains. *Ann. Rev. Ecol. Syst.* **8**, 83–107.

Grubb, P. J. and Whitmore T. C. (1966) A comparison of montane and lowland rain forest in Ecuador, 2: The climate and its effects on the distribution and physiognomy of the forests. *Journal of Ecology* **54**, 303–333.

Haffer, J. (1997) Alternative models of vertebrate speciation in Amazonia: An overview. *Biodiversity and Conservation* **6**, 451–476.

Haffer, J. T. and Prance, G. T. (2001) Climatic forcing of evolution in Amazonia during the Cenozoic: On the refuge theory of biotic differentiation. *Amazoniana—Limnologia et Oecologia regionalis systemae fluminis Amazonias* **16**, 579–605.

Hastenrath, S. (1968) Certain aspects of the three-dimensional distribution of climate and vegetation belts in the mountains of central America and southern Mexico. *Colloquium Geogr.* **9**, 122–130.

Hope, G. S. (1986) Development of present day biotic distributions in the New Guinea mountains. In: B. A. Barlow (ed.), *Flora and Fauna of Alpine Australasia: Ages and Origins* (pp. 129–145). C.S.I.R.O., Melbourne.

Jansen, M. A. K., Gaba, V., and Greenberg, B. M. (1998) Higher plants and UV-B radiation: Balancing damage, repair and acclimation. *Trends in Plant Science* **3**, 131–135.

Kats, L. B. and Ferrer R. P. (2003) Alien predators and amphibian declines: Review of two decades of science and the transition to conservation. *Diversity and Distributions* **9**, 99–110.

Landsberg, H. E., Lippmans, H., Paffen, K. H., and Troll, C. (1966) *World Maps of Climatology.* Springer-Verlag, Berlin.

Leach, D. G. (1962) *Rhododendrons of the World.* Allen & Unwin, London.

Leavitt, P. R., Vinebrooke, R. D., Donald, D. B., Smol, J. P., and Schindler, D. W. (1997) Past ultraviolet radiation environments in lakes derived from fossil pigments. *Nature* **388**, 457–459.

Lee, D. W. and Lowry, J. B. (1980a) Solar ultraviolet on tropical mountains: Can it affect plant speciation? *The American Naturalist* **115**, 880–883.

Lee, D. W. and Lowry, J. B. (1980b) Young leaf anthocyanin and solar ultraviolet. *Biotropica* **12**, 75–76.

Lindoo, S. J. and Caldwell, M. M. (1978) Ultraviolet-B radiation-induced inhibition of leaf expansion and promotion of anthocyanin production. *Plant Physiology* **61**, 278–282.

Lodish, H., Berk, A., Zipursky, S. L., Matsudaira, P., Baltimore, D., and Darnell, J. (2000) *Molecular Cell Biology.* Freeman, New York.

Madronich, S., McKenzie, R. L., Caldwell, M. M., and Bjorn, L. O. (1995) Changes in ultraviolet light reaching the Earth's surface. *Ambio* **24**, 143–152.

Martin, P. S. (1963) *The Last 10,000 Years: A Fossil Pollen Record of the American Southwest.* University of Arizona Press, Tucson, AZ.

Monteith, J. L. (1973) *Principles of Environmental Physics.* Arnold, London.

Morley, R. J. (2000) *Origin and Evolution of Tropical Rain Forests.* John Wiley & Sons, Chichester, U.K.

Murali, N. S. and Teramura A. H. (1986a) Intraspecific differences in *Cucumis sativus* sensitivity to ultraviolet-B radiation. *Physiol. Plant* **68**, 673–677.

Murali, N. S. and Teramura, A. H. (1986b) Effectiveness of UV-B radiation on the growth and physiology of field grown soybean modified by water stress. *Photochemistry and Photobiology* **44**, 215–219.

Murali, N. S. and Teramura, A. H. (1986c) Effects of supplemental ultraviolet-B radiation on the growth and physiology of field-grown soybean. *Environmental and Experimental Botany* **26**, 233–242.

Pearson, P. N., Ditchfield, P. W., Singano, J., Harcourt-Brown, K. G., Nicholas, C. J., Olsson, R. K., Shackleton, N. J., and Hall, M. A. (2001) Warm tropical sea surface temperatures in the Late Cretaceous and Eocene epochs. *Nature* **413**, 481–487.

Pietras, J. T., Carroll, A. R., Singer, B. S., and Smith, M. E. (2003) 10 Kyr depositional cyclicity in the early Eocene: Stratigraphic and Ar-40/Ar-39 evidence from the lacustrine Green River Formation. *Geology* **31**, 593–596.

Salomans, J. B. (1986) Paleoecology of volcanic soils in the Colombian Central Cordillera (Parque Nacional de los Nevados). *Dissertationes Botanicae* **95**, 1–212.

Schroeter, C. (1908) *Das Pflanzenleben der Alpen: Eine Schilderung der Hochgebirgsflora*. Verlag von Albert Raustein, Zurich, Switzerland [in German].

Shindell, D. T., Rind, D., and Lonergan, P. (1998a) Increased polar stratospheric ozone losses and delayed eventual recovery owing to increasing greenhouse-gas concentrations. *Nature* **392**, 589–592.

Shindell, D. T., Rind, D., and Lonergan, P. (1998b) Climate change and the middle atmosphere, Part IV: Ozone response to doubled CO_2. *J. Clim.* **11**, 895–918.

Sleumer, H. (1966) *An Account of Rhododendron in Malesia*. Noordhoff, Groningen, The Netherlands.

Smith, P. M. and Warr, K. (eds.) (1991) *Global Environmental Issues* (295 pp.). Hodder & Stoughton/Open University, London.

Son, K. C., Kim, H., and Park, Y. S. (2002) Effects of DIF and temperature drop on the growth and flowering of *Begonia* × *hiemalis*. *Journal of the Korean Society for Horticultural Science* **43**, 492–496.

Stolarski, R., Bojkov, R., Bishop, L., Zerefos, C., Staehelin, J., and Zawodry, J. (1992) Measured trends in atmospheric ozone. *Science* **256**, 342–349.

Storfer, A. (2003) Amphibian declines: Future directions. *Diversity and Distributions* **9**, 151–163.

Stuart, S. N., Chanson, J. S., Cox, N. A., Young, B. E., Rodrigues, A. S. L., Fischman, D. L., and Waller, R. W. (2004) Status and trends of Amphibian declines and extinctions worldwide. *Science* **306**, 1783–1786.

Sullivan, J. H., Teramura, A. H., and Ziska, L. H. (1992) Variation in UV-B sensitivity in plants from a 3,000-m elevational gradient in Hawaii. *American Journal of Botany* **79**, 737–743.

Teramura, H. (1983) Effects of ultraviolet-B radiation on the growth and yield of crop plants. *Physiol. Plant* **58**, 415–427.

Troll, C. (1959) *Die tropischen Gebirge: Ihre dreidimensionale klimatische und pflanzengeographische Zonierung* (93 pp.). Dummlers, Bonn [in German].

van de Staaij, J. W. M., Bolink, E., Rozema, J., and Ernst, W. H. O. (1997) The impact of elevated UV-B (280–320 nm) radiation on the reproduction biology of a highland and a lowland population of *Silene vulgaris*. *Plant Ecology* **128**, 172–179.

van der Hammen, T. (1974) The Pleistocene changes of vegetation and climate in tropical South America. *Journal of Biogeography* **1**, 3–26.

van Steenis, C. G. G. J. (1934–36) On the origin of the Malaysian mountain flora. *Bulletin du Jardin Botanique Buitenzorg Series III*: Part I **13**, 135–262; Part II **13**, 289–417; Part III **14**, 56–72.

van Steenis, C. G. G. J. (1972) *The Mountain Flora of Java*. E. J. Brill, Leiden, The Netherlands.

Visscher, H., Looy, C. V., Collinson, M. E., Brinkhuis, H., Cittert, J. H. A. V. K. V., Kurschner, W. M., and Sephton, M. A. (2004) Environmental mutagenesis during the end-Permian ecological crisis. *Proc. Nat. Acad. Sci. U.S.A.* **101**, 12952–12956.

Walker, D. and Flenley, J. R. (1979) Late Quaternary vegetational history of the Enga District of upland Papua New Guinea. *Phil. Trans. R. Soc. B* **286**, 265–344.

Wilf, P., Cuneo, N. R., Johnson, K. R., Hicks, J. F., Wing, S. L., and Obradovich, J. D. (2003) High plant diversity in Eocene South America: Evidence from Patagonia. *Science* **300**, 122–125.

Willis, K. J. and Niklas, K. J. (2004) The role of Quaternary environmental change in plant macroevolution: The exception or the rule? *Phil. Trans. R. Soc. B* **359**, 159–172.

Zachos, J., Pagani, M., Sloan, L. *et al.* (2001) Trends, rhythms and aberrations in global climate 65 Ma to present. *Science* **292**, 686–693.

9

Climate change and hydrological modes of the wet tropics

J. Marengo

9.1 INTRODUCTION

The Amazon Basin is the world's largest drainage system. In fact, 1,100 rivers make up the Amazon system. The Amazon River carries one-fifth of all the river water in the world. The source of the Amazon can be traced to the Apurimac River, located at 5,200 m above sea level. The 1,100 tributaries flow through nine South American countries: Brazil, Bolivia, Peru, Ecuador, Colombia, Venezuela, Guyana, Surinam, and French Guiana. The Amazon River and its tributaries drain most of the area of heavy rainfall. The area is called "Amazonia" and most of it is a sparsely populated rainforest. Most of the Amazon Basin is in Brazil (Amazonia Legal). The Amazon River represents 16% of annual global river runoff (Shiklomanov, 2001).

The Amazon River system is the single largest source of freshwater on Earth and its flow regime is subject to interannual and long-term variability represented as large variations in downstream hydrographs (Richey *et al.*, 1989; Vörösmarty *et al.*, 1996; Marengo *et al.*, 1998a; Marengo and Nobre, 2001; Marengo 2004a, b). A better understanding of rainfall and river variability will depend on the physical mechanisms related to regional and large-scale atmospheric–oceanic–biospheric forcings that impact the temporal and spatial variability of the hydrometeorology of the Amazon Basin. The impacts could be felt on various timescales.

The implementation of field experiments in the region during the last 20 years—such as the ABRACOS (Anglo Brazilian Amazon Climate Observational Study) during the 1980s, the LBA (Large Scale Biosphere Atmosphere experiment in Amazonia), and the SALLJEX (South American Low Level Jet field experiment) during the late 1990s and early 2000s—has allowed for the development of new knowledge on climate and hydrology in the Amazon Basin, including the interaction between land surface processes in rainfall, and the development of regional and global climate models tuned with more realistic representations of physical processes for the region (Gash and Nobre, 1997; Silva Dias *et al.*, 2002; Vera *et al.*, 2006). Moisture transport into and out of the Amazon Basin has also been studied, and regional circulation features responsible for this transport and its variability in time and space have been detected and studied using observations from these field experiments and other global data sets (Marengo *et al.*, 2002, 2004a, b; Vera *et al.*, 2006).

On the basis of what is now known on climate variability in Amazonia and the moisture transport in and out of the basin based on observational studies and model simulations, the question that arises is: What are the possible impacts on the Amazon ecosystem of regional-scale deforestation or the increase of greenhouse gas (GHG) concentrations in the atmosphere and subsequent global warming. The issue of deforestation has been explored in various numerical experiments since the 1980s using atmospheric global climate models—general circulation models (GCMs)—all of which show that the Amazon will become drier and warmer (see reviews in Marengo and Nobre, 2001; Voldoire and Royer, 2004). Even though there are no clear signs of trends for reduction of rainfall in the basin due to deforestation—as suggested by climate models on deforestation scenarios—one study (Costa *et al.*, 2003) has detected changes in the Tocantins River discharges as a result of land-use changes in its upper basin following the construction of the city of Brasilia in the 1960s.

Furthermore, since the early 2000s new developments in atmosphere–ocean–biosphere coupled models—by the Hadley Centre for Climate Research and Prediction in the U.K., the Institute Pierre et Simon Laplace at the University of Paris in France, and the Frontier Research Center for Global Change in Japan—have allowed for better simulation of future climate change scenarios. The new models include interactive vegetation schemes that more realistically represent the water vapor, carbon, and other gas exchange between the vegetation and the atmosphere. Projections for future climate change from the Hadley Centre model have shown that an increase in the concentration of greenhouse gases in the atmosphere will produce changes in vegetation such that Amazonia will become a savanna by the 2050s, and the region will become drier and warmer with most of the moisture coming from the tropical Atlantic such that normally produced rainfall in the region will not find the environment to condense above the savanna vegetation by 2050, and the moist air stream will move to southeastern South America producing more rainfall in those regions. Therefore, after 2050, the Amazon Basin may well behave as a "source of moisture" rather than a sink (like its present day climate) (Cox *et al.*, 2000, 2004; Betts *et al.*, 2004; Huntingford *et al.*, 2004).

Therefore, this chapter is focused on the role of the Amazon Basin in the functioning and modulation of the regional climate and hydrology, in both present and future climates, by means of (a) description of hydrological regimes and maintenance of humidity and import/export of moisture, (b) variability of climate and hydrology in various timescales; and (c) sensitivity of the Amazon system to changes in land use or climate change due to an increase in the concentration of GHG in the atmosphere.

9.2 RAINFALL AND HYDROLOGICAL REGIMES IN THE AMAZON BASIN

9.2.1 Rainfall distribution and seasonal rainfall and river seasonal variability

The Amazon River drains an area of $6.2 \times 10^6 \, km^2$ and discharges an average of $6,300 \, km^3$ of water to the Atlantic Ocean annually. The annual cycle of rainfall in the region has been extensively described in Rao and Hada (1990), Marengo (1992), Figueroa and Nobre (1999), Marengo and Nobre (2001), Liebmann and Marengo (2001), Marengo (2004b). The spatial distribution of rainfall shows three centers of abundant precipitation in the Amazon Basin. One is located in northwest Amazonia, with more than 3,600 mm per year. Another region with abundant rainfall is the central part of Amazonia around 5°S with 2,400 mm per year. A third center is found close to the mouth of the Amazon River near Belém, with more than 2,800 mm per year. In the Rio Negro basin area, in northwestern Amazonia, the rainfall is abundant throughout the year reaching its maximum in April–June, while southern Amazonia's rain peaks earlier (January–March). The extreme high and localized values of precipitation in narrow strips along the eastern side of the Andean slopes are thought to be due to upglide condensation and a rain shadow effect on the lee side, so the localized maximum is due to the easterly winds being lifted when they flow over the Andes. The coastal maximum is caused by nocturnal convergence between the trade winds and the land breeze. In the central-north and south-southeast sections, rainfall is lower—in the region of 1,500 mm. The peak of the rainy season occurs earlier (December–February) in southern Amazonia, while northern and central Amazonia experience maximum rain in March–May (Figure 9.1, see color section).

The seasonal variability in rainfall in Figure 9.1 can be better understood by viewing it with Figure 9.2, which shows the mean seasonal distribution of rainfall. The austral summer—that is, December to February—season is characterized by the peak of the rainy season in southern Amazonia and the dry season in the Amazon region north of the equator, with less than 360 mm in the entire season. March–May represents the peak of the rainy season in central Amazonia, all the way from western Amazonia to the mouth of the Amazon River, and June–August represents the dry season over most of the region, with less than 180 mm in the entire season in southern and eastern Amazonia and less than 360 mm in the entire season in central Amazonia, while the extreme north of Amazonia experiences its wet season.

Following the annual cycle of rainfall, river discharge peaks first in southern and eastern Amazonia (January–March)—as in the Tocantins-Araguaia and Madeira Rivers in Figure 9.3—while the Negro and Amazon River peak in March–May. Measurements taken at Óbidos integrate the contributions of the Solimêes River (southern and western Amazonia) and the Rio Negro (northern Amazonia). Chu (1982) shows that almost 70% of the contribution to Óbidos measurements comes from the waters of the Solimêes River. The records of the Amazon, Negro, Xingú and Tocantins Rivers (Marengo et al., 1998b) are displayed for two El Niño years (1982–83, 1986–87) and La Niña years (1975–76, 1988–89). Extreme years were chosen since they may better show the associations between El Niño and rain in their basins. The

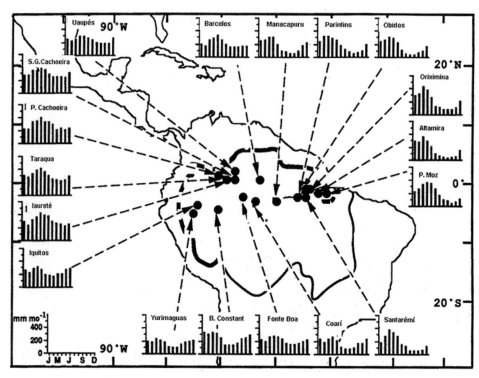

Figure 9.2. Seasonal cycle of rainfall in various stations across the Amazon Basin. Location of stations is indicated by a small square, and units are in mm month^{-1}. Scale is shown at the lower left side of the panel. Modified from Marengo (1992).

year before an El Niño peak, Amazon River discharges at Óbidos are anomalously high, while in actual El Niño years the discharges are lower than average. During La Niña years, the discharges at Óbidos are more than 7,000 m^3 s^{-1} above the normal.

River discharge peaks earlier in southern and eastern Amazonia (Tocantins and Madeira Rivers) than in northern Amazonia (Rio Negro) (Figure 9.2). Discharges of the Amazon River measured at Óbidos (200 km inland from the mouth of the Amazon) do not represent the true amount of water that reaches the mouth of the Amazon, since they do not include the waters of the Xingú and Tocantins Rivers (Marengo et al., 1994). Mean discharge at the Óbidos gauging station is 175,000 m^3 s^{-1} (or 2.5 mm day^{-1}), while the correct value at the mouth of the Amazon (Roads et al., 2002; Marengo, 2004b) is 210,000 m^3 s^{-1} (or 2.9 mm day^{-1}).

9.2.2   Atmospheric and hydrological water balance

The hydrological cycle of the Amazon region is of great importance since the region plays an important role in the functioning of regional and global climate. Variations in its regional water and energy balances at year-to-year and longer timescales are of

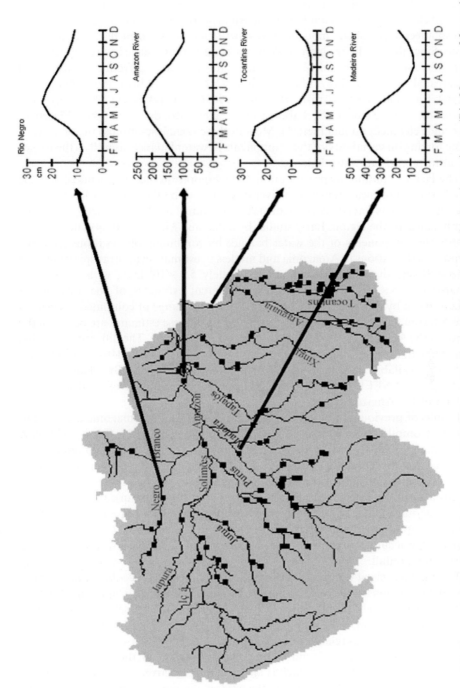

Figure 9.3. Seasonal cycle of river discharges and levels in northern and southern Amazonia. Levels are in cm (Rio Negro at Manaus). Discharges from the Amazon, Madeiras, and Tocantins Rivers are in $10^3 \text{ m}^3 \text{ s}^{-1}$. *Source:* M. Costa and M. Coe.

special interest, since alterations in circulation and precipitation can translate ultimately to changes in the streamflow of the Amazon River. In addition, these changes
can also affect atmospheric moisture transport from the Amazon region to adjacent
regions. Since the late 1970s, large-scale water budget studies have been conducted for
this region using a variety of observational data sets varying from radiosondes to
global reanalyses (Salati, 1987; Matsuyama, 1992; Eltahir and Bras, 1994; Marengo *et
al.*, 1994; Rao *et al.*, 1996; Vörösmarty *et al.*, 1996; Costa and Foley, 1999; Curtis and
Hastenrath, 1999; Zeng, 1999; Labraga *et al.*, 2000; Roads *et al.*, 2002; Marengo,
2004b and references quoted therein). Most of these reanalyses discuss the impacts of
remote forcing on variability of the components of water balance, as well as the role of
evapotranspiration in water balance.

The lack of continuous precipitation and evaporation measurements across the
entire basin and of measurements of river discharge along the Amazon River and its
main tributaries has forced many scientists to use indirect methods for determining the
water balance for the region. Early studies by Salati and Marques (1984) attempted to
quantify the components of the water balance by combining observations from the
few radiosonde stations in Amazonia and models to estimate evapotranspiration. The
Amazon Rivers drains an area of approximately $5.8 \times 10^6 \, \text{km}^2$, with an average
discharge of $5.5 \times 10^{12} \, \text{m}^3 \, \text{yr}^{-1}$. However, different estimates of the area of the
Amazon Basin by different authors have led to a wide range of computed discharges
of the Amazon River during the last 25 years. Most of these estimates are based on the
records of the Amazon at Óbidos, and are shown in Table 9.1. The differences are due
to the different areas considered, and more recently (Marengo, 2004b) the discharges
at Óbidos (available since the mid-1970s) were corrected by the Brazilian National
Water Agency ANEEL so that they would be more representative of the discharge at
the delta of the Amazon River.

Results of previous studies of the annual water budget in Amazonia are listed in
Table 9.2 (compiled from Matsuyama, 1992; Marengo *et al.*, 1994; Costa and Foley,
2000; Marengo, 2004b). Main differences in results are due to the different areas
considered for the basin that translate to different discharge and derived run-off,
different precipitation networks and methods of assessment (mostly based on gridded
rainfall data or from rain gauges distributed irregularly in the basin), and the methods
used to determine annual water balance, where evapotranspiration *ET* is estimated as
residual precipitation *P* and discharge *R*. However, the equation $P = ET + R$ does
not guarantee accurate estimates of the possible role of the tropical forest in recycling
moisture for rainfall.

The annual cycle of water balance terms shows some differences between northern
and southern sections of the basin (Figure 9.3). There is seasonality in *R* and *P*: with *R*
peaking between 3 and 4 months after *P*. The E/P ratio of the dry season is larger than
that of the rainy season, indicating that the role of evaporation (and evapotranspiration) on the water cycle is relatively more important in the dry season than in the rainy
season. The largest E/P is found during the dry season in the southern region,
reaching values greater than 1, which is larger than respective values in the northern
region (Marengo, 2004b). In general, in the present climate the Amazon Basin can be
considered as a moisture sink ($P > E$).

Table 9.1. Observed river discharge for the Amazon River at Óbidos (Matsuyama, 1992; Marengo *et al.*, 1994; Marengo and Nobre, 2001; Marengo, 2004b).

Study	Amazon River discharge $(10^3 \text{ m}^3 \text{ s}^{-1})$
Leopold (1962)	113.2
UNESCO (1971)	150.9
Nace (1972)	175.0
UNESCO (1974)	173.0
Baumgartner and Reichel (1975)	157.0
Villa Nova *et al.* (1976)	157.0
Milliman and Meade (1983)	199.7
Nishizawa and Tanaka (1983)	160.0
Oki *et al.* (1995)	155.1
Matsuyama (1992)	155.1
Russell and Miller (1990)	200.0
Vörösmarty *et al.* (1989)	170.0
Sausen *et al.* (1994)	200.0
Marengo *et al.* (1994)	202.0
Costa and Foley (1998a)	162.0
Zeng (1999)	205.0
Leopoldo (2000)*	160.0
Leopoldo (2000)**	200.0
Roads *et al.* (2002)	224.0
Marengo (2004b)*	175.0
Marengo (2004b)**	210.0

*   Measured at Óbidos.
** Measured (corrected) at the mouth.

Estimates of the water balance in the Amazon region exhibit some uncertainities, derived mainly from the use of different rainfall data sets (either gridded or station data), use of streamflow data at the Óbidos gauge site (corrected or uncorrected), and use of global reanalyses produced by some meteorological centers in the U.S. and Europe. The National Centers for Environmental Prediction (NCEP) and the European Center for Medium Range Weather Forecast (ECMWF) have carried out retrospective analyses (reanalyses) over the last decade using a single model and data assimilation to represent climate evolution from as early as World War II. These reanalyses can highlight characteristic features of circulation and water balance. However, while data assimilation should in principle provide for a description of the water flux field, there are no guarantees that this description will be superior to that obtained from objective analysis and radiosonde observations alone, especially over continental regions. There is a need for the level of uncertainty to be identified in the measurement or estimation of the components of the water budget.

Table 9.3 shows the annual values of the water budget components for Amazonia in its entirety, giving the mean and two extremes of interannual variability: El Niño

Table 9.2. The annual water budget of the Amazon Basin. P = Precipitation; ET = Evapotranspiration; R = Streamflow—all in mm y^{-1}. In this table the water balance equation $ET = P - R$ is used, P and R are measured, and ET is obtained as a residual. Marengo (2004b) used the water balance equation that considers the non-closure of the Amazon Basin (Marengo and Nobre, 2001; Marengo, 2004b).

Study	P	ET	R
Baumgartner and Reichel (1975)	2,170	1,185	985
Villa Nova et al. (1976)	2,000	1,080	920
Jordan and Heuveldop (1981)	3,664	1,905	1,759
Leopoldo et al. (1982)	2,076	1,676	400
Franken and Leopoldo (1984)	2,510	1,641	869
Vörösmarty et al. (1989)	2,260	1,250	1,010
Russell and Miller (1990)	2,010	1,620	380
Nishizawa and Koike (1992)	2,300	1,451	849
Matsuyama (1992)	2,153	1,139	849
Marengo et al. (1994)	2,888	1,616	1,272
Costa and Foley (2000)	2,166	1,366	1,800
Marengo (2004b)	2,117	1,570	1,050

1982–83, and El Niño 1997–98 and La Niña 1988–89. Reduced precipitation (P), run-off (R), and moisture convergence (C) are found during these two strong El Niño events while values larger than normal are found during La Niña 1988/89, and in all cases $P > E$, suggesting that the Amazon region is an atmospheric moisture sink. The difference between these two El Niño events in Amazonia is that during 1997/98 large-scale circulation anomalies over the Atlantic sector did not allow for much convergence of moisture. In the long term $P > E$, during La Niña events it is shown that $P > E$, and during El Niño 1982–83 and 1997/98 $P > E$, even though the difference is smaller than the mean and La Niña years.

Table 9.3. Climatological water budget 1970–99 for the Amazon Basin. Comparisons are made for the 1982–83 El Niño and the 1988–89 La Niña. P is derived from observations (Marengo, 2004b), E and C are derived from NCEP/NCAR reanalyses, and R is run-off from the historical discharge records of the Amazon River at Óbidos. Units are in mm day^{-1}. +C = Moisture convergence (Marengo, 2004b).

Component	Mean	El Niño 1982–83	El Niño 1997–98	La Niña 1988–89
P	5.8	4.6	5.2	6.7
E	4.3	4.5	4.1	4.4
R	2.9	2.1	2.5	2.9
C	1.4	1.3	1.3	3.1
$P - E$	+1.5	+0.4	+0.9	+2.3
$P - E - C$	+0.1	−0.9	−0.1	−0.8
Imbalance $= [((C/R) - 1)]$	51%	38%	52%	6%

Table 9.4. Water budget 1970–99 of the entire Amazon Basin. P is derived from several data sources: Global Historical Climatology Network (GHCN), Xie and Arkin (CMAP), GPCP, NCEP, Legates–Wilmott (LW), Climate Research Unit (CRU) and from observations by Marengo (2004a). E and C are derived from NCEP/NCAR reanalyses, and R = Corrected run-off from the historical discharge records of the Amazon River at Óbidos. Units are in mm day^{-1}. $+C$ = Moisture convergence (Marengo, 2004b).

Component	GHCN	CMAP	GPCP	NCEP	LW	CRU	Marengo (2004b)
P	8.6	5.6	5.2	6.4	5.9	6.0	5.8
E	4.3	4.3	4.3	4.3	4.3	4.3	4.3
R	2.9	2.9	2.9	2.9	2.9	2.9	2.9
C	1.4	1.4	1.4	1.4	1.4	1.4	1.4
$P - E$	4.3	1.3	0.9	2.1	1.6	1.6	1.5
$P - E - C$	+2.9	−0.1	−0.5	+0.7	+0.2	+0.3	+0.1

The rain gauge based rainfall estimates used by Marengo (2004b) produced a mean of 5.8 mm day^{-1}, which is close to the values obtained in similar studies using different rainfall gridded data sets (CMAP, CRU, GHCN, and GPCP). The observed R at the mouth of the Amazon has been estimated as 2.9 mm day^{-1} (or 210,000 m^3 s^{-1} for a basin area of 6.1 million square kilometers), and this represents the combination of Amazon discharges at Óbidos and those of the Xingú and Tocantins Rivers. Table 9.4 shows that, depending on the rainfall observational data set used, the results for water balance in the region can vary and the $P - E$ difference can get as high as 4.3 mm day^{-1} (GHCN) or as low as 0.9 mm day^{-1} (GPCP).

9.2.3 Maintenance of humidity and import/export of moisture in the Amazon Basin

Under the present climate the Amazon Basin behaves as a moisture sink ($P > E$), and therefore the basin receives moisture from sources such as the tropical rainforest by means of intense recycling and by transport from the tropical Atlantic by near-surface easterly flows or trade winds. The former has generated plenty of concern due to the possible impact of deforestation on the hydrological cycle of the basin (see Section 9.2.5). In this context, water that evaporates from the land surface is lost to the system if it is advected out of the prescribed region by atmospheric motion, but recycled in the system if it falls again as precipitation (Brubaker et al., 1993). Studies on recycling of water in the hydrological cycle of the Amazon Basin have been performed since the mid-1970s by Molion (1975), Lettau et al. (1979), Salati et al. (1979), Salati and Voce (1984), Salati (1987), and Eltahir and Bras (1993) among others. All indicate the active role of evapotranspiration from the tropical forest in the regional hydrological cycle. During highly active precipitation episodes, moisture convergence can account for 70 to 80% of precipitation. However, on the monthly or longer term, mean evapotranspiration is responsible for approximately 50% of the precipitation, especially in the southeastern parts of the basin.

Bosilovich *et al.* (2002) adapted the passive tracer methodology developed by Koster *et al.* (1986) to the NASA GEOS climate model in order to identify the sources of precipitation in various continental-scale basins, among them the Amazon Basin. They found that the largest contribution to rainfall in the Amazon Basin comes from the South American continent (45.5%) and the tropical Atlantic contributes 37%. The continental sources for Amazon precipitation are large throughout the year, but the oceanic sources vary with the seasonal change of the easterly flow over the tropical Atlantic. In the Amazon, significant amounts of water are not transported from very long distances, contrasting with the Mackenzie River in Canada that receives a significant contribution from Asian sources, or the North American sources for rainfall in the Baltic Sea region in Europe.

In the context of regional circulation in South America, the role of moisture transport from the tropical North Atlantic to the Amazon region has been documented in previous studies (see reviews in Hastenrath, 2001), and the interannual variability of rainfall anomalies in the region has been linked to variability in moisture transport and the intensity of the trade winds in the tropical Atlantic sector. Furthermore, the Amazon Basin is a region that provides moisture to regions in subtropical South America—such as southern Brazil and the La Plata River Basin—as has been shown in various studies.

A relevant feature of South American low-level circulation during the wet warm season is a poleward warm and moist air stream immediately to the east of the Andes often referred to as a low-level jet, because of its resemblance to the U.S. Great Plains Low-Level Jet east of the Rocky Mountains. This moist air current is referred as the "South American Low-Level Jet east of the Andes" or SALLJ—a component of the seasonal low-level circulation in the region that is detected all year long but mostly during the warm season (Berbery and Barros, 2002; Marengo *et al.*, 2004a). Figure 9.4 shows a conceptual model of the SALLJ. It illustrates moisture transport reaching the Amazon Basin by means of tropical North Atlantic easterly trade winds. The moisture transport typical of austral summer time is enriched by evapotranspiration from the Amazon Basin. Once the trade winds reach the Andes they are deflected by the mountains, changing the near-surface flow from northeast to southeast. During winter, subtropical Atlantic highs move towards southern Brazil and northern Argentina, and the winds from the northwest—at the western flank of this anticyclone—seem to replace the northwestern flow from Amazonia typical of summer. This winter flow carries less moisture than the summer flow even though it can sometimes be stronger. In both seasons, this northwest stream at the exit region of the jet converges with air masses from the south, which can favor the development of convective activity and rain at the exit region of the jet in southeastern South America over the La Plata River Basin. The conceptual model also shows the effects of topography in the SALLJ through dry and moist processes, the impact of the energy balance terms (sensible and latent heat) released from the Bolivian Plateau, while the near-surface heat low is important in terms of the impacts of transients (cold fronts, cyclogenesis, etc.) on the SALLJ (see reviews in Nogues-Paegle *et al.*, 2002; Marengo *et al.*, 2004a).

SALLJ variability in time and space is relatively poorly understood because of the limited upper-air observational network in South America east of the Andes, which

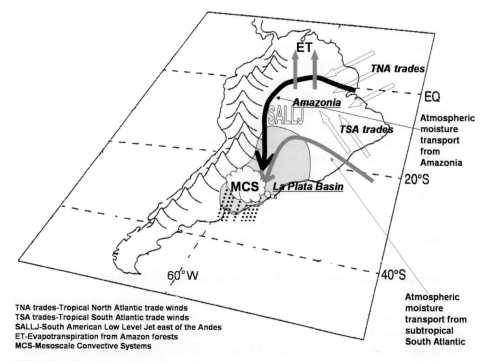

TNA trades-Tropical North Atlantic trade winds
TSA trades-Tropical South Atlantic trade winds
SALLJ-South American Low Level Jet east of the Andes
ET-Evapotranspiration from Amazon forests
MCS-Mesoscale Convective Systems

Figure 9.4. Conceptual model of the South American Low Level Jet (SALLJ) east of the Andes. *Source*: Marengo *et al.* (2004a).

seems to be unsuitable to capture the occurrence of the low-level jet, its horizontal extension and intensity, or temporal variability. Regarding time variability, SALLJ events seems to occur all year long, being more intense in terms of wind speed and moisture transport during the austral summer. As shown in Figure 9.4, more frequent wintertime SALLJs are related to the intensity and position of the subtropical South Atlantic anticyclone, and the source of moisture is the tropical–subtropical South Atlantic. During the summertime, the most important source of moisture is the tropical Atlantic–Amazon Basin system, when northeast winds coming from the tropical North Atlantic are deflected to the southeast by the Andes and in doing so get enriched by moisture from the Amazon Basin. This SALLJ, which brings tropical moisture from the Amazon to southern Brazil and northern Argentina, is more frequent in the warm season.

From the moisture budget calculations of Saulo *et al.* (2000), using regional models, a net convergence of moisture flux is found over an area that includes the La Plata Basin, with a maximum southward flux through the northern boundary at low levels that represents the moisture coming from Amazonia via the SALLJ. While there is evidence to suggest that this model provides a realistic description of the local circulation, it is emphasized that observational data are needed to gain further understanding of the behavior of the South American Low-Level Jet and its role in the regional climate.

Another feature of the low-level circulation in South America is the semi-permanent South Atlantic Convergence Zone (SACZ). The SACZ is influenced by SST anomalies over the southwestern tropical Atlantic, has a strong impact on the rainfall regime over southern northeast Brazil, southeast and southern Brazil, and contributes to modulate underlying SSTs over the southwest tropical Atlantic (Chaves and Nobre, 2004). There is evidence that the phases and location of the SACZ respond to Rossby wave activity (Liebmann *et al.*, 1999) and to Madden Julian Oscillation (MJO) (Carvalho *et al.*, 2004). On the other hand, its intensity depends on moisture coming from the Amazon region during summertime. Analyses performed by Nogues-Paegle (2002), Herdies *et al.* (2002), and Marengo *et al.* (2004a) suggest that there is an out-of-phase relationship between the SALLJ and SACZ. Moisture transport and possibly rainfall downstream of the jet or in the SACZ show a contrasting pattern, with enhanced convection and rainfall due to enhanced SALLJ consistent with periods of weak SACZ and *vice versa*. The SALLJ and SACZ are components of the South American Monsoon System (SAMS).

9.2.4   Interannual variability: El Niño and tropical Atlantic impacts

Rainfall variability in various timescales in Amazonia has been the subject of several studies regarding physical causes that could include remote and local forcings. At seasonal and interannual timescales, remotely forced seasonal variations are usually linked to SST anomalies in the tropical Pacific and Atlantic Oceans. The Southern Oscillation (SO) and its extremes—linked to anomalies in the tropical Pacific (El Niño or La Niña at interannual scales), and to sea surface temperature (SST) anomalies and meridional contrasts in the tropical Atlantic—have been associated with rainfall anomalies in the Amazon Basin. Various papers have been devoted to studies on the impact of regional and global SST anomalies in the tropical Pacific and Atlantic Oceans on rainfall anomalies in the region (Ropelewski and Halpert, 1987, 1989; Aceituno, 1988; Richey *et al.*, 1989; Rao and Hada, 1990; Marengo, 1992, 2004a; Meggers, 1994; Nobre and Shukla, 1996; Rao *et al.*, 1996; Guyot *et al.*, 1997; Marengo *et al.*, 1998a, b; Uvo *et al.*, 1998; Fu *et al.*, 1999, 2001; Botta *et al.*, 2002; Foley *et al.*, 2002; Ronchail *et al.*, 2002).

The low SO phase, which is associated with the El Niño phenomenon, is related to negative rainfall anomalies in northern and central Amazonia and anomalously low river levels in the Amazon River, while the high SO phase (related to the La Niña phenomenon) features anomalously wet seasons in northern and central Amazonia. Below-average southern summer rainfall throughout the Amazon Basin during seasons with weaker northeast trades, due to reduced moisture flux from Amazonia, has been identified during extreme El Niño years. In fact, a tendency towards drier rainy seasons and lower Rio Negro levels was detected during El Niño events in 1925–26, 1982–83, and more recently during 1997–98, while wetter conditions were observed during La Niña years in 1988–89 and 1995–96 (Figure 9.5).

The drought of 1998 in north and central Amazonia is generally considered as the most intense of the last 118 years. Kirchoff and Escada (1998) described the "wildfire of the century'" in 1998 as one of the most tragic that have ever occurred in Brazil.

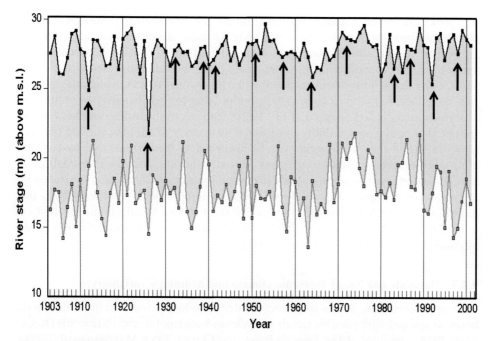

Figure 9.5. Time series of annual maximum (*top*) and annual minimum (*bottom*) river stage (in cm) of the Rio Negro at Manaus, Amazonas. *Source*: Williams *et al.* (2005). Arrows indicate occurrence of the El Niño phenomenon. The 1903–2004 long-term mean stage is 23.22 cm.

However, Williams *et al.* (2005) have suggested that the most severe drought in tropical South America during the 20th century occurred in 1926 during the El Niño of 1925–26. They established that dryness in the northern portion of the Rio Negro basin in 1925 also contributed to the major drought in 1926, through both depletion of soil moisture and possibly a negative feedback on rainfall from the abundant smoke aerosol (see below for elaboration). Annual rainfall deficits are broadly consistent with the reduction in annual discharge for 1926—estimated as 30–40%. The reduction in peak discharge during 1926 is closer to 50%. Sternberg (1987) describes an unparalleled drop in the high-water levels of the Rio Negro at Manaus during the El Niño event in 1926, during a severe dry season in which a great fire blazed for over a month, scorching the vegetation along the main channel. The drought also affected the Orinoco Basin with widespread and drought-related fires in the savannas. Evidence for a dry year in 1912 is also apparent in the Amazon discharge record in Figure 9.5, as is the minimum discharge prior to 1926, 1983, and 1998.

Regional forcing is associated with the systematic buildup of planetary boundary layer moisture and can affect the onset of the rainy season (Fu *et al.*, 1999; Marengo *et al.*, 2001). The atmosphere in southern Amazonia is quite stable; therefore, a great amount of surface heating is required to force the transition between dry and wet seasons, with soil moisture seeming to play a role in the predictability of the rainy season in southern Amazonia (Koster *et al.*, 2000; Goddard *et al.*, 2003; Marengo *et al.*, 2003). Near the equator, stability during the dry season is weaker and the transition from the dry to the wet season is more dependent on adjacent SST anomalies. Recent studies have demonstrated that the transition between the dry and wet seasons in southern Amazonia is also dependent on the presence of biogenic aerosol and aerosol produced by biomass burning in the region, which play a direct role in surface and tropospheric energy budget due to their capacity to scatter and absorb solar radiation (Artaxo *et al.*, 1990). This aerosol can also influence atmospheric thermodynamic stability in as much as it tends to cool the surface (by scattering radiation that would otherwise be absorbed at the surface) and by warming atmospheric layers above by absorption. Recent modeling and observational results have indicated that the aerosol plume produced by biomass burning at the end of the dry season is transported to the south and may interact with frontal systems, thus indicating a possible feedback to the precipitation regime by affecting the physics of rainfall formation (Freitas *et al.*, 2004) through the radiative forcing of cloud microphysical processes (Silva Dias *et al.*, 2002).

9.2.5   Decadal- and longer timescale climate and hydrology variability

On longer timescales, studies have documented long-term variations in rainfall in the basin, associated with possible trends or cycles in both rainfall and river levels (Rocha *et al.*, 1989; Chu *et al.*, 1994; Dias de Paiva and Clarke, 1995; Marengo *et al.*, 1998a; Chen *et al.*, 2001; Zhou and Lau, 2001), as well as climatic tendencies in water balance and moisture transport in the basin (Costa and Foley, 1998; Curtis and Hastenrath, 1999; Marengo, 2004b). Botta *et al.* (2002) and Foley *et al.*, (2002) found a 3–4-year peak in Amazonian rainfall which was related to the typical mode of variability of El Niño events (2–7 years). They also found a 24–28-year oscillation which has been discussed previously by Marengo (2004b) and Zhou and Lau (2001), and is indicative of decadal-mode variability. Coe *et al.* (2002) also identify an ~28-year mode variability in the Amazon Basin that drives spatial and temporal variability in river discharge and flooded areas throughout the Amazon/Tocantins River. Previous studies by Marengo (1995) and Callede *et al.* (2004) have also identified positive trends in water levels of the Rio Negro in Manaus and the reconstructed series of discharge for the Amazon River at Óbidos, respectively. The latter author measured increases of 9% in mean annual discharges and 10% in floods between 1903 and 1999.

Recently, Marengo (2004a) performed an analysis of decadal and long-term patterns of rainfall, and—using a combination of rain gauge and gridded rainfall data sets for 1929–98—identified slightly negative rainfall trends for Amazonia in its entirety. However, the most important findings were the presence of decadal timescale variations in rainfall in Amazonia, with periods of relatively drier and wetter conditions, contrasting between northern and southern Amazonia. Northern Amazonia exhibits a weak non-significant negative trend, while southern Amazonia exhibits a positive trend statistically significant at the 5% level (Figure 9.6).

Shifts in the rainfall regime in both sections of the Amazon Basin were identified in the mid-1940s and 1970s (Figure 9.6). After 1975–76, northern Amazonia exhibited less rainfall than before 1975. Changes in the circulation and oceanic fields after 1975

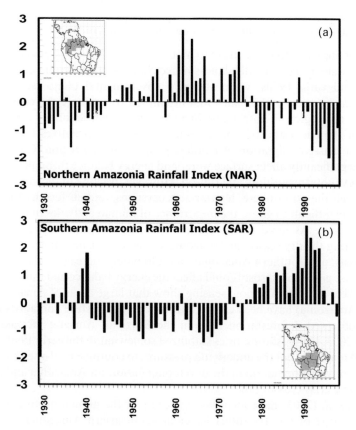

Figure 9.6. Time series of normalized departures of rainfall in northern Amazonia (NAR index) and southern Amazonia (SAR index). The domains of the regions appear in the map inside each panel. *Source*: Marengo (2004a).

suggest the important role played by warming of the tropical central and eastern Pacific on decreasing rainfall in northern Amazonia, due to more frequent/intense strong El Niño events during the relatively dry period 1975–98. This decadal-scale variability has also been detected in other regions in South America, and it seems that other regions—such as the La Plata Basin—also experience these shifts. They have been linked to phases of the Pacific Decadal Oscillation (PDO) (Zhang *et al.*, 1997). The positive PDO phase started in the mid-1970s and apparently ended in the early 2000s. A negative PDO phase started in the mid-1940s and extended until 1975–76, a period with more frequent and intense El Niño events.

Callede *et al.* (2004) suggested a reduction in the interannual variability of mean discharge during 1927–70 that is consistent with the positive phase of the PDO, and positive/negative rainfall departures in northern/southern Amazonia as shown by Marengo (2004a). Their analysis of mean annual discharge and average annual rainfall during 1945–98 demonstrates an increase in flow in relation to rainfall, which could be, according to the authors, the consequence of Amazonian deforestation.

9.2.6   Changes in land use and impacts on Amazon climate

A variety of human activities can act to modify various aspects of climate and surface hydrologic systems. Historically, land-surface changes in Amazonia got intensified in the mid- and early 1970s, when strategic governmental plans—such as Brazil's *Programa de Integração Nacional*—first attempted to promote the economic development of the region. Those plans included the construction of extensive roads throughout the basin and the implementation of fiscal incentives for new settlers, triggering a massive migration of landless people into the region. Changes in land cover can significantly affect surface water and energy balance through changes in net radiation, evapotranspiration, and run-off. However, because of the intricate relationships between the atmosphere, terrestrial ecosystems, and surface hydrological systems, it is still difficult to gauge the importance of human activities in the Amazonian hydrologic cycle. As indicated in Section 9.2.2, aerosol and smoke from biomass burning during the dry season in Amazonia seems to have an impact on the onset of the rainy season in southern Amazonia, and ultimately increase in the concentration of greenhouse gases and aerosol could affect the energy balance and thus climate of the region. Recent data from remote sensing show that large areas of Amazonia (mostly Brazilian Amazonia) have been changed from forest to pasture and agricultural land and that observed deforestation rates in the Brazilian Amazon increased in 2004 relative to 2003. Deforestation rates stabilized somewhat in the early 1990s, mainly in Brazilian Amazonia, but the underlying pressures to continue land-use change are still present: a growing population in the developing nations of Amazonia and plans for a road network criss-crossing the region.

Costa *et al.* (2003) have identified increases in the annual mean and high-flow season discharge of the Tocantins River in southeastern Amazonia since the late 1970s, even though rainfall has not increased. They suggest that changes in the land cover in the basin for agricultural purposes and urban development have altered the hydrological cycle of the basin. Callede *et al.* (2004) suggest that increases in the mean annual discharge of the reconstructed series of the Amazon River at Óbidos during 1945–98 could be the consequence of Amazon deforestation.

The construction of reservoirs for hydroelectric generation in Amazonia has an impact on the hydrological regime as well as on biodiversity and water quality (Tundisi *et al.*, 2002), depending on the size of the inundated area in the tropical rainforest. Brazil has five reservoirs dedicated to hydroelectricity generation (Coaracy Nunes, Curua-Una, Tucurui, Balbina, and Samuel) and a further six are planned to be built (Manso, Cachoeira, Ji-Parana, Karanaô, Barra do Peixe, and Couto Magalhães). Studies on Samuel, Balbina, and Tucurui show that there are different degrees of biomass degradation, and that it is more advanced in Tucurui because it is older. Land-use changes have also been reported near the site of the reservoir due to human settlements in the region.

In an attempt to investigate the possible impact of Amazon deforestation on regional climate and hydrology, global climate model simulations of land-use changes—where forest is replaced by grassland throughout the whole basin—have suggested a possible change in regional and global climate as a result of tropical

Table 9.5. Comparison of climate simulation experiments of Amazon deforestation from global climate models. Results show the differences between deforested and control runs. E = Change in evapotranspiration (mm day^{-1}); T = Change in surface air temperature ($^{\circ}$K); P = Change in precipitation (mm day^{-1}); R = Run-off, calculated as the difference between P and E ($R = P - E$). Modified from Marengo and Nobre (2001).

Experiment	E	T	P	R
Dickinson and Henderson-Sellers (1988)	−0.5	+3.0	0.0	+0.5
Dickinson and Kennedy (1992)	−0.7	+0.6	−1.4	−0.7
Henderson-Sellers et al. (1993)	−0.6	+0.5	−1.6	−1.0
Hahmann and Dickinson (1995)	−0.4	+0.8	−0.8	−0.4
Zeng et al. (1996)	−2.0	—	−3.1	−1.1
Hahmann and Dickinson (1997)	−0.4	+1.0	−1.0	−0.6
Costa and Foley (2000)	−0.6	+1.4	−0.7	−0.1
Lean and Warrilow (1989)	−0.9	+2.4	−1.4	−0.5
Lean and Warrilow (1991)	−0.6	+2.0	−1.3	−0.7
Lean and Rowntree (1993)	−0.6	+1.9	−0.8	−0.3
Lean and Rowntree (1997)	−0.8	+2.3	−0.3	+0.5
Lean et al. (1996)	−0.8	+2.3	−0.4	+0.4
Manzi and Planton (1996)	−0.3	−0.5	−0.4	−0.1
Nobre et al. (1991)	−1.4	+2.5	−1.8	−0.4
Shukla et al. (1990), Nobre et al. (1991)	−1.4	+2.5	−1.8	−0.4
Dirmeyer and Shukla (1994)	−0.4	—	−0.7	−0.3
Sud et al. (1990)	−1.2	+2.0	−1.5	−0.3
Sud et al. (1996b)	−1.0	+3.0	−0.7	+0.3
Walker et al. (1995)	−1.2	—	−1.5	−0.3
Polcher and Laval (1994a)	−2.7	+3.8	+1.0	+3.7
Folcher and Laval (1994b)	−0.4	+0.1	−0.5	−0.1
Zhang et al. (2001)	−0.4	+0.3	−1.1	−0.01
Voldoire and Royer (2004)	−0.6	−0.1	−0.4	—

deforestation (see reviews in Salati and Nobre, 1991; Marengo and Nobre, 2001). Under a hypothesized Amazon Basin deforestation scenario, almost all models show a significant reduction in precipitation and evapotranspiration (Table 9.5), and most found a decrease in streamflow and precipitation and increases in air temperature. Deforestation results in increased surface temperature, largely because of decreases in evapotranspiration. The combined effect of deforestation and a doubling of CO_2, including interactions between processes, is projected to increase temperature in the order of +1.4°C.

However, such predictions disagree with results found by mesoscale models, which have been consistently predicting the establishment of enhanced convection—and potential rainfall—above sites of fragmented deforestation. Modeling studies have tried to reproduce that effect, and it was noted that mesoscale circulations between forested and deforested patches may significantly affect the timing and formation of clouds, potentially altering both intensity and distribution of precipitation (Chen and Avissar, 1994). It was estimated that, at the mesoscale, a landscape with a relatively large discontinuity tends to produce more precipitation than a homogenous domain, inducing a negative feedback that ultimately tends to eliminate the discontinuity (Avissar and Liu, 1996). In some cases, the thermal circulation induced may get as intense as that of a sea breeze—such as that over domains with extended areas of unstressed, dense vegetation bordering areas of bare soil. The horizontal scale of such landscape heterogeneities is another factor that may affect the establishment of precipitation (Pielke, 2001), while the optimum scale for triggering convection seems to depend on the air humidity level (Avissar and Schmidt, 1998). A strong enough synoptic (or background) wind-field may also interact with the induced circulation, possibly masking its existence at times (Segal *et al.*, 1988). It was noted that a mild background wind of $5\,\mathrm{m\,s^{-1}}$ may be sufficient to virtually remove all thermal impacts generated by land-surface discontinuities (Avissar and Schmidt, 1998), although more recent studies have revealed that a strong background wind may only advect instabilities elsewhere rather than disperse them (Baidya Roy and Avissar, 2003). The results by Baidya Roy and Avissar (2003) and Weaver and Baidya Roy (2002) were derived from high-resolution mesoscale models simulating the effect of land-surface and land-use changes in Rondonia, southern Amazonia, during the LBA-WET AMC field campaign in 1999 (Silva Dias *et al.*, 2002). They found that coherent mesoscale circulations were triggered by surface heterogeneity and that synoptic flow did not eliminate these circulations but advected them away. These circulations affect the transport of moisture, heat the synoptic scale, and can affect climate.

All these projected changes in Amazonia (Table 9.5) may have climatic, ecological, and environmental implications for the region, the continent, and the globe. A sound knowledge of how the natural system functions is thus a prerequisite to defining optimal development strategies. The complex interactions between the soil, vegetation, and climate must be measured and analyzed so that the limiting factors to vegetation growth and soil conservation can be established. New knowledge and improved understanding of the functioning of the Amazonian system as an integrated entity and of its interaction with the Earth system will support development of national and regional policies to prevent exploitation trends from bringing about irreversible changes in the Amazonian ecosystem. Such knowledge, in combination with enhancement of the research capacities and networks between Amazonian countries, will stimulate land managers and decision-makers to devise sustainable, alternative land-use strategies along with forest preservation strategies.

9.3   PROJECTIONS OF FUTURE CHANGES IN CLIMATE AND HYDROLOGY OF THE AMAZON BASIN

9.3.1   Detected changes in air temperature

According to the IPCC *Third Assessment Report* (IPCC, 2001), observed air temperature trends have shown a warming of $0.6 \pm 0.2°C$ during the 20th century. Climate reconstructions have shown the 20th century as being the warmest of the last 1,000 years. Specifically, the 1990s was the warmest decade of the millennium, with 1998 as the warmest (an El Niño year), followed by 2002, 2003, and 2004. Warming since 1976 was of the order of 0.19°C/decade—higher than the warming of the period 1910–1945 (0.14°C/decade). The warmest years since 1976 (in decreasing order) were: 1998, 2002, 2003, 2004, 2001, 1995, 1997, 1990, 1999, 1991, and 2000.

For the Amazon region, few studies have identified long-term trends in air temperature. Sansigolo *et al.* (1992) and Marengo (2004b) detected warming in many of the major cities in Brazil, including Manaus and Belém in the Amazon Basin, but this warming was linked to the urbanization and heat island effects in big cities. At a regional level, Victória *et al.* (1998) detected an observed warming of +0.56°C/100 years until 1997, while Marengo (2003) updated this warming to +0.85°C/100 years until 2002. Recently, Vincent *et al.* (2005) have identified some positive trends in extreme air temperatures and negative trends in the diurnal temperature range at quite a few stations in South America. Similarly, positive trends were identified in nighttime and daytime air temperatures at some stations in Amazonia.

9.3.2   Future projections of changes in air temperature

Regarding future climate change projections in the region, different IPCC models for A2 and B2 scenarios were assessed by Marengo and Soares (2003), who detected a warming of different magnitudes as a result of their use. The largest warming was approximately 4–6°C in central Amazonia during austral winter in the A2 scenario and a 2–3°C increase in the B2 scenario of the Hadley Centre model for the year 2100. At the level of time slices, Marengo and Soares (2003) show that the HadCM3 model shows a larger warming in Amazonia after 2080 than those expected in 2020 and 2050—especially during austral spring (September–November)—reaching up to 11°C in the A2 scenario and 8°C for the B2. The other models show similar tendencies with warming a bit lower, with the exception of the CSIRO model that actually shows cooling in Amazonia.

Figure 9.7 shows a time series of mean annual air temperature up to 2100 from six IPCC models for the A2 scenario: HadCM3 from the U.K., the CCCma from Canada, the CSIRO from Australia, R30 from the U.S., and the CCSR/NIES from Japan. All IPCC models exhibit warming up to 2100, with the warmest projections being from the HadCM3 (almost 10°C warmer than the 1961–90 long-term mean) and the CCSR/NIES (7°C warmer than the 1961–90 long-term mean). The other models exhibit warming of the order of 3°C or less. The figure shows consistent warming in all models but with a large spread among models after 2070.

9.3.3   Detected and projected changes in precipitation

The IPCC *Third Assessment Report* (IPPC, 2001) has shown no clear signs of negative trends in Amazonia due to increased deforestation as one would expect (Section 9.2.3). Section 9.2.5 describes trends and long-term variability in observed mean seasonal and annual rainfall and discharges. The magnitude and size of the trends depend on rainfall data sets, length of records, etc. and the uncertainty is high since

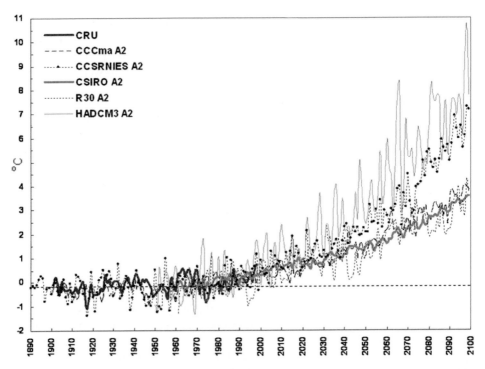

Figure 9.7. Air temperatures in the Amazon region as simulated by five IPCC models: R30 (GFDL, U.S.A.), CCCMa (Canada), HadCM3 (U.K.), CSIRO (Australia), CCSR/NIES (Japan). Anomalies are in °C from the 1961–1990 long-term mean. Observations are from the CRU rainfall data set for the 1903–1998 period. Air temperatures are for the A2 SRES.

various studies (Section 9.2.4) found trends that vary in direction when different length periods are used. What is clear is the presence of decadal-scale rainfall variability, with wetter periods during 1945–76 and relatively drier periods during 1977–2000 in northern Amazonia. Analysis of extreme rainfall events in Amazonia (Haylock *et al.*, 2006) has also shown this, with some degree of spatial coherence and uncertainty due to the reduced number of stations used for the region. Using the records at Manaus and Belém, since 1961 an increase in rainfall exceeding the 99th and 95th percentiles has been detected, suggesting a tendency for more intense and extreme events during the last 40 years, even though the annual totals may have not exhibited a significant positive trend.

9.3.4 Future projections of changes in precipitation

For the future, the different IPCC models for the A2 scenario (Hulme and Sheard, 1999; Marengo and Soares, 2003) show negative rainfall anomalies in most of the Amazon Basin (up to 20% reduction), while for the lower emission scenario (B2) this reduction was 5%. In addition to these emission scenarios, there are climate change scenarios for the 21st century: new developments in dynamic vegetation schemes and coupled climate–carbon models (Cox *et al.*, 2000, 2004; Betts *et al.*, 2004) have shown an effect—named "die-back of the Amazon forest"—by means of which rising atmospheric CO_2 is found to contribute to a 20% rainfall reduction and to more than a 30% increase in surface temperatures in the Amazon Basin, through physiological forcing of stomatal closure. They also show an increase in rainfall in southern Brazil and northern Argentina. These projections for drought in Amazonia after 2040 also show systematic warming in the tropical Pacific indicative of an El Niño-like mode of variability becoming persistent after 2040. However, the likelihood of this extended El Niño or more frequent/intense El Niño mode scenarios in a global warming world is still an open issue. Furthermore, these studies do not give any information on the likelihood of change in extreme rainfall events.

If we consider the climate change scenarios discussed by Marengo and Soares (2003) and the model results from Cox *et al.* (2004) and Betts *et al.* (2004), increase in the concentration of greenhouse gases in the atmosphere may result in increased warming and rainfall reduction rates in Amazonia, implying more intense respiration and the closing of stomata, leading to the decline of Amazon tropical forests. These projections also suggest that vegetation will be more susceptible to fires due to the drying of Amazonia. Figure 9.6 shows the observed increase in rainfall in southern Amazonia contrasting with the slight decrease of rainfall in northern Amazonia. Systematic increases in rainfall in southern Amazonia and southeastern South America, as detected by Obregon and Nobre (2003), are consistent with a tendency for more intense/frequent transport of moisture from Amazonia to the La Plata Basin by the South American Low-Level Jet (SALLJ) east of the Andes (Marengo *et al.*, 2004c). Thus, it could be hypothesized that—following Betts' results from the HadCM3 model after the mid-2050s—the drying of the Amazon Basin and humidification of the southern Brazil and northern Argentina region in an extended El Niño mode predicted by this model could be explained by changes in regional circulation, with an increase in SALLJ frequency and/or intensity in a global warming world. However, not much can be said regarding changes in extreme rainfall events.

A reduction in rainfall will also have impacts on river and water levels, as well as on rainfall distribution in them, with longer dry periods and intense rainfall events concentrated on a few days. However, the uncertainty level is still high since these experiences were obtained from a model—in this case, the HadCM3 model—and other models are being tested now to see if they also simulate Amazon die-back. It is clear that models must improve or include the representation of natural processes, such as clouds and the impacts of aerosol.

9.3.5 Changes in the hydrology of the Amazon River Basin

Macroscale hydrological models, which model the land-surface hydrological dynamics of continental-scale river basins, have rapidly developed during the last decade (Russell and Miller, 1990; Marengo *et al.*, 1994; Miller *et al.*, 1994; Nijssen *et al.*, 1997, 2001). These models can act as links between global climate models and water resource systems on large spatial scales and long-term timescales. Predictions of river discharge in the Amazon Basin for present climates and $2 \times CO_2$ future scenarios have been calculated by Russell and Miller (1990) and Nijssen et al. (2001) using global models. Some problems in the parameters of the model—or perhaps the availability of suitable run-off data for validation—indicate that in most models rainfall and run-off in Amazonia are underestimated (Marengo et al., 2003). This also generates uncertainty in the projected values of run-off in the future, forced either by increase in GHGs or in changes in land use and land cover. More recent simulations by Coe et al. (2002), using a terrestrial ecosystem model, have been successful in simulating inter-annual and seasonal run-off variability in Amazonia, and—even though the discharge is consistently underestimated—the model captures climate variability and the impacts of El Niño events since the early 1950s.

Land-surface changes are accompanied by alterations in climate, and consequently in the hydrological cycle. Water flux anomalies associated with these changes have already occurred in many parts of the globe and have been detected, for example, over tropical and subtropical basins—such as the Yangtze (Yin and Li, 2001; Yang et al., 2002), Mekong (Goteti and Lettenmaier, 2001), Amazon River Basin (see reviews in Marengo and Nobre, 2001)—as well as on several catchment areas within the African continent (Calder et al., 1995; Hetzel and Gerold, 1998). Recently, major land-surface changes have been observed in various parts of the tropics (Aldhous, 1993), and Amazonia—which holds more than 40% of all remaining tropical rainforests in the world—has been the focus of many studies about the impact of these changes on hydrological dynamics.

Increasing trends in discharge and precipitation were observed at all but the eastern parts of the Amazon Basin between the late 1950s and the early 1980s, and—despite contentions that these trends were associated with upstream deforestation (Gentry and Lopez-Parodi, 1980)—most time series retreated to their long-time means by the end of the 2000 period. These are more indicative of decadal tendencies—rather than of any unidirectional trends—as a response to fluctuations over the Tropical Pacific associated with ENSO events, and not deforestation. These findings thus support the idea that the atmospheric fluctuations induced by remote forcings (Richey et al., 1989; Fu et al., 2001) can potentially offset or overshadow the effects of deforestation (Chen et al., 2001). The existence of trends in additional terms of the hydrological cycle in Amazonia has also been tested, and the swings and tendencies of significant changes on spatial averages for the input and output fluxes of water vapor (decreasing) vary according to the type and length of time series used. The use of spatially aggregated point data may not be appropriate for the detection of trends, due to the inevitable "dilution" of the signal during the upscaling process, while the use of gridded data sets may also create artificial trends. What has also been observed is decadal timescale variability, more so than any unidirectional trend towards systematic drying or moistening of the Amazon region in the long term. However, these tendencies do not provide information on whether significant changes in precipitation extremes would occur in the future.

The deforestation experiments listed in Table 9.6 show that most models simulate a decrease in rainfall and run-off due to the large-scale removal of forests in the Amazon Basin. However, these experiments did not alter at all the concentration of

Table 9.6. Natural and anthropogenic forcing, climatic tendencies, and human and ecosystem dynamics in Amazon countries.

Forcing (natural or anthropogenic)	Impacts on water resources	Consequences
El Niño and tropical Atlantic sea surface temperature anomalies	Changes in rainfall distribution in the Amazon region	Drought in northern Amazon region; problems in transportation due to low river water levels; high risk of forest fires at seasonal level; impacts on natural river ecosystems; impacts on agriculture; impacts on water storage for hydroelectric generation
Climate change due to increase in concentration of GHG	Possible changes in the hydrological cycle; changes in the energy balance and warming; changes in biodiversity and natural ecosystems	Dynamics of vegetation affected; Amazon forests die and become savanna; drying of the Amazon region; floods or extremely low water levels likely to occur; more frequent forest fires; impacts on water storage for hydroelectric generation
Deforestation and land-use change	Possible changes in hydrological and energy cycles; changes in water quality and chemistry due to deforestation in the east flank of the Andes (upper-Amazon countries)	Regional rainfall reduction; regional warming; erosion; sedimentation along the main channel and accumulation of sediments in reservoirs; water quality and biodiversity may be affected
Biomass burning (natural and man-made)	Changes in water and energy cycles; changes in air quality	Impacts on the onset of the rainy season and physics of rainfall; impacts on air quality and sensitivity to warming due to release of large amounts of GHG and aerosol

GHGs or aerosol in the atmosphere. Later experiments on $2 \times CO_2$ scenarios have produced reductions in streamflows of the Amazon River and major rivers in the tropics as compared to present day streamflows, even though present day streamflows were somewhat over- or under-estimated in the past.

Table 9.6 lists natural and anthropogenic forcing that affects climate and hydrology in the Amazon region, the expected impacts on water resources of the region, and the possible consequences on the components of the climate system. The major forcings can be natural or human-induced and their impact could be on various timescales: intraseasonal (onset of the rainy season in the region), interannual (drought or flood associated with El Niño), and long-term (changes in vegetation, soil hydrology, etc.). The table can serve as a starting point in the assessment of possible impacts of future climate change in Amazonia.

9.4   UNCERTAINTIES IN CLIMATE AND HYDROLOGY VARIABILITY, LONG-TERM TRENDS, AND CLIMATE CHANGE IN THE AMAZON REGION

As can be seen from the analysis of observed hydrometeorological fields in previous sections, no significant trends have been detected in the rainfall regime of the Amazon region as a whole, or on the discharge of the Amazon River and its tributaries, that could be attributed to deforestation and human-induced land-use changes, or to increase in the concentration of greenhouse gases. The detected contrasting rainfall trends in northern and southern Amazonia on decadal timescales are somewhat surprising, since it would suggest the dominance of natural climate variability on rainfall variability on both sides of the basin, linked to El Niño or El Niño like modes. Some previous studies have detected upward discharge trends in Peruvian Amazon rivers in the mid-1980s, but these trends were more a consequence of changes in the main channel or in the measurement techniques, and not of rainfall changes. However, rainfall records do not go back more than 50 years, and there are large sections of the basin with no data, so there is considerable uncertainty about climate trends in Amazonia. River data exhibit longer time records, but there is uncertainty in climate trends detected using river series or on short-time rainfall series, since some of the river data could be affected by non-climatic signals, such as the back-water effect in the Rio Negro series in Manaus, Brazil. Gridded rainfall data sets have helped in solving problems on regional coverage, but they may have added even more uncertainties since there are differences among data sets. National hydrometeorological networks are deteriorating, and there are fewer stations in 2000 than there were in the 1970s. Automatic hydrological stations are being installed in some countries, but the operational costs are high and not all Amazon countries have adopted the same operational system, and problems with calibration and data intercomparison have been reported.

The choice of rainfall data sets has an impact on results regarding the water budget. Uncertainties exist in these results and they are sensitive to rainfall network density—when using station data—and to numerical interpolation techniques and use of satellite data to fills gaps in time and space across the basin—when using gridded data. The lack of continuous upper-air observations in the basin makes it difficult to estimate moisture transport into and out of the basin, and we have to rely on model data for this—such as reanalyses. Furthermore, the use of uncorrected discharge under-estimates freshwater discharge, and thus adds to the uncertainty in closing the water budget. Therefore, it can be stated that estimation of water vapor fluxes in the Amazon Basin is another source of uncertainty in water balance calculations. In addition, there are very few observations of evaporation in some isolated parts of the basin. Perhaps there is need for a unique rainfall data set for the entire region, with rainfall information coming from rainfall stations in the Amazonian areas of each country on a continuous basis. Some Amazon countries are implementing networks of automatic weather and hydrological stations in remote areas of the basin to solve the problem of areas devoid of data, but they need calibration and are expensive to implement and maintain. In addition, data have only recently started to be collected.

Finally, some of the studies reviewed for the writing of this chapter suggest that increases in the concentrations of GHG and aerosol in the atmosphere, as well as changes in land cover for agriculture, have already affected the hydrology of the Amazon Basin. Some uncertainties can be attached to these results due to model limitations, and to the lack of continuous and long-term observational series of climate and hydrological variables. Large uncertainties were also identified in anthropogenic aerosol forcing and response, and changes in land use leading to biomass burning and their impacts on rainfall in the basin. Analyses of the few available long-term rainfall and river series suggest an absence of significant, unidirectional trends towards drier or wetter conditions in the basin. However, more evidence of variability at interannual and decadal timescales is apparent, and these observed trends may be due to natural climate variability and observed climate shifts, while no signs of land-use changes have yet appeared in the long-term climate and hydrologic variability of the region.

Acknowledgements

The contents of this chapter were derived from various LBA (Large Scale Atmosphere Biosphere Experiment in Amazonia) studies and projects funded by the National Aeronautics and Space Administration (NASA), Inter American Institute for Global Change (IAI), and European and Brazilian agencies since the late 1990s. In particular, I am grateful for funding from the Research Foundation of the State of Sao Paulo (FAPESP grant 01/13816-1) and from the Brazilian Research Council (CNPq).

9.5 REFERENCES

Aceituno, P. (1988) On the functioning of the Southern Oscillation in the South American sector, Part I: Surface climate. *Mon. Wea. Rev.* **116**, 505–524.

Aldous, P. (1993) Tropical deforestation: Not just a problem in Amazonia. *Science* **259**, 1390–1390.

Artaxo, P., Maenhaut, W., Storms. H., and Van Grieken, R. (1990) Aerosol characteristics and sources for the Amazon Basin during the wet season. *J. Geophys. Res.* **95**, 16971–16985.

Avissar, R. and Liu, Y. (1996) Three-dimensional numerical study of shallow convective clouds and precipitation induced by land surface forcings. *J. Geophys. Res.* **101**, 7499–7518.

Avissar, R. and Schmidt, T. (1998) An evaluation of the scale at which ground-surface heat flux patchiness affects the convective boundary layer using large-eddy simulations. *J. Atmos. Sci.* **55**, 2666–2689.

Baumgartner, A. and Riechel, R. (1975) *The World Water Balance* (464 pp.). Elsevier, New York.

Baidya Roy, S. and Avissar, R. (2002) Impact of land use/land cover change on regional hydrometeorology in Amazonia. *J. Geophys. Res.* **107**, doi:10.1029/2000JD000266.

Baidya Roy, S., Weaver, C. P., Nolan, D., and Avissar, R. (2003) A preferred scale for landscape forced mesoscale circulations? *J. Geophys. Res.* **108**, doi:10.1029/2002JD003097.

Berbery, E. H. and Barros, V. (2002) The hydrological cycle of the La Plata Basin in South America. *J. Hydromet.* **3**, 630–645.

Betts, R., Cox, P., Collins, M., Harris, P., Huntingford, C., and Jones, P. (2004) The role of ecosystem–atmosphere interactions in simulated Amazonian precipitation decrease and forest dieback under global change warming. *Theoretical and Applied Climatology* **78**, 157–175.

Bosilovich, M., Sud, Y., Schubert, S., and Walker, G. (2002) GEWEX CSE sources of precipitation using GCM water vapor tracers. *GEWEX News* **12**, 1–7.

Botta, A., Ramankuttym, N., and Foley, J. A. (2002) Long-term variations of climate and carbon fluxes over the Amazon Basin. *Geophysical Research Letters* **29**, doi:10.1029/2001GL013607.

Brubacker, K. L., Entekhabi, D., an dEagleson, P. S. (1993) Estimation of continental precipitation recycling. *J. Climate* **6**, 1077–1089.

Callede, J., Guyot, J. L., Rocnchail, J., L'Hote, Y., Niel, H., and De Oliveira E. (2004) Evolution du debit de l'Amazone à Obidos de 1903 à 1999. *Hydrological Processes* **49**, 85–97 [in French].

Carvalho, L. M. V., Jones, C., and Liebmann, B. (2004) The South Atlantic convergence zone: Intensity, form, persistence and relationships with intraseasonal to interannual activity and extreme rainfall. *J. Climate* **17**, 88–118.

Chaves, R. R. and P. Nobre (2004) Interactions between the South Atlantic Ocean and the atmospheric circulation over South America. *Geophys. Res. Lett.* **31**, L03204, doi:10.1029/2003GLO18647.2004.

Chen, F. and Avissar, R. (1994) Impact of land-surface moisture variability on local shallow convective cumulus and precipitation in large-scale models. *J. Appl. Met.* **33**, 1382–1401.

Chen, T.-C., Yoon, J., St. Croix, K. J., and Takle, E. S. (2001) Suppressing impacts of Amazonian deforestation by global circulation change. *Bull. Amer. Met. Soc.* **82**, 2209–2215.

Chu, P. S. (1982) Diagnostics of climate anomalies in tropical Brazil. Ph.D. dissertation, Department of Meteorology, University of Wisconsin, Madison, WI.

Chu, P. S., Yu, P., and Hastenrath, S. (1994) Detecting climate change concurrent with deforestation in the Amazon basin: Which way has it gone? *Bull. Amer. Met. Soc.* **75**, 579–583.

Coe, M., Costa, M. H., Botta, A., and Birkett, C. (2002) Long term simulations of discharge and floods in the Amazon river. *J. Geophys Res.* **107**, 11-1/11-17.

Costa, M. H. and Foley, J. A. (1998) A comparison of precipitation datasets for the Amazon Basin. *Geophys. Res. Lett.* **25**, 155–158.

Costa, M. H. and Foley, J. A. (1999) Trends in the hydrologic cycle of the Amazon Basin. *J. Geophys. Res.* **104**, 14189–14198.

Costa, M. H. and Foley, J. A. (2000) Combined effects of deforestation and doubled atmospheric CO_2 concentrations on the climate of Amazonia. *J. Climate* **13**, 18–34.

Costa, M. H., Botta, A., and Cardille, J. A. (2003) Effects of large-scale changes in land cover on the discharge of the Tocantins River, Southeastern Amazonia. *J. Hydrol.* **283**, 206–217.

Cox, P., Betts, R., Jones, C., Spall, S., and Totterdell, T. (2000) Acceleration of global warming due to carbon-cycle feedbacks in a coupled climate model. *Nature* **408**, 184–187.

Cox, P., Betts, R., Collins, M., Harris, P., Huntingford, C., and Jones, C. (2004) Amazonian forest dieback under climate-carbon cycle projections for the 21th century. *Theoretical and Applied Climatology* **78**, 137–156.

Curtis, S. and Hastenrath, S (1999) Trends of upper-air circulation and water vapor over equatorial South America and adjacent oceans. *Int. J. Climatol.* **19**, 863–876.

Dias de Paiva, E. M. C. and Clarke, R. (1995) Time trends in rainfall records in Amazonia. *Bull. Amer. Met. Soc.* **75**, 579–583.

Dickinson, R. and Henderson-Sellers, A. (1988) Modeling tropical deforestation: A study of GCM land-surface parameterization. *Quart. J. Roy. Meteor. Soc.* **114**, 439–462.

Dickinson, R. and Kennedy, P. (1992) Impacts on regional climate of Amazon deforestation. *Geophys. Res. Letters* **19**, 1947–1950.

Dirmeyer, P. and Shukla, J. (1994) Albedo as a modulator of climate response to tropical deforestation. *J. Geophys. Res.* **99**, 20863–20877.

Eltahir, E. A. B. and Bras, R. L. (1993) On the response of the tropical atmosphere to large-scale deforestation. *Quart. J. Roy. Meteor. Soc.* **119**, 779–783.

Eltahir, E. A. B. and Bras, R. L. (1994) Precipitation recycling in the Amazon Basin. *Quart. J. Roy. Met. Soc.* **120**, 861–880.

Figueroa, N. and Nobre, C. (1990) Precipitation distribution over central and western tropical South America. *Climanálise* **5**, 36–40.

Foley, J. A., Botta, A., Coe, M. T., and Costa, M. H. (2002) The El Niño-Southern Oscillation and the climate, ecosystem and rivers of Amazonia. *Global Biogeochem. Cycles*, doi:10.1029/2002GB001872.

Franken, W. and Leopoldo, P. (1984) Hydrology of catchment areas in Central-Amazonian forest streams. In: H. Soili (ed.), *The Amazon: Limnology and Landscape Ecology of a Mighty Tropical River and its basin* (pp. 501–519).

Freitas, S. R., Longo, K. M., Silva Dias, M. A. F., Silva Dias, P. L., Recuero, F. S., Chatfield, R., Prins, E., and Artaxo, P. (2004) Monitoring the transport of biomass burning emissions in South America. *Environmental Fluid Mechanics* in press.

Fu, R., Zhu, B, and Dickinson, R. E. (1999) How do atmosphere and land surface influence seasonal changes of convection in the tropical Amazon? *J. Climate* **12**, 1306–1321.

Fu, R, Dickinson, R. E., Chen, M., and Wang, H. (2001) How do tropical sea surface temperatures influence the seasonal distribution of precipitation in the equatorial Amazonia? *J. Climate* **14**, 4003–4026.

Gash, J. and Nobre, C. (1997) Climatic effects of Amazonian deforestation: Some results from ABRACOS. *Bull. Amer. Met. Soc.* **78**, 823–830.

Gentry, A. H. and Lopez-Parodi, J. (1980) Deforestation and increased flooding of the upper Amazon. *Science* **210**, 1354–1356.

Goddard, L., Barnston, A., and Mason, S. (2003) Evaluation of the IRI's "net assessment" seasonal climate forecasts: 1997–2001. *Bull. Amer. Met. Soc.* **84**, 1761–1781.

Goteti, G. and Lettenmaier, D. P. (2001) Effects of streamflow regulation and land cover change on the hydrology of the Mekong river basin. Master thesis, Department of Civil and Environmental Engineering. University of Washington, WA (98 pp.).

Guyot, J-L., Callede, J., Molinier, M., Guimaraes, W., and de Oliveira, E. (1997), La variabilité hydrologique actuelle dans le bassin de l'Amazone. *Seminario Internacional Consecuencias Climáticas e Hidrológicas del Evento El Niño a Escala Regional y Local, Memorias Técnicas, 26–29 Noviembre, Quito, Ecuador* (pp. 285–293) [in French].

Hahmann, A. and Dickinson, R. (1995) Performance and sensitivity of the RCCM2/BATS model to tropical deforestation over the Amazon Basin. *Conference Proceedings of the XXI IUGG General Assembly, Boulder, CO.*

Hahmann, A. and Dickinson, R. (1997) RCCM2 BATS model over tropical South America: Application to tropical deforestation. *J. Climate* **10**, 1944–1964.

Hastenrath, S. (2001) Interannual and longer-term variability of upper air circulation in the Northeast Brazil–Tropical Atlantic sector. *J. Geophys. Res.* **105**, 7327–7335.

Haylock, M. R., Peterson, T., Abreu de Sousa, J. R., Alves, L. M., Ambrizzi, T., Baez, J., Barbosa de Brito, J. I., Barros, V. R., Berlato, M. A., Bidegain, M. *et al.* (2006) Trends in

total and extreme South American rainfall 1960–2000 and links with sea surface temperature. *Journal of Climate* **19**, 1490–1512.

Henderson-Sellers, A., Dickinson, R., Durbidge, T., Kennedy, P., McGuffie, K., and Pitman,
A. (1993) Tropical deforestation: Modeling local to regional scale climate change. *J.
Geophys. Res.* **98**, 7289–7315.

Herdies, D. L., Da Silva, A., Silva Dias, M. A., and Nieto-Ferreira, R. (2002) Moisture budget
of the bimodal pattern of the summer circulation over South America. *J. Geophys. Res.*
107, 42/1–42/10.

Hetzel, F. and Gerold, G. (1998) The water cycle of a moist deciduous rainforest and a cocoa
plantation in Cote d'Ivoire. *Water Resources Variability in Africa during the XXth Century.
Proceedings of the Abidjan '98 Conference,Abidjan, Côte d'Ivoire, November 1998* (IAHS
Publ. 216, pp. 411–418). International Association of Hydrological Sciences, Wallingford,
U.K.

Hulme, M. and Sheard N. (1999) *Cenários e Alteraçêes Climáticas para o Brasil* (6 pp.). Climate
Research Unit, Norwich, U.K.

Huntingford, C., Harris, P., Gedney, P., Cox, P., Betts, R., Marengo, J., and Gash, J. (2004)
Using a GCM analogue model to investigate the potential for Amazon Forest dieback.
Theoretical and Applied Climatology **78**, 177–186.

IPCC (2001) *Climate Change 2001: The Scientific Basis* (contribution of Working Group I to the
Third Assessment Report of the Intergovernmental Panel on Climate Change (IPCC), 944
pp.). Cambridge University Press, Cambridge, U.K.

Jordan, C. and Heuveldop, J. (1981) The water balance of an Amazonian rain forest. *Acta
Amazônica* **11**, 87–92.

Kirchoff, V. W. J. H. and Escada, P. A. S. (1998) *O Megaincêndio do Século [The Wildfire of the
Century]* (86 pp.). Transect Editorial Press, São Jose dos Campos, São Paulo, Brazil [in
Portuguese].

Koster, R., Jouzel, J., Suozzo, R., Russell, G., Broecker, W., Rind, D., and Eagleson, P. (1986)
Global sources for local precipitation as determined by the NASA GISS GCM. *Geophys.
Res. Lett.* **13**, 121–124.

Koster, R., Suarez, M., and Heiser, M. (2000) Variance and predictability of precipitation at
seasonal-to-interannual timescales. *J. Hydromet.* **1**, 26–46.

Labraga, J. C., Frumento, O., and Lopez, M. (2000) The atmospheric water vapor in South
America and the tropospheric circulation. *J. Climate* **13**, 1899–1915.

Lean, J. and Rowntree, P. (1993) A GCM simulation of the impact of Amazon deforestation on
climate using an improved canopy representation. *Quart. J. Roy. Met. Soc.* **119**, 509–530.

Lean, J. and Rowntree, P. (1997) Understanding the sensitivity of a GCM simulation of
Amazonian deforestation to specification of vegetation and soil characteristics. *J.
Climate* **6**, 1216–1235.

Lean, J. and Warrilow, D. (1989) Climatic impact of Amazon deforestation. *Nature* **342**, 311–
313.

Lean, J. and Warrilow, D. (1991) Climatic impact of Amazon deforestation. *Nature* **342**, 311–
313.

Lean, J., Bunton, C., Nobre, C., and Rowntree, P. (1996) The simulated impact of Amazonian
deforestation on climate using measured ABRACOS vegetation characteristics. In: J.
Gash, C. Nobre, J. Roberts, and R. Victória (eds.), *Amazonian Deforestation and
Climate* (pp. 549–576).

Leopold, L. (1962) Rivers. *American Scientist* **50**, 511–537.

Leopoldo, P. (2000) O ciclo hidrológico em bacias experimentais da Amazonia central. In: E.
Salati, M. L. Absy, and R. L. Victória (eds.), *Amazonia: Um Ecossistema em Transforma-
çãão* (pp. 87–117). INPA, Manaus, Brazil [in Portuguese].

Leopoldo, P., Franken, W., Matsui, E., and Roibeiro, M. (1982) Estimativa da evapotranspir-
ação da floresta Amazônica de terra firme. *Acta Amazônica* **12**, 23–28 [in Portuguese].

Lettau, H., Lettau, K., and Molion, L. (1979) Amazonia's hydrologic cycle and the role of
atmospheric recycling in assessing deforestation effects. *Mon. Wea. Rev.* **107**, 227–238.

Liebmann, B. and Marengo, J. A. (2001) Interannual variability of the rainy season and rainfall
in the Brazilian Amazonia. *J. Climate* **14**, 4308–4318.

Liebmann, B., Kiladis, G., Marengo, J. A., Ambrizzi, T., and Glick, J (1999) Submonthly
convective variability over South America and the South Atlantic Convergence Zone. *J.
Climate* **12**, 1877–1891.

Manzi, O. and Planton, S. (1996) Calibration of a GCM using ABRACOS and ARME data and
simulation of Amazonian deforestation. In: J. H. C. Gash (eds.), *Amazonian Deforestation
and Climate*. John Wiley & Sons, New York.

Marengo, J. A. (1992) Interannual variability of surface climate in the Amazon basin. *Int. J.
Climatol.* **12**, 853–863.

Marengo, J. A. (1995) Variations and change in South American streamflow. *Climate Change*
31, 99–117.

Marengo, J. A. (2003) Condições climaticas e recursos hídricos no norte do Brasil. *Clima e
Recursos Hídricos* (No. 9, pp. 117–156). Associação Brasileira de Recursos Hídricos/
FBMC-ANA, Porto Alegre, Brasil [in Portuguese].

Marengo, J. A. (2004a) Interdecadal variability and trends of rainfall across the Amazon basin.
Theoretical and Applied Climatology **78**, 79–96.

Marengo, J. A. (2004b) On the characteristics and variability of the water budget in the Amazon
Basin. *Climate Dynamics* in press.

Marengo, J. A. and Nobre, C. A. (2001) The hydroclimatological framework in Amazonia. In:
J. Richey, M. McClaine, and R. Victoria (eds.), *Biogeochemistry of Amazonia* (pp. 17–42).
Oxford University Press, Oxford, U.K.

Marengo, J. A. and Soares, W. (2003) Impacto das modificações da mudança climática—sintese
do terceiro relatório do IPCC. Condições climaticas e recursos hídricos no norte do Brasil.
Clima e Recursos Hídricos (No. 9, pp. 209–233). Associação Brasileira de Recursos
Hídricos/FBMC-ANA, Porto Alegre, Brasil [in Portuguese].

Marengo, J. A., Miller, J. A., Russell, G., Rosenzweig, C., and Abramopoulos, F. (1994)
Calculations of river-runoff in the GISS GCM: Impact of a new land surface parameterization and runoff routing on the hydrology of the Amazon River. *Climate Dynamics* **10**,
349–361.

Marengo, J. A., Tomasella, J., and Uvo, C. (1998a) Long-term streamflow and rainfall
fluctuations in tropical South America: Amazonia, Eastern Brazil and Northwest Peru.
J. Geophys. Res. **103**, 1775–1783.

Marengo, J., Nobre, C. A., and Sampaio, G. (1998b) On the associations between hydro-
meteorological conditions in Amazonia and the extremes of the Southern Oscillation.
*Seminario Internacional Consecuencias Climáticas e Hidrológicas del Evento El Niño a
Escala Regional y Local, Memorias Tecnicas, 26–29 Noviembre 1997, Quito, Ecuador*
(Extended Abstracts, pp. 257–266).

Marengo, J. A., Liebmann, B., Kousky, V., Filizola, N., and Wainer, I. (2001) On the onset and
end of the rainy season in the Brazilian Amazon Basin. *Journal of Climate* **14**, 833–852.

Marengo, J. A., Douglas, M., and Silva Dias, P. L. (2002) The South American Low-Level Jet
East of the Andes during the LBA-TRMM and WET AMC campaign of January–April
1999. *J. Geophys. Research* **107**, 47/1–47/12

Marengo, J. A., Cavalcanti, I. F. A., Satyamurty, P., Nobre, C. A., Bonatti, J. P., Manzi, A.,
Trosnikov, I., Sampaio, G., Camargo, H., Sanches, M. B. *et al.* (2003) Ensemble simulation
of regional rainfall features in the CPTEC/COLA atmospheric GCM: Skill and predictability assessment and applications to climate predictions. *Clim. Dyn.* **21**, 459–475.

Marengo, J., Soares, W., Saulo, C., and Nicolini, M. (2004a) Climatology of the LLJ east of the Andes as derived from the NCEP reanalyses. *Journal of Climate* **17**, 2261–2280.

Marengo, J., Fisch, G., Vendrame, I., Cervantes, I., and Morales, C. (2004b) On the diurnal and day-to-day variability of rainfall in Southwest Amazonia during the LBA-TRMM and LBA-WET AMC campaigns of summer 1999. *Acta Amazônica* in press.

Marengo, J., Liebmann, B., Vera, C., Paegle, J., and Baez, J. (2004c) Low frequency variability of the SALLJ. *CLIVAR Exchanges* **9**, 26–27.

Matsuyama, H. (1992) The water budget in the Amazon River basin during the FGGE period. *J. Meteorol. Soc. Jap.* **70**, 1071–1083.

Meggers, B. J. (1994) Archeological evidence for the impact of mega-Niño events on Amazonia during the past two millenia. *Climatic Change* **28**, 321–338.

Miller, J., Russell, G., and Caliri, G. (1994) Continental scale river flow in climate models. *J. Climate* **7**, 914–928.

Milliman, J. and Meade, R. (1983) World-wide delivery of river sediment to the oceans. *J. Geol.* **91**, 1–21.

Molion, L. C. B. (1975) A climatonomic study of the energy and moisture fluxes of the Amazon basin with considerations of deforestation effects. Ph.D. thesis, University of Wisconsin, Madison, WI.

Nace, R. I. (1972) World hydrology: Status and prospects. *World Water Balance* (Vol. 1, pp. 1–10). IAHS/UNESCO/WMO.

Nijssen, B., Letenmaier, D., Liang, X., Wetzel, S., and Wood, E. (1997) Streamflow simulation for continental-scale river basins. *Water Resource Res.* **33**, 711–724.

Nijssen, B., O'Donnell, G., and Lettenmaier, D. (2001) Predicting the discharge of global rivers. *J. Climate* **14**, 3307–3323.

Nishizawa, T. and Koike, Y. (1992) *Amazon: Ecology and Development* (221 pp.). Iwanami, Tokyo.

Nobre, P. and Shukla, J. (1996) Variations of sea surface temperature, wind stress, and rainfall over the tropical Atlantic and South America. J Climate 9, 2464–2479.

Nobre, C., Sellers, P., and Shukla, J. (1991) Amazonian deforestation and regional climate change. *J. Climate* **4**, 957–988.

Nogués-Paegle, J. and Mo, K.-C. (1997) Alternating wet and dry conditions over South America during summer. Mon. Wea. Rev. 125, 279–291.

Nogues-Paegle, J., Mechoso, C., Fu, R., Berbery, H., Winston, C., Chao, T., Cook, K., Diaz, A., Enfield, D., Ferreira, R. *et al.* (2002) Progress in pan American CLIVAR research: Understanding the South American Monsoon. *Meteorologica* **27**, 1–30.

Oki, T., Musiake, K., Matsuyama, H., and Masuda, K. (1995) Global atmposheric water balance and runoff from large river basins. *Hydrological Processes* **9**, 655–678.

Pielke, R. A., Sr. (2001) Influence of the spatial distribution of vegetation and soils on the prediction of cumulus convective rainfall. *Rev. Geophys.* **39**, doi:10.1029/1999RG000072.

Polcher, J. and Laval, K. (1994a) The impact of African and Amazonian deforestation on tropical climate. *J. Hydrology* **155**, 389–405.

Polcher, J. and Laval, K. (1994b) A statistical study of the regional impact of deforestation on climate in the LMD GCM. *Climate Dynamics* **10**, 205–219.

Rao, V. B. and Hada, K. (1990) Characteristics of rainfall over Brasil: Annual variations and connections with the Southern Oscillation. *Theor. Appl. Climatol.* **42**, 81–91.

Rao, V. B., Cavalcanti, I., and Hada, K. (1996) Annual variation of rainfall over Brazil and water vapor characteristics over South America. *J. Geophys. Res.* **101**, 26539–26551.

Richey, J. E., Nobre, C., and Deser, C. (1989) Amazon river discharge and climate variability: 1903 to 1985. *Science* **246**, 101–103.

Roads, J., Kanamitsu, M., and Stewart, R. (2002) CSE water and energy budgets in the NCEP-DOE reanalyses. *J. Hydromet.* **3**, 227–248.

Rocha, H., Nobre, C., and Barros, M. (1989) Variabilidade natural de longo prazo no ciclo hidrológico da Amazonia. *Climanálise* **4**(12), 36–42 [in Portuguese].

Ronchail, J., Cochonneau, G., Molinier, M., Guyot, J. L., Chaves, A. G. M., Guimarães, V., and Oliveira, E. (2002) Interannual variability in the Amazon basin and sea-surface temperature in the equatorial Pacific and the tropical Atlantic Oceans. *Int. J. Climatol.* **22**, 1663–1686.

Ropelewski, C. and Halpert, M. (1987) Global and regional scale precipitation patterns associated with the El Niño-Southern Oscillation. *Mon. Wea. Rev.* **115**, 1606–1626.

Ropelewski, C. and Halpert, M. (1989) Precipitation patterns associated with the high index of the Southern Oscillation. *J. Climate* **2**, 268–284.

Russell, G. and Miller, J. (1990) Global river runoff calculated from a global atmospheric general circulation model. *J. Hydrol.* **117**, 241–254.

Salati, E. (1987) The forest and the hydrological cycle. In: R. E. Dickinson (ed.), *The Geophysiology of Amazonia: Vegetation and Climate Interactions* (pp. 273–296). John Wiley & Sons, New York.

Salati, E. and Marques, J. (1984) Climatology of the Amazon region. In: H. Sioli (ed.), *The Amazon: Limnology and Landscape Ecology of a Mighty Tropical River and Its Basin.* W. Junk, Dordrecht, The Netherlands.

Salati, E. and Nobre, C. A. (1991) Possible climatic impacts of tropical deforestation. *Climatic Change* **19**, 177–196.

Salati, E. and Voce, P. (1984) Amazon basin: A system in equilibrium. *Science* **225**, 128–138.

Salati, E., Dall'Ollio, A., Matsui, E., and Gat, J. (1979) Recycling of water in the Amazon basin: An isotopic study. *Water Resource Res.* **15**, 1250–1258.

Sansigolo, C., Rodrigues, R., and Etchichury, P. (1992) Tendências nas temperaturas médias do Brasil. *Anais do VII Congresso Brasileiro de Meteorologia* (Vol. 1, pp. 367–371) [in Portuguese].

Saulo, C., Nicolini, M.. and Chou, S. C. (2000) Model characterization of the South American low-level flow during the 1997–98 spring–summer season. *Clim. Dyn.* **16**, 867–881.

Sausen, R., Schubert, S., and Dumenil, L. (1994) A model of river runoff for use in coupled atmosphere–ocean models. *J. Hydrol.* **155**, 337–352.

Segal, M., Avissar, R., McCumber, M. C., and Pielke, R. A. (1988) Evaluation of vegetation effects on the generation and modification of mesoscale circulations. *J. Atmos. Sci.* **45**, 2268–2292.

Shiklomanov, I. (ed.) (2001) *World Water Resources and Their Use.* UNESCO, Paris (available online from *www.unesco.org*).

Shukla, J., Nobre, C., and Sellers, P. (1990) Amazonia deforestation and climate change, *Science* **247**, 1322–1325.

Silva Dias, M., Rutledge, S., Kabat, P., Silva Dias, P., Nobre, C., Fisch, G., Dolman, H., Zipser, E., Garstang, M., Manzi, A. *et al.* (2002) Clouds and rain processes in a biosphere atmosphere interaction context in the Amazon Region. *J. Geophys. Research* **107**, 39/1–39-18.

Sternberg, H. R. (1987) Aggravation of floods in the Amazon River as a consequence of deforestation? *Geografiska Annaler* **69A**, 201–219.

Sud, Y., Yang, R., and Walker, G. (1996a) Impact of in situ deforestation in Amazonia on the regional climate: General circulation model simulation study. *J. Geophys. Res.* **101**, 7095–7109.

Sud, Y., Walker, G., Kim, J.-H., Liston, G., Sellers, P., and Lau, W. (1996b) Biogeophysical consequences of a tropical deforestation scenario: A GCM simulation study. *J. Climate* **9**, 3226–3247.

Tundisi, J. G., Tundisi, T. M., and Rocha, O. (2002) Ecossistemas de aguas interiores. In: A. Rebouças, B. Braga, and J. G. Tundisi (eds.), *Aguas Doces do Brasil: Capital Ecológico,*

Uso e Conservação (2nd edn., pp. 153–192). Escrituras Editora, São Paulo, Brazil [in Portuguese].

UNESCO (1971) *Discharge of Selected Rivers of the World* (Vols. 2/3). UNESCO, Paris.

UNESCO (1974) *Discharge of Selected Rivers of the World* (Vols. 3/2). UNESCO, Paris.

Uvo, C. R. B., Repelli, C. A., Zebiak, S., and Kushnir, Y. (1998) The relationship between tropical Pacific and Atlantic SST and Northeast Brazil monthly precipitation. *J. Climate* **11**, 551–562.

Vera, C., Higgins, W, Gutzler, J., Marengo, J. A., Garreaud, R., Amador, J., Gochis, D., Nogues-Paegle, J., Zhang, C., Ambrizzi, T., Mechoso, C., and Lettenmaier, D. (2006) A unified vision of the American monsoon systems. *J. of Climate* (in press).

Victoria, R., Martinelli, L., Moraes, J., Ballester, M. V., Krushche, A., Pellegrino, G., Almeida, R., and Richey, J. (1998) Surface air temperature variations in the Amazon region and its border during this century. *J. Climate* **1**, 1105–1110.

Villa Nova, N., Salati, E., and Matsui, E. (1976) Estimativa da evapotranspiraçãão na bacia Amazônica. *Acta Amazônica* **6**, 215–228 [in Portuguese].

Vincent, L. A., Peterson, T. C. Barros, V. R., Marino, M. B., Rusticucci, M., Carrasco, G., Ramirez, E., Alves, L. M., Ambrizzi, T., Berlato, M. A. *et al.* (2005) Observed trends in indices of daily temperature extremes in South America 1960–2000. *J. Climate* **18**, 5011–5023.

Voldoire, A. and Royer, J. F. (2004) Tropical deforestation and climate variability. *Climate Dynamics* **22**, 857–874.

Vörösmarty, C. J., Moore, B., Gildea, M. P., Peterson, B., Melillo, J., Kicklighter, D., Raich, J., Rastetter, E., and Steudler, P. (1989) A continental-scale model of water balance and fluvial transport: Application to South America. *Global Biogeochemical Cycles* **3**, 241–265.

Vörösmarty, C., Willmott, C., Choudhury, B., Schloss, A., Stearns, T., Robertson, S., and Dorman, T. (1996) Analysing the discharge regime of a large tropical river trough remote sensing, ground climatic data, and modeling. *Water Resource Res.* **32**, 3137–3150.

Walker, G., Sud, Y., and Atlas, R. (1995) Impact of the ongoing Amazonian deforestation on local precipitation: A GCM simulation study. *Bull. Am. Met. Soc.* **76**, 346–361.

Weaver, C. and Baidya-Roy, S. (2002) Sensitivity of simulated mesoscale atmospheric circulations resulting from landscape heterogeneity to aspects of model configuration. *J. Geophys. Res.* **107**(D20), 59/1–59/21.

Williams, E., Dall'Antonia, A., Dall'Antonia, V., de Almeida, J., Suarez, F., Liebmann, B., and Malhado, A. (2005) The Drought of the Century in the Amazon Basin: An analysis of the regional variation of rainfall in South America in 1926. *Acta Amazônica* **35**(2), 231–238.

Yang, S. L., Zhao, Q. Y., and Belkin, I. M. (2002) Temporal variation in the sediment load of the Yangtze River and the influence of human activities. *J. Hydrol.* **263**, 56–71.

Yin, H. F. and Li, C. (2001) Human impact on flood and flood disasters on the Yangtze River. *Geomorphology* **41**, 105–109.

Zeng, N. (1999) Seasonal cycle and interannual variability in the Amazon hydrologic cycle. *J. Geophys. Res.* **104**, 9097–9106.

Zeng, N., Dickinson, R., and Zeng, X. (1996) Climatic impact of Amazon deforestation: A mechanistic study. *J. Climate* **9**, 859–883.

Zhang, Y., Wallace, J. M., and Battisti, D. (1998) ENSO-like interdecadal variability: 1900–93. *J. Climate* **10**, 1004–1020.

Zhang, H., Henderson-Sellers, A., and McGuffie, K. (2001) The compounding effects of tropical deforestation and greenhouse warming on Climate. *Climatic Change* **49**, 309–338.

Zhou, J. and Lau, K. M. (2001) Principal modes of interannual and decadal variability of summer rainfall over South America. *Int. J. Climatol.* **21**, 1623–1644.

10

Plant species diversity in Amazonian forests

M. R. Silman

10.1 INTRODUCTION

Looking at the Amazonian landscape from space one sees unbroken forest stretching from the eastern lowlands of Colombia south through Peru to Bolivia, and east from the Andes to the Atlantic Ocean. The expanse of trees—covering over half a billion hectares—is draped over a relatively flat landscape broken only by large rivers and human-induced habitat modification (Figure 10.1a, see color section).

Amazonian forests are the most diverse on the planet, estimated to harbor 30,000 species of vascular plants, with 5,000–10,000 species of trees alone (Henderson *et al.*, 1991; Thomas, 1999; Myers *et al.*, 2000). While we have a picture of Neotropical forests as spectacularly diverse, how diversity is distributed across the landscape is less well-known. The monotonous green sameness belies large changes in diversity and species composition of the forest. A hectare of Amazonian forest may harbor anywhere from 30 to 300 species of trees, and may be dominated by one of more than a dozen different plant families (Gentry, 1988). As one moves across the Amazon Basin, forest composition and diversity changes at all spatial scales (Campbell, 1994). Species lists of trees from an eastern Amazonian site on poor soils are bewilderingly unfamiliar to a botanist trained in western Amazonia. Go from upper Amazonian forest on predominantly rich soils to the poor soils of eastern Amazonia and the character and composition of the forests will be as distinct as those of the major forest classes in temperate areas. Changes in community composition on different soils at local scales can be equally dramatic (Balslev *et al.*, 1987; Duivenvoorden, 1996; Lips and Duivenvoorden, 1996; Tuomisto *et al.*, 2002; Phillips *et al.*, 2003; Masse, 2005). In a new set of 1-ha plots in western Amazonia, rich- and poor-soil plots have <5% overlap in species (N. Pitman, unpublished). Ecological interactions have been shown to underlie interspecific trade-offs in performance on distinct soils, and the diversification of certain clades shows repeated evolution of edaphic specialization (Fine *et al.*, 2004, 2005).

Looking further into the environmental variables underlying changes in composition and diversity, it becomes clear that climates in Amazonian forests range from highly variable both within and among years to near absolute monotony, and that forests are underlain by geologies ranging from bleached and nutrient-poor metamorphics and granites from the earliest parts of the Earth's history to young sediments weathered from the Andes.

This suite of features of the Amazonian landscape—its immense size, variable climates, heterogenous geology and soils—has been pinpointed as the reason for its high plant diversity in the ecological literature (Willis, 1922; Terborgh, 1973; Rosenzweig, 1995), even though the low diversity in the vast expanses of Siberian forests or the high diversity of geographically-restricted South African plant communities provide poignant counterexamples. Moving beyond the simple assumption that large areas, long timespans, and high environmental heterogeneity equal high species diversity, how each of those factors—gradients in soils, climate, and geological history—influence forest diversity at sites (alpha diversity), how diversity in a landscape accrues as one moves across sites (beta diversity), and how these combine to form regional (gamma) diversity is only recently becoming known and even more rarely integrated into explanations of diversity that treat both local and regional scales (e.g., Pitman *et al.*, 2002; Ricklefs, 2004) .

Nearly 20 years ago, Alwyn Gentry (1988) wrote the first quantitative paper on Neotropical forest diversity and floristics, examining four environmental gradients: latitude, precipitation, soils, and elevation. Though providing empirical hypotheses for patterns of diversity across gradients, Gentry's paper was based on just 38 lowland and 11 montane 0.1-ha plots, and just a few preliminary hectare inventories in upper Amazonia. Since 1988, empirical work on Amazonian flora has exploded, including over 400 ha of tree inventories in the Amazonian lowlands (Ter Steege *et al.*, 2003; ATDN, 2006). New climatological tools, such as satellite-based measurement of rainfall and compilations of station data, paint a more complex picture of precipitation amount and variability—both within and among years—than previously known (Malhi and Wright, 2004; Marengo, 2004; Chiu *et al.*, 2006). In many cases the details of the gradients Gentry (1988) treated are only now becoming known due to advances in remote sensing and ongoing revisions of our understanding of Amazonian historical geology, climatology, and paleoecology. Geological surface and soil data were even sparser in the 1980s, particularly in western Amazonia, where ages were poorly constrained, and large expanses of terrain had never been visited by geologists. Consequently, soils were often classified simply through educated guesses based on aerial photos and satellite images. Andean uplift timing was relatively unconstrained, and the modern understanding of the complex history of the formation of the modern Amazon River from the Miocene through the present was a little-known hypothesis. How these new data on environmental gradients affect our understanding of diversity remains unexplored.

This chapter looks at patterns of within-site (alpha), between-site (beta), and regional (gamma) diversity in Amazonian forests. A goal is to update patterns of floristic composition and diversity given the current understanding of Amazonian climatology, geology, and history, revisiting Gentry's ideas in the light of new data.

10.2 STUDY SITE: OVERVIEWS OF AMAZONIAN GEOGRAPHY, GEOLOGY, AND CLIMATE

An overview of Amazonian geography is central to understanding the environmental gradients that influence Amazonian plant communities. The following sections give an overview of current empirical results and hypotheses regarding environmental gradients in the Amazon Basin.

10.2.1 Amazonian geography, geology, and soils

The Amazon Basin is largely in the southern hemisphere, with its bulk lying south of $\sim 3°$N in the west and $\sim 1°$N in the east. The Amazon Basin proper has its western border in the Andean highlands of Bolivia, Peru, Ecuador, and Colombia, where large white-water rivers—the Madeira, Ucayali, Marañon, Napo, and Caqueta—descend onto the large alluvial fans of upper Amazonia (Figure 10.1b, see color section). West of Manaus, Brazil, the *terra firme* habitats along the Amazon are a complex mix of young (Late Tertiary and Quaternary) sediments derived from changing depositional environments (Räsänen *et al.*, 1992; Hoorn, 1994a, b; Potter, 1997; Hovikoski *et al.*, 2005; Rossetti *et al.*, 2005). Soils in the western lowlands range from rich soils on the higher elevations of the alluvial fans to a mosaic of rich and poor soils along the main stem of the Amazon from Iquitos eastward (Davis *et al.*, 1997). The northern limit of the Amazon Basin lies $\sim 3°$N on the highly weathered Paleozoic sandstones and Proterozoic/Archean metamorphic series of the Guianan Shield with soils that are largely podzolized and deeply acidic (Ducke and Black, 1956; Davis *et al.*, 1997; Johnson *et al.*, 2001; Malhi *et al.*, 2004). These ancient and nutrient-poor landscapes are drained in the west by the black-water rivers of the Río Negro system: the Vaupés, upper Negro, and Branco. The southern and southeastern limits of the Amazon Basin lie on Proterozoic metamorphic rocks of the Brazilian Shield and are drained by the clear water Tapajos and Xingú river systems (for a general overview see Sioli, 1984; Schenk *et al.* 1997). As in Guianan Shield areas, soils are largely poor, particularly when compared with those on upper-Amazonian alluvium (Figure 10.1c, see color section).

The geology of the basin is such that there are relatively fertile soils on young alluvium and recently exposed rock formations of Andean orogeny in the western Amazon Basin, which are replaced by alluvial deposits weathered from the Guianan and Brazilian Shields along the main stem of the Amazon, and poor soils on some of the oldest rocks on Earth at the northern and southern limits of the Amazon (Schenk *et al.*, 1997) (Figure 10.1c). This pattern of origin leads to a broad correlation between substrate age and fertility in the Amazon, with soils derived from younger geological formations generally more fertile than those derived from old (Malhi *et al.*, 2004).

Studies on Amazonian historical geology show the basin to be dynamic, with modern Amazon drainage being a relatively recent (Late Miocene to Early Pliocene) geologic feature whose origin is concurrent with, or even later than, the divergence of many animal and plant taxa (Potter, 1997; Rossetti *et al.*, 2005). From the Paleozoic through the Cretaceous the non-shield areas of the Amazon were depositional environments (Potter, 1997). To the east of Manaus, deposition largely stopped in the Cretaceous, save for the Miocene fluvial deposits of the Barreiras Formation and Quaternary deposits along river banks (Irion *et al.*, 1995; Potter, 1997; Rossetti *et al.*, 2005). From the Paleozoic through the Miocene, waters drained west from shield areas onto a continental margin and then, with the uplift of the northern Andes in the Miocene, as series of lacustrine and transitional marine habitats. Late Tertiary tectonic changes in land elevation caused the Amazon to change flow from westward to eastward, adopting its current bed through the depression between the Guianan and Brazilian Shield areas at Santarem (Hoorn, 1994b; Hoorn *et al.*, 1995; Potter, 1997).

Direct correlations between plant diversity as derived from inventory data and geology require high-resolution maps of soils and surface geology. While several excellent country-wide campaigns are currently underway (e.g., the Peruvian INGENMET and Brazilian CPRM geological survey campaigns), geological maps of the Amazon Basin have spotty coverage (e.g., Rosetti *et al.*, 2005). The problem is compounded by nomenclatural differences in geological formations that span several countries.

Gentry (1988) hypothesized that diversity would be positively correlated with soil fertility, but that the effect would be subsidiary to precipitation amount and predictability. Subsequent research has found a large effect of soil fertility on productivity in Amazonian forests (Malhi *et al.*, 2004), but not diversity (Clinebell *et al.*, 1995). The present analysis looks at diversity as it relates to depositional age (which is correlated with substrate type) using a compilation of South American geology from the United States Geological Survey (Schenk *et al.*, 1997), updated with information from Brazil's CPRM and Rosetti *et al.* (2005).

10.2.2   Elevation

Except for its western and northern extremities, the Amazon Basin is marked in its flatness; 85% of the Amazon Basin, including its Andean headwaters, are under 500 m in elevation. Ninety-seven percent of rainforests lie below 500 m (S. Saatchi, pers. commun.). The main stem of the Amazon drops only 215 m from where the Marañón passes the final Andean foothill at the Pongo de Manseriche to its mouth 3,200 km away at Ilha de Marajó (Figure 10.1b). To the north and south of the Marañón, alluvial fans of Andean sediments rise to ~400 m at the base of the Andean foothills. The northern limit of Amazonia rises up to ~2,000 m in the Tepuis of the Guianan Shield, and the southern limit of the basin gradually rises up to ~1,000 m on the Brazilian Shield. The western border of the basin is found on the east Andean flank, with forests rising to 4,800 m in the *Polylepis* woodlands of the high Andes (e.g., Hoch and Korner, 2005). Given the measured moist air lapse rate of $5.6°C\,km^{-1}$ (Bush *et al.*, 2004), elevation *per se* will have little effect on lowland Amazonian forests, save for the transition from lowland to Andean forests at the basin's western margin.

10.2.3   Precipitation

The Amazon River discharges roughly 20% of Earth's river water that reaches the ocean, even though its basin occupies just 2% of continental land area. The spectacular discharge rate is due to the deep convection that forms over the Amazon Basin, the moisture brought in from the Atlantic by easterly winds, and the huge amounts of orographic rainfall generated as water vapor from the Atlantic is forced upwards by the Andean massif. While the mean annual rainfall is $\sim$2,400 mm yr^{-1}, precipitation is not uniformly distributed over the Amazon Basin, either in space or in time (Sombroek, 2001; Malhi and Wright, 2004; Marengo, 2004). The number of dry months (months averaging <100-mm precipitation) at Amazonian sites ranges from 0 to 8, and annual precipitation over closed-canopy humid forest and rainforest varies from 1,200 mm yr^{-1} in the southeastern Amazon Basin to >8,000 mm yr^{-1} in the Andean foothills.

Total rainfall in the Amazon Basin is high in a large region of upper Amazonia in northeastern Peru, eastern Ecuador, eastern Colombia, and western Amazonas state in Brazil, as well as near the mouth of the Amazon in eastern Pará and Amapá states. High-rainfall areas are also found along the base of the Andes to from $\sim$5°N to $\sim$16°S. Rainfall amounts drop rapidly south of $\sim$6° to 8° in the western and central Amazon. A region of lower rainfall in central Pará state, the Roraima territory of Brazil, and adjacent Guayana and Surinam is known as the transverse dry belt (Nimer, 1977; Pires and Prance, 1977, Davis *et al.*, 1997), though total annual rainfall in this area is as high as much of the upper-Amazonian forests in southeast Peru and Acre, Brazil.

Precipitation varies temporally in the Amazon Basin on all timescales, even though the effects of super-annual variation on diversity have been little studied. Variability ranges from daily cycles and anomalies in rainfall rate due to local climatic processes, to intra-annual variation in rainfall, to among-year variation in rainfall driven by changes in sea surface temperature—such as ENSO and the North Atlantic Oscillation (Marengo *et al.*, 2001; Pezzi and Cavalcanti, 2001; Marengo, 2004). On even longer timescales, precipitation changes are driven by the orbital parameters of Earth and their interaction with proximate climate drivers—such as the South American Low-Level Jet and the South American Convergence Zone (Baker *et al.*, 2001; Cruz *et al.*, 2005). Far from being static, detailed paleoclimatological reconstructions show that the main atmospheric and oceanic features influencing climate in the Amazon Basin show large variations through time. Indeed, modern studies correlating these drivers with historical weather patterns show that temporal variation also translates into differential spatial effects across the Amazon Basin (Giannini *et al.*, 2001; Cruz *et al.*, 2005; Vuille and Werner, 2005).

Previous empirical hypotheses of tree diversity suggested that tree diversity increased with precipitation amount and decreased with the number of months having <100-mm precipitation, a figure where evapotranspiration exceeds precipitation in a typical lowland ecosystem (Gentry, 1988; Clinebell *et al.*, 1995; ter Steege *et al.*, 2003). Annual precipitation amount and dry-season length are also inversely correlated with each other (Figure 10.2a).

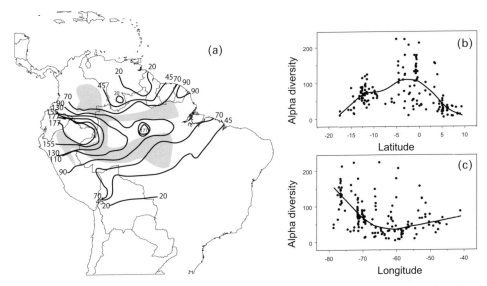

Figure 10.2. (a) Local (alpha) diversity derived from 423 1-ha Amazonian tree plots. Isoclines drawn from universal kriging fit for 1° grid-cells. Figure adapted from *Amazon Tree Diversity Network*. Shaded regions show areas with no inventory plots. (b) Local diversity versus latitude. (c) Local diversity versus longitude.

Precipitation regimes have traditionally been characterized in terms of monthly and yearly averages interpolated from existing gauges, with varying degrees of quality and duration (e.g., Sombroek, 2001). Much of the Amazon Basin remains wilderness, and large areas are without rain gauges, particularly in upper Amazonia. In the present analysis we use satellite measurements of total monthly precipitation on a 0.25° × 0.25° grid collected over 7 years by the Tropical Rainfall Monitoring Mission (Chiu *et al.*, 2006). These data give complete temporal and spatial coverage for precipitation across the tropics in grid-cells just less than 30 km per side, or for every ~900 km^2, and allow examination of precipitation trends in the Amazon with unprecedented clarity.

From these data we derive four measures of precipitation amount and variability: average annual precipitation; average dry-season length, taken as the average number of months with precipitation below 100 mm (Clinebell *et al.*, 1995; Sombroek, 2001; Ter Steege *et al.*, 2003); inter-annual variability, taken as the coefficient of variation of rainfall among years; and total variability, taken as the coefficient of variation among months over the 7-year period.

10.2.4   Species diversity and environmental gradients

To look at the influence of these environmental gradients on species diversity, we looked at alpha and gamma diversity in Amazonia (Whittaker, 1972). Alpha diversity was measured as both species density (N ha^{-1}) and also Fisher's alpha (Fisher *et al.*,

1943). Fisher's alpha has been used extensively in comparisons of Neotropical forest diversity as it is relatively independent of plot size (Condit *et al.*, 1996; Leigh, 1999; Ter Steege *et al.*, 2003). Gamma diversity was estimated as the total number of species that include a given latitude and elevation within their range (Silman *et al.*, unpublished).

Data on tree species diversity from individual locales were taken from compilations of Ter Steege *et al.* (2000) and Pitman *et al.* (1999, 2001), as well as unpublished data of J. Terborgh, P. Núñez and N. Pitman. Basin-wide estimates of tree alpha diversity were taken from universal kriging estimates based on 423 plots in moist *terra firme* forest in Amazonia and the Guianan Shield (Ter Steege *et al.*, 2003, as updated by the ATDN, 2004). Gamma diversity estimates for upper Amazonia and the adjacent eastern Andes are derived from 263,000 collections of vascular plants from 0 to 23°S in Ecuador, Peru, and Bolivia, housed in the Missouri Botanical Garden TROPICOS database (Silman *et al.*, unpublished).

Many of the analyses are simple correlations or graphical comparisons of empirical patterns of diversity with environmental gradients. In cases where multiple factors are compared quantitatively, relationships were modeled non-parametrically using generalized additive models (GAMs; Hastie and Tibshirani, 1990). GAMs make no assumptions about the form of the relationship among variables, letting the data present empirical hypotheses.

10.3   RESULTS

10.3.1   Precipitation: patterns

Patterns of precipitation amount and variability in Amazonia change with both latitude and longitude. Analysis of TRMM data shows average annual precipitation in the Amazonian lowlands peaks at ~2–4°S latitude, with a broad plateau from ~4°N to 4°S of the equator. Precipitation decreases steadily to the south and north (Figure 10.3e). Precipitation at low latitudes can be either high or low, while sites to the north and south have more uniformly low rainfall. Precipitation variability, measured as dry-season length and total variability, reaches a minimum between ~1°N and 4°S (Figures 10.3a and 100.3c, respectively). Large areas of upper Amazonia have almost no variability in rainfall from month to month and year to year. Near the equator at Yasuní, Ecuador, maximum monthly rainfall has occurred in every month of the year (Pitman *et al.*, 2001). Precipitation variability increases as one moves away from the equator, with dry season length increasing through 20°S, and precipitation variability reaching a peak ~15°S (Figures 10.3a, c).

In contrast to latitude, mean annual precipitation shows little relationship to longitude, though areas of high rainfall are more common in western Amazonia, causing a slight decrease in average rainfall as one moves from west to east (Figure 10.3f). Dry season length is on average shorter in western Amazonia and increases to the east, though the main cause of the trend is the absence of areas with no dry season in eastern Amazonia (Figure 10.3d). Total variability in rainfall is lowest in western Amazonia and increases consistently as one moves east (Figure 10.3b).

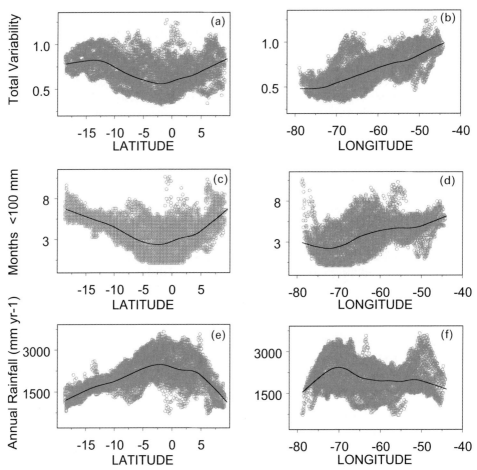

Figure 10.3. Latitudinal and longitudinal variation in Amazonian precipitation variability (a) and (b), dry-season length (c) and (d), and total amount (e) and (f) derived from TRMM measurements. Average dry-season length and precipitation variability reach a minimum centered at ~2–3°S latitude. Dry-season length and precipitation variability are lowest in western Amazonia and incease to the east. Precipitation also peaks ~2–3°S, but shows little trend from west to east.

As noted by other authors, precipitation variables are correlated (e.g., Gentry, 1988; Clinebell *et al.*, 1995; Ter Steege *et al.*, 2003). Mean dry-season length decreases with increasing rainfall, though areas with the highest rainfall can have dry-season lengths that vary from 0 to 4 months (Figure 10.4a). Total variability in precipitation is on average lowest in areas with high rainfall and increases with decreasing rainfall down to ~1,800 mm yr^{-1}, below which it is approximately constant (Figure 10.4b). At precipitation levels above ~1,800 mm yr^{-1}, however, one can find sites with either high or low total variability in precipitation. Variability among years shows little trend with annual precipitation above 1,500 mm yr^{-1}, save for a slight tendency for areas

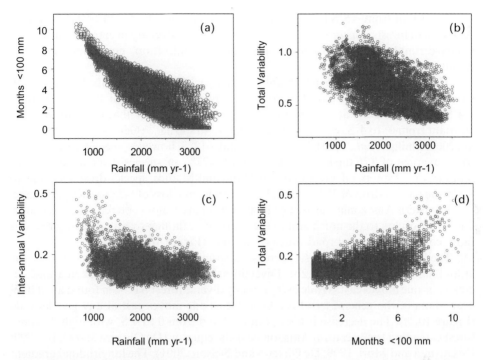

Figure 10.4. Amazonian rainfall and three measures of variation in rainfall derived from 7 years (1998–2004) of satellite measurements from the Tropical Rainfall Monitoring Mission (TRMM). (a) Total precipitation versus length of dry season as measured by months having <100-mm precipitation (Spearman's rho, $r_s = -0.85$). (b) Total variability in rainfall (measured as the coefficient of variation) among 94 months of precipitation versus accumulated rainfall ($r_s = 0.60$). (c) Inter-annual variability versus accumulated rainfall ($r_s = -0.17$). (d) Total variability versus dry-season length (Spearman's rho, $r_s = 0.86$). In all cases measures of variability in precipitation decrease with increasing rainfall. There is substantial variability in the trend, however, with sites at any given precipitation having short or long dry seasons, and predictable or unpredictable rainfall amounts from month to month or year to year.

with high annual precipitation to have fewer sites with high inter-annual variability. Total variability in precipitation is highly correlated with dry-season length ($r_s = 0.80$), but the increase in variability with dry-season length is slight until dry seasons exceed 5 months in length.

10.3.2   Latitudinal and longitudinal gradients in Amazonian diversity

Even within the tropical forests of South America there are large gradients in plant species diversity (Gentry, 1988, figs. 2a–c). For both tree alpha diversity and vascular plant gamma diversity, diversity has a broad peak between ∼1°N and 5°S. South of 5°S, tree alpha diversity decreases steeply and nearly linearly with latitude (Figure 10.2b). Regional (gamma) diversity of all vascular plants peaks at ∼4°S and remains fairly constant from ~5°S to 10°S, decreasing rapidly south of ~14°S. To the north, Amazonian forests *sensu stricto* end at ~3°N. Even so, diversity in extralimital forests, which certainly are Amazonian in taxonomic composition, continues to decline rapidly.

The pattern of decrease in tree alpha diversity with increasing latitude is clearest in western Amazonia. Tree plot data from four inventories—Yasuni at ~0°, Loreto at ~4°S, Madre de Dios at ~12°S, and Madidi at ~15°—show that diversity remains constant from 0° to 4°S, with plots averaging ~240 spp ha^{-1} (Gentry, 1988; Pitman *et al.*, 2001; Phillips *et al.*, 2003). Between 4°S and ~12°S, however, species diversity falls to an average of 176 spp ha^{-1}, a decrease of ~27% in species diversity over 8° of latitude, or ~7 spp per degree. Between Manu and Madidi, species diversity drops to ~118 spp ha^{-1}, a drop of 36% over 2.5° of latitude, or a loss of ~23 species per degree.

In eastern Amazonia the area of highest diversity is not centered on the equator, but rather occurs to the north and to the south of it, though overall diversity remains fairly constant across a broad range of latitude (Figure 10.2a).

The longitudinal (west–east) decrease in species diversity is nearly as great as the latitudinal change (Figure 10.2b). Diversity is highest in upper Amazonia and decreases as one moves east. While exceptionally diverse forest has been found at ~60°W near Manaus, the general trend is a decrease as one moves east at any give latitude (Figure 10.2c). The decrease is highest in the band from 0 to 4°S, with high-diversity forest extending to the central Amazon near the equator (De Oliveira and Daly, 1999; De Oliveira and Mori, 1999; De Oliveira and Nelson, 2001). The longitudinal gradient in species diversity becomes much less pronounced beyond 6°N or S of the main stem of the Amazon River (Figure 10.2a).

An important result is that areas central to understanding the gradients of tree diversity in Amazonia remain largely unsampled (Figure 10.2a; see also Ter Steege *et al.*, 2003). In addition to the large areas without inventory shown in Figure 10.2a, certain habitats—such as swamp forest (Ancaya, 2000), bamboo forest (Nelson *et al.*, 1994; Griscom and Ashton, 2003; Silman *et al.*, 2003), and upper Amazonian dry forest (e.g., Gentry, 1995; Pennington, 2000, p. 265)—remain poorly sampled, reinforcing the point that this chapter presents empirical hypotheses of patterns of diversity that can be tested and updated with more sampling.

10.3.3   Diversity in relation to precipitation and geology

In the current analysis, alpha diversity shows broad correlations with both precipitation patterns and geology (Figure 10.5a–d, see color section). Total precipitation amount is positively correlated with diversity (Figure 10.5a) and average length of dry season negatively correlated (Figure 10.5d). Both reproduce the basic pattern of high average diversity in central and western Amazonia, and along the main stem of the Amazon, though with notable discrepancies. Strong gradients in alpha diversity are found across areas of high rainfall, particularly in northwest Amazonia, and, conversely, areas of low diversity are found in areas of high average annual rainfall

(Figure 10.5a). Manaus, in particular, has diversity as high as western Amazonian forests, but has much lower rainfall. High rainfall areas at the mouth of the Amazon and in Guianan forests show similar diversity to much drier forests in the western Amazon. Average dry-season length, while having its minimum at areas of highest diversity in western Amazonia, also has strong gradients in diversity across areas of short average dry-season length, particularly in northwest Amazonia (Figure 10.5c).

Total variability in monthly precipitation and diversity shows a closer correspondence to the basic patterns of alpha diversity (Figure 10.5b). Isoclines in total variability largely follow isoclines in alpha diversity, particularly in southern and northwest Amazonia. The transverse dry belt, separating central and western Amazonian forests from Guianan forests, is clearly visible in these images (Figures 10.5a–c) as an area of not only relatively low rainfall, but also an area of highly variable rainfall. Species diversity in this area, and to the east, is correspondingly lower than one would expect from a forest with similar total precipitation, but lower intra- and inter-annual variability.

Geologic age also shows broad correspondence to patterns of alpha diversity, with the highest diversity forests all falling on the Tertiary and Quaternary sediments west of Manaus, and the area of high diversity extending east along a narrowing tongue of Miocene and younger-aged Tertiary sediments (Figure 10.5d). Forests on the Proterozoic- and Archean-aged rocks of the Guianan and Brazilian Shields have relatively uniformly low diversity, even in areas of high rainfall and stable precipitation regime. Geologic age remains correlated with diversity, even after accounting for the association between total variability in precipitation and geologic age (residuals from local regression fit of total variability in precipitation and alpha diversity versus geologic age, Kruskal–Wallis $\chi^2 = 14.5509$, $df = 3$, $p = 0.002$).

Because of the relatively homogeneous Middle- to Late Cenozoic sediments, patterns of diversity in central and western Amazonia present a test of precipitation's influence on diversity while minimizing variability in geologic age. Focusing on areas $\geq 60°W$, where Amazonian tree plot density is highest, average alpha diversity of forest trees shows a broad peak from the equator south to $\sim 5°$ with a nearly linear decrease to the south and a steeper decline to the north (Figure 10.6). Looking at the maximum diversity (*sensu* Ter Steege *et al.*, 2003) one sees broadly the same pattern, the exception being a slower rate of decrease to the north of the equator. Both of these patterns correspond well to patterns of dry-season length and total precipitation variability in the southern hemisphere. North of the equator, maximum diversity follows precipitation, while average diversity decreases more rapidly than precipitation amount or either measure of precipitation variability. Another notable feature of this figure is the paucity of plots from 0 to 5°N and 6 to 10°S (see also Figure 10.2a).

Southern hemisphere vascular plant gamma (regional) diversity increases to $\sim 4°S$, then decreases slowly to $\sim 12°S$, falling off rapidly as one moves farther south. Gamma diversity remains high much farther south than alpha (local) diversity (Figure 10.6). While this trend appears to be discordant with predictions based on precipitation, it is understandable in terms of how underlying species abundances change as their ranges cross the precipitation gradient. Gamma diversity in this analysis is based on range data with a species only having to include a particular

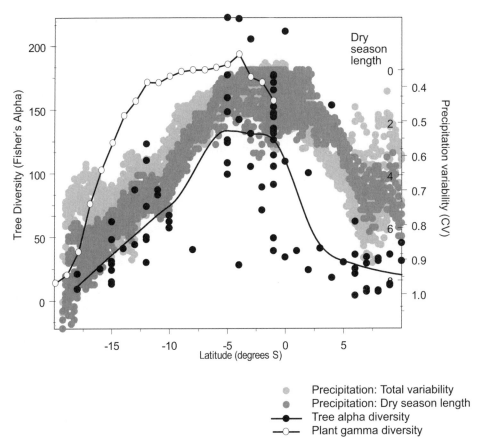

Figure 10.6. Change in species richness ($N_{species}$) with elevation in four Neotropical inventories
(Andes: Gentry, 1988; Costa Rica: Lieberman *et al.*, 1996; Mexico: Vazquez and Givnish, 1998;
western South America: Silman *et al.*, unpublished). (a) 0.1-ha diversity results for woody plants
>2.5 cm dbh. Mexican results are from a single elevational transect, while Andean plots range
from 11°N to 16°S. Mexico and the Andes show similar rates of change in diversity with
elevation (Mexico ~70 spp km^{-1}; Andes ~60 spp km^{-1}), with a distinct number of species in the
source pools (intercepts) in the two regions. (b) Results for gamma diversity (all species whose
ranges include a given elevation) in western Amazonian and Andean vascular plants with ≥10
collections (*left axis*) and tree alpha (local) diversity from a transect of 1-ha plots in Costa Rica
(*right axis*; replotted from Lieberman *et al.*, 1996, $r^2 = 0.99$).

latitude/elevation combination in its range to be counted. The species can be present in
the landscape, but be at low abundance near the edge of its range and therefore not
likely to contribute to alpha diversity (*sensu* Holt *et al.*, 1997). This influence on
diversity will be particularly true when species outliers are found in local areas of
suitable habitat outside its central range (Levin, 1995; Holt and Keitt, 2000).

10.3.4   Elevation and diversity

There are not enough hectare inventories from the Andes to look at changes in diversity with elevation, much less patterns of changes in diversity along the elevational gradient with respect to latitude, geology, and climatic variability (though see Boyle, 1996). However, data from Neotropical inventories of all woody plants ≥2.5 cm d.b.h. (diameter at breast height) in both the Andes (Gentry, 1988) and Mexico (Vazquez and Givnish, 1998) show a linear decrease in diversity with elevation (Figure 10.7a). Inventories of tree alpha diversity from 1-ha plots in Costa Rica (Lieberman *et al.*, 1996) and gamma diversity in South American collection data also show a linear decrease in species number with elevation (Figure 10.7b).

10.4   DISCUSSION

10.4.1   Precipitation and diversity

Diversity is clearly correlated with climatic stability, and the plateau in Amazonian diversity from 1°N to 5°S falls squarely on the area with no predictable dry season, and the lowest variability in climate over time (Figures 10.5b, c). Both local and regional diversity in all data sets analyzed showed this trend. Diversity may also appear anomalously low in areas with a shorter dry season as calculated through the number of months that average below 100 mm, especially if the timing of rainfall is unpredictable and that area is subject to episodic super-annual drought that may not change average monthly figures in a systematic way. The correlation of diversity with seasonality and climate variability across latitude is seen clearly in western Amazonian plots, with the close correspondence between tree alpha diversity and climate standing in stark contrast to the patterns of vertebrate species diversity in western Amazonia, which remains nearly constant to ~14°S (Mares, 1992; Stotz *et al.*, 1996; Symula *et al.*, 2003). In addition to the latitudinal pattern, the longitudinal decrease in tree diversity in Amazonia also correlates well with both seasonality and climate variability. These data suggest that species are limited by physiological tolerances to drought, with areas of constant, wet climate decreasing water-use efficiency constraints and allowing species to occupy understory light environments that would be unprofitable in dry environments (Pitman *et al.*, 2002). Comparisons of forests along climatic stability and rainfall gradients support this hypothesis, with species diversity of forests in wet areas increasing disproportionately rapidly among understory taxa, and with certain families being much more diverse in wet areas (Gentry and Emmons, 1987; Pitman *et al.*, 2002). Givnish (1999) hypothesized that stable climate would increase distance- and density-dependence caused by moisture-loving plant pathogens. Though this complementary hypothesis remains untested, it would not necessarily predict the observed increase in specific plant families, or in understory plants. Indeed, it is unclear whether Janzen–Connell effects are stronger in the tropics than in the temperate zone (HilleRisLambers *et al.*, 2002).

Previous results have shown both absolute amount of precipitation and average dry-season length to be related to diversity (Gentry, 1988; Clinebell *et al.*, 1995; Ter

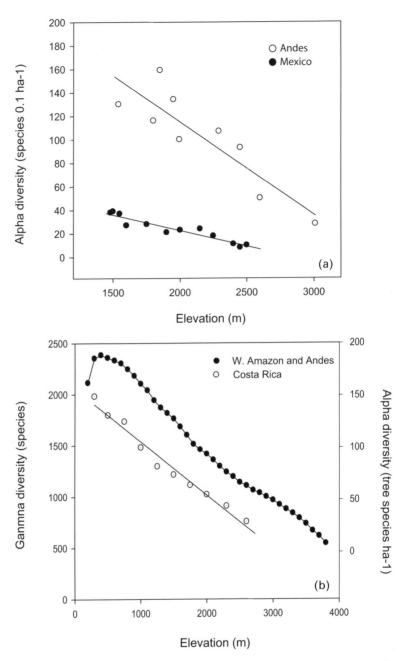

Figure 10.7. Changes in dry-season length, precipitation variability, tree alpha diversity, and vascular plant gamma diversity with latitude for western Amazonian forests. Dry-season length and precipitation variability are lowest ~2°N to ~5°S. Tree alpha diversity shows a plateau from the equator to ~5°S. Gamma diversity, measured for the southern hemisphere, peaks at ~4°S with a slow decline south to ~11°S, with a steep decline farther south.

Steege *et al.*, 2003). Gentry (1988) presented evidence that showed diversity to be strongly correlated with absolute rainfall. He noted, however, that it was likely seasonality, as evidenced by the strength of the dry season, which was the true cause of the increase in diversity, rather than precipitation *per se*. Clinebell *et al.* (1995; and others subsequently) demonstrated an inverse correlation between diversity and the length of the dry season, taken as the number of months with average rainfall below either 65 mm or 100 mm.

Because rainfall in the Amazon is variable on multiple timescales, precipitation stability may be a better indicator of moisture regime as it affects plants over their life-cycles. Places with high average monthly rainfall but subject to episodic drought may cause decreases in diversity belied by measures of climate means. Areas with the same "average" dry-season length can have a wide range of variability, from a very consistent dry season from year to year, to one that is highly variable among years (Figures 10.3a–d).

The main oceanic climate forcings on Amazonian forests—El Niño/Southern Oscillation and the North Atlantic Oscillation/South Atlantic Oscillation dipole—do not affect Amazonian forests equally (Pezzi and Cavalcanti, 2001). For example, the warm phase of ENSO causes drought in the northeastern Amazon and Bolivia, and increased raininess in the western Amazon in Ecuador and northern Peru and central and southern Brazil. Central and southern Peru and the central Amazon are transition areas for ENSO correlations, and can have either positive or negative rainfall anomalies depending on the event. The main effect of oceanic forcings on Amazonian precipitation is to change the length of the rainy season while not changing the daily rain rate (Marengo *et al.*, 2001). Because of this, the metric "dry-season length" varies across the Amazon. In areas of low variability the mean is representative of the long-term trend, while in areas of high variability the mean will over- or under-estimate dry-season length in any one year. Whether the variability affects forest structure and species composition awaits further study.

10.4.2 Geology and diversity

The current study shows that, for South American tropical forests, tree alpha diversity is also associated with geological age of substrate. While subsidiary to rainfall effects, alpha diversity tended to be higher on younger substrates, which in turn were more fertile substrates, even after accounting for precipitation variability (Figure 10.5d). This is almost certainly not an effect of age *per se*, but rather due to the correlation between age and fertility in Amazonian surface geology. This effect can be seen most clearly in Amazonas state of Venezuela and adjacent Brazil and Colombia, where forests on shield areas have much lower alpha diversity than those on adjacent Tertiary sediments, even though total precipitation and precipitation variability are similar. The result that the highest diversity forests in Amazonia are on some of the youngest substrates emphasizes the role of assembly through plant migration rather than *in situ* diversification (Wilf *et al.*, 2003; Ricklefs, 2004) (though Richardson *et al.*, 2001 present an interesting counterexample in the genus *Inga*). If tropical forests are museums of diversity, they are museums where the exhibits are constantly re-arranged.

Substrate also has a large effect on the floristic composition of Amazonian tree communities, and these effects are conserved at higher phylogenetic levels. At a local spatial scale, at our upper-Amazonian sites, tree plots can be reliably classified to floodplain or *terra firme* forest even if stems are only identified to family, and that result holds generally at localities across the Amazon Basin (e.g., Terborgh *et al.*, 1996). Terborgh and Andresen (1998) showed that at larger spatial scales, however, adjacent *terra firme* and floodplain sites were more similar to each other than either was to the same habitat type at a more distant site. Floodplain sediments by and large reflect local to regional sediment transport and suggest that trees are responding to soil characteristics. Another explanation would be that trees are highly dispersal-limited, and that floristic differences among regions reflect *in situ* evolutionary differences (Campbell, 1994). While biogeographic explanations cannot be completely discounted, collecting expeditions to areas of similar geologies, even areas separated by hundreds to thousands of kilometers, have surprisingly similar floras (Schulenberg and Awbrey, 1997; Neill, 1999; Holst, 2001; P. Núñez, unpublished). Conversely, juxtaposed areas of distinct geologies show distinct species compositions. At the southern limit of the Amazon Basin in Bolivia, rainforest in Madidi National Park shows more similarity to forests derived from Andean sediments ~2,000 km away in Ecuador than the Brazilian Shield forests at the same latitude in Noel Kempff Mercado National Park, a distance of ~500 km (Pitman *et al.*, 2001; Macia and Svenning, 2005; Silman *et al.*, 2006).

Soils and their underlying parent materials affect diversity in two ways. The first is in the total diversity at a point, or alpha diversity. Gentry (1988) presented empirical data that suggested that alpha diversity is highest on rich soils, though the difference in soil fertility was much less important than precipitation amount and seasonality. This result was borne out by Clinebell *et al.* (1995) and the results of the current study. A second way that soils and geology influence diversity is through their effects on species distributions through niche relations. In this case taxa have preferences for soil types, leading to different species composition among soil types, with this beta diversity increasing the total (gamma) diversity of a region. Thus, edaphic effects on diversity can act through ecological processes at the hectare scale, and through floristic (distributional) effects at larger spatial scales.

Changes in community composition and plot-to-plot similarity with geological substrate are well-known in both the Neotropics and Paleotropics (e.g., Duivenvoorden, 1995; Clark *et al.*, 1998; Potts *et al.*, 2002; Phillips *et al.*, 2003; Tuomisto *et al.*, 2003; Palmiotto *et al.*, 2004; Valencia *et al.*, 2004; Masse, 2005; Russo *et al.*, 2005). Tests of substrate effects will need to be carried out at multiple spatial scales because substrate geology can affect plots; both through species-level ecophysiological effects (the niche) and larger-scale and longer-term effects on local species pools. Physiological and related ecological niche effects are likely due to direct effects of nutrient status or water holding capacity on species ability to maintain a positive population growth rate on a certain substrate, or indirect effects of natural enemies mediated through physiological effects (e.g., Givnish, 1999; Fine *et al.*, 2004).

The role of the dynamic pre-Quaternary geological history of the area encompassed by modern Amazonia in generating modern patterns of plant diversity is only recently being integrated into studies of floristics and diversity (Fine *et al.*, 2005). However, phylogeographic explanations in light of the Andean orogeny and its associated effects on the continental margin and shield areas have become standard explanations of animal diversification in Amazonia. Much of the lower Amazon was under water until 2-5 Myr BP due to high Miocene and Pliocene sea levels, and western Amazonia was a sequence of depositional centers very different from today, with a mosaic of shallow lakes and seas (Potter, 1997; Kronberg *et al.*, 1998; Rossetti *et al.*, 2005). The rapid uplift of the Bolivian Andes, rising ~3 km in elevation between ~10.3 and 6.8 Myr BP (Lamb, 2004; Ghosh *et al.*, 2006), and the dynamism of the Amazonian forelands throughout western Amazonia (Räsänen *et al.*, 1992; Kronberg *et al.*, 1998; Hovikoski *et al.*, 2005; Roddaz *et al.*, 2005) certainly had profound effects on Andean and Amazonian phytogeography, though they remain unexplored.

10.4.3 Gamma diversity and range limits along environmental gradients

In the current analysis gamma diversity remains high much farther south in upper Amazonia than alpha diversity. When thought of in terms of species ranges, the result is not surprising. A general macroecological pattern is that species are more abundant, or frequently encountered, at the center of their ranges than at their periphery (Brown. 1984, 1995). Decreases in evenness in species abundances, whether due to local or regional effects, would decrease alpha diversity, even if local plant communities were assembled through random sampling from a landscape species pool. Rarer species would still occur in areas of the landscape that fit their ecological requirements, but would be included in communities less frequently, leading to lower alpha diversity. Examples of this are frequently observed when working along elevational gradients in the Andes where, on geological formations, tree taxa common in the lowlands are found infrequently over a kilometer above their "usual" range (R. Foster and P. Nuñez, pers. communs.; Silman, unpublished).

10.4.3.1 Beta diversity

The data in this chapter do not treat how habitat specificity translates patterns of local (alpha) diversity into regional (gamma) diversity. However, previous research suggests that, although substrate strongly influences floristic composition at a variety of spatial scales, it does not create a large number of absolute habitat specialists (Pitman *et al.*, 1999). If one samples enough area in *terra firme* forests, species common in other habitat types occur. This has led to an unproductive argument in the literature about the role of soils in influencing community similarity in Amazonian forests (Pitman *et al.*, 1999, 2001; Tuomisto *et al.*, 2003). The fact that species can be widely distributed and relatively abundant on different substrates is not incompatible with these same substrates causing substantial changes in community composition from site to site. Because most species occur at frequencies of one or two individuals in even the largest Amazonian inventories, both quantitative measures and presence–absence measures of community similarity can show apparently high beta diversity, while the nature of the forest from site to site remains largely similar. Understanding the effects of performance differences across climatic and edaphic gradients on species distributions, and the origins of these differences, will not be solved by correlative studies, but rather require experimental approaches (e.g., Fine *et al.*, 2004.)

10.4.4   Elevation and diversity

Gentry found that species diversity decreased nearly linearly with elevation in a small sample of 0.1-ha plots scattered at different elevations from 9°N to 22°S. This linear decrease in species diversity was also found in a comparable Central American sample taken in Mexico, and—when plotted on a log scale—shows that the rates of species loss with elevation are nearly the same in the two locales, but with the South American samples being nearly an order of magnitude more diverse (Figure 10.5a).

   Data from the Andes to test this hypothesis remain scarce. In the current study, both the collections of all Andean vascular plants and Costa Rican hectare tree inventories show a linear decrease with elevation (Figure 10.5b). We have no comprehensive explanation of the decrease in diversity with elevation, and both niche-based hypotheses and biogeographically-based hypotheses are compelling. If one makes the simple prediction that diversity is related to area, the trends will basically agree, as area decreases regularly with elevation in the Andes up to ~3,600 m. Above that, however, area increases on the broad highland plateaus of the Andes, while diversity does not. The decrease in area with increasing elevation is accentuated when allowance is made for the relatively recent occurrence of the Andean uplift. However, western Amazonian lowland landscapes were undergoing large changes in the Late Tertiary as well (Hoorn, 1994b; Potter, 1997; Costa *et al.*, 2001; Rossetti *et al.*, 2005), yet they harbor the consistently highest diversity in the Amazon Basin.

   More compelling are the large and steep environmental gradients across elevation in the Andes. Temperature changes with elevation at ~ 1°C per 175 m. Rainfall gradients along the Andean flank can be correspondingly steep. In southeastern Peru, rainfall varies from 1,700 mm yr^{-1} near the border with Bolivia and Brazil, to 6,900 mm yr^{-1} in the Andean foothills. Much of that change comes over just 20 km. That these factors and others correlated with them are known to drive plant distributions points to the importance of explanations for observed diversity centered on niche relationships and species interactions.

   Importantly, no mid-elevational bulge is apparent in either the inventories of all vascular plants, or of trees. A mid-elevational bulge in epiphytes, both vascular and non-vascular, was expected due to the reliance of many of these taxa on persistent cloud, particularly when the spectacular radiation of Andean orchids is included. Still, the absence of a mid-elevational bulge is compelling in these data, and, if it does exist and was missed by this analysis, would occur low in the foothills (<1,700 m) where local climatic conditions allow Amazonian families to coexist with what are traditionally considered Andean families.

10.4.5   Long-term climate change and Amazonian diversity: a Holocene minimum in western Amazonia?

While Amazonia conjures up images of vast lowlands, Amazonian forest extends up the east slope of the Andes, gradually losing its lowland elements. Because modern temperatures are exceptionally warm compared with most of the Quaternary, forest tree species found today on the Andean flank up to ~1,700 m were likely members of lowland Amazonian forest communities for much of the last 2 million years, with many showing large altitudinal migrations with changing climate (Colinvaux *et al.*, 1996; Bush *et al.*, 2004). The non-equilibrium model of community structure makes the prediction that increasing regional species diversity would also increase within-site diversity (Hubbell, 2001). A prediction that emerges is that diversity, at least in upper-Amazonian forests, was even higher during the cooler periods of the Quaternary when lowland floras included taxa that are currently thought of as montane. Whether the Holocene is a diversity minimum for upper-Amazonian forests awaits further study.

10.4.6   The mid-domain effect

The idea that simple geometry explains many patterns of diversity (Colwell and Lees, 2000) has been advanced for taxa as disparate as small mammals (McCain, 2004), corals (Connolly *et al.*, 2003), and Andean epiphytes (Kessler, 2001). However, this study found no evidence for the mid-domain effect in Amazonian or Andean flora. Ter Steege *et al.* (2003) also suggested that the mid-domain effect was responsible for the peak in Amazonian diversity ~5° south of the equator. Results from the current study show that this peak falls in areas where climate is wet and stable, with a high degree of predictability both within and among years. Additionally, forests in Amazonia are clearly most diverse in central and western parts, with the highest consistent diversity falling in western South America (Figure 10.2a). Geometric constraints do not predict this pattern, with environmental effects and historical explanations providing more plausible and biologically satisfying answers (Hawkins *et al.*, 2005).

10.4.7   Discounting migration

This chapter has implicitly discounted dispersal limitation as a major factor limiting the distribution of tree species in the Amazon Basin, even though it has been demonstrated for small mammals and certain bird taxa. Both theoretical data on species dispersal ranges and empirical data on the paleo-distributions of species argue that in Amazonia migration *per se* has not limited the majority of plant taxa (Clark *et al.*, 1999, 2001, 2003; Higgins *et al.*, 2003). This same conclusion has been reached for North American forests, and we expect it to hold true for South American forests as well. At the continental scale, Pennington and Dick (2004) report that up to 20% of the taxa in Amazonian forests are likely long-distance immigrants from other continents. While these results do not discount the effect of dispersal limitation as being important in some taxa, particularly autochorous taxa, it does refocus investigations of species distributions towards edaphic and climatic explanations and the way these factors interact with processes of species formation and extinction. Given recent ecophysiological results on habitat limitation, the role of longer-term climate variability might leave imprints on species distributions that become clear when looking at community level data.

10.4.8 Future efforts

While tree inventory work in the Amazon has exploded, with the installation of large numbers of tree plots and the recent publication of several excellent local floras (e.g., Vásquez Martínez, 1997; Ribeiro *et al.*, 1999), large areas of the Amazon remain uninventoried. Ter Steege *et al.* (2003) used the imprecision of their extrapolations of Amazonian diversity to suggest where floristic inventories should be focused. In more simple terms, Figure 10.2a shows that large areas of Amazonia remain uninventoried, including important geological transitions, rainfall gradients, and areas of rapid change in tree diversity. Completing Amazonian forest inventories—such as those of the Amazon Tree Diversity Network—are imperative for an overview of Amazonian diversity. In addition to siting these plots on areas of high uncertainty in diversity from the predictions of extrapolative and interpolative models, plots should be stratified to include areas of changing geology with similar climate, and similar geology with changing climates. Indeed, areas with strong or weak dry seasons can have either predictable or unpredictable climates on either a within-year or among-year basis, giving one the ability to tease apart the effects of dry-season *per se* versus climatic variability at longer timescales. Doing so would allow us to start asking questions that get at historical factors influencing tropical forest diversity.

Though we have focused on diversity–environment correlations in explaining why diversity varies among Amazonian forests, these diversity–environment correlations are underlain by ecological, physiological, and historical mechanisms (Wright, 2002; Leigh *et al.*, 2004; Ricklefs, 2004). The results of this study show that even concepts like "dry-season length," though seemingly concrete, may have complicated links to diversity depending on the degree to which modern forest community membership reflects whether species are in equilibrium with climate and substrate and are simply limited by their physiological and ecological tolerances, or whether forests bear longer-term imprints of climatological variability.

10.5 REFERENCES

Ancaya, E. J. (2000) *Diversity and Floristics of Upper-Amazonian Swamp Forests Biology* (182 pp.). Wake Forest University, Winston-Salem, NC.

ATDN (2006) *Amazon Tree Diversity Network*. Available online at *http://www.bio.uu.nl/ ∼ herba/guyana/amazon_plot_network/index.htm*

Baker, P. A., Seltzer, G. O., Fritz, S. C., Dunbar, R. B., Grove, M. J., Tapia, P. M., Cross, S. L., Rowe, H. D., and Broda, J. P. (2001) The history of South American tropical precipitation for the past 25,000 years. *Science* **291**, 640–643.

Balslev, H., Luteyn, J., Ällgaard, B., and Holm-Nielsen, L. B. (1987) Composition and structure of adjacent unflooded and floodplain forest in Amazonian Ecuador. *Opera Botanica* **92**, 37–57.

Boyle, B. F. (1996) *Changes on Altitudinal and Latitudinal Gradients in Neotropical Montane Forests*. Washington University, St. Louis.

Brown, J. H. (1984) On the relationship between abundance and distribution of species. *American Naturalist* **124**, 255–279.

Brown, J. H. (1995) *Macroecology*. University of Chicago Press, Chicago.

Bush, M. B., Silman, M. R., and Urrego, D. H. (2004) 48,000 Years of climate and forest change in a biodiversity hot spot. *Science* **303**, 827–829.

Campbell, D. G. (1994) Scale and patterns of community structure in Amazonian forests. In: P. J. Edwards, R. M. May, and N. R. Webb (eds.), *Large-scale Ecology and Conservation Biology* (pp. 179–194). Blackwell Scientific, Oxford, U.K.

Chiu, L., Liu, Z., Rui, H., and Teng, W. L. (2006) Tropical Rainfall Measuring Mission (TRMM) data and access tools. In: J. J. Qu, W. Gao, M. Kafatos, R. E. Murphy, and V. V. Salomonson (eds.), *Earth Science Satellite Remote Sensing*. Springer-Verlag and Tsinghua University Press.

Clark, D. B., Clark, D. A., and Read, J. M. (1998) Edaphic variation and the mesoscale distribution of tree species in a neotropical rain forest. *Journal of Ecology* **86**, 101–112.

Clark, J. S., Silman, M., Kern, R., Macklin, E., and HilleRisLambers, J. (1999) Seed dispersal near and far: Patterns across temperate and tropical forests. *Ecology* **80**, 1475–1494.

Clark, J. S., Lewis, M., and Horvath, L. (2001) Invasion by extremes: Population spread with variation in dispersal and reproduction. *American Naturalist* **157**, 537–554.

Clark, J. S., Lewis, M., Mclachlan, J. S., and HilleRisLambers, J. (2003) Estimating population spread: What can we forecast and how well? *Ecology* **84**, 1979–1988.

Clinebell, R. R., Phillips, O. L., Gentry, A. H., Stark, N., and Zuuring, H. (1995) Prediction of neotropical tree and liana species richness from soil and climatic data. *Biodiversity and Conservation* **4**, 56–90.

Colinvaux, P. A., Liu, K. B., De Oliveira, P., Bush, M. B., Miller, M. C., and Kannan M. S. (1996) Temperature depression in the lowland tropics in glacial times. *Climatic Change* **32**, 19–33.

Colwell, R. K. and Lees, D. C. (2000) The mid-domain effect: Geometric constraints on the geography of species richness. *Trends in Ecology and Evolution* **15**, 70–76.

Condit, R., Hubbell, S. P., Lafrankie, J. V., Sukumar, R., Manokaran, N., Foster, R. B., and Ashton, P. S. (1996) Species-area and species-individual relationships for tropical trees: A comparison of three 50-ha plots. *Journal of Ecology* **84**, 549–562.

Connolly, S. R., Bellwood, D. R., and Hughes, T. P. (2003) Indo-Pacific biodiversity of coral reefs: Deviations from a mid-domain model. *Ecology* **84**, 2178–2190.

Costa, J. B. S., Bemerguy, R. L., Hasui, Y., and Borges, M. D. (2001) Tectonics and paleo-geography along the Amazon River. *Journal of South American Earth Sciences* **14**, 335–347.

Cruz, F. W., Burns, S. J., Karmann, I., Sharp, W. D., Vuille, M., Cardoso, A. O., Ferrari, J. A., Dias, P. L. S., and Viana, O. (2005) Insolation-driven changes in atmospheric circulation over the past 116,000 years in subtropical Brazil. *Nature* **434**, 63–66.

Davis, S. D., Heywood, V. H., Herrera-Macbryde, O., Villa-Lobos, J., and Hamilton, A. (1997) *Centres of Plant Diversity: A Guide and Strategy for Their Conservation, Vol. 3: The Americas* (available online at *http://www.nmnh.si.edu/botany/projects/cpd/*). IUCN Publications Unit, Cambridge, U.K.

De Oliveira, A. A. and Daly D. C. (1999) Geographic distribution of tree species occurring in the region of Manaus, Brazil: Implications for regional diversity and conservation. *Biodiversity and Conservation* **8**, 1245–1259.

De Oliveira, A. A. and Mori S. A. (1999) A central Amazonian *terra firme* forest, I: High tree species richness on poor soils. *Biodiversity and Conservation* **8**, 1219–1244.

De Oliveira, A. A. and Nelson B. W. (2001) Floristic relationships of *terra firme* forests in the Brazilian Amazon. *Forest Ecology and Management* **146**, 169–179.

Ducke, A. and Black, G. A. (1956) Phytogeographical notes on the Brazilian Amazon. *Anais da Academia Brasileira de Ciências* **25**, 1–46.

Duivenvoorden, J. F. (1995) Tree species composition and rain forest–environment relationships in the Middle Caqueta Area, Colombia, NW Amazonia. *Vegetatio* **120**, 91–113.

Duivenvoorden, J. F. (1996) Patterns of tree species richness in rain forests of the Middle Caqueta Area, Colombia, NW Amazonia. *Biotropica* **28**, 142–158.

Fine, P. V. A., Mesones, I., and Coley, P. D. (2004) Herbivores promote habitat specialization by trees in Amazonian forests. *Science* **305**, 663–665.

Fine, P. V. A., Daly, D. C., Muñoz, G. V., Mesones, I., and Cameron K. M. (2005) The contribution of edaphic heterogeneity to the evolution and diversity of Burseraceae trees in the western Amazon. *Evolution* **59**, 1464–1478.

Fisher, R. A., Corbet, A. S., and Williams, C. B. (1943) The relation between the number of species and the number of individuals in a random sample of an animal population. *Journal of Animal Ecology* **12**, 42–57.

Gentry, A. H. (1988) Changes in plant community diversity and floristic composition on environmental and geographical gradients. *Annals of the Missouri Botanical Garden* **75**, 1–34.

Gentry, A. H. and Emmons, L. H. (1987) Geographical variation in fertility, phenology and composition of the understory of neotropical forests. *Biotropica* **19**, 216–227.

Ghosh, P., Garzione, C. N., and Eiler, J. M. (2006) Rapid uplift of the altiplano revealed through ^{13}C–^{18}O bonds in paleosol carbonates. *Science* **311**, 511–515.

Giannini, A., Chiang, J. C. H., Cane, M. A., Kushnir, Y., and Seager, R. (2001) The ENSO teleconnection to the tropical Atlantic Ocean: Contributions of remote and local SSTs to rainfall variability in the tropical Americas. *Journal of Climate* **14**, 4530–4544.

Givnish, T. J. (1999) On the causes of gradients in tropical tree diversity. *Journal of Ecology* **87**, 193–210.

Griscom, B. and Ashton, P. M. S. (2003) Bamboo control of forest succession: *Guadua sarcocarpa* in southeastern Peru. *Forest Ecology and Management* **175**, 445–454.

Hastie, T. J. and Tibshirani, R. J. (1990) *Generalized Additive Models*. Chapman & Hall, New York.

Hawkins, B. A., Diniz, J. A. F., and Weis, A. E. (2005) The mid-domain effect and diversity gradients: Is there anything to learn? *American Naturalist* **166**, E140–E143.

Henderson, A., Churchill, S. P., and Luteyn, J. L. (1991) Neotropical plant diversity. *Nature* **351**, 21–22.

Higgins, S. I., Clark, J. S., Nathan, R., Hovestadt, T., Schurr, F., Fragoso, J. M. V., Aguiar, M. R., Ribbens, E., and Lavorel, S. (2003) Forecasting plant migration rates: Managing uncertainty for risk assessment. *Journal of Ecology* **91**, 341–347.

HilleRisLambers, J., Clark, J. S., and Beckage, B. (2002) Density-dependent mortality and the latitudinal gradient in species diversity. *Nature* **417**, 732–735.

Hoch, G. and Korner, C. (2005) Growth, demography and carbon relations of *Polylepis* trees at the world's highest treeline. *Functional Ecology* **19**, 941–951.

Holst, B. K. (2001) Vegetation of an outer limestone hill in the Central–East Cordillera Vilcabamba region, Peru. In: L. E. Alonso, A. Alonso, T. S. Schulenberg, and F. Dallmeier (eds.), *Biological and Social Assessment of the Cordillera de Vilcabamba, Peru* (pp. 80–84). Conservation International, Washington, DC.

Holt, R. D. and Keitt, T. H. (2000) Alternative causes for range limits: A metapopulation perspective. *Ecology Letters* **3**, 41–47.

Holt, R. D., Lawton, J. H., Gaston, K. J., and Blackburn, T. M. (1997) On the relationship between range size and local abundance: Back to basics. *Oikos* **78**, 183–190.

Hoorn, C. (1994a) An environmental reconstruction of the Palaeo-Amazon River System (Middle–Late Miocene, NW Amazonia). *Palaeogeography, Palaeoclimatology, Palaeoecology* **112**, 187–238.

Hoorn, C. (1994b) Fluvial paleoenvironments in the Intracratonic Amazonas Basin (Early Miocene–early Middle Miocene, Colombia). *Palaeogeography, Palaeoclimatology, Palaeoecology* **109**, 1–54.

Hoorn, C., Guerrero, J., Sarmiento, G. A., and Lorente, M. A. (1995) Andean tectonics as a cause for changing drainage patterns in Miocene northern South America. *Geology* **23**, 237–240.

Hovikoski, J., Räsänen, M., Gingras, M., Roddaz, M., Brusset, S., Hermoza, W., and Pittman L. R. (2005) Miocene semidiurnal tidal rhythmites in Madre de Dios, Peru. *Geology* **33**, 177–180.

Hubbell, S. P. (2001) *The Unified Neutral Theory of Biodiversity and Biogeography* (375 pp.). Princeton University Press, Princeton, NJ.

Irion, G., Muller, J., Demello, J. N., and Junk, W. J. (1995) Quaternary Geology of the Amazonian lowland. *Geo-Marine Letters* **15**, 172–178.

Johnson, C. M., Vieira, I. C. G., Zarin, D. J., Frizano, J., and Johnson, A. H. (2001) Carbon and nutrient storage in primary and secondary forests in eastern Amazonia. *Forest Ecology and Management* **147**, 245–252.

Kessler, M. (2001) Patterns of diversity and range size of selected plant groups along an elevational transect in the Bolivian Andes. *Biodiversity and Conservation* **10**, 1897–1921.

Kronberg, B. I., Fralick, P. W., and Benchimol, R. E. (1998) Late Quaternary sedimentation and palaeohydrology in the Acre Foreland Basin, SW Amazonia. *Basin Research* **10**, 311–323.

Lamb, S. (2004) *Devil in the Mountain: A Search for the Origin of the Andes* (340 pp.). Princeton University Press, Princeton, NJ.

Leigh, E. G., Jr. (1999) *Tropical Forest Ecology: A View from Barro Colorado Island.* Oxford University Press, Oxford, U.K.

Leigh, E. G., Davidar, P., Dick, C. W., Puyravaud, J. P., Terborgh, J., Ter Steege, H., and Wright, S. J. (2004) Why do some tropical forests have so many species of trees? *Biotropica* **36**, 447–473.

Levin, D. A. (1995) Plant outliers: An ecogenetic perspective. *American Naturalist* **145**, 109–118.

Lieberman, D., Lieberman, M., Peralta, R., and Hartshorn, G. S. (1996) Tropical forest structure and composition on a large-scale altitudinal gradient in Costa Rica. *Journal of Ecology* **84**, 137–152.

Lips, J. M. and Duivenvoorden, J. F. (1996) Regional patterns of well drained upland soil differentiation in the Middle Caquetá Basin of Colombian Amazonia. *Geoderma* **72**, 219–257.

Macia, M. J. and Svenning J. C. (2005) Oligarchic dominance in western Amazonian plant communities. *Journal of Tropical Ecology* **21**, 613–626.

Malhi, Y. and Wright, J. (2004) Spatial patterns and recent trends in the climate of tropical rainforest regions. *Philosophical Transactions of the Royal Society of London, Series B: Biological Sciences* **359**, 311–329.

Malhi, Y., Baker, T. R., Phillips, O. L., Almeida, S., Alvarez, E., Arroyo, L., Chave, J., Czimczik, C. I., Di Fiore, A., Higuchi, N. *et al.* (2004) The above-ground coarse wood productivity of 104 neotropical forest plots. *Global Change Biology* **10**, 563–591.

Marengo, J. A. (2004) Interdecadal variability and trends of rainfall across the Amazon Basin. *Theoretical and Applied Climatology* **78**, 79–96.

Marengo, J. A., Liebmann, B., Kousky, V. E., Filizola, N. P., and Wainer, I. C. (2001) Onset and end of the rainy season in the Brazilian Amazon Basin. *Journal of Climate* **14**, 833–852.

Mares, M. A. (1992) Neotropical mammals and the myth of Amazonian biodiversity. *Science* **255**, 976–979.

Masse, D. (2005) The effects of distance and geomorphology on the floristic composition of lowland tropical tree communities. *Biology* (67 pp.). Wake Forest University, Winston-Salem, NC.

McCain, C. M. (2004) The mid-domain effect applied to elevational gradients: Species richness of small mammals in Costa Rica. *Journal of Biogeography* **31**, 19–31.

Myers, N., Mittermeier, R. A., Mittermeier, C. G., Da Fonseca, G. A. B., and Kent, J. (2000) Biodiversity hotspots for conservation priorities. *Nature* **403**, 853–858.

Neill, D. A. (1999) Introduction: Geography, geology, paleoclimates, climates and vegetation of Ecuador. In: P. M. Jorgensen and S. León-Yáñez (eds.), *Catalogue of the Vascular Plants of Ecuador* (pp. 2–25). *Monographs in Systematic Botany from the Missouri Botanical Garden* **75**, 1–1181

Nelson, B. W., Kapos, V., Adams, J. B., Oliveira, W. J., Braun, O. P. G., and Doamaral, I. L. (1994) Forest disturbance by large blowdowns in the Brazilian Amazon. *Ecology* **75**, 853–858.

Nimer, E. (1977). Clima. *Geografia do Brasil*, Vol. I. *Regio Norte* (pp. 39–58). IBGE, Rio de Janeiro [in Portuguese].

Palmiotto, P. A., Davies, S. J., Vogt, K. A., Ashton, M. S., Vogt, D. J., and Ashton, P. S. (2004) Soil-related habitat specialization in dipterocarp rain forest tree species in Borneo. *Journal of Ecology* **92**, 609–623.

Pennington, R. T. and Dick, C. W. (2004) The role of immigrants in the assembly of the South American rainforest tree flora. *Philosophical Transactions of the Royal Society of London, Series B: Biological Sciences* **359**, 1611–1622.

Pezzi, L. P. and Cavalcanti, I. F. A. (2001) The relative importance of ENSO and tropical Atlantic sea surface temperature anomalies for seasonal precipitation over South America: A numerical study. *Climate Dynamics* **17**, 205–212.

Phillips, O. L., Vargas, P. N., Monteagudo, A. L., Cruz, A. P., Zans, M. E. C., Sánchez, W. G., Yli-Halla, M., and Rose, S. (2003) Habitat association among Amazonian tree species: A landscape-scale approach. *Journal of Ecology* **91**, 757–775.

Pires, J. M. and Prance, G. T. (1977). The Amazon forest: A natural heritage to be preserved. In: G. T. Prance and T. S. Elias (eds.), *Extinction Is Forever: Threatened and Endangered Species of Plants in the Americas and Their Significance in Ecosystems Today and in the Future* (pp. 158–194). New York Botanical Garden, New York.

Pitman, N. C. A., Terborgh, J., Silman, M. R., and Nuñez, P. (1999) Tree species distributions in an upper Amazonian forest. *Ecology* **80**, 2651–2661.

Pitman, N. C. A., Terborgh, J. W., Silman, M. R., Nuñez, P., Neill, D. A., Ceron, C. E., Palacios, W. A., and Aulestia, M. (2001) Dominance and distribution of tree species in upper Amazonian *terra firme* forests. *Ecology* **82**, 2101–2117.

Pitman, N. C. A., Terborgh, J. W., Silman, M. R., Nuñez, P., Neill, D. A., Ceron, C. E., Palacios, W. A., and Aulestia, M. (2002) A comparison of tree species diversity in two upper Amazonian forests. *Ecology* **83**, 3210–3224.

Potter, P. E. (1997) The Mesozoic and Cenozoic paleodrainage of South America: A natural history. *Journal of South American Earth Sciences* **10**, 331–344.

Potts, M. D., Ashton, P. S., Kaufman, L. S., and Plotkin, J. B. (2002) Habitat patterns in tropical rain forests: A comparison of 105 plots in northwest Borneo. *Ecology* **83**, 2782–2797.

Räsänen, M. E., Neller, R., Salo, J., and Jungner, H. (1992) Recent and ancient fluvial deposition systems in the Amazonian Foreland Basin, Peru. *Geological Magazine* **129**, 293–306.

Ribeiro, J. E. L. S., Hopkins, M. J. G., Vicentini, A., Sothers, C. A., Costa, M. A. S., De Brito, J. M., De Souza, M. A. D., Martins, L. H. P., Lohmann, L. G., Assunção, P. A. C. L *et al.* (1999) *Guia de Identificação das plantas vasculares de uma floresta de terra-firme na Amazônia Central* (773 pp.). INPA-DFID, Manaus, Brazil [in Portuguese].

Richardson, J. E., Pennington, R. T., Pennington, T. D., and Hollingsworth, P. M. (2001) Rapid diversification of a species-rich genus of neotropical rain forest trees. *Science* **293**, 2242–2245.

Ricklefs, R. E. (2004) A comprehensive framework for global patterns in biodiversity. *Ecology Letters* **7**, 1–15.

Roddaz, M., Baby, P., Brusset, S., Hermoza, W., and Darrozes, J. M. (2005) Forebulge dynamics and environmental control in western Amazonia: The case study of the Arch of Iquitos (Peru). *Tectonophysics* **399**, 87–108.

Rosenzweig, M. L. (1995) *Species Diversity in Space and Time.* Cambridge University Press, Cambridge, U.K.

Rossetti, D. D., De Toledo, P. M., and Goes, A. M. (2005) New geological framework for Western Amazonia (Brazil) and implications for biogeography and evolution. *Quaternary Research* **63**, 78–89.

Russo, S. E., Davies, S. J., King, D. A., and Tan, S. (2005) Soil-related performance variation and distributions of tree species in a Bornean rain forest. *Journal of Ecology* **93**, 879–889.

Schenk, C. J., Viger, R. J., and Anderson, C. P. (1997) *Geologic Provinces of the South America Region.* U.S. Geological Survey, Washington, D.C.

Schulenberg, T. S. and Awbrey, K. (eds.) (1997) *The Cordillera del Condor Region of Ecuador and Peru: A Biological Assessment* (232 pp.). Conservation International, Washington, D.C.

Silman, M. R., Ancaya, E. J., and Brinson, J. (2003) Bamboo forests of western Amazonia. In: N. C. A. Pitman, R. Leite, and P. Alvarez (eds.), *Alto Purus: Biodiversity, Conservation, and Management.* Center for Tropical Conservation Press, Durham, NC.

Silman, M. R., Araujo Murakami, A., Pariamo, H., Bush, M., and Urrego, D. (2006) Changes in tree community structure at the southern limits of Amazonia: Manu and Madidi. *Ecología en Bolivia* **40**(3), 443–452.

Sioli, H. (1984) The Amazon and its main affluents: Hydrology, morphology of the river courses, and river types. In: H. Sioli (ed.), *The Amazon: Limnology and Landscape Ecology of a Mighty Tropical River and Its Basin* (pp. 127–165). Dr. W. Junk, Dordrecht, The Netherlands.

Sombroek, W. (2001) Spatial and temporal patterns of Amazon rainfall. *Ambio* **30**, 388–396.

Stotz, D., Fitzpatrick, J., Parker, T., and Moskovitz, D. (1996) *Neotropical Birds: Ecology and Conservation* (482 pp.). University of Chicago Press, Chicago.

Symula, R., Schulte, R., and Summers, K. (2003) Molecular systematics and phylogeography of Amazonian poison frogs of the genus *Dendrobates*. *Molecular Phylogenetics and Evolution* **26**, 452–475.

Ter Steege, H., Sabatier, D., Castellanos, H., Van Andel, T., Duivenvoorden, J., De Oliveira, A. A., Ek, R., Lilwah, R., Maas, P., and Mori S. (2000) An analysis of the floristic composition and diversity of Amazonian forests including those of the Guiana Shield. *Journal of Tropical Ecology* **16**, 801–828.

Ter Steege, H., Pitman, N., Sabatier, D., Castellanos, H., Van Der Hout, P., Daly, D. C., Silveira, M., Phillips, O., Vásquez, R., Van Andel, T. *et al.* (2003) A spatial model of tree alpha-diversity and tree density for the Amazon. *Biodiversity and Conservation* **12**, 2255–2277.

Terborgh, J. (1973) On the notion of favorableness in plant ecology. *American Naturalist* **107**, 481–501.

Terborgh, J. and Andresen, E. (1998) The composition of Amazonian forests: Patterns at local and regional scales. *Journal of Tropical Ecology* **14**, 645–664.

Terborgh, J., Foster, R. B., and Núñez, P. (1996) Tropical tree communities: A test of the nonequilibrium hypothesis. *Ecology* **77**, 561–567.

Thomas, W. W. (1999) Conservation and monographic research on the flora of tropical America. *Biodiversity and Conservation* **8**, 1007–1015.

Tuomisto, H., Ruokolainen, K., Poulsen, A. D., Moran, R. C., Quintana, C., Canas, G., and Celi, J. (2002) Distribution and diversity of pteridophytes and Melastomataceae along edaphic gradients in Yasuni National Park, Ecuadorian Amazonia. *Biotropica* **34**, 516–533.

Tuomisto, H., Ruokolainen K., and Yli-Halla, M. (2003) Dispersal, environment, and floristic variation of western Amazonian forests. *Science* **299**, 241–244.

Valencia, R., Foster, R. B., Villa, G., Condit, R., Svenning, J. C., Hernández, C., Romoleroux, K., Losos, E., Magard, E., and Balslev, H. (2004) Tree species distributions and local habitat variation in the Amazon: Large forest plot in eastern Ecuador. *Journal of Ecology* **92**, 214–229.

Vásquez Martínez, R. (1997) *Flórula de las Reservas Biológicas de Iquitos, Perú* (1046 pp.). Missouri Botanical Garden Press, St. Louis [in Spanish].

Vazquez, J. A. and Givnish, T. J. (1998) Altitudinal gradients in tropical forest composition, structure, and diversity in the Sierra de Manantlan. *Journal of Ecology* **86**, 999–1020.

Vuille, M. and Werner M. (2005) Stable isotopes in precipitation recording South American summer monsoon and ENSO variability: Observations and model results. *Climate Dynamics* **25**, 401–413.

Whittaker, R. H. (1972) Evolution and measurement of species diversity. *Taxon* **21**, 213–251.

Wilf, P., Cuneo, N. R., Johnson, K. R., Hicks, J. F., Wing, S. L., and Obradovich, J. D. (2003) High plant diversity in Eocene South America: Evidence from Patagonia. *Science* **300**, 122–125.

Willis, J. C. (1922) *Age and Area*. Cambridge University Press, Cambridge, U.K.

Wright, S. J. (2002) Plant diversity in tropical forests: A review of mechanisms of species coexistence. *Oecologia* **130**, 1–14.

11

Nutrient-cycling and climate change in tropical forests

M. E. McGroddy and W. L. Silver

11.1 INTRODUCTION

Increased inputs of greenhouse gases have altered the composition of the atmosphere over the past 150 years (IPCC, 2001), resulting in shifts in temperature and precipitation around the globe. The scientific community has put an enormous effort into understanding the causes of these changes, and predicting future climate and the interactions between climate and the biosphere that may moderate or accelerate current trends. Most of the research on climate change, both ongoing and predicted, has focused on boreal and north temperate ecosystems where temperature shifts are predicted to be the largest (IPCC, 2001), and deep organic soils present the potential for a strong positive feedback to climate change (Vourlitis and Oechel, 1997; Oechel *et al.*, 1998; Hobbie *et al.*, 2002).

Tropical forests play an important role in the global carbon cycle, accounting for an estimated 43% of global net primary production, and storing over 25% of the C found in forest soils (Brown and Lugo, 1982; Melillo *et al.*, 1993). There has been considerable controversy over the potential effects of elevated CO_2 and climate change on productivity and C cycling in tropical forests (Körner, 1998; Silver, 1998; Chambers *et al.*, 2001; Chambers and Silver, 2004; Clark, 2004; Körner, 2004; Cramer *et al.*, 2004; Lewis *et al.*, 2004; Ometto *et al.*, 2005). Long-term forest inventory plots in Amazonia show increasing tree biomass over the past 50 years, coupled with parallel trends in tree recruitment and, to a lesser degree, mortality. While this has been suggested to result from increased atmospheric CO_2 concentrations (Phillips *et al.*, 1998; Baker *et al.*, 2004) it is equally possible that there are other causal factors—such as changes in disturbance regime or climate variation (Chambers and Silver, 2004; Lewis *et al.*, 2004). Greater productivity, whatever the ultimate cause, could slightly moderate the rate of increase of atmospheric CO_2 (i.e., Brown *et al.*, 1993; Wang and Polglase, 1995; McKane *et al.*, 1995; van Noordwijk *et al.*, 1997; Tian *et al.*, 1998; Körner 2004). Almost all efforts to model or predict the forest response to elevated atmospheric CO_2 and climate change are qualified by the caveat that nutrient limitations may restrict the ability of vegetation to respond with increased productivity (Lynch and St. Clair, 2004). The low C use efficiency of tropical trees has led to the hypothesis that productivity is likely to be limited by nutrients and/or water (Chambers *et al.*, 2004).

In attempting to understand the effects of predicted climate change and atmospheric CO_2 enrichment on nutrient-cycling in tropical forests and their implications for the future of these ecosystems we must keep in mind that atmospheric CO_2 concentrations have doubled over the past century and mean annual temperatures in most tropical forest regions have shown a strong warming trend over the past 40 years (Hulme *et al.*, 2001; Malhi and Wright, 2004). Inter-annual and multidecadal rainfall variability in the tropics is quite substantial and patterns in rainfall are less clear than those of temperature. Current measures of nutrient dynamics reflect a system that is already responding to climate changes. Long-lived perennial plants, adapted to fairly stable climates—such as those characteristic of tropical forest regions—may have limited plasticity to respond to the changes in atmospheric composition and climate they are experiencing. Thus, nutrient cycle processes dominated by plant physiology might have a limited range of responses while those dominated by microbes may be more rapidly and effectively shifting to match current conditions. Based on data from experimental manipulations we suggest that nutrient cycles in pre-industrial tropical forests may well have reflected more C-efficient physiological strategies with less C allocated to nutrient acquisition by plants (i.e., fine root biomass, root exudates including phosphatases and organic acids, mycorrhizal associations, etc.). Carbon conservation may have resulted in higher N mineralization rates, through N-based microbial metabolism. The combination of slightly cooler temperatures and reduced C inputs to below-ground processes might have resulted in lower rates of decomposition, P mineralization, and plant nutrient uptake.

Over the next century, mean annual temperature in tropical regions is predicted to increase anywhere from 1 to 5°C (IPCC, 2001; Hulme *et al.*, 2001; Lal *et al.*, 2002; Cramer *et al.*, 2004; Table 11.1). Tropical climates are generally characterized by warm temperatures with little seasonal variation in temperature (Holdridge, 1967). Near-constant warm temperatures throughout the year are likely to result in little temperature-related stress to organisms relative to that seen in temperate and boreal ecosystems, but may also make tropical systems particularly sensitive to even small

Table 11.1. Magnitude of predicted changes in regional climates in tropical regions. Data are from Cramer *et al.* (2004), based on output from four different climate models for the period 2081–2100 and are presented as anomalies relative to the period 1969–1998.

Region	Temperature (%)	Rainfall (%)
Neo-Tropics	+3 to +5.5	−17 to +8
Africa	+3.7 to +7.7	−17 to +10
Asia	+3.1 to +5.6	+7.5 to +23

changes in climate (Townsend *et al.*, 1992; Silver, 1998). Rainfall is more variable in tropical forests, ranging from 1 to 8 m/yr. Climate change induced patterns in precipitation are difficult to predict, and may vary significantly across regions within the tropics; this is in addition to predicted changes in the volume of rainfall changes in the seasonality or magnitude of rainfall events, which may also be important on the regional scale (IPCC, 2001; Lal *et al.*, 2001, 2002; Hulme *et al.*, 2001). Changes in precipitation that lead to increased drought, drenching rains, or soil saturation are likely to have a significant impact on tropical forest form and function. Approximately 42% of tropical forests currently experience significant drought during part of the year (Brown and Lugo, 1982). In these ecosystems, litterfall, decomposition, and nutrient uptake are synchronized with the timing and quantity of rainfall (Jaramillo and Sanford, 1995; Martinez-Yrizar, 1995). Changes in the frequency and severity of drought in seasonal and aseasonal forests will feed back on the amount and distribution of above- and below-ground NPP, nutrient mineralization rates, and the frequency and severity of fires (Mueller-Dombois and Goldammer, 1990).

In this chapter we discuss the potential effects of climate change on nutrient-cycling and explore the possibility for nutrient limitation to alter ecosystem response to elevated CO_2 and climate change. We briefly review the basic attributes of tropical soils, and then focus our analysis on soil phosphorus (P) and nitrogen (N), internal ecosystem fluxes (litterfall, decomposition, plant nutrient uptake), and C and N trace gas emissions. In tropical forests on highly weathered soils, P is thought to be the primary limiting nutrient to NPP, and thus the most likely to impact ecosystem response to global changes. In contrast, N is rarely thought to be limiting, but climate change coupled with increasing anthropogenic N deposition in tropical regions could have a significant impact on gaseous N losses, N leaching, and associated cation-leaching. We focus on short-term (less than 100 yr) effects of climate change on tropical forest biogeochemical cycling. Longer-term changes are more likely to include significant shifts in species composition and forest structure (Pimm and Sugden, 1994; Condit *et al.*, 1996; Bazzaz, 1998; Hilbert *et al.*, 2001; Enquist, 2002; McLaughlin *et al.*, 2002; Chambers and Silver, 2004; Jensen, 2004) confounding the effects of nutrient availability.

11.2 TROPICAL FOREST SOILS

The warm, and generally moist climate of the tropics combined with the lack of large-scale disruptions—such as the periodic glaciations of the boreal and temperate regions—has allowed soil development to continue undisturbed for millions of years. While most soil orders are represented in the tropics (Sanchez, 1976), the most common and extensive soil orders are the highly weathered Oxisols and Ultisols (McGill and Cole, 1981; Jordan, 1985; Vitousek and Sanford, 1986). These soils are generally characterized by fine textures, low charge density in the mineral fraction, and the dominance of variable, pH-dependent charge (Sanchez, 1976; Sollins *et al.*, 1988). Changes in precipitation can stimulate shifts in pH, which in turn affect cation and anion retention in soils. For example, dry-season irrigation in a moist forest in

Panama resulted in increased permanent charge and cation retention in surface soils (Yavitt and Wright, 2002).

With advanced weathering, most of the primary minerals in soils are absent, and Fe and Al oxides and hydroxides predominate. These minerals impart properties to soils that can decrease their nutrient-holding capacity in the mineral fraction, increase organic matter storage and associated nutrient retention, and lower pH (Uehara, 1995). In highly weathered tropical soils, organic coatings on mineral surfaces control a significant proportion of nutrient availability and cation exchange capacity (Tiessen *et al.*, 1994). Thus, factors that influence rates of production and decomposition can feed back on nutrient-cycling directly through mineralization of organically bound nutrients, and indirectly through changes in nutrient retention and storage associated with soil organic matter.

Old tropical soils tend to have low, exchangeable P pools (Cross and Schlesinger, 1995). In terrestrial ecosystems, P is primarily derived from the weathering of parent material; ecosystem P pools—and especially the available P fraction—reach a maximum early in ecosystem development (Walker and Syers, 1976; Crews *et al.*, 1995). Phosphorus availability generally declines over geologic time as fresh weathering inputs diminish and losses remain constant or increase (Walker and Syers, 1976). Geochemical reactions with Fe and Al oxides provide an additional sink for P which can function as a loss from the perspective of the biota. If P becomes occluded with Fe and Al, it forms secondary minerals that may require hundreds to thousands of years to weather into plant-available forms (Walker and Syers, 1976; Tate, 1985; Stevenson and Cole, 1999). Supply of residual primary P through weathering usually occurs well below the surface layers of the soil where most of the active roots occur. The low total P pools, lack of primary mineral P, slow weathering of secondary P minerals, and low rates of P deposition provide the basis for potential P limitation to NPP in tropical forests (McGill and Cole, 1981; Galloway *et al.*, 1982; Vitousek, 1984; Vitousek and Sanford, 1986; Andreae *et al.*, 1990; Williams *et al.*, 1997).

However, the extent and importance of P limitation has recently come under scrutiny (Johnson *et al.*, 2003; Davidson *et al.*, 2004b; Chacon *et al.*, 2006), and some of the mechanisms proposed are likely to be sensitive to climate change. For example, P availability increases under short-term anaerobic conditions. Some humid tropical forest soils experience fluctuating redox in surface horizons due to rapid rates of oxygen consumption that exceed diffusive resupply (Silver *et al.*, 1999). As soil oxygen declines, oxidized Fe forms are reduced, releasing bound P and decreasing the bonding efficiency for new P (Peretyazhko and Sposito, 2005; Chacon *et al.*, 2006). Increased temperature and/or rainfall may result in more frequent anaerobic events in humid tropical soils, potentially increasing labile soil P pools (Silver *et al.* 1999). In contrast, increased drought in humid regions could enhance the proportion of oxidized Fe forms and decrease plant P availability through strong P sorption and occlusion.

Elevated CO_2 could indirectly impact soil P availability in a number of ways. Increased plant demand for P due to a CO_2 fertilization response could result in increased plant production of phosphatases and/or organic acids, resulting in increased mineralization of organically bound P. Additionally, increased plant produc-

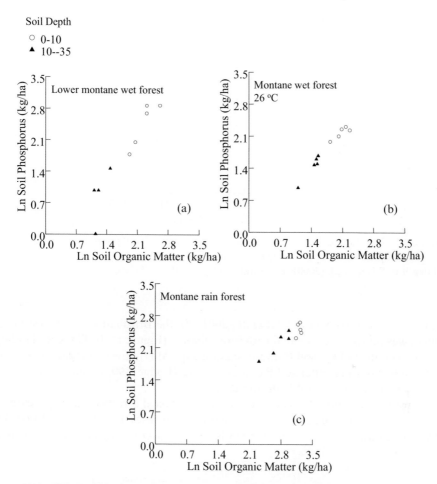

Figure 11.1. Relationships between soil organic matter content and exchangeable P in the 0–10- and 10–35-cm depths. Data are log-transformed and are from Silver *et al.* (1999). (a) For the lower montane wet forest $r_2 = 0.89$, $P < 0.01$; (b) for the montane wet forest $r_2 = 0.96$, $P < 0.01$; and (c) for the montane rain forest $r_2 = 0.82$, $P < 0.01$.

tion and soil organic matter density could result in decreased P sorption due to organic matter coating of Fe and Al oxide minerals (Lloyd *et al.*, 2001). At the landscape scale, soil organic matter was positively correlated with exchangeable P in soils along a rainfall and temperature gradient in a subtropical forest in Puerto Rico (Figure 11.1) and in a Bornean rainforest (Burghouts *et al.*, 1998). Similarly, labile P in mineral soils was strongly positively correlated with soil C pools along a localized soil texture gradient in moist tropical forest in Brazil (Figure 11.2). Elevated CO_2 or climate changes that decrease the mass of soil organic matter could indirectly impact soil P pools, and feed back to decrease NPP. It is also possible, however, that the strong correlations between P and soil organic matter along natural gradients result from other factors that affect soil P availability and in turn stimulate organic matter

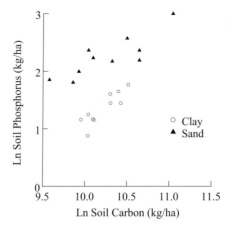

Figure 11.2. The relationship between total C and exchangeable P pools in sand and clay soils in a moist tropical forest in Brazil. Data are log-transformed and are taken from the 0–10-cm soil depth in Silver *et al.* (2000). For sandy soils, $r_2 = 0.71$, $P < 0.01$; for clay soils $r_2 = 0.77$, $P < 0.01$.

production and P content (Lloyd *et al.*, 2001). In the Brazilian forest, forest floor P content was positively correlated with forest floor C (Figure 11.3). This could indicate top-down control of the soil C and P relationship. Alternatively, fungal colonization of the forest floor may alter soil P concentrations (Lodge, 1993), leading to a narrow range of C : P ratios favorable during decay.

In contrast to many temperate ecosystems, tropical forests on highly weathered soils tend to have adequate to high N availability (Vitousek and Howarth, 1991). This is likely due both to N accumulation via N fixation over geological timescales (Riley

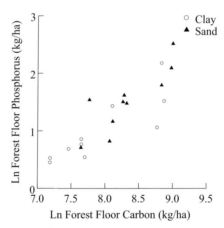

Figure 11.3. The relationship between forest floor C and forest floor P pools in sand and clay soils in a moist tropical forest in Brazil. Data are log-transformed and are from Silver *et al.* (2000). For sandy and clay soils combined, $r_2 = 0.71$, $P < 0.01$.

and Vitousek, 1995; Crews *et al.*, 2001), and internal N retention mechanisms that limit losses (Silver *et al.*, 2001, 2005b). Even when it is abundant, N-cycling in tropical forests can impact the availability of other nutrient cations and anions. Variable charge soils in high rainfall environments can experience ammonium and associated anion-leaching (Sollins *et al.*, 1988). High nitrification rates coupled with abundant rainfall can stimulate nitrate and associated cation-leaching from tropical soils (Silver and Vogt, 1993).

Factors that affect decomposition (see below) are likely to influence N mineralization rates. In temperate and boreal ecosystems, net N mineralization rates tend to increase with mean annual temperature and moisture, and generally increase with *in situ* warming (Rustad *et al.*, 2001). There have been few studies looking at soil N cycling in tropical forests in response to changes in temperature or moisture. In laboratory studies, net N mineralization rates were insensitive to temperature changes along a tropical montane forest elevation gradient (Marrs *et al.*, 1988), as were soils from a Colombian cloud forest (Cavelier *et al.*, 2000). In a recent review, Silver *et al.* (2006) found no apparent effect of temperature on net or gross N mineralization and nitrification rates along tropical elevation and temperature gradients.

Nitrogen-cycling is likely to be sensitive to changes in precipitation, particularly at the arid or very wet extremes. In dry tropical forests, pulses of precipitation can result in rapid rates of N mineralization, nitrification, and nitric oxide emissions (Davidson *et al.*, 1993; Lodge *et al.*, 1994). In wet forests, increased precipitation can enhance element-leaching (Sollins and Radulovich, 1988), lower soil redox (Silver *et al.*, 1999), and lead to increased N losses via denitrification (Silver *et al.*, 2001). If increased precipitation lowers soil redox it could also decrease rates of nitrification, a strictly aerobic process, providing a potential negative feedback to cation-leaching and gaseous N losses.

Agricultural and urban development in the tropics is dramatically increasing the rate of N deposition to tropical ecosystems (Galloway *et al.*, 1994, 1995). Current N deposition theory argues that N-rich environments should experience large and rapid increases in N losses with N deposition (Aber *et al.*, 1989; Matson *et al.*, 1999). The interaction of N deposition and climate change, particularly increased duration or intensity of rainfall, could significantly increase nutrient losses from these ecosystems (Figure 11.4). Nitrogen fertilization in N-rich Hawaiian forests led to increased nitric oxide (NO) and nitrous oxide (N_2O) fluxes (Hall and Matson, 1999), and stimulated N-leaching (Lohse and Matson, 2005). Increased N-leaching could enhance cation and anion losses, leading to nutrient limitation to NPP. Nitrogen-leaching also pollutes streams and groundwater sources (Vitousek *et al.*, 1997). The interactions of N deposition, elevated CO_2, and climate change are poorly understood for tropical forests. This clearly should be a high priority for future research.

11.3 LITTER INPUTS

Alterations to the amount and timing of rainfall in tropical forests may significantly affect nutrient-cycling via litter production, although the relationship between climate

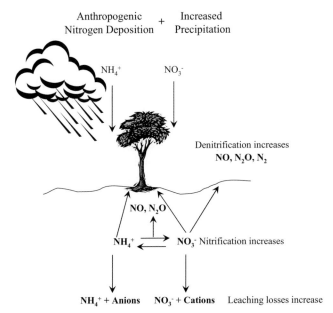

Figure 11.4. The potential effects of anthropogenic nitrogen deposition and increased precipitation in tropical forests. Increased inputs of ammonium could stimulate ammonium-leaching in variable charge soils, and lead to associated anion losses. Ammonium deposition could stimulate nitrification, enhancing nitric and nitrous oxide fluxes. Increased nitrate pools via nitrification or direct deposition could stimulate nitrate- and cation-leaching and denitrification. If soil redox declines with climate change, then nitrification rates could decrease, partially offsetting the effects of nitrogen deposition.

and litter dynamics is likely to be complex. Litter P concentrations have been found to correlate positively with rainfall seasonality and with inter-annual rainfall in moist and dry tropical rainforests (Read and Lawrence, 2003; Wood et al., 2005), but not spatially along larger-scale elevation and rainfall gradients (Silver, 1998). The observed increase in litterfall P content during the wet season or wet years in seasonal environments may be due to increased soil P availability and/or decreased demand for the retranslocation of P during leaf senescence (Wood et al., 2005). Similarly, there may be significant changes in plant P demand associated with seasonal phenological activity (Lal et al., 2001). Although foliar N may increase per unit leaf mass with increasing precipitation (Wright et al., 2001; Wright and Westoby, 2002; Santiago and Mulkey, 2005), leaf litter N concentrations do not appear to vary with precipitation in moist (Santiago and Mulkey, 2005; Read and Lawrence, 2003) or humid (Silver, 1998; Wood et al., 2005) tropical forests. Elevated atmospheric CO_2 has also been shown in some cases to increase the litter C:N and C:P ratios, though again the effect is not consistent (Kanowski, 2001; Santiago and Mulkey, 2005).

Climate change impacts on litter inputs may also come from shifts in timing or amount of litterfall. In seasonal tropical forests, litterfall and nutrient uptake are synchronized to the annual patterns in precipitation (Jaramillo and Sanford, 1995).

Seasonality of litterfall has been shown to be negatively correlated to mean annual precipitation in gradient studies (Read and Lawrence 2003; Santiago and Mulkey, 2005). Shifts in the duration of the dry season may lead to temporal separation between plant demands and nutrient availability (Silver, 1998). Increased duration and severity of droughts may also lead to an increase in drought deciduousness, or a decrease in the leaf area index of the canopy (Nepstad *et al.*, 2002). This could decrease litter inputs and/or disrupt the synchronicity between nutrient inputs and plant demand (Lodge *et al.*, 1994). Over the long term, a shift to a significantly wetter or drier environment could lead to a species composition shift with associated changes in both litter quality and quantity (Condit, 1998; Santiago *et al.*, 2005).

Below-ground litter inputs are difficult to study and few data are available. Elevated CO_2 may increase allocation to root tissues in tropical forests if nutrients are limiting (Arnone and Körner, 1995), although this could be offset by increased root mortality and turnover. Thus far, the data from field and greenhouse experiments are very mixed (Norby and Jackson, 2000). At a global scale, there were no strong relationships between root turnover (below-ground NPP/fine root standing stocks) and temperature or precipitation in forested ecosystems (Gill and Jackson, 2000). In temperate forests, fine root growth cycles appear to be regulated by temperature, resulting in strong annual signals of productivity and mortality (Pregitzer *et al.*, 2000). In contrast, tropical soils experience little variation in soil temperature seasonally, suggesting that any patterns in turnover are more likely to be dominantly controlled by soil moisture, nutrient supply, or internal regulation of root : shoot ratios. Dry-season irrigation did not change overall root phenology in a moist tropical forest, but did alter the timing of root growth and mortality by increasing the longevity of new roots while simultaneously increasing the mortality of older roots (Yavitt and Wright, 2001).

11.4   DECOMPOSITION

Actual evapotranspiration is one of the strongest predictors of decomposition on a global scale (Aerts, 1997), although its explanatory power is quite weak ($r_2 = 0.14$). Regionally, the relationship between rainfall and decomposition appears even weaker. Using elevation gradient studies within tropical forests, Silver (1998) found no predictive relationship between rainfall and decomposition rates. A 50% reduction in precipitation reaching the forest floor had no effect on litter decomposition rates in a partial throughfall exclusion experiment in a moist forest in Amazonia (Nepstad *et al.*, 2002), and dry-season irrigation resulted in only a small increase in decomposition rates of the forest floor in a moist forest in Panama (Wieder and Wright, 1995). In Hawaii, leaves decayed faster in moist forests than in wet forests (Schuur, 2001), but a common substrate showed a weaker trend, suggesting an important interaction of plant characteristics with climate or site conditions. Given this, the direct effects of either an increase or a decrease in mean annual precipitation on decomposition would probably be quite small. There are, however, indirect effects of climate change that might significantly affect the rate of both decomposition and mineralization of essential nutrients. In tropical sites, litter chemistry and in particular $C:P$, lignin$:P$, $C:N$, and lignin$:N$ ratios are often inversely related to decomposition rates (Ostertag and Hobbie, 1999; Hobbie and Vitousek, 2000). As discussed above, both decreased rainfall and/or elevated atmospheric CO_2 might lead to increased $C:$ nutrient ratios in litter. Significant immobilization of nutrients during the early stages of decomposition is commonly observed in tropical forest ecosystems (Ostertag and Hobbie, 1999; McGroddy et al., 2004; but see Kitayama et al., 2004). Increased $C:$ nutrient ratios in leaf litter could increase immobilization of nutrients in the microbial biomass during decomposition, could lower mineralization rates, and could reduce plant available nutrient pools.

Below-ground decomposition appears to be less sensitive to climate factors than above-ground litter, with tissue quality playing a key role in regulating the rate of decay (Silver and Miya, 2001). Matamala and Schlesinger (2000) found no effect of elevated CO_2 on root decomposition or tissue quality in a young temperate loblolly pine forest, though previous studies have found decreased root tissue N content under elevated CO_2 treatments (Crookshanks et al., 1998).

11.5   NUTRIENT UPTAKE

Not surprisingly, the impact of climate change on below-ground ecology has received far less attention than above-ground effects. In order to stimulate more research, Norby and Jackson (2000) produced an excellent review of the data currently available on root responses to elevated CO_2 and climate changes. Plant uptake of essential nutrients is highly sensitive to soil temperature, moisture, and aeration, and thus is likely to be affected by climate change, but measured changes in nutrient uptake kinetics have not been strongly correlated with changes in productivity (Chapin, 1974; Bassirirad, 2000). The rate of plant nutrient uptake depends on the amount of active root surface, the movement of nutrients to the active root surface, nutrient availability, and the transport rate across membranes into both the root itself and finally into the xylem system. All of these factors may respond to climate change in ways that magnify or mitigate the impacts.

Tropical forests are often characterized by high below-ground biomass (Jackson et al., 1996) and root turnover (Gill and Jackson, 2000; Trumbore et al., 2006). Soil temperature and soil moisture both have the potential to affect root architecture and root growth (Gill and Jackson, 2000), and can be positively correlated to fine root length and root uptake (Bassirirad, 2000; Pregitzer et al., 2000), though field studies have found phenological and other constraints often weaken these relationships (Fitter et al., 1998). Under elevated CO_2 some species increased their below-ground biomass, suggesting that the additional available C is used to increase fine root volume and nutrient uptake, particularly in deciduous forests (Norby et al., 1999). Other studies, however, have shown no effect or increased root mortality, resulting in higher turnover of fine root pools and no measurable change in stocks (Arnone et al., 2000; Tingey et al., 2000; Pregitzer et al., 2000). Based on work with crop plants and the regulatory role of simple sugars on cell division, Pritchard and

Rogers (2000) suggested that under elevated CO_2 roots will be larger, more highly branched, but less efficient at nutrient and water uptake. They go on to suggest that this might amplify the impact of warmer, drier soil conditions predicted for some tropical regions—such as equatorial Africa.

Fine roots are only one component of plant uptake, and for P uptake, in particular, mycorrhizal associations are vital components (Bolan, 1991; Miyasaka and Habte, 2001). Mycorrhizal hyphae increase the amount of soil volume explored and exude phosphatases and organic acids to increase P uptake (Bolan, 1991). Mycorrhizal associations are ubiquitous in tropical forest soils, where up to 90% of tree species are thought to support associations with endomycorrhizae (Bolan, 1991). If elevated CO_2 increased C allocation below ground and, in particular, to mycorrhizal associations, this could in turn increase plant nutrient uptake. Studies thus far, however, have not found support for this (Fitter et al., 2000; Zak et al., 2000a; Gavito et al., 2003; Staddon et al., 2004). It appears that elevated CO_2 on its own has little or no impact on mycorrhizal infection rates or production of mycorrhizal tissue, at least under experimental conditions (Staddon and Fitter, 1998). Mycorrhizae do appear to respond positively to increased soil temperature, though most studies have been done on temperature ranges more typical of temperate regions (Braunberger et al., 1997; Fitter et al., 2000; Staddon et al., 2004), and it is not clear if the relationship will continue to hold at the warmer temperatures typical of the tropics. Mycorrhizal infections are inversely related to soil P availability (Janos, 1983), and if plant P demand were to increase it could result in increased mycorrhizal infection rates. Responses of mycorrhizae to changes in soil temperature and moisture often appear to be species-specific; thus, any shift in climate that results in a significant shift in mycorrhizal community composition has potentially wide-reaching implications for nutrient cycles (Fitter et al., 2000).

The nutrients held in the microbial biomass represent a small but very rapidly cycled pool. Though soil microbial communities are widely considered to be C-limited, recent work in highly weathered tropical sites suggests that microbial activity may be limited by P availability or co-limited by C and P (Cleveland et al., 2002). As with plants, it is likely that microbial communities in highly weathered tropical soils are more limited by P than by N (Vitousek and Matson, 1988). The rate of mineralization of organically bound nutrients is closely linked to the rate of microbial activity and turnover of the microbial biomass. Field studies have found slightly positive, but highly variable responses in microbial biomass or microbial respiration to elevated atmospheric CO_2 in temperate forests (Zak et al., 2000b). Under low-nutrient conditions, microbial immobilization of limiting nutrients increased in response to elevated CO_2 (Berntson and Bazzaz, 1997, 1998). This could lead to significant negative feedbacks to productivity in nutrient-limited ecosystems.

Drought may also strongly affect microbial biomass and activity including nutrient immobilization and mineralization. In seasonal tropical forests, microbial biomass and P pools have been found to increase in the dry season, presumably in response to decreased plant uptake (McGroddy et al., submitted). As discussed above for litter inputs, there is a synchronous flush of nutrients released from the microbial biomass at the onset of the wet season in these seasonal forests, coinciding with the onset of new root growth and increased plant demand (Perrott *et al.*, 1990). In contrast, aseasonal tropical forests do not show intra-annual variations in microbial P pools (Yavitt and Wright, 1996; Luizão *et al.*, 1992). Shifts in the timing and length of dry seasons in seasonal forests and the introduction of drought into aseasonal forests could alter the competitive balance between microbial biomass and plant demand for nutrients. Drought or shifts in temperature can also lead to a shift in microbial community composition and function (Papatheodorou *et al.*, 2004; Sowerby *et al.*, 2005) with potential long-term effects on rates of nutrient-cycling and productivity. Several studies have shown the direct effects of elevated CO_2 on the composition and function of the microbial community, presumably through increased or altered carbon exudates into the rhizosphere, though this response may be moderated by soil nutrient or C availability (Zak *et al.*, 2000a).

11.6   TRACE GAS EMISSIONS

Tropical forests are important sources and sinks of greenhouse gases, particularly CO_2, N_2O, and methane (CH_4). The high NPP typical of moist and humid tropical forests is a significant component of the global C cycle (Melillo *et al.*, 1993; Clark *et al.*, 2003). Tropical forests also have the highest rates of soil respiration, which releases CO_2 back to the atmosphere (Raich and Schlesinger, 1992). Drought, fire, clearing, and disease—all factors that rapidly increase rates of tree mortality—can alter microbial activity or community structure, resulting in pulses of CO_2 and other greenhouse gases (Schimel and Gulledge, 1998). Methane is produced under anaerobic conditions and tropical forests have generally been considered a weak sink (Keller and Reiners, 1994; Steudler *et al.*, 1996). However, upland tropical forest soils have recently been identified as a significant source of CH_4 at local (Keller *et al.*, 1986; Silver *et al.*, 1999; Teh *et al.*, 2005) and regional (Frankenberg *et al.*, 2005) scales.

Humid tropical forests are the largest natural source of N_2O globally (Lashof and Ahuja, 1990). Nitrous oxide is produced via nitrification and denitrification. Factors that stimulate nitrification—such as the death of N-rich tissues and anthropogenic N deposition—can increase N_2O emissions (Hall and Matson, 1999; Silver *et al.*, 2005a); similarly, if soils become reduced, denitrification rates to N_2O and dinitrogen (N_2) may increase (Firestone *et al.*, 1980; Silver *et al.*, 2001). In the humid tropics, denitrification is likely to be the dominant source of N_2O emissions from soils. The potential effects of climate change on the ratio of $N_2O : N_2$ are poorly understood. Theory suggests that denitrification to N_2 is favored under low-NO_3^- and low-redox conditions, so it is possible that if increased rainfall and temperature lead to more strongly reducing conditions less NO_3^- will be available for denitrification and N_2 production will be favored. It is extremely difficult to accurately estimate N_2 fluxes from ecosystems, although this is an active area of research.

There have been few controlled experiments looking at the effects of climate changes on greenhouse gas production in tropical forests. Throughfall exclusion experiments—mimicking enhanced drought conditions—decreased N_2O emissions from seasonal forests in tropical Brazil (Cattanio *et al.*, 2002; Nepstad *et al.*, 2002;

Table 11.2. Summary of potential effects of climate change on nutrient cycling in tropical forests.

	Increased temperature	Altered rainfall (+/−)	Other effects	Research directions
Soil nutrient availability	Increase in mineralization rates for both N and P	+Increased P availability due to shift in redox conditions, increased N losses via nitrification, denitrification, and leaching	Increased soil organic matter due as a result of elevated CO_2 could increase nutrient retention capacity and organic coatings could decrease P sorption capacity	What is the impact of increases in nutrient deposition due to urbanization and intensification of agriculture in addition to climate shifts and elevated CO_2 in these systems?
Litter nutrient inputs and decomposition		Data show both + and − correlations between rainfall and litter nutrient concentrations Shifts in seasonality could decouple decomposition of litter from nutrient demand	Some evidence for increased C:nutrient ratios under elevated CO_2 potentially leading to decreased decomposition rates and/or increased nutrient immobilization during decomposition	What are the effects of climate change on below-ground nutrient inputs and decomposition?
Nutrient uptake	Root uptake kinetics are positively related to soil temperature. Mycorrhizal infection appears to be positively correlated to soil temperature	Stronger drought/dry-season dynamics may reduce fine root and microbial biomass in surface soils, thus reducing nutrient uptake	Shifts in relative and absolute below-ground plant biomass under elevated CO_2 appear to be species-specific, thus hard to predict in a diverse tropical forest	Are changes in nutrient uptake dynamics translated into shifts in net productivity?

(continued)

Table 11.2 (*cont.*)

	Increased temperature	Altered rainfall (+/−)	Other effects	Research directions
Trace gas emissions		N_2O emissions are sensitive to soil moisture conditions and have been found to positively correlate to rainfall. Under wet conditions upland tropical forests can switch from a slight sink to a net source of CH_4		Will trace gas dynamics in tropical forests under wetter and warmer conditions lead to positive feedbacks for the global climate?

Davidson *et al.*, 2004a). Similar results were found during an induced drought in a tropical megacosm experiment (van Haren *et al.*, 2005). Observational data along a tropical montane rainfall gradient showed increased N_2O emissions with increasing rainfall (Keller *et al.*, 1986) and decreased soil O_2 availability (Silver *et al.*, 1999).

11.7 SUMMARY AND FUTURE DIRECTIONS

In summary, nutrient-cycling in tropical forests is likely to be sensitive to current and future global changes, particularly changes in the amount and distribution of precipitation (Table 11.2). Existing research has focused on the effects of increased drought, which impacts P- and N-cycling primarily through changes in microbial processes and plant litter dynamics. From this review it is clear that predictions of the response of tropical forests to climate change and elevated atmospheric CO_2 are based primarily on extrapolation from other regions, observational changes along climate gradients, and a few direct experiments. Extrapolation from other regions—particularly, north temperate and boreal ecosystems—can be problematic because tropical forests: occur on highly weathered, P-limited soils; support very diverse microbial, plant, and animal communities; and have high mean annual temperatures with low temporal variability. Similarly, while climate gradient research can provide insights into systematic changes across plant communities and habitats (Silver, 1998; Schuur, 2001), they are generally poorly suited to explore climate changes within a given community or site. There have been some significant strides forward with the establishment of large-scale *in situ* manipulations in the tropics such as the throughfall exclusion experiment in eastern Amazonia (Nepstad *et al.*, 2002), and the irrigation of a seasonally dry forest in Panama (Wright, 1992; Yavitt and Wright, 1996). Clearly, more experimentation is needed to determine the sensitivity of tropical forest flora, fauna, and biogeochemical cycles to elevated CO_2 and climate change.

Future research should be designed to capture ecosystem responses within a realistic range of temperature and/or precipitation change, and over a long enough time interval to determine whether self-regulation or equilibration to a new condition occurs. As mentioned above, tropical regions are characterized by mean annual temperatures near the biological optimum. Under these conditions a small shift in temperature may have a stronger or different effect than a large increase in ecosystems that typically experience wide seasonal shifts in temperature. Collaborative research should be a key component of future climate change experiments, to better understand the linkages among population, community, and ecosystem processes. Consideration of other human impacts on the environment, such as land-use change, anthropogenic inputs such as N deposition, and invasive species will also improve our ability to predict future conditions.

11.8   REFERENCES

Aber, J. D., Nadelhoffer, K. J., Steudler, P., and Melillo, J. M. (1989) Nitrogen saturation in northern forest ecosystems. *BioScience* **39**, 378–386.

Aerts, R. (1997) Climate, leaf litter chemistry, and leaf litter decomposition in terrestrial ecosystems: A triangular relationship. *OIKOS* **79**, 439–449.

Andreae, M. O., Talbot, R. W., Beresheim, H., and Beecher, K. M. (1990) Precipitation chemistry in Central Amazonia. *Journal of Geophysical Research* **95**, 16987–16999.

Arnone III, J. A. and Körner, C. (1995) Soil and biomass carbon pools in model communities of tropical plants under elevated CO_2. *Oecologia* **104**, 61–71.

Arnone, J. A., Zaller, J. G., and Spehn, E. M. (2000) Dynamics of root systems in native grasslands: Effects of elevated atmospheric CO_2. *New Phytologist* **147**, 73–86.

Baker, T., Phillips, O. L., Malhi, Y., Almeida, S., Arroyo, L., Di Fiore, A., Erwin, T., Higuchi, N., Killeen, T. J., Laurance, S. G. *et al.* (2004) Increasing biomass in Amazonian forest plots. *Philosophical Transactions of the Royal Society of London, Series B: Biological Sciences* **359**, 353–365.

Bassirirad, H. (2000) Kinetics of nutrient uptake by roots: Responses to global change. *New Phytologist* **147**, 155–169.

Bazzaz, F. A. (1998) Tropical forests in a future climate: Changes in biological diversity and impact on the global carbon cycle. *Climatic Change* **39**, 317–336.

Berntson, G. and Bazzaz, F. (1997) Nitrogen cycling in microcosms of yellow birch exposed to elevated CO_2: Simultaneous positive and negative below-ground feedbacks. *Global Change Biology* **3**, 247–258.

Berntson, G. and Bazzaz, F. (1998) Regenerating temperate forest mesocosms in elevated CO_2: Belowground growth and nitrogen cycling. *Oecologia* **113**, 115–125.

Bolan, N. S. (1991) A critical review on the role of mycorrhizal fungi in the uptake of phosphorus by plants. *Plant and Soil* **134**, 189–207.

Braunberger, P. G., Abbott, L. K., and Robson, A. D. (1997) The effect of rain in the dry-season on the formation of vesicular–arbuscular mycorrhizas in the growing season of annual clover-based pastures. *New Phytologist* **127**, 107–114.

Brown, S. and Lugo, A. E. (1982) The storage and production of organic matter in tropical forests and their role in the global carbon cycle. *Biotropica* **14**, 161–187.

Brown, S., Hall, C. A. S., Knabe, W., Raich, J., Trexler, M. C., and Woomer, P. (1993) Tropical forests: Their past, present and potential future in the terrestrial carbon budget. *Water Air and Soil Pollution* **70**, 71–94.

Burghouts, T. B. A., van Straalen, N. M., and Bruijnzeel, L. A. (1998) Spatial heterogeneity of element and litter turnover on a Bornean rain forest. *Journal of Tropical Ecology* **14**, 477–506.

Cattanio, J. H., Davidson, E. A., Nepstad, D. C., Verchot, L. V., and Ackerman, I. L. (2002) Unexpected results of a pilot throughfall exclusion experiment on soil emissions of CO_2, CH_4, N_2O, and NO in eastern Amazonia. *Biology and Fertility of Soils* **36**, 102–108.

Cavelier, J., Tanner, E., and Santamaria, J. (2000) Effect of water, temperature and fertilizers on soil nitrogen net transformations and tree growth in an elfin cloud forest of Colombia. *Journal of Tropical Ecology* **16**, 83–99.

Chacon, N., Silver, W. L., Dubinsky, E. A., and Cusack, D. F. (2006) Iron reduction and soil phosphorus solubilization in humid tropical forest soils: The roles of labile carbon pools and an electron shuttle compound. *Biogeochemistry* **78**, 67–84.

Chambers, J. Q. and Silver, W. L. (2004) Some aspects of ecophysiological and biogeochemical responses of tropical forests to atmospheric change. *Philosophical Transactions of the Royal Society of London, Series B: Biological Sciences* **359**, 463–476.

Chambers, J. Q., Higuchi N., Tribuzy, E. S., and Trumbore, S. E. (2001) Carbon sink for a century. *Nature* **410**, 429.

Chambers, J. Q., Tribuzy, E. S., Toledo, L. C., Crispim, B. F., Higuchi, N., dos Santos, J., Araújo, A. C., Kruijt, B., Nobre, A. D., and Trumbore, S. E. (2004) Respiration from a tropical forest ecosystem: Partitioning of sources and low carbon use efficiency. *Ecological Applications* **14**, S72–S88.

Chapin, F. S. (1974) Phosphate absorption capacity and acclimation potential in plants along a latitudinal gradient. *Science* **183**, 521–523.

Clark, D. A. (2004) Tropical forests and global warming: Slowing it down or speeding it up? *Frontiers in Ecology and the Environment* **2**, 73–80.

Clark, D. A., Piper, S. C., Keeling, C. D., and Clark, D. B. (2003) Tropical rain forest tree growth and atmospheric carbon dynamics linked to interannual temperature variation during 1984–2000. *Proceedings of the National Academy of Sciences* **100**, 5852–5857.

Cleveland, C. C., Townsend, A. R., and Schmidt, S. K. (2002) Phosphorus limitation of microbial processes in moist tropical forests: Evidence from short-term laboratory incubations and field studies. *Ecosystems* **5**, 680–691.

Condit, R. (1998) Ecological implications of changes in drought patterns: Shifts in forest composition in Panama. *Climatic Change* **39**, 413–427.

Condit, R., Hubbell, S. P., and Foster, R. B. (1996) Changes in a tropical forest with a shifting climate: Results from a 50 hectare permanent census plot at Barro Colorado Island in Panama. *Journal of Tropical Ecology* **12**, 231–256.

Cramer, W., Bondeau, A., Schaphoff, S., Lucht, W., Smith, B., and Stich, S. (2004) Tropical forests and the global carbon cycle: Impacts of atmospheric carbon dioxide, climate change and rate of deforestation. *Phil. Trans. R. Soc. London B* **359**, 331–343.

Crews, T. E., Kitayama, K., Fownes, J. H., Riley, R. H., Herbert, D. A., Mueller-Dombois, D., and Vitousek, P. M. (1995) Changes in soil phosphorus fractions and ecosystem dynamics across a long chronosequence in Hawaii. *Ecology* **76**, 1407–1424.

Crews, T. E., Kurina, L. M., and Vitousek, P. M. (2001). Organic matter and nitrogen accumulation and nitrogen fixation during early ecosystem development in Hawaii. *Biogeochemistry* **52**, 259–279.

Crookshanks, M., Taylor, G., and Broadmeadow, M. (1998) Elevated CO_2 and tree root growth: Contrasting responses in *Fraxinus excelsior*, *Quercus petraea* and *Pinus sylvestris*. *New Phytologist* **138**, 241–250.

Cross, A. F. and Schlesinger, W. H. (1995) A literature review and evaluation of the Hedley fractionation: Applications to the biogeochemical cycle of soil phosphorus in natural ecosystems. *Geoderma* **64**, 197–214.

Davidson, E. A., Matson, P. A., Vitousek, P. M., Riley, R., Dunkin, K., Garcia-Mendez, G., and Maass, J. M. (1993) Processes regulating soil emissions of NO and NO₂ in a seasonally dry tropical forest. *Ecology* **74**, 130–139.

Davidson, E. A., Ishida, F. Y., and Nepstad, D. C. (2004a) Effects of an experimental drought on soil emissions of carbon dioxide, methane, nitrous oxide, and nitric oxide in a moist tropical forest. *Global Change Biology* **10**, 718–730.

Davidson, E. A., de Carvalho, C. J. R. Viera, I. C. G., Figueiredo, R. de O., Moutinho, P., Ishida, F. Y., dos Santos, M. T. P., Guerrero, J. B., Kalif, K., and Sabá, R. T. (2004b) Nitrogen and phosphorus limitation of biomass growth in tropical secondary forest. *Ecological Applications* **14**, S150–S163.

Enquist, C. A. F. (2002) Predicted regional impacts of climate change on the geographical distribution and diversity of tropical forests in Costa Rica. *Journal of Biogeography* **29**, 519–534.

Firestone, M. K., Firestone, R. B., and Tiedje, J. M. (1980) Nitrous-oxide from soil denitrification: Factors controlling its biological production. *Science* **208**, 749–751.

Fitter, A. H., Graves, J. D., Self, G. K., Brown, T. K., Bogie, D. S., and Taylor, K. (1998) Root production, turnover and respiration under two grassland types along an altitudinal gradient: Influence of temperature and solar radiation. *Oecologia* **114**, 20–30.

Fitter, A. H., Heinemeyer, A., and Staddon, P. L. (2000) The impact of elevated CO₂ and global climate change on arbuscular mycorrhizas: A myocentric approach. *New Phytologist* **147**, 179–187.

Frankenberg, C., Meirink, J. F., van Weele, M., Platt, U., and Wagner, T. (2005) Assessing methane emissions from global space-borne observations. *Science* **308**, 1010–1014.

Galloway, J. N., Likens, G. E., Keene, W. C., and Miller, J. M. (1982) The composition of precipitation in remote areas of the world. *Journal of Geophysical Research* **87**, 8771–8786.

Galloway, J. N., Levy, H., and Kashibhatla, P. S. (1994) Year 2020: Consequences of population growth and development on deposition of oxidized nitrogen. *Ambio* **23**, 120–123.

Galloway, J. N., Schlesinger, W. H., Levy II, H., Michaels, A., and Schnoor, J. L. (1995) Nitrogen fixation: Anthropogenic enhancement–environmental response. *Global Biogeochemical Cycles* **9**, 235–252.

Gavito, M. E., Schweiger, P., and Jakobsen, I. (2003) P uptake by arbuscular mycorrhizal hyphae: Effect of soil temperature and atmospheric CO₂ enrichment. *Global Change Biology* **9**, 106–116.

Gill, R. A. and Jackson, R. B. (2000) Global patterns of root turnover for terrestrial ecosystems. *New Phytologist* **147**, 13–31.

Hall, S. J. and Matson, P. A. (1999) Nitrogen oxide emissions after nitrogen additions in tropical forests. *Nature* **400**, 152–155.

Hilbert, D. W., Ostendorf, B., and Hopkins, M. S. (2001) Sensitivity of tropical forests to climate change in the humid tropics of north Queensland. *Australian Ecology* **26**, 590–603.

Hobbie, S. E. and Vitousek, P. M. (2000) Nutrient limitation of decomposition in Hawaiian forests. *Ecology* **81**, 1867–1877.

Hobbie, S. E., Nadelhoffer, K. J., and Hogberg, P. (2002) A synthesis: The role of nutrients as constraints on carbon balances in boreal and arctic regions. *Plant and Soil* **242**, 163–170.

Holdridge, L. R. (1967) *Life Zone Ecology* (206 pp.). Tropical Science Center, San Jose, Costa Rica.

Hulme, M., Doherty, R., Ngara, T., New, M., and Lister, D. (2001) African climate change: 1900–2100. *Climate Research* **17**, 145–168.

IPCC (2001) *Climatic Change 2001: The Scientific Basis*. Cambridge University Press, Cambridge, U.K.

Jackson, R. B., Canadell, J., Ehleringer, J. R., Mooney, H. A., Sala, O. E., and Schulze, E. D. (1996) A global analysis of root distributions for terrestrial biomes. *Oecologia* **108**, 389–411.

Janos, D. P. (1983) Tropical mycorrhizas, nutrient cycles and plant growth. In: S. L. Sutton, T. C. Whitemore, and Chadwick, A. C. (eds.), *Tropical Rain Forest: Ecology and Management* (pp. 327–345). Blackwell Scientific, Oxford, U.K.

Jaramillo, V. J. and Sanford Jr., R. L. (1995) Nutrient cycling in tropical deciduous forests. In: S. H. Bullock, H. A. Mooney, and E. Medina (eds.), *Seasonally Dry Tropical Forests* (pp. 346–361). Cambridge University Press, Cambridge, U.K.

Jensen, M. N. (2004) Climate warming shakes up species. *BioScience* **54**, 722–729.

Johnson, A. H., Frizano, J., and Vann, D. R. (2003) Biogeochemical implications of labile phosphorus in forest soils determined by the Hedley fractionation procedure. *Oecologia* **135**, 487–499.

Jordan, C. F. (1985) *Nutrient Cycling in Tropical Forest Ecosystems: Principles and Their Application in Management and Conservation*. John Wiley & Sons. New York.

Kanowski, J. (2001) Effects of elevated CO_2 on the foliar chemistry of seedlings of two rainforest trees from north-east Australia: Implications for folivorous marsupials. *Australian Ecology* **26**, 165–172.

Keller, M. and Reiners, W. A. (1994) Soil atmosphere exchange of nitrous oxide, nitric oxide, and methane under secondary succession of pasture to forest in the Atlantic lowlands of Costa Rica. *Global Biogeochemical Cycles* **8**, 399–410.

Keller, M., Kaplan, W. A., and Wofsy, S. C. (1986) Emissions of N_2O, CH_4, and CO_2 from tropical forest soils. *Journal of Geophysical Research* **91**, 11791–11802.

Kitayama, K., Aiba, S. I., Takyu, M., Majalap, N., and Wagai, R. (2004) Soil phosphorus fractionation and phosphorus-use efficiency of a Bornean tropical montane rain forest during soil aging with podzolization. *Ecosystems* **7**, 259–274.

Körner, C. (1998) Tropical forests in a CO_2-rich world. *Climatic Change* **39**, 297–315.

Körner, C. (2004) Through enhanced tree dynamics carbon dioxide enrichment may cause tropical forests to lose carbon. *Philosophical Transactions of the Royal Society of London, Series B: Biological Sciences* **359**, 493–498.

Lal, C. B., Annapurna, C., Raghubanshi, A. S., and Singh, J. S. (2001a) Foliar demand and resource economy of nutrients in dry tropical forest species. *Journal of Vegetation Science* **12**, 5–14.

Lal, M., Nozawa, T., Emori, S., Harasawa, H., Takahashi, K., Kimoto, M., Abe-Ouchi, A., Nakajima, T., Takemura T., and Numaguti, A. (2001b) Future climate change: Implications for Indian summer monsoon and its variability. *Current Science* **81**, 1196–1207.

Lal, M., Harasawa, H., and Takahashi, K. (2002) Future climate change and its impact over small island states. *Climate Research* **19**, 179–192.

Lashof, D. A. and Ahuja, D. R. (1990) Relative contributions of greenhouse gas emissions to global warming. *Nature* **344**, 529–531.

Lewis, S. L., Malhi, Y., and Phillips, O. L. (2004) Fingerprinting the impacts of global change on tropical forests. *Philosophical Transactions of the Royal Society of London, Series B: Biological Sciences* **359**, 437–462.

Lloyd, J., Bird, M. I., Veenendaal, E. M., and Kruijt, B. (2001) Should phosphorus availability be constraining moist tropical forest responses to increasing CO_2 concentrations? In: E. D. Schulze (ed.), *Global Biogeochemical Cycles in the Climate System* (pp. 95–114). Academic Press, San Diego.

Lodge, D. (1993) Nutrient cycling by fungi in wet tropical forests. In: S. Isaac, J. C. Frankland, R. Watling, and A. J. S. Whalley (eds.), *Aspects of Tropical Mycology*. Cambridge University Press, Cambridge, U.K.

Lodge, D. J., McDowell, W. H., and McSwiney, C. P. (1994) The importance of nutrient pulses in tropical forests. *Trends in Ecology and Evolution* **9**, 384–387.

Lohse, K. A. and Matson, P. (2005) Consequences of nitrogen additions for soil losses from wet tropical forests. *Ecological Applications* **15**, 1629–1648.

Luizão, R. C. C., Bonde, T. A., and Rosswell, T. (1992) Seasonal variation of soil microbial biomass: The effects of clearfelling a tropical rainforest and establishment of a pasture in the central Amazon. *Soil Biology and Biochemistry* **24**, 805–813.

Lynch, J. P. and St. Clair, S. B. (2004) Mineral stress: The missing link in understanding how global climate change will affect plants in real world soils. *Field Crops Research* **90**, 101–115.

Malhi, Y. and Wright, J. (2004) Spatial patterns and recent trends in the climate of tropical rainforest regions. *Philosophical Transactions of the Royal Society of London, Series B: Biological Sciences* **359**, 311–329.

Marrs, R. H., Proctor, J., Heaney, A., and Mountford, M. D. (1988) Changes in soil nitrogen-mineralization and nitrification along an altitudinal transect in tropical rain forest in Costa Rica. *Journal of Ecology* **76**, 466–482.

Martinez-Yrizar, A. (1995) Biomass distribution and primary productivity of tropical dry forests. In: S. H. Bullock, H. A. Mooney, and E. Medina (eds.), *Seasonally Dry Tropical Forest* (pp. 326–345). Cambridge University Press, Cambridge U.K.

Matamala, R. and Schlesinger, W. H. (2000) Effects of elevated atmospheric CO_2 on fine root production and activity in an intact temperate forest ecosystem. *Global Change Biology* **6**, 967–979.

Matson, P. A., McDowell, W. H., Townsend, A. R., and Vitousek, P. M. (1999) The globalization of N deposition: Ecosystem consequences in tropical environments. *Biogeochemistry* **46**, 67–83.

McGill, W. B. and Cole, C. V. (1981) Comparative aspects of cycling of organic C, N, S and P through soil organic matter. *Geoderma* **26**, 267–286.

McGroddy, M. E., Silver, W. L., and de Oliveira, R. C. (2004) The effect of phosphorus availability on decomposition dynamics in a seasonal lowland Amazonian forest. *Ecosystems* **7**, 172–179.

McKane, R. B., Rastetter, E. B., Melillo, J. M., Shaver, G. R., Hopkinson, C. S., Fernandes, D. N., Skole, D. L., and Chomentowski, W. H. (1995) Effects of global change on carbon storage in tropical forests of South America. *Global Biogeochemical Cycles* **9**, 329–350.

McLaughlin, J. F., Hellmand, J. J., Boggs, C. L., and Ehrlich, P. R.. (2002) Climate change hastens population extinctions. *Proceedings of the National Academy of Sciences of the United States of America* **99**, 6070–6074.

Melillo, J. M., McGuire, A. D., Kicklighter, D. W., Moore III, B., Vörösmarty, C. J., and Schloss, A. L. (1993) Global climate change and terrestrial net primary production. *Nature* **363**, 234–240.

Miyasaka, S. C. and Habte, M. (2001) Plant mechanisms and mycorrhizal symbioses to increase phosphorus uptake efficiency. *Communications in Soil Science and Plant Analysis* **32**, 1101–1147.

Mueller-Dombois, D. and Goldammer, J. (1990) Fire in tropical ecosystems and global environmental change: An introduction. In: J. Goldammer (ed.), *Fire in the Tropical Biota: Ecosystem Processes and Global Challenges* (pp. 1–10). Springer-Verlag, Berlin.

Nepstad, D. C., Moutinho, P., Dias-Filho, M. B., Davidson, E., Cardinot, G., Markewitz, D., Figueiredo, R., Vianna, N., Chambers, J., Ray, D. *et al.* (2002) The effects of partial throughfall exclusion on canopy processes, aboveground production and biogeochemistry of an Amazon forest. *Journal of Geophysical Research*, doi:10.1029/2001JD000360.

Norby, R. J. and Jackson, R. B. (2000) Root dynamics and global change: Seeking an ecosystem perspective. *New Phytologist* **147**, 3–12.

Norby, R. J., Wullschleger, S. D., Gunderson, C. A., Johnson, D. W., and Ceulemans, R. (1999) Tree responses to rising CO_2: Implications for the future forest. *Plant, Cell and Environment* **22**, 683–714.

Oechel, W. C., Vourlitis, G. L., and Hastings, S. J. (1998) The effects of water table manipulation on the net CO_2 flux of wet sedge tundra ecosystems. *Global Change Biology* **4**, 77–90.

Ometto, J. P. H. B., Nobre, A. D., Rocha, H. R., Artaxo, P., and Martinelli, L. A. (2005) Amazonia and the modern carbon cycle: Lessons learned. *Oecologia* **143**, 483–500.

Ostertag, R. and Hobbie, S. E. (1999) Early stages of root and leaf decomposition in Hawaiian forests: Effects of nutrient availability. *Oecologia* **121**, 564–573.

Papatheodorou, E. M., Stamou, G. P., and Giannotaki, A. (2004) Response of soil chemical and biological variables to small and large scale changes in climatic factors. *Pedobiologia* **48**, 329–338.

Peretyazhko, T. and Sposito, G. (2005) Iron(III) reduction and phosphorous solubilization in humid tropical forest soils. *Geochimica et Cosmochimica Acta* **69**, 3643–3652.

Perrott, K. W., Sarathchandra, S. U., and Waller, J. E. (1990) Seasonal storage and release of phosphorus and potassium by organic matter and the microbial biomass in a high-producing pastoral soil. *Australian Journal of Soil Research* **28**, 593–608.

Phillips, O. L., Malhi, Y., Higuchi, N., Laurence, L. F., Nuñez, V. P., Vasqueth, M. R., Laurence, S. G., Ferreira, L. V., Stern, M., Brown, S., and Grace, J. (1998) Changes in the carbon balance of tropical forests: Evidence from long-term plots. *Science* **282**, 439–442.

Pimm, S. L. and Sugden, A. M. (1994) Tropical diversity and global change. *Science* **263**, 933–934.

Pregitzer, K. S., King, J. S., Burton, A. J., and Brown, S. E. (2000) Responses of tree fine roots to temperature. *New Phytologist* **147**, 105–115.

Pritchard, S. G. and Rogers, H. H. (2000) Spatial and temporal deployment of crop roots in CO_2-enriched environments. *New Phytologist* **147**, 55–71.

Raich, J. W. and Schlesinger, W. H. (1992) The global carbon dioxide flux in soil respiration and relationship to vegetation and climate. *Tellus* **44B**, 81–99.

Read, L. and Lawrence, D. (2003) Litter nutrient dynamics during succession in dry tropical forests of the Yucatan: Regional and seasonal effects. *Ecosystems* **6**, 747–761.

Riley, R. H. and Vitousek, P. M. (1995) Nutrient dynamics and nitrogen trace gas flux during ecosystem development in montane rain forest. *Ecology* **76**, 292–304.

Rustad, L., Campbell, J., Marion, G., Norby, R., Mitchell, M., Hartley, A., Cornelissen, J. Gurevitch, J. and GCTE-NEWS (2001) A meta-analysis of the response of soil respiration, net nitrogen mineralization, and aboveground plant growth to experimental ecosystem warming. *Oecologia* **126**, 543–562.

Sanchez, P. A. (1976) *Properties and Management of Soils in the Tropics*. John Wiley & Sons, New York.

Santiago, L. S. and Mulkey, S. S. (2005) Leaf productivity along a precipitation gradient in lowland Panama: Patterns from leaf to ecosystem. *Trees—Structure and Function* **19**, 349–356.

Santiago, L. S., Schuur, E. A., and Silvera, K. (2005) Nutrient cycling and plant–soil feedbacks along a precipitation gradient in lowland Panama. *Journal of Tropical Ecology* **21**, 461–470.

Schimel, J. P. and Gulledge, J. (1998) Microbial community structure and global trace gases. *Global Change Biology* **4**, 745–758.

Schuur, E. A. G. (2001) The effect of water on decomposition dynamics in mesic to wet Hawaiian montane forests. *Ecosystems* **4**, 259–273.

Silver, W. L. (1998) The potential effects of elevated CO_2 and climate change on tropical forest soils and biogeochemical cycling. *Climatic Change* **39**, 337–361.

Silver, W. L. and Miya, R. (2001) Global patterns in root decomposition: Comparisons of climate and litter quality effects. *Oecologia* **129**, 407–419.

Silver, W. L. and Vogt, K. A. (1993) Fine root dynamics following single and multiple disturbances in a subtropical wet forest ecosystem. *Journal of Ecology* **8**, 729–738.

Silver, W. L., Lugo, A. E., and Keller, M. (1999) Soil oxygen availability and biogeochemistry along rainfall and topographic gradients in upland wet tropical forest soils. *Biogeochemistry* **44**, 301–328.

Silver, W. L., Neff, J., McGroddy, M., Veldkamp, E., Keller, M., and Cosme, R. (2000) Effects of soil texture on belowground carbon and nutrient storage in a lowland Amazonian forest ecosystem. *Ecosystems* **3**, 193–209.

Silver, W. L., Herman, D. J., and Firestone, M. K. (2001) Dissimilatory nitrate reduction to ammonium in upland tropical forest soils. *Ecology* **82**, 2410–2416.

Silver, W. L., Thompson, A. W., McGroddy, M. E., Varner, R. K., Dias, J. D., Silva, H., Crill, P. M., and Keller, M. (2005a) Fine root dynamics and trace gas fluxes in two lowland tropical forest soils. *Global Change Biology* **11**, 290–306.

Silver, W. L., Thompson, A. W., Reich, A., Ewel, J. J., and Firestone, M. K. (2005b) Nitrogen cycling in tropical plantation forests: Potential controls on nitrogen retention. *Ecological Applications* **15**, 1604–1614.

Silver, W. L., Thompson, A. W., Herman, D. J., and Firestone, M. K. (2006) In: L. S. Hamilton and P. Bubb (eds.), *Forests in the Mist: Science for Conserving and Managing Tropical Montane Cloud Forests.* University of Hawaii Press, Honolulu, HI.

Sollins, P. and Radulovich, R. (1988) Effects of soil physical structure on solute transport in a weathered tropical soil. *Soil Science Society of America Journal* **52**, 1168–1173.

Sollins, P., Robertson, G. P., and Uehara, G. (1988) Nutrient mobility in variable- and permanent-charge soils. *Biogeochemistry* **6**, 181–199.

Sowerby, A., Emmett, B., Beier, C., Tietema, A., Penuelas, J., Estiarte, M., van Meeteren, M. J. M., Hughes, S., and Freeman, C. (2005) Microbial community changes in heathland soil communities along a geographical gradient: Interaction with climate change manipulations. *Soil Biology and Biochemistry* **37**, 1805–1813.

Staddon, P. L. and Fitter, A. H. (1998) Does elevated atmospheric carbon dioxide affect arbuscular mycorrhizas? *Trends in Ecology and Evolution* **13**, 455–458.

Staddon, P. L., Gregersen, R., and Jakobsen, I. (2004) The response of two *Glomus* mycorrhizal fungi and a fine endophyte to elevated atmospheric CO_2, soil warming and drought. *Global Change Biology* **10**, 1909–1921.

Steudler, P. A., Melillo, J. M., Feigl, B. J., Neill, C., Piccolo, M. C., and Cerri, C. C. (1996) Consequences of forest-to-pasture conversion on CH_4 fluxes in the Brazilian Amazon Basin. *Journal of Geophysical Research* **101**, 18547–18554.

Stevenson, F. J. and Cole, C. V. (1999) The phosphorus cycle. In: F. J. Stevenson and C. V. Cole (eds.), *Cycles of Soil: Carbon, Nitrogen, Phosphorus, Sulfur, Micronutrients* (2nd Edn.). John Wiley & Sons, New York.

Tate, K. R. (1985) Soil phosphorus. In: D. Vaughn and R. E. Malcolm (eds.), *Soil Organic Matter and Biological Activity* (Developments in Plant and Soil Sciences Series). Marinus Nijhoff/Dr. W. Junk, Boston.

Teh, Y. A., Silver, W. L., and Conrad, M. E. (2005) Oxygen effects on methane production and oxidation in humid tropical forest soils. *Global Change Biology* **11**, 1283–1297.

Tian, H. J., Melillo, J. M., Kicklighter, D. W., McGuire, A. D., Helfrich III, J. V. K., Moore III, B., and Vörösmarty, C. J. (1998) Effect of interannual climate variability on carbon storage in Amazonian ecosystems. *Nature* **396**, 664–667.

Tiessen, H., Cuevas, E., and Chacon, P. (1994) The role of soil organic matter in sustaining soil fertility. *Nature* **371**, 783–785.

Tingey, D. T., Phillips, D. L., and Johnson, M. G. (2000) Elevated CO_2 and conifer roots: Effects on growth, life span and turnover. *New Phytologist* **147**, 87–103.

Townsend, A. R., Vitousek, P. M., and Holland, E. A. (1992) Tropical soils could dominate the short-term carbon cycle feedbacks to increased global temperatures. *Climatic Change* **22**, 293–303.

Trumbore, S., Salazar da Costa, E., Nepstad, D. C., de Camargo, P. B., Martinelli, L. A., Ray, D., Restom, T., and Silver, W. (2006) Dynamics of fine root carbon in Amazonian tropical ecosystems and the contribution of roots to soil respiration. *Global Change Biology* **12**, 217–229.

Uehara, G. (1995) Management of isoelectric soils of the humid tropics. In: R. Lal, J. Kimble, E. Levine and B. A. Stewart (eds.), *Soil Management and the Greenhouse Effect: Advances in Soil Science* (pp. 271–278). CRC Press, Boca Raton, FL.

van Haren, J. L. M., Handley, L. L., Biel, K. Y., Kudeyarov, V. N., McLain, J. E. T., Martens, D. A., and Colodner, D. C. (2005) Drought-induced nitrous oxide flux dynamics in an enclosed tropical forest. *Global Change Biology* **11**, 1247–1257.

van Noordwijk, M., Cerri, C., Woomer, P. L., Nugroho, K., and Bernoux, M. (1997) Soil carbon dynamics in the humid tropical forest zone. *Geoderma* **79**, 187–225.

Vitousek, P. M. (1984) Litterfall, nutrient cycling and nutrient limitation in tropical forests. *Ecology* **65**, 285–298.

Vitousek, P. M. and Howarth, R. W. (1991) Nitrogen limitation on land and in the sea: How can it occur? *Biogeochemistry* **13**, 87–115.

Vitousek, P. M. and Matson, P. A. (1988) Nitrogen transformations in a range of tropical forest soils. *Soil Biology and Biochemistry* **20**, 361–367.

Vitousek, P. M. and Sanford Jr., R. L. (1986) Nutrient cycling in moist tropical forest. *Annual Review of Ecology and Systematics* **17**, 137–167.

Vitousek, P. M., Aber, J., Howart, R. W., Likens, G. E., Matson, P. A., Schindler, D. W., Schlesinger, W. H., and Tilman, G. D. (1997) Human alteration of the global nitrogen cycle: Causes and consequences. *Issues in Ecology* **1**, 1–15.

Vourlitis, G. L. and Oechel, W. C. (1997) Landscape-scale CO_2, H_2O vapor and energy flux of moist–wet coastal tundra ecosystems over two growing seasons. *Journal of Ecology* **85**, 575–590.

Walker, T. W. and Syers, J. K. (1976) The fate of phosphorus during pedogenesis. *Geoderma* **15**, 1–19.

Wang, Y. P. and Polglase, P. J. (1995) Carbon balance in the tundra, boreal forest and humid tropical forest during climate-change: Scaling-up from leaf physiology and soil carbon dynamics. *Plant, Cell and the Environment* **18**, 1226–1244.

Wieder, R. K. and Wright, S. J. (1995) Tropical forest litter dynamics and dry season irrigation on Barro-Colorado Island, Panama. *Ecology* **76**, 1971–1979.

Williams, M. R., Fisher, T. R., and Melack, J. H. (1997) Chemical composition and deposition of rain in the central Amazon, Brazil. *Atmospheric Environment* **31**, 207–217.

Wood, T. E., Lawrence, D., and Clark, D. A. (2005) Variation in leaf litter nutrients of a Costa Rican rain forest is related to precipitation. *Biogeochemistry* **73**, 417–437.

Wright, I. J. and Westoby, M. (2002) Leaves at low versus high rainfall: Coordination of structure, lifespan and physiology. *New Phytologist* **155**, 403–416.

Wright, I. J., Reich, P. B., and Westoby, M. (2001) Strategy shifts in leaf physiology, structure and nutrient content between species of high- and low-rainfall and high- and low-nutrient habitats. *Functional Ecology* **15**, 423–434.

Wright, S. J. (1992) Seasonal drought, soil fertility and the species density of tropical forest plant communities. *Trends in Ecology and Evolution* **7**, 260–263.

Yavitt, J. B. and Wright, S. J. (1996) Temporal patterns of soil nutrients in a Panamanian moist forest revealed by ion-exchange resin and experimental irrigation. *Plant and Soil* **183**, 117–129.

Yavitt, J. B. and Wright, S. J. (2001) Drought and irrigation effects on fine root dynamics in a tropical moist forest, Panama. *Biotropica* **33**, 421–434.

Yavitt, J. B. and Wright, S. J. (2002) Charge characteristics of soil in a lowland tropical moist forest in Panama in response to dry-season irrigation. *Australian Journal of Soil Research* **40**, 269–281.

Zak, D. R., Pregitzer, K. S., Curtis, P. S., and Holmes, W. E. (2000a) Atmospheric CO_2 and the composition and function of soil microbial communities. *Ecological Applications* **10**, 47–59.

Zak, D. R., Pregitzer, K. S., King, J. S., and Homes, W. E. (2000b) Elevated atmospheric CO_2, fine roots and the response of soil microorganisms: A review and hypothesis. *New Phytologist* **147**, 201–222.

12

The response of South American tropical forests to contemporary atmospheric change

O. L. Phillips, S. L. Lewis, T. R. Baker, and Y. Malhi

12.1 INTRODUCTION

Ecosystems worldwide are changing as a result of anthropogenic activities. Processes such as deforestation are physically obvious, but others, such as hunting and surface fires, are subtler but affect biodiversity in insidious ways (cf. Lewis *et al.*, 2004a; Laurance, 2004). Increased rates of nitrogen deposition and increases in air temperatures and atmospheric CO_2 concentrations are altering the environment of even the largest and most well-protected areas (e.g., Prentice *et al.*, 2001; Galloway and Cowling, 2002; Malhi and Wright, 2004). Anthropogenic atmospheric change will become more significant during this century, as CO_2 concentrations reach levels unprecedented for the last 20 million or perhaps even 40 million years (Retallack, 2001; Royer *et al.*, 2001). Nitrogen-deposition rates and climates are predicted to move far beyond Quaternary envelopes (Prentice *et al.*, 2001; Galloway and Cowling, 2002). Moreover, the rate of change in all these basic ecological drivers is likely to be without precedent in the evolutionary span of most species on Earth today (Lewis *et al.*, 2004a). This then is the Anthropocene: we are living through truly epoch-making times (Crutzen, 2002).

Given the scale of the anthropogenic experiment with the atmosphere–biosphere system, it is now self-evident that all ecosystems on Earth are affected by human activities in some sense. Recent research (Malhi and Phillips, 2005) suggests that tropical forests far from areas of deforestation are indeed undergoing profound shifts in structure, dynamics, productivity, and function. Here we synthesize recent results from a network of long-term monitoring plots across tropical South America that indicate how these forests are changing.

Changes in tropical forests are of societal importance for three reasons. First, tropical forests play an important role in the global carbon cycle and hence the rate of climate change, as ~40% of terrestrial vegetation carbon stocks lie within tropical forests (Malhi and Grace, 2000). Second, as tropical forests house at least half of all

Earth's species, changes will have a large impact on global biodiversity (Groombridge and Jenkins, 2003). Finally, as different plant species vary in their ability to store and process carbon, both climate and biodiversity changes are potentially linked by feedback mechanisms (e.g., Cox et al., 2000).

Evidence suggests that the remaining Amazonian rainforest is currently a global carbon sink (Malhi and Grace, 2000). The evidence is from long-term monitoring plots which show that forest stands are increasing in above-ground biomass (Phillips et al., 1998, 2002a; Baker et al., 2004a), and from inverse modeling of atmospheric CO_2 concentrations that indicate tropical ecosystems may contribute a carbon sink of 1–3 Gt (1 gigatonne = 1 billion metric tonnes) per year (e.g., Rayner et al., 1999; Rodenbeck et al., 2003). The existence of a substantial tropical carbon sink is consistent with modeling and laboratory studies that imply changes in the productivity of tropical forests in response to increasing CO_2 (e.g., Lloyd and Farquhar, 1996; Norby et al., 1999; Lewis et al., 2004a). Although these interpretations are still being debated (see Lewis et al., 2004a), efforts to overcome limitations in each line of evidence have generally confirmed the presence of a sink. Thus, it is reasonable to suggest that tropical forests are providing a substantial buffer against global climate change. Indeed, the results from long-term forest-monitoring plots suggest that intact Amazonian forests have increased in biomass by ~0.3–0.5% per year, and hence sequester carbon at approximately the same rate that the European Union (in January 2004) emits it by burning fossil fuels (Phillips et al., 1998; Malhi and Grace, 2000; Baker et al., 2004a).

Increasing atmospheric CO_2 concentrations and rising air temperatures will alter fundamental ecological processes and in turn will likely effect changes in tropical biodiversity. Changes in biodiversity as a consequence of anthropogenic climate change have in fact already been noted in better-studied temperate areas (e.g., Parmesan and Yohe, 2003) and in a well-studied old-growth tropical forest landscape in Brazil (Laurance et al., 2004). The interactive "balance" among tens of thousands of tropical plant species and millions of tropical animal species is certain to shift, even within the largest and best-protected forest ecosystems, which are traditionally thought of as "pristine" wilderness. These areas are vital refugia—where global biodiversity may most easily escape the current extinction crisis—as they are large enough to allow some shifts in the geographic ranges of species in response to global changes, and are afforded some protection from industrial development, such as logging and agriculture. However, how most tropical forest taxa will respond to rising temperatures and CO_2 concentrations, among other global changes, is currently unknown (Thomas et al., 2004).

Biodiversity change has inevitable consequences for climate change because different plant species vary in their ability to store and process carbon. One example of this is how shifts in the proportion of faster-growing light-demanding species may alter the carbon balance of tropical forests. Long-term plots suggest that mature humid Neotropical forests are a net carbon sink of ~0.6 gigatonnes per year (Phillips et al., 1998; Baker et al., 2004a). However, tree mortality rates have increased substantially in recent decades, so causing a likely increase in the frequency of tree-fall gap formation (Phillips and Gentry, 1994; Phillips et al., 2004). A shift in forests towards gap-favoring, light-demanding species with high growth rates, at the expense of more shade-tolerant species, is plausible (Körner, 2004). Such fast-growing species generally have lower wood-specific gravity, and hence lower carbon content (West *et al.*, 1999), than do shade-tolerant trees. An Amazon-wide decrease in mean wood specific gravity of just 0.4% would cancel out the carbon sink effect apparently caused by accelerated plant productivity. Whether such changes are occurring is currently poorly understood, but it is clear that the biodiversity and climate-change issues are closely linked and merit further study.

   In this chapter we present a summary of the latest findings from permanent plots monitored by a large network of Amazon forest researchers, known as "RAINFOR" (*Red Amazónica de Inventarios Forestales*, or Amazon Forest Inventory Network; *http://www.geog.leeds.ac.uk/projects/rainfor/*). Here we summarize findings from old-growth forests in terms of (a) structural change, (b) dynamic-process change, and (c) functional change, over the past two decades.

12.2   THE PLOT NETWORK

For these analyses, we define a plot as an area of forest where all trees above 10-cm diameter at breast height (d.b.h., measured at 1.3-m height or above any buttress or other deformity) are tracked individually over time. All trees are marked with a unique number, measured, mapped, and identified. Periodically (generally every 5 years), the plot is revisited, and all surviving trees are re-measured, dead trees are noted, and trees recruited to 10-cm d.b.h. are uniquely numbered, measured, mapped, and identified. This allows calculation of: (i) the cross-sectional area that tree trunks occupy (termed "basal area"), which can be used with allometric equations to estimate tree biomass (Baker *et al.*, 2004a); (ii) tree growth (the sum of all basal-area increments for surviving and newly recruited stems over a census interval); (iii) the total number of stems present; (iv) stem recruitment (number of stems added to a plot over time); and (v) mortality (either the number or basal area of stems lost from a plot over time). We present results from 50 to 91 plots, depending upon selection criteria for different analyses (most critically, the number of census intervals from a plot and whether only stem-count data or the full tree-by-tree data set is available). The number of plots used for stem-density changes is more than that used in the biomass study because full tree-by-tree data are required to calculate biomass (using Baker *et al.*'s, 2004a methods), whereas stem-change data can often be obtained from published studies. The plots span the Amazonian forests of northern South America (Figure 12.1), including Bolivia, Brazil, Ecuador, French Guiana, Peru, and Venezuela. Most are 1 ha in size and comprise ~600 trees of ≥10-cm d.b.h. The smallest are 0.4 ha and the largest is 9 ha, all large enough to avoid undue influence by the behavior of an individual tree (Chave *et al.*, 2003). Many plots have been monitored for more than a decade, although they range in age from 2 to 25 years. The earliest plot inventory was started in 1971, the latest in 2002. Details of the exact plot locations, inventory and monitoring methods, and issues relating to collating and analysing plot data are omitted from this chapter for reasons of space, but are discussed in detail

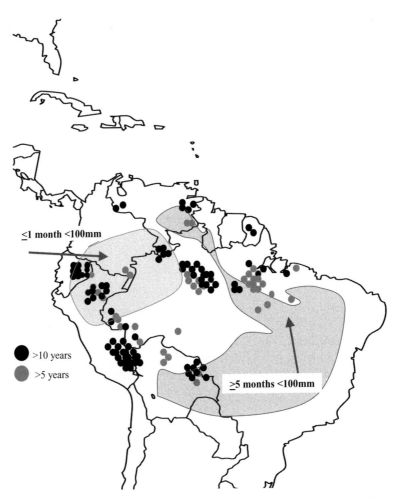

Figure 12.1. Plot locations used in this study. Symbols represent approximate locations of each plot; grey circle for plots monitored for 5–10 years, black for those with >10 years of monitoring. The approximate extent of seasonal and highly seasonal areas of tropical South America are indicated.

elsewhere (Phillips *et al.*, 2002a, b, 2004; Baker *et al.*, 2004a, b; Malhi *et al.*, 2002, 2004; Lewis *et al.*, 2004b).

12.3   STRUCTURAL CHANGES

Among 59 plots monitored in old-growth Amazon forests with full tree-by-tree data, there has been a significant increase in above-ground biomass between the first and last time they were measured. Over approximately the last 20 years, the increase has been 0.61 ± 0.22 tonnes of carbon per hectare per year, or a relative increase of $0.50 \pm 0.17\%$ per year (mean $\pm 95\%$ confidence interval; Baker *et al.*, 2004a). Across all 59 plots, the above-ground biomass change is normally distributed and shifted to the right of 0 (Figure 12.2a). The estimate of a net increase of 0.61 ± 0.22 t C ha^{-1} yr^{-1} is slightly higher than the 0.54 ± 0.29 t C ha^{-1} yr^{-1} estimated earlier for the lowland Neotropics by Phillips *et al.* (1998) using 50 sites up to 1996.

We estimate the magnitude of the South American carbon sink by multiplying 0.61 tonnes per hectare per year by the estimated area of mature Neotropical humid-forest cover (c. 8,705,100 km^2; FAO, 1990), which yields a value of about 0.5 giga-tonnes of carbon per year. If we further assume that the ratio of above-ground to below-ground biomass is 3 : 1 (cf. Phillips *et al.*, 1998), and that below-ground biomass is increasing in proportion to above-ground biomass, then the sink increases to 0.71 ± 0.26 gigatonnes of carbon per year. If other biomass components—such as small trees, lianas, and coarse woody debris—are also increasing in biomass, then the sink will be fractionally larger still. However, these estimates depend critically on (i) how representative the 59 tree-by-tree plots are of South American forests; (ii) assumptions about the extent of mature, intact forest remaining in South America; (iii) the extent to which we have sampled the regional-scale matrix of natural disturbance and recovery.

Clark (2002) raised two concerns about the original findings of Phillips *et al.* (1998) that Amazon biomass was increasing, suggesting that (i) some floodplain plots that Phillips *et al.* considered mature may still be affected by primary succession, and that (ii) large buttress trees in some plots may have been measured in error—that is, not above the buttress, as protocols dictate, but around them. However, Baker *et al.* (2004a) showed that the carbon sink remains when plots on old floodplain substrates and those that may have buttress problems are removed from the analysis.

Consideration of all 91 RAINFOR plots shows a small increase in stem density between the first and last time they were measured, of 0.84 ± 0.77 stems per hectare per year (Figure 12.2b; paired t-test, $t = 2.12$, $P = 0.037$), or a $0.15 \pm 0.13\%$ per year increase (Phillips *et al.*, 2004). Across all plots, stem-change rates are approximately normally distributed and slightly shifted to the right of 0 (Figure 12.2b). The same test using 59 plots (from the Baker *et al.* 2004a study) shows a similar increase in stem density ($0.16 \pm 0.15\%$ per year), while a smaller but longer-term data set (50 plots from Lewis *et al.*, 2004b) shows a slightly larger increase ($0.18 \pm 0.12\%$ per year). While still significant, these changes in stem density are proportionally not as great as the biomass changes.

12.4 DYNAMIC CHANGES

An alternative way of examining forest change is to look for changes in processes (growth, recruitment, death), as well as in structure (biomass, stem density). Are these forests getting more active or simply gaining mass? We measure the dynamics of forests in two ways. First, we can examine changes in stem population dynamics. By

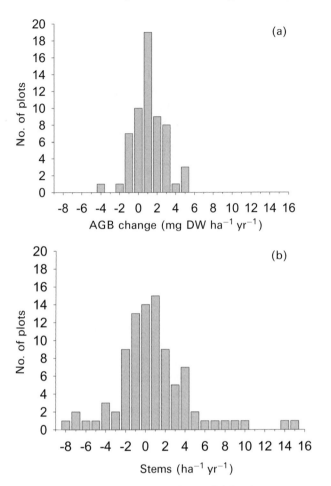

Figure 12.2. (a) Above-ground biomass change (dry weight) of trees greater than 10-cm diameter in 59 Amazon plots, based on initial and final stand-biomass estimates calculated using an allometric equation relating individual tree diameter to biomass, and incorporating a correction factor to account for variation in wood density among species (from Baker *et al.*, 2004a). As would be expected in a random sample of small plots measured for a finite period, some sites show a decline in biomass during that period indicating that at that particular point in space and time tree mortality has exceeded tree growth. However, the mean and median are shifted significantly to the right ($P < 0.01$). (b) Stem number change in 91 plots from across South American tropical forests. Stems were counted during the first and final censuses of each plot (plots are the same as those used by Phillips *et al.*, 2004). The mean and median are shifted significantly to the right ($P < 0.05$).

convention (Phillips and Gentry, 1994) we estimate stem turnover between any two censuses as the mean of annual mortality and recruitment rates for the population of trees $\geq$10-cm diameter. Second, we examine changes in biomass fluxes of the forest— in terms of growth of trees and the biomass lost with mortality events. These stand-

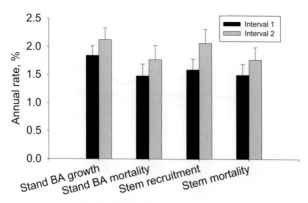

Figure 12.3. Annualized rates of stand-level basal-area growth, basal-area mortality, stem recruitment, and stem mortality from plots with two consecutive census intervals (i.e., the subset
of RAINFOR sites that have been inventoried on at least three successive occasions), each
giving the mean from 50 plots with 95% confidence intervals. Paired t-tests show that all of the
increases are significant. The average mid-year of the first and second censuses was 1989 and
1996, respectively (from Lewis *et al.*, 2004b).

level rates of "biomass growth" and "biomass loss" should be approximately proportional to the rate at which surviving and recruiting trees gain basal area and the rate at
which basal area is lost from the stand through tree death (Phillips *et al.*, 1994).

Among 50 old-growth plots across tropical South America with at least three
censuses (and therefore at least two consecutive monitoring periods that can be
compared), we find that all of these key ecosystem processes—stem recruitment, mortality, and turnover, and biomass growth, loss, and turnover—are increasing significantly (Figure 12.3), between the first and second halves of the monitoring period
(Lewis *et al.*, 2004b). Thus, over the past two decades, these forests have become, on
average, faster growing and more dynamic. Notably, the increases in the rate of
dynamic fluxes (growth, recruitment, and mortality) are about an order of magnitude
larger than are the increases in the structural pools (above-ground biomass and stem
density; Lewis *et al.*, 2004b).

These and similar results can be demonstrated graphically in a number of ways. In
Figure 12.4, we plot the across-site mean values for stem recruitment and mortality as
a function of calendar year. This shows that the increase has not been short-term (e.g.,
the result of a spike around a year with unusual weather), that recruitment rates have
on average consistently exceeded mortality rates, and that mortality appears to lag
recruitment (Phillips *et al.*, 2004).

Using data for the 50 plots with two consecutive census intervals, we can also
separate them into two groups: one faster growing and more dynamic (mostly western
Amazonian), and one slower growing and less dynamic (mostly eastern and central
Amazonian). Both groups showed increased stem recruitment, stem mortality, stand
basal-area growth, and stand basal-area mortality, with larger absolute increases in
rates in the faster growing and more dynamic sites than in the slower-growing and less
dynamic sites (Figure 12.5; Lewis *et al.*, 2004b). However, the proportional increases

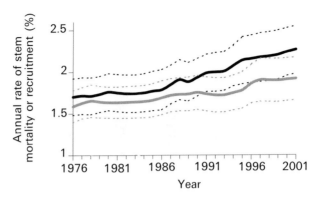

Figure 12.4. Mean and 95% confidence intervals for stem recruitment and mortality rates against calendar year, for plots arrayed across Amazonia. Rates for each plot were corrected for the effects of differing census-interval lengths, for "site-switching", and for "majestic-forest bias". A detailed justification methodology for these corrections is given in Phillips *et al.* (2004); all trends are robust and hold equally if these corrections are not applied. *Black* indicates recruitment, *grey* indicates mortality, *solid lines* are means, and *dots* are 95% confidence intervals (from Phillips *et al.*, 2004).

in rates were similar, and statistically indistinguishable, across both forest types (Lewis *et al.*, 2004b). This shows that increasing growth, recruitment, and mortality rates are occurring proportionately similarly across different forest types and geographically widespread areas.

12.5 FUNCTIONAL CHANGES

Changes in the structure and dynamics of tropical forests are likely to be accompanied by changes in species composition and function. There is, moreover, no *a priori* reason to expect that large changes in Amazon forests should be restricted to trees. Phillips *et al.* (2002b) studied woody climbers (structural parasites on trees, also called "lianas"), which typically contribute 10–30% of forest leaf productivity but are ignored in almost all monitoring studies except in most of our western Amazonian sites. Across the RAINFOR plots of western Amazonia there has been a concerted increase in the density, basal area, and mean size of lianas (Figure 12.6; Phillips *et al.*, 2002b). Over the last two decades of the 20th century, the density of large lianas relative to trees increased here by 1.7–4.6% per year. This was the first direct evidence that intact tropical forests are changing in terms of their composition and function. A long-term monitoring study from beyond Amazonia (Barro Colorado Island in Panama) has since reported a substantial increase in absolute and relative liana leaf-fall rates since the 1980s, indicating that lianas are both increasing and becoming more dominant there (Wright *et al.*, 2004). There is some experimental evidence (Granados and Körner, 2002) for a very strong response of tropical lianas to elevated atmospheric CO_2 concentrations, much stronger than the normal experimental response of trees.

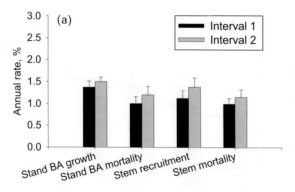

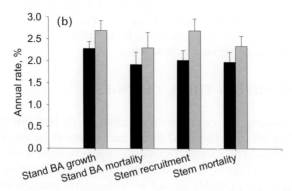

Figure 12.5. Annualized rates of stand-level basal-area growth, basal-area mortality, stem recruitment, and stem mortality over consecutive census intervals for plots grouped into "slower growing less-dynamic" (a) and "faster growing more-dynamic" (b) forests. Of the slower-dynamics group, 20 of 24 plots are from eastern and central Amazonia, whereas just two are from western Amazonia. Of the faster-dynamics group, 24 of 26 plots are from western Amazonia, with just one from central Amazonia. The remaining three plots are from Venezuela and outside the Amazon drainage basin. Changes have occurred across the South American continent, and in both slower- and faster-dynamic forests (from Lewis et al., 2004b).

Finally, a recent paper from a cluster of plots in central Amazonia has shown consistent changes in tree species composition over the past two decades (Laurance et al., 2004). Many faster-growing genera of canopy and emergent stature trees increased in basal area or density, whereas some slower-growing genera of sub-canopy or understory trees decreased in density. Laurance et al. (2004) provide evidence of pervasive changes in central Amazonian forests: growth, mortality, and recruitment all increased significantly over two decades (basal area also increased, but not significantly so), with faster-growing genera showing much larger absolute and relative increases in growth, relative to slower-growing genera. Further studies are urgently needed to determine whether comparably large shifts in tree communities are occurring throughout the tropics.

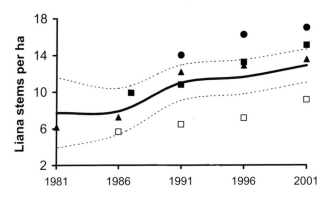

Figure 12.6. Five-year running means (*solid line*) with 95% confidence intervals (*dashed lines*) of liana stem density per hectare (>10-cm diameter at breast height), with values plotted separately for northern Peru (*filled squares*), southern Peru (*filled triangles*), Bolivia (*filled circle*), and Ecuador (*unfilled squares*) (adapted from Phillips *et al.*, 2002b; see that paper for full details of field and analytical methodology).

12.6   WHAT IS DRIVING THESE CHANGES?

What could be causing the continent-wide changes in tree growth, recruitment, mortality, stem density, and biomass? Many factors could be invoked, but there is only one parsimonious explanation. The results appear to show a coherent fingerprint of increasing growth—that is, increasing net primary productivity (NPP)—across tropical South America, probably caused by a long-term increase in resource availability (Lewis *et al.* 2004a, b). According to this explanation, increasing resource availability increases NPP, which then increases stem growth rates. Faster individual growth rates account for the increase in stand basal-area growth and stem recruitment rates, and the fact that these factors show the "clearest" signals (statistically most significant) in the analyses (Lewis *et al.*, 2004b). Because of increased growth, competition for limiting resources—such as light, water, and nutrients—increases. Over time some of the faster-growing, larger trees die, as do some of the "extra" recruits (the accelerated growth percolates through the system), resulting in increases in the rate of biomass mortality and stem mortality. Thus, the system gains biomass and stems, while the losses lag some years behind, causing an increase in above-ground biomass and stems. Overall, this suite of changes may be qualitatively explained by a long-term increase in a limiting resource.

   The changes in composition can also be explained by increasing resource availability, as the rise in liana density may be either a direct response to rising resource supply rates, or a response to greater disturbance caused by higher tree-mortality rates. The changing tree composition in central Amazonian plots (Laurance *et al.*, 2004) is also consistent with increasing resource supply rates, as experiments show that faster-growing species are often the most responsive, in absolute terms, to increases in resource levels (Coomes and Grubb, 2000).

   What environmental changes could be increasing the growth and productivity of tropical forests? While there have been widespread changes in the physical, chemical, and biological environment of tropical trees (Lewis *et al.*, 2004a), only increasing atmospheric CO_2 concentrations (Prentice *et al.*, 2001), increasing solar radiation inputs (Wielicki *et al.*, 2002), and rising air temperatures (Malhi and Wright, 2004) have been documented across Amazonia and could be responsible for increased growth and productivity. For none of these three changes, however, do we have overwhelming evidence that the driver has both actually changed and that such a change must accelerate forest growth (Lewis *et al.*, 2004a).

The increase in atmospheric CO_2 is our leading candidate, because of the undisputed long-term increase in CO_2 concentrations, the key role of CO_2 in photosynthesis, and the demonstrated positive effects of CO_2 fertilization on plant growth rates, including experiments on whole temperate-forest stands (Hamilton *et al.*, 2002; Norby *et al.*, 2002; Lewis *et al.*, 2004a). At present, however, no experiments have assessed the effects of increasing CO_2 availability on intact, mature tropical-forest stands, and this interpretation is still contested by some (e.g., Chambers and Silver, 2004).

Air-temperature increases are also undisputed, and could conceivably be causing the changes we document. However, many authors expect that the $0.26°C$ per decade air-temperature increase (Malhi and Wright, 2004) would actually reduce, not increase, forest growth, as respiration costs are likely to increase with temperature. However, increased air temperatures will also increase soil temperatures, which could in turn increase soil mineralization rates and thus nutrient availability (see review by Lewis *et al.*, 2004a). Whether recent rises in air temperature have increased or decreased tropical forest NPP requires further study.

Recent satellite data suggest an increase in incoming solar radiation across the tropics between the mid-1980s and late 1990s as a result of reduced cloudiness (Wielicki *et al.*, 2002). However, because stem turnover has increased across the tropics since the 1950s (Phillips and Gentry, 1994; Phillips, 1996), increasing solar radiation since the mid-1980s may not have occurred over a long enough period of time to explain the trends in forest-plot data, at least in terms of stem turnover. Furthermore, as the *difference* between stand-level basal-area growth and mortality was similar at the start (1980s) and end (1990s) of the study by Lewis *et al.* (2004b), the factor causing changes in growth, recruitment, and mortality was probably operating before the onset of the study, and hence before the observed increase in incoming solar radiation. Finally, the evidence for increased insolation in Amazonia is not strong and there is a possibility that forest NPP may be greater under cloudy conditions (e.g., Roderick *et al.*, 2001: cloudiness increases the fraction of radiation that is indirect, which penetrates further into the canopy and could thus have a positive impact on whole forest NPP since canopy leaves overheat in midday tropical sun and may already be saturated with respect to light).

Determining which environmental change, or changes, has caused the trends we document across South American tropical forest is very difficult. However, each environmental change is expected to leave a unique signature, or fingerprint, in forest data, as different environmental changes initially impact different processes, have different distributions in time and space, and may affect some forests more than others (e.g., depending upon soil fertility). Future analyses of forest-plot data at finer spatially and temporally resolved scales should therefore allow a further narrowing of potential causes underlying rising productivity across South American tropical forests (Lewis *et al.*, 2004a).

12.7 THE FUTURE

For those concerned about future biodiversity losses and global climate change, our analyses suggest both worrying trends and some apparently "good news". The Amazon, the world's largest remaining tract of tropical forest, has shown concerted changes in forest dynamics over the past two decades. Such unexpected and rapid alterations—apparently in response to anthropogenic atmospheric change—raise concerns about other possible surprises that might arise as global changes accelerate in coming decades. Tropical forests are evidently very sensitive to changes in incoming resource levels and may show large structural and dynamic changes in the future, as resource levels alter further and temperatures continue to rise (Lewis *et al.*, 2004a). The implication of such rapid changes for the world's most biodiverse region is unknown, but could be substantial.

Moreover, old-growth Amazonian forests are evidently helping to slow the rate at which CO_2 is accumulating in the atmosphere, thereby acting as a buffer to global climate change—certainly "good news" for the moment. The concentration of atmospheric CO_2 is rising at a rate equivalent to 3.2 gigatonnes of carbon per year; this would be significantly greater without the tropical South American carbon sink of 0.5 to 0.8 gigatonnes of carbon per year. However, this subsidy from nature could be a relatively short-lived phenomenon. Mature Amazonian forests may either (i) continue to be a *carbon sink* for decades (Chambers *et al.*, 2001, Cramer *et al.*, 2001), or (ii) soon become a *small carbon source* because of changes in functional and species composition (Cramer *et al.*, 2001; Phillips *et al.*, 2002b; Körner, 2004; Laurance *et al.*, 2004), or (iii) become a *mega-carbon source*, possibly in response to climate change (Cox *et al.*, 2000; Cramer *et al.*, 2001). Given that a 0.3% annual increase in Amazon forest biomass is roughly equivalent to the entire fossil-fuel emissions of the European Union (in January 2004), a switch of tropical forests from a moderate carbon sink to even a moderate carbon source would have profound implications for global climate, biodiversity, and human welfare.

Finally, it is important to emphasize that climate-based models that project the future carbon balance in Amazonia (and future climate-change scenarios) have made no allowance for changing forest composition. This omission is likely to lead to erroneous conclusions. For example, lianas contribute little to forest biomass but kill trees and suppress tree growth (Schnitzer and Bongers, 2002), and their rapid increase suggests that the tropical carbon sink might shut down sooner than current models suggest. Large changes in tree communities could undoubtedly lead to net losses of carbon from tropical forests (Phillips and Gentry, 1994; Körner, 2004). The potential scope for such impacts of biodiversity changes on carbon storage is highlighted by Bunker *et al.* (2005) who explored various biodiversity scenarios based on the tree species at Barro Colorado Island. When slower-growing tree taxa are lost from an accelerated, liana-dominated forest, as much as one-third of the carbon storage capacity of the forest could be lost. Clearly, projections of future carbon fluxes will need to account for the changing composition and dynamics of tropical forests.

Acknowledgements

The results summarized here depended on contributions from numerous field assistants, rural communities, and field-station managers in Brazil, Bolivia, Ecuador, French Guiana, Peru, and Venezuela, and more than 50 grants from funding agencies in Europe and the U.S.A. This support is acknowledged in our earlier publications. Simon Lewis and Yadvinder Malhi are supported by Royal Society University Research Fellowships. Timothy Baker is supported by NERC grant NE/B503384/1 and a Roberts Fellowship at the University of Leeds. We thank M. Alexiades, S. Almeida, L. Arroyo, S. Brown, J. Chave, J. A. Comiskey, C. I. Czimczik, A. Di Fiore, T. Erwin, N. Higuchi, T. Killeen, C. Kuebler, S. G. Laurance, W. F. Laurance, J. Lloyd, A. Monteagudo, H. E. M. Nascimento, D. A. Neill, P. Núñez Vargas, J. Olivier, W. Palacios, S. Patiño, N. C. A. Pitman, C. A. Quesada, M. Saldias, J. N. M. Silva, J. Terborgh, A. Torres Lezama, R. Vásquez Martínez, B. Vinceti for contributing data and discussions to earlier papers on which this overview is based.

12.8   REFERENCES

Baker, T. R., Phillips, O. L., Malhi, Y., Almeida, S., Arroyo, L., Di Fiore, T., Higuchi, N., Killeen, T., Laurance, S. G., Laurance, W. F. *et al.* (2004a) Increasing biomass in Amazonian forest plots. *Philosophical Transactions of the Royal Society, Series B* **359**, 353–365.

Baker, T. R., Phillips, O. L., Malhi, Y., Almeida, S., Arroyo, L., Di Fiore, T., Killeen, T., Laurance, S. G., Laurance, W. L., Lewis, S. L. *et al.* (2004b) Variation in wood density determines spatial patterns in Amazonian forest biomass. *Global Change Biology* **10**, 545–562.

Bunker, D. E., De Clerck, F., Bradford, J. C., Colwell, R. K., Perfecto, I., Phillips, O. L., Sankaran, M., and Naeem, S. (2005) Species loss and above-ground carbon storage in a tropical forest. *Science* **310**, 1029–1031.

Chambers, J. Q. and Silver, W. L. (2004) Some aspects of ecophysiological and biogeochemical responses of tropical forests to atmospheric change. *Philosophical Transactions of the Royal Society, Series B* **359**, 463–476.

Chambers, J. Q., Higuchi, N., Tribuzy, E. S., and Trumbore, S. E. (2001) Carbon sink for a century. *Nature* **410**, 429.

Chave, J., Condit, R., Lao, S., Caspersen, J. P., Foster, R. B., and Hubbell, S. P. (2003) Spatial and temporal variation of biomass in a tropical forest: Results from a large census plot in Panama. *Journal of Ecology* **91**, 240–252.

Clark, D. A. (2002) Are tropical forests an important carbon sink? Reanalysis of the long-term plot data. *Ecological Applications* **12**, 3–7.

Coomes, D. A. and Grubb, P. J. (2000) Impacts of competition in forests and woodlands: A theoretical framework and review of experiments. *Ecological Monographs* **200**, 171–207.

Cox, P. M., Betts, R. A., Jones, C. D., Spall, S. A., and Totterdell, I. J.. (2000) Acceleration of global warming due to carbon-cycle feedbacks in a coupled climate model. *Nature* **408**, 184–187.

Cramer, W., Bondeau, A., Woodward, F. I., Prentice, I. C., Betts, R. A., Brovkin, V., Cox, P. M., Fisher, V., Foley, J. A., Friend, A. D. *et al.* (2001) Global response of terrestrial ecosystem structure and function to CO_2 and climate change: Results from six dynamic global vegetation models. *Global Change Biology* **7**, 357–373.

Crutzen, P. J. (2002) Geology of mankind. *Nature* **415**, 23.

Galloway, J. N. and Cowling, E. B. (2002) Reactive nitrogen and the world: 200 years of change. *Ambio* **31**, 64–71.

Granados, J. and Körner, C. (2002) In deep shade: Elevated CO_2 increases the vigour of tropical climbing plants. *Global Change Biology* **8**, 1109–1117.

Groombridge, B. and Jenkins, M. D. (2003) *World Atlas of Biodiversity*. University of California Press, Berkeley, CA.

Hamilton, J. G., DeLucia, E. H., George, K., Naidu, S. L., Finzi, A. C., and Schlesinger, W. H. (2002) Forest carbon balance under elevated CO_2. *Oecologia* **131**, 250–260.

Körner, C. (2004) Through enhanced tree dynamics carbon dioxide enrichment may cause tropical forests to lose carbon. *Philosophical Transactions of the Royal Society, Series B* **359**, 493–498.

Laurance, W. F. (2004) Forest–climate interactions in fragmented tropical landscapes. *Philosophical Transactions of the Royal Society, Series B* **359**, 345–352.

Laurance, W. F., Oliveira, A. A., Laurance, S. G., Condit, R., Nascimento, H. E. M., Sanchez-Thorin, A. C., Lovejoy, T. E., Andrade, A., D'Angelo, S., Ribeiro, J. E. *et al.* (2004) Pervasive alteration of tree communities in undisturbed Amazonian forests. *Nature* **428**, 171–174.

Lewis, S. L., Malhi, Y., and Phillips, O. L. (2004a) Fingerprinting the impacts of global change on tropical forests. *Philosophical Transactions of the Royal Society, Series B* **359**, 437–462.

Lewis, S. L., Phillips, O. L., Baker, T. R., Lloyd, J., Malhi, Y., Almeida, S., Higuchi, N., Laurance, W. F., Neill, D. A., Silva, J. N. M. *et al.* (2004b) Concerted changes in tropical forest structure and dynamics: Evidence from 50 South American long-term plots. *Philosophical Transactions of the Royal Society, Series B* **359**, 421–436.

Lloyd, J. and Farquhar, G. D. (1996) The CO_2 dependence of photosynthesis, plant growth responses to elevated atmospheric CO_2 concentrations and their interaction with plant nutrient status. *Functional Ecology* **10**, 4–32.

Malhi, Y. and Grace, J. (2000) Tropical forests and atmospheric carbon dioxide. *Trends in Ecology and Evolution* **15**, 332–337.

Malhi, Y. and Phillips, O. L. (2005) *Tropical Forests and Global Atmospheric Change* (260 pp.). Oxford University Press, Oxford, U.K.

Malhi, Y. and Wright, J. (2004) Spatial patterns and recent trends in the climate of tropical rainforest regions. *Philosophical Transactions of the Royal Society, Series B* **359**, 311–329.

Malhi, Y., Phillips, O. L., Baker, T. R., Almeida, S., Frederiksen, T., Grace, J., Higuchi, N., Killeen, T., Laurance, W. F., Leaño, C. *et al.* (2002) An international network to understand the biomass and dynamics of Amazonian forests (RAINFOR). *Journal of Vegetation Science* **13**, 439–450.

Malhi, Y., Baker, T. R., Phillips, O. L., Almeida, S., Alvarez, E., Arroyo, L., Chave, J., Czimczik, C., Di Fiore, A., Higuchi, N. *et al.* (2004). The above-ground coarse woody productivity of 104 neotropical forest plots. *Global Change Biology* **10**, 563–591.

Norby, R. J., Wullschleger, S. D., Gunderson, C. A., Johnson, D. W., and Ceulemans, R. (1999) Tree responses to rising CO_2 in field experiments: Implications for the future forest. *Plant Cell and Environment* **22**, 683–714.

Norby, R. J., Hanson, P. J., O'Neill, E. G., Tschaplinski, T. J., Weltzin, J. F., Hansen, R. A., Cheng, W. X., Wullschleger, S. D., Gunderson, C. A., Edwards, N. T. *et al.* (2002) Net primary productivity of a CO_2-enriched deciduous forest and the implications for carbon storage. *Ecological Applications* **12**, 1261–1266.

Parmesan, C. and Yohe, G. (2003) A globally coherent fingerprint of climate change impacts across natural systems. *Nature* **421**, 37–42.

Phillips, O. L. (1996) Long-term environmental change in tropical forests: Increasing tree turnover. *Environmental Conservation* **23**, 235–248.

Phillips, O. L. and Gentry, A. H. (1994) Increasing turnover through time in tropical forests. *Science* **263**, 954–958.

Phillips, O. L., Hall, P., Gentry, A. H., Sawyer, S. A., and Vásquez, R. (1994) Dynamics and species richness of tropical forests. *Proceedings of the National Academy of Sciences (U.S.A.)* **91**, 2805–2809.

Phillips, O. L., Malhi, Y., Higuchi, N., Laurance, W. F., Nuñez, P. V., Vásquez, M. R., Laurance, S. G., Ferriera, L. V., Stern, M., Brown, S. *et al.* (1998) Changes in the carbon balance of tropical forest: Evidence from long-term plots. *Science* **282**, 439–442.

Phillips, O. L., Malhi, Y., Vinceti, B., Baker, T., Lewis, S. L., Higuchi, N., Laurance, W. F., Vargas, P. N., Martínez, R. V., Laurance, S. G. *et al.* (2002a) Changes in the biomass of tropical forests: Evaluating potential biases. *Ecological Applications* **12**, 576–587.

Phillips, O. L., Martínez, R. V., Arroyo, L., Baker, T. R., Killeen, T., Lewis, S. L., Malhi, Y., Mendoza, A. M., Neill, D., Vargas, P. N. *et al.* (2002b) Increasing dominance of large lianas in Amazonian forests. *Nature* **418**, 770–774.

Phillips, O. L., Baker, T. R., Arroyo, L., Higuchi, N., Killeen, T., Laurance, W. F., Lewis, S. L., Lloyd, J., Malhi, Y., Monteagudo, A. *et al.* (2004) Pattern and process in Amazon tree turnover, 1976–2001. *Philosophical Transactions of the Royal Society, Series B* **359**, 381–407.

Prentice, I. C., Farquhar, G. D., Fasham, M. J. R., Goulden, M. L., Heimann, M., Jaramillo, V. J., Kheshgi, H. S., Le Quéré, C., Scholes, R. J., Wallace, D. W. R. *et al.* (2001) The carbon cycle and atmospheric carbon dioxide. *Climate Change 2001: The Scientific Basis* (Intergovernmental Panel on Climate Change Third Assessment Report, pp. 183–237). Cambridge University Press, Cambridge, U.K.

Rayner, P. J., Enting, I. G., Francey, R. J., and Langenfelds, R. L. (1999). Reconstructing the recent carbon cycle from atmospheric CO_2, $\delta^{13}C$ and O_2/N_2 observations. *Tellus* **51B**(2), 213–232.

Retallack, G. J. (2001) A 300-million-year record of atmospheric carbon dioxide from fossil plant cuticles. *Nature* **411**, 287–290.

Rodenbeck, C., Houweling, S., Gloor, M., and Heimann, M. (2003) CO_2 flux history 1982–2001 inferred from atmospheric data using a global inversion of atmospheric transport. *Atmospheric Chemistry and Physics* **3**, 1919–1964.

Roderick, M. L., Farquhar, G. D., Berry, S. L., and Noble, I. R. (2001) On the direct effect of clouds and atmospheric particles on the productivity and structure of vegetation. *Oecologia* **129**, 21–30.

Royer, D. L., Wing, S. L., Beerling, D. J., Jolley, D. W., Koch, P. L., Hickey, L. J., and Berner, R. A. (2001) Paleobotanical evidence for near present-day levels of atmospheric CO_2 during part of the Tertiary. *Science* **292**, 2310–2313.

Schnitzer, S. A. and Bongers, F. (2002) The ecology of lianas and their role in forests. *Trends in Ecology and Evolution* **17**, 223–230.

Thomas, C. D., Cameron, A., Green, R. E., Bakkenes, M., Beaumont, L. J., Collingham, Y. C., Erasmus, B. F. N., Ferreira de Siqueira, M., Grainger, A., Hannah, L. *et al.* (2004) Extinction risk from climate change. *Nature* **427**, 145–148.

West, G. B., Brown, J. H., and Enquist, B. J. (1999) A general model for the structure and allometry of vascular plant systems. *Nature* **400**, 664–667.

Wielicki, B. A., Wong, T., Allan, R. P., Slingo, A., Kiehl, J. T., Soden, B. J., Gordon, C. T., Miller, A. J., Yang, S. K., Randall, D. A. *et al.* (2002) Evidence for large decadal variability in tropical mean radiative energy budget. *Science* **295**, 841–844.

Wright, S. J., Calderon, O., Hernandez, S., and Paton, S. (2004) Are lianas increasing in importance in tropical forests? A 17-year record from Panama. *Ecology* **85**, 484–489.

13

Ecophysiological response of lowland tropical plants to Pleistocene climate

S. A. Cowling

13.1 INTRODUCTION

Climate changes associated with the Last Glacial Maximum (LGM, 21 kyr) are probably the most extreme that terrestrial vegetation, including tropical lowland ecosystems, have been forced to respond in over the past 100,000 years. The degree of tropical cooling can be reconstructed by paleoproxies that generally indicate a minimal cooling of 3°C and a maximum cooling of approximately 7°C (Guilderson et al., 1994; Stute et al., 1995; Mix et al., 1999; Behling and Negrelle, 2001; Mora and Pratt, 2001; Behling, 2002; Urrego et al., 2005). Some debate surrounds the degree of tropical decreases in glacial precipitation, primarily because precipitation patterns are strongly regional, and thus wide discrepancies in paleoprecipitation trends occur between different reconstructions. Despite this, a value of approximately 20% decrease in LGM rainfall is typically reconstructed from pollen-proxies in tropical catchments—such as the Amazonian Basin (Bush and Silman, 2004). Paleoclimate simulations of the South American monsoon during the LGM indicate an annual reduction in rainfall across Amazonia of between 25–35% relative to today (Cook and Vizy, 2006). Research also shows that glacial decreases in rainfall likely occurred in wet as opposed to dry seasonal months (Bush and Silman, 2004).

Less ambiguous with respect to paleoclimate reconstruction is the decline in atmospheric CO_2 that occurred at the LGM. Direct measurement of CO_2 gas trapped in Antarctic and Greenland polar ice provides us with a record of CO_2 for the past 420,000 years (Indermuhle et al., 1999; Monnin et al., 2001). Ice core studies indicate that atmospheric CO_2 at the LGM was on average 200 parts per million by volume (p.p.m.V) (relative to modern day values of >380 p.p.m.V), indicating an over 40% reduction in atmospheric levels during glacial periods.

A combination of climate cooling, decreasing precipitation, and low atmospheric CO_2 surely promoted changes in equatorial vegetation form and function, but in what way and to what extent is still a matter of discussion. There are far fewer palynological sites located in tropical regions relative to those in temperate North America and Eurasia, simply due to geological history. Discovery of new coring sites with good paleoecological reconstruction potential, however, is increasing (Bush, 2002; Mayle *et al.*, 2000).

As a result of the current shortage of tropical palynological records, dynamic global vegetation models (DGVMs) play a key role in filling knowledge gaps where no tropical pollen profiles currently exist (Harrison and Prentice, 2003; Cowling *et al.*, 2004). Vegetation models can also be important for elucidating the underlying mechanisms of vegetation change because models are built upon fundamental principles of plant physiology, biochemistry, and ecosystem ecology. Models differ in the way they emphasize or parameterize particular processes, but they address fundamental biophysical mechanisms nonetheless.

Research effort towards better understanding past vegetation changes is essential, for if we can't explain ecological changes that we know to have occurred within past climates, then how can we place confidence on our estimations about ecological responses to future changes in climate? By outlining knowledge on how plants physiologically and biochemically respond to different abiotic stresses, we can begin to hypothesize about how lowland vegetation may have looked in the past, and how it might be influenced in the future. Equipped with this knowledge, we are able to make more informed decisions concerning aspects of tropical conservation, as well as developing mitigation strategies before a time when unwanted ecological changes may occur.

In this chapter I introduce some of the biochemical and physiological plant processes (mechanisms) that are important in terms of ecosystem-level responses to climate change, focusing on those most influenced by Pleistocene climate. Due to the ability of models to deconvolve co-varying responses, I address the independent versus interactive effects of Pleistocene precipitation, temperature, and atmospheric CO_2 on lowland plant ecology. Whether or not C_4 plants (mostly subtropical and tropical grasslands) experienced widespread proliferation during glacials will be a topic of discussion, one that includes recent research highlighting the potential for over-prediction of C_4 plant abundance due to caveats associated with stable carbon isotope analyses. The latter sections of this review will contain more speculative discussions of the stratification of tropical lowland forests based on the results of different modeling experiments, and of the possible response of tropical soil processes to Pleistocene climate change.

13.2   ECOPHYSIOLOGICAL PRIMER

Natural vegetation can be divided into two broad categories (C_3 and C_4), named according to the number of carbon atoms in the first organic intermediate of photosynthesis. Most plants exhibit the C_3 photosynthetic pathway and include a vast range of herbaceous, woody, and grass species. On the other hand, C_4 species are nearly all grasses typically found in seasonally moist and semi-arid regions of the tropics (Sage *et al.*, 1999).

In C_3 plants the enzyme that is responsible for initiating biochemical reactions involving CO_2 (called "carboxylation") is called "Rubisco" (ribulose-1,5-bisphosphate carboxylase-oxygenase). Rubisco is reactive with both CO_2 and O_2 at rates determined by concentrations of O_2 and CO_2, the catalytic properties of Rubisco, as well as leaf temperature (Pearcy and Ehleringer, 1984). Photorespiration (the oxygenation of Rubisco) causes an overall loss of fixed carbon in the leaf, and is a process that is highly sensitive to changes in leaf temperature. High temperatures reduce the affinity of Rubisco for CO_2 relative to O_2 (called the "specificity factor"), thereby elevating rates of photorespiration (Brooks and Farquhar, 1985).

Carboxylation in C_4 species occurs within structures surrounded by bundle sheath cells (called "Kranz morphology") (Pearcy and Ehleringer, 1984). The bundle sheath functions as a type of "CO_2 pump" in that it actively transports CO_2 across cell membranes to concentrate CO_2 at the site of carboxylation. The primary carboxylating enzyme in C_4 plants, phosphoenolpyruvate (PEP) carboxylase, is non-reactive with O_2, so that photorespiration is far less limiting to plant carbon balance than in C_3 species (Ehleringer and Bjorkman, 1977).

Plant carbon and water relations are tightly linked via the functioning of leaf pores called "stomata". The concentration of CO_2 in intercellular spaces (C_i) is sensed by plants, promoting subsequent feedbacks onto the shape and size of stomata as mediated by changes in stomatal conductance (Farquhar and Sharky, 1982). The ratio of the amount of CO_2 gained relative to the amount of water lost, called "water-use efficiency" (WUE), is a good indicator of plant–water balance and can have a strong influence on overall plant productivity.

^{13}C is a naturally-occurring stable isotope in the atmosphere. The diffusion of $^{12}CO_2$ through stomata is faster than $^{13}CO_2$ simply because $^{12}CO_2$ is a lighter compound. The concentration of ^{13}C in C_3 and C_4 plants, however, differs because ^{13}C is differentially discriminated against, primarily because of the photosynthetic pathway, but also from the plant–water status (Farquhar et al., 1989). C_3 plants discriminate against ^{13}C more than do C_4 plants because Rubisco has a higher affinity for $^{12}CO_2$. Within C_3 plants themselves, the ratio of ^{13}C to ^{12}C (denoted $\delta^{13}C$) varies as a result of differences in plant–water balance, which can be initiated by changes in soil moisture content or atmospheric vapor pressure deficit (VPD) (Lloyd and Farquhar, 1994). When C_3 plants become water-stressed, they are less able to discriminate against ^{13}C, therefore become enriched with the heavier ^{13}C isotope.

Most of what has been speculated about how paleovegetation may have responded to Pleistocene conditions (primarily to low atmospheric CO_2) is based on the results of empirical research involving modern day plant genotypes treated with low atmospheric CO_2, either in isolation or in combination with other environmental stresses (Sage and Reid, 1992; Johnson et al., 1993; Mayeux et al., 1997; Sage and Cowling, 1999; Anderson et al., 2001; Gill et al., 2002). Experiments involving modern day genotypes could be biased because plants may be able to genetically adapt to climatic variations, in specific to long-term changes in atmospheric CO_2 (Ward and Strain, 1997; Ward et al., 2000). In other words, we may be forcing plants adapted to relatively higher CO_2 concentrations to respond to lower CO_2, resulting in an over-sensitivity of plant processes to experimental reductions in the level of CO_2. In contrast, some have suggested that modern day plants may still be exhibiting adaptations to the lower atmospheric CO_2 levels of the past (Sage and Cowling, 1999), thus minimizing potential evolutionary bias.

Generally, however, many of the studies involving plant responses to low CO_2 can be characterized by the following features: (a) experimental species that are annuals and highly genetically modified (i.e., crop species), (b) experimental designs that involve closed-environment chambers, and (c) short-term exposure to CO_2 or other climate treatment. Together, these factors could limit the extent to which empirical knowledge is transferable to growth-limiting environmental controls during the Pleistocene.

13.3   INDEPENDENT VERSUS INTERACTIVE CLIMATE EFFECTS

13.3.1   Global cooling

Paleoecological reconstructions of montane vegetation under glacial climate reveal a sensitivity of vegetation growing along altitudinal gradients (Colinvaux *et al.*, 1996a, b; Street-Perrott *et al.*, 1997; Olago *et al.*, 1999). Paleorecords of vegetation along the Andean flank and within the highlands of Central America and eastern Africa indicate that, in general, altitudinal vegetation ranges shifted down the mountainside in response to glacial cooling. Unique tropical ecosystems, containing non-analogous admixtures of highland and lowland forest species, have been reconstructed from tropical paleosediments (Street-Perrott *et al.*, 1997; Bush *et al.*, 2004).

The influence of glacial cooling may have not been as profound for tropical lowlands as it was for vegetation in montane regions. The average temperature optimum (the temperature at which photosynthesis is the highest) for C_3 plants is around 22–23°C (Collatz *et al.*, 1998). Adaptation to temperatures above the C_3 thermal optimum is possible, many examples of which can be found in plants thriving in very hot regions—such as desert (>40°C) and some tropical grasslands/croplands (>30°C; Crafts-Brandner and Salvucci, 2000; Salvucci and Crafts-Brandner, 2004). Not many studies have investigated the *in situ* thermal optimum of forest species growing in tropical lowlands. A survey of experimental and field data on thermal response of photosynthesis to changes in temperature indicate that woody species tend to exhibit less species-to-species variability in thermal optima than do herbaceous species (Medlyn *et al.*, 2002). Eddy covariance measurements of net ecosystem exchange (NEE) in an old-growth Amazonian forest near Pará (Brazil) show that hour-to-hour changes in temperature result in reduced photosynthetic uptake (Goulden *et al.*, 2004), showing that the thermal optimum of tropical forest species can be exceeded on a daily basis. Based on extrapolations from studies that have been performed (i.e., on seedlings or crop species grown in controlled environments; see Amthor, 1991), some lowland plant species are probably growing near or above their photosynthetic optimum in today's climate. Very little is known about the tolerance levels of tropical lowland forest to changes in temperature extremes; studies show that thermal extremes are more important for plant survival (growth) than monthly or annual climate means (Asner et al., 2000; Tian et al., 2000).

Supporting evidence that tropical forests are growing near their thermal optima comes in part from the results of modeling experiments that are able to isolate the effects of warm temperatures from those of dry atmospheres (i.e., climatic variables that often co-vary). These studies indicate a substantial increase in forest productivity when monthly temperatures are reduced from modern day (Kutzbach et al., 1998; Cowling et al., 2001, 2004; Harrison and Prentice, 2003). A decline in ambient temperature, particularly in the range of the last glacial maximum (between -3 and $-5°C$) probably had little *direct* influence on lowland vegetation, except to perhaps force compositional changes depending on the ability of plant species to either adapt to changing climate or to alter species range. The aforementioned modeling studies show that tropical cooling during glacials could have been a benefit to lowland vegetation. A reduction of photorespiration in response to lowering ambient temperature promotes conservation of carbon in plants. In an already carbon-starved environment (i.e., due to low atmospheric CO_2; Ward et al., 2005) conserved carbon could in turn be re-allocated to organs (i.e., roots, seeds, shoots) or to enable stress tolerance mechanisms.

13.3.2 Decreased glacial precipitation

Vegetation modeling studies performed for African tropical forest indicate that the reduction in potential evapotranspiration (PET) following reduced tropical temperatures could have acted to compensate for small decreases in glacial precipitation (Kutzbach et al., 1998). This mechanism provides an explanation for why some paleovegetation modeling studies show that the area of tropical forest remains the same as today, or even increases, when only glacial temperature and precipitation values are included in LGM simulations (Harrison and Prentice, 2003).

Modern day tropical field and modeling studies, however, indicate that drought is responsible for inter-annual reductions in tropical ecosystem carbon balance, and that this decline is related to a complex interaction of temperature warming, drought, and drought-induced increases in fire frequency (Laurance and Williamson, 2001). Analyses of remotely sensed data for the 1980s in the Amazonian region of northern Brazil indicate that forest stem density is significantly reduced during ENSO years (Batista et al., 1997; Asner et al., 2000) because of the heightened occurrence of drought (Williamson et al., 2000; Laurance et al., 2001; Potter et al., 2001a; Nepstad et al., 2004), with concomitant increases in forest fire frequency (Laurance et al., 2001; Barlow and Peres, 2004).

The accuracy of scaling small-scale hydrological responses to the ecosystem level is problematic for modern day hydrological cycling, and thus becomes even more uncertain in paleoenvironments. The response of lowland forest to drought may not be linear because of interactions with other ecosystem-level processes. Analysis of observed rates of rainfall, runoff, and evaporation in a seasonal forest in Amazonia indicate that tropical rainforest is able to take up and store excess water from year to year, showing lag-times in the response of rainforest to drought (Zeng, 1999). Eddy covariance studies of forest near Santarem (Brazil) indicate that ecosystem carbon—that is, net ecosystem exchange (NEE)—is lost in the wet season and gained in the dry season (Saleska et al., 2003), a response opposite to intuition, but is likely the result of ecosystem-scale responses superimposed on long-term disturbance history.

13.3.3   Low atmospheric CO_2

A decrease in atmospheric CO_2 generally decreases C_3 photosynthesis because the concentration of CO_2 as a photosynthetic substrate is reduced and the competition between CO_2 and O_2 for active binding sites on Rubisco is greater. Reductions in atmospheric CO_2 also promote increases in stomatal conductance in C_3 plants, causing an increase in the rate of water lost to transpiration (Farquhar and Sharkey, 1982).

Closed-chamber experiments using a variety of crop species (wheat, oat, mustard, and bean) indicate that water-use efficiency (WUE) is significantly lower (up to 50%) at low (~200 p.p.m.V) relative to ambient (~360 p.p.m.V) CO_2 (Polley et al., 1993, 1995; Cowling and Sage, 1998). In experimental low-CO_2 treatments, the combination of decreased photosynthesis and lower WUE often results in an overall decrease in C_3 plant productivity (Sage, 1995; Tissue et al., 1995; Mayeaux et al., 1997; Sage and Coleman, 2001). On the other hand, responses of C_4 plants to changes in atmospheric CO_2 are different than C_3 responses because of dissimilarities in morphology and biochemistry (Pearcy and Ehleringer, 1984). Due to the insensitivity of PEP carboxylase to O_2 and because of C_4's active pumping of CO_2 into the bundle sheath, stomatal conductance in C_4 plants responds much less to CO_2 fluctuations than C_3. Consequently, C_4 plants exhibit substantially much higher water-use efficiencies than C_3, even when exposed to sub-ambient atmospheric CO_2 levels.

Leaf-level changes in WUE can be scaled to the level of plant and ecosystem. WUE averaged over the life-span of the plant is estimated from fossil organic matter using stable carbon isotopes (Farquhar et al., 1989). WUE estimated from fossil leaves dating to the mid-Holocene (Araus and Buxo, 1993) and over the last two and a half centuries (Bert et al., 1997) are significantly lower than in modern plant material, likely as a result of elevations in atmospheric CO_2 since the start of the Industrial Revolution. In summary, low atmospheric CO_2 results in low productivity because of low rates of photosynthesis and induced decreases in plant water-use efficiency.

13.3.4   Interactive effects of Pleistocene climate and atmospheric CO_2

Modeling studies can address research questions requiring the separation of interactive multiple factors because of the modular construction of models, a research objective that cannot be accomplished in "real-life" vegetation experiments. Several modeling studies show that it is specifically the *interactive* effects of Pleistocene climate and CO_2 that were likely responsible for most of the reconstructed changes in glacial vegetation form and function (Cowling, 1999; Harrison and Prentice, 2003). Perhaps the most profound illustration of this is found in Harrison and Prentice's (2003) study that incorporates 17 different paleoclimate reconstructions of the LGM in connection with one dynamic global vegetation model. The authors conducted factorial experiments promoting vegetation responses to a combination of climate stressors and were clearly able to show that the interactive effects of low atmospheric CO_2 are most important for vegetation dynamics in equatorial relative to high-latitude regions.

13.4　ECOLOGICAL RESPONSES TO PLEISTOCENE CLIMATE CHANGE

13.4.1　Expansion of C_4 grasslands

The distribution of modern day C_3 and C_4 plants is primarily governed by temperature, but may be modified by changes in atmospheric CO_2 (Ehleringer *et al.*, 1991, 1997; Sage *et al.*, 1999). C_4 plants, with their intrinsically higher WUE, have a strong competitive advantage in hot and arid regions, but the associated energetic costs of the C_4-syndrome (i.e., CO_2-pump) renders them less competitive in regions experiencing moderate to low temperatures. The point at which C_4 plant abundance drops below 50% is commonly referred to as the C_4–C_3 transition temperature. On average, transition temperature is between 20°C and 28°C in a variety of C_3 and C_4 plants surveyed across North America, South America, Asia, Australia, Africa, Europe, and Central America (Ehleringer *et al.*, 1997), and is mechanistically simulated in vegetation models (Cerling *et al.*, 1998; Collatz *et al.*, 1998).

Decreases in atmospheric CO_2, like that during the LGM, are thought as favoring C_4 grasses over C_3 herbaceous plants because of low CO_2 induced increases in C_3 photorespiration (Ehleringer *et al.*, 1997). Robinson (1994) argues that glacial reductions in global temperature would have restricted C_4 plant ranges more than low atmospheric CO_2 would have favored expansion; however, Collatz *et al.* (1998) demonstrate that low CO_2 causes a lowering of the C_4–C_3 transition temperature. A decrease in transition temperature could have expanded the geographical extent to which C_4 grasses are competitive against C_3 herbaceous species.

Analyses of $\delta^{13}C$ of organic fossil sediments can be used to determine the relative proportion of C_3 and C_4 abundance in the past because C_3-dominated soils contain less of the heavier isotope ($\delta^{13}C = -28$ per mil) than C_4 ($\delta^{13}C = -14$ per mil). Unfortunately, very few carbonate sediments in tropical lowlands dating to the LGM have been analyzed for relative C_3 versus C_4 presence, with most of these types of studies being performed in tropical highlands (Aucour *et al.*, 1994; Liu *et al.*, 1996; Sukumar *et al.*, 1995; Street-Perrott *et al.*, 2004). Because the $\delta^{13}C$ of C_3 plants under water stress is similar to that of unstressed C_4 vegetation, Liu *et al.* (2005) caution that reconstructed abundance of LGM C_4 grasslands may be over-estimated. This is because research shows that some tropical C_3 grasslands (e.g., monsoonal China; Liu *et al.*, 2005) were drought-stressed during the LGM, and could have indicated a characteristic C_4–$\delta^{13}C$ sediment profile.

Haberle and Maslin's (1999) analysis of pollen found in ocean sediments at the mouth of the Amazon River, and Kastner and Goni's (2003) study of organic matter composition of Amazonian deep-sea sediments support the hypothesis that C_4 or C_3 grasslands did not extensively expand their range in tropical lowlands such as

Amazonia. Study of pollen cores concentrated in a region in western Brazil indicates uninterrupted lowland forest cover from the last glaciation to the modern day (Bush *et al.*, 2004). On the other hand, some tropical areas seem to have experienced substantial C_4 grassland expansion (e.g., South Africa; Scott, 2002). Wide-scale encroachment of C_4 grasslands in the tropics, therefore, appears to be a regional and not necessarily tropical glacial phenomenon.

13.4.2   Rainforest versus seasonal forest

Based strictly on the physiological effect of low CO_2 in promoting low plant water-use efficiency, one might predict that the occurrence of seasonal forest (with adaptations to regularly-occurring drought) would have been more prominent during glacials relative to today, and that any expansion of seasonal forest may have occurred at the expense of tropical evergreen forest. In support, several palynological studies demonstrate the heightened presence of seasonal forest during glacials, in areas that are covered with rainforest today (Mayle, 2004). Modern day biogeographical studies of forest distributions in central Amazonia (Chaco Forest) indicate that seasonally dry forest was significantly more abundant in the past than at present (Pennington *et al.*, 2000), although species composition was probably much different than today (Mayle, 2004). A comprehensive collection of African pollen data dating to the LGM, for example, indicate that—although data for the lowlands are sparse—modern day tropical rainforest was largely replaced by tropical seasonal forest, while areas that are today dominated by seasonal forest were typically encroached upon by savanna (Elenga *et al.*, 2000). Modeling studies are in agreement with paleo-proxy data. Mayle and Beerling (2004) compare various paleodata with results from dynamic vegetation model simulations, and conclude that southern Amazonia was likely covered by deciduous (seasonal) forest rather than evergreen rainforest.

   Some researchers suggest that the occurrence of fire was greater during the last glacial than at present, not only because of decreased moisture, but also because seedlings may not have been able to grow fast or tall enough to avoid death by fire (i.e., due to low plant carbon resources) (Bond *et al.*, 2003). The increased occurrence of seasonal forest, particularly seasonal forest having a thinner canopy, may have contributed to the ability of tropical lowland ecosystems to overcome persistent fire events. Research by Marod *et al.* (2004) in a seasonal forest in Thailand shows that seedling survival rates for plants grown in canopy gaps (i.e., experiencing high light levels) were greater than those established under dense canopy (i.e., shade). Thus, increased occurrence of open-canopied, seasonal forest at savanna–forest ecotones during glacial periods may have contributed to a decreased probability of grasslands invading forest, despite the catalyst of increased fire frequency.

13.4.3   Vertical stratification of glacial forests

The structure and stratification of tropical lowland forest may have been substantially different during the Pleistocene relative to today (Cowling, 2004). Canopy density, the vertical and horizontal stratification of forest canopies, implicitly encompasses various architectural traits—such as branching, crown depth, and tree height (Meir *et al.*, 2000; Clark *et al.*, 2004; Lalic and Mihailovic, 2004). The silvicultural and vegetation modeling communities tend to parameterize canopy density in terms of an index called "leaf area index" (LAI)—that is, one-sided leaf surface area (m^2) relative to ground cover (m^2). In general, LAI values $<3\,m^2\,m^{-2}$ are found in association with open vegetation types like tropical grasslands, with values between 4 and $8\,m^2\,m^{-2}$ associated with closed-canopied rainforests (Chapin *et al.*, 2002).

A recent simulation of tropical Africa indicates that LAI declines from 5.2 to $4.1\,m^2\,m^{-2}$ in response to LGM climate relative to modern day (Cowling *et al.*, 2004). Harrison and Prentice (2003) model a global decrease in LAI from 4.28 to approximately $4.03\,m^2\,m^{-2}$ as a result of cooler and drier LGM climate and lower atmospheric CO_2 conditions. Most vegetation–climate simulations indicate that vegetation canopy density was probably lower under glacial climate, although models disagree on the actual degree of reductions relative to today (Levis *et al.* 1999; Cowling *et al.*, 2001; Harrison and Prentice, 2003). Modern day observations in the eastern, central, and southwestern regions of Amazonia indicate large differences in forest structure between permanent study plots (Vieira *et al.*, 2004). Thus, landscape heterogeneity (i.e., of tree height and canopy density) may have been even more profound during glacials, where low concentrations of atmospheric CO_2 could have promoted further variations in canopy thickness and tree height.

An opened canopy in response to Pleistocene climate may have permitted a more developed understory layer of grass and herbaceous plants, although the functional types and dominant species were likely much different than today. A thinning canopy allows more solar radiation to reach the ground, causing surface temperatures to rise, but also cause greater diurnal variations in temperature (Leopoldo *et al.*, 1993; Potter *et al.*, 2001b; Thery, 2001; Montgomery, 2004). Species that are light-intolerant, or sensitive to wide fluctuations in temperature, would have been detrimentally affected by such canopy changes. Alternatively, understory seedlings adapted to high light environments and tolerant of relatively wider temperature ranges would have thrived in the new glacial forest (canopy) microclimate.

Thinner canopies also result in less moisture retention within forest strata because of the greater mixing of intra-canopy air, as well as from reduced rates of evapo-transpiration (Costa and Foley, 1997; Albertson *et al.*, 2001). Understory fern species adapted to relatively high light levels and to wide-ranging atmospheric moisture levels, for example, may have been favored in open-canopied glacial forests. Tropical forest epiphytes that require high canopy humidity may have been severely disadvantaged in glacial forest microclimates. The tight coupling between canopy density and canopy microclimate holds strong implications for species phylogenies. Reconstructed low population size and narrow distribution of some species, for example, found in combination with larger population size and widening distributions of other species, may be indicators of changes in paleoforest structure.

13.5   SOIL PROCESSES AND PLEISTOCENE CLIMATE CHANGE

Within discussions of how tropical ecosystems may have adjusted to climate change during glacial periods, very little is said of below-ground processes, which is surprising considering the importance attributed to soil processes in modern day and future ecosystem studies (Ball and Drake, 1997; Ross *et al.*, 2002; Lenton and Huntingford, 2003; Sotta *et al.*, 2004; Trumbore, 2006). Whether or not soil processes act as a negative or positive feedback to climate change depends on the balance of the response of net primary production of terrestrial vegetation, and of soil respiration and litter decomposition, to changes in temperature, soil moisture, and atmospheric CO_2 (Kirschbaum, 2000).

Tropical climate cooling may have reduced rates of heterotrophic respiration. Glacial changes in leaf tissue chemistry—such as plant litter with relatively more nitrogen-based complex molecules than carbon-based (i.e., one of the observed physiological effects of low CO_2)—may have resulted in slower rates of litter decomposition. The quantity of tannins (nitrogen-based compounds) is shown to relate directly to decomposition in *Cecropia* species in secondary forest in central Amazonia (Mesquita *et al.*, 1998). Changes in litter chemistry, therefore, could have promoted a continual build-up of tropical soil carbon throughout the duration of a glacial cycle. Most of the carbon stored in modern day tropical ecosystems is found in above-ground structures, but this may have been reversed during periods of glacial advance.

The effect of reduced plant productivity for lowering quantity of leaf litter has traditionally been considered the most important variable in modifying soil carbon, although there are few experimental data to verify this statement. A C_3 perennial shrub subjected to a range of CO_2 levels, from sub- (200 p.p.m.V) to super-ambient (550 p.p.m.V) CO_2 (and no other changes in climate), shows only a marginal (11%) reduction in bulk soil carbon over 4 years (Gill *et al.*, 2002). Rates of respiration, translocation, and nitrate reduction are observed in soybean grown in low-CO_2 growth chambers (Bunce, 2004). Modeling studies tend to indicate decreases in soil carbon storage at the LGM, primarily because model calculation of soil carbon is strongly dependent on the type of ecosystem present, as well as the productivity (NPP) of that ecosystem (Friedlingstein *et al.*, 1992; Prentice *et al.*, 1993; Kubatzki and Claussen, 1999; Levis *et al.*, 1999; Kaplan *et al.*, 2002). Kirschbaum (2000) highlights the problems associated with assuming that there is no difference in soil carbon storage for past and present biome types.

Another factor that may have influenced processes modifying soil carbon storage during glacials is the response of groundwater and water-table levels to decreases in sea level. Faure *et al.* (2002) believe that the 120-m glacial drop in current sea level could have increased hydrostatic pressure on continental water sources, causing an increase in groundwater flow to coastal regions, and a subsequent decline in regional water-table levels (Clapperton, 1993). Drier surface soils could have reduced microbial decomposition and contributed to an increase in tropical soil carbon storage during glacial periods. Without a better grasp of the relative strength of soil carbon processes to changes in climate, we will be unable to confidently predict how below-ground processes may have been altered during glacial and interglacial cycles of the Pleistocene.

13.6 CONCLUSION

It is specifically the *interaction* of low atmospheric CO_2 and global cooling and aridity that is critical for understanding the potential response of plants to changes in

Pleistocene climate. The combination of modeling and palynological data should put to rest the notion that lowland tropical forests were severely restricted in glacial times by expanding C_4-dominated grasslands. Rather, evidence indicates that seasonal forest likely encroached on rainforest in lowland regions and that grasslands expanded at forest–grassland boundaries. Perhaps one of the potentially largest differences in the character of lowland tropical forests in the Pleistocene relative to today involves changes to canopy density, and subsequent changes in species composition of canopy strata. One of the greatest uncertainties with respect to tropical ecosystem responses to Pleistocene climate change involves below-ground soil processes. If we assume that decreases in vegetation productivity had the dominant effect on soil processes, then soil carbon storage was lower during glacials than present. Alternatively, if we assume that the temperature and moisture response of soil processes drives below-ground carbon storage, then tropical soil storage increased during glacials relative to today. As with modern day ecosystem research, much has to be learnt about the response of below-ground processes to biotic and abiotic stress in order to predict how they might have changed in the past, or how they might change in the future.

13.7 REFERENCES

Albertson, J., Katul, G., and Wiberg, P. (2001) Relative importance of local and regional controls on coupled water, carbon and energy fluxes. *Advances in Water Resources* **24**, 1103–1118.

Amthor, J. S. (2001) Effects of atmospheric CO_2 concentration on wheat yield: Review of results from experiments using various approaches to control CO_2 concentration. *Field Crops Research* **73**, 1–34.

Anderson, L., Maherali, H., Johnson, H., Polley, H., and Jackson, R. (2001) Gas exchange and photosynthetic acclimation over sub-ambient to elevated CO_2 in C_3–C_4 grassland. *Global Change Biology* **7**, 693–707.

Araus, J. and Buxo, R. (1993) Changes in carbon isotope discrimination in grain cereals from the north-western Mediterranean basin during the past seven millennia. *Australian Journal of Plant Physiology* **20**, 117–128.

Asner, G. P., Townsend, A. R., and Braswell, B. H. (2000) Satellite observation of El Niño effects on Amazon forest phenology and productivity. *Geophysical Research Letters* **27**, 981–984.

Aucour, A. M., Hillaire-Marcel, C., and Bonnefille, R. (1994) Late Quaternary biomass changes from C-13 measurements in a highland peatbog from equatorial Africa (Burundi). *Quaternary Research* **41**, 225–233.

Ball, A. S. and Drake, B. G. (1997) Short-term decomposition of litter produced by plants grown in ambient and elevated atmospheric CO_2 concentrations. *Global Change Biology* **3**, 29–35.

Barlow, J. and Peres, C. A. (2004) Ecological responses to El Niño-induced surface fires in central Brazilian Amazonia: Management implications for flammable tropical forests. *Philosophical Transactions of the Royal Society, London (B)* **359**, 367–380.

Batista, G. T., Shimabukuro, Y. E., and Lawrence, W. T. (1997) The long-term monitoring of vegetation cover in the Amazonian region of northern Brazil using NOAA-AVHRR data. *International Journal of Remote Sensing* **18**, 3195–3210.

Behling, H. (2002) South and southeast Brazilian grasslands during Late Quaternary times: A synthesis. *Palaeogeography, Palaeoclimatology, Palaeoecology* **177**, 19–27.

Behling, H. and Negrelle, R. R. B. (2001) Tropical rain forest and climate dynamics of the Atlantic lowland, southern Brazil, during the Late Quaternary. *Quaternary Research* **56**, 383–389.

Bert, D., Leavitt, S., and Dupouey, J. (1997) Variations of wood delta C-13 and water-use efficiency of *Abies alba* during the last century. *Ecology* **78**, 1588–1596.

Bond, W. J., Midgley, G. F., and Woodward, F. I. (2003) The importance of low atmospheric CO_2 and fire in promoting the spread of grasslands and savannas. *Global Change Biology* **9**, 973–982.

Brooks, A. and Farquhar, G. (1985). Effect of temperature on the CO_2/O_2 specificity of ribulose-1,5-bisphosphate carboxylase/oxygenase and the rate of respiration in the light. *Planta* **165**, 397–406.

Bunce, J. A. (2004) A comparison of the effect of carbon dioxide concentration and temperature on respiration, translocation and nitrate reduction in darkened soybean leaves. *Annals of Botany* **93**, 665–669.

Bush, M. B. (2002) Distributional change and conservation on the Andean flank: A palaeo-ecological perspective. *Global Ecology and Biogeography* **11**, 463–473.

Bush, M. B. and Silman, M. R. (2004) Observations on late-Pleistocene cooling and precipitation in the lowland Neotropics. *Journal of Quaternary Science* **19**, 677–684.

Bush, M. B., de Oliveira, P. E., Colinvaux, P. A., Miller, M. C., and Moreno, J. E. (2004) Amazonian paleoecological histories: One hill, three watersheds. *Palaeoegeography, Palaeoeoclimatology, Palaeoecology* **214**, 359–393.

Cerling, T. E., Ehleringer, J. R., and Harris, J. M. (1998) Carbon dioxide starvation, the development of C_4 ecosystems, and mammalian evolution. *Philosophical Transactions of the Royal Society, London (B)* **353**, 159–171.

Chapin, F. S., Matson, P. A., and Mooney, H. A. (2002) *Principles of Terrestrial Ecosystem Ecology*. Springer-Verlag, New York.

Clapperton, C. M. (1993) Nature of environmental changes in South America at the Last Glacial Maximum. *Palaeoegeography, Palaeoeoclimatology, Palaeoecology* **101**, 189–208.

Clark, M. L., Clark, D. B., and Roberts, D. A. (2004) Small-footprint lidar estimation of sub-canopy elevation and tree height in a tropical rain forest landscape. *Remote Sensing of Environment* **91**, 68–89.

Colinvaux, P. A., De Oliveira, P. E., Moreno, J. E., Miller, M. C., and Bush, M. B. (1996a) A long pollen record from lowland Amazonia: Forest and cooling glacial times. *Science* **274**, 85–89.

Colinvaux, P. A., Liu, K. B., De Oliveira, P., Bush, M. B., Miller, M. C., and Kannan, M. S. (1996b) Temperature depression in the lowland tropics in glacial times. *Climatic Change* **32**, 19–33.

Collatz, G., Berry, J., and Clark, J (1998). Effects of climate and atmospheric CO_2 partial pressure on the global distribution of C_4 grasses: Past, present and future. *Oecologia* **114**, 441–454.

Cook, K. H. and Vizy, E. K. (2006) South American climate during the Last Glacial Maximum: Delayed onset of the South American monsoon. *Journal of Geophysical Research: Atmospheres* **111**, article #D02110.

Costa, M. H. and Foley, J. A. (1997) Water balance of the Amazon Basin: Dependence on vegetation cover and canopy conductance. *Journal of Geophysical Research: Atmospheres* **102**(D20), 23973–23989.

Cowling, S. A. (1999) Simulated effects of low atmospheric CO_2 on structure and composition of North American vegetation at the Last Glacial Maximum. *Global Ecology and Biogeography* **8**, 81–93.

Cowling, S. A. (2004) Tropical forest structure: A missing dimension to Pleistocene landscapes. *Journal of Quaternary Science* **19**, 733–743.

Cowling, S. A. and Sage, R. F. (1998) Interactive effects of low atmospheric CO_2 and elevated temperature on growth, photosynthesis and respiration in *Phaseolus vulgaris*. *Plant Cell and Environment* **21**, 427–435.

Cowling, S. A., Maslin, M. A., and Sykes, M. T. (2001). Paleovegetation simulations of lowland Amazonia and implications for neotropical allopatry and speciation. *Quaternary Research* **55**, 140–149.

Cowling, S. A., Betts, R. A., Cox, P. M., Ettwein, V. J., Jones, C. D., Maslin, M. A., and Spall, S. A. (2004) Contrasting simulated past and future responses of the Amazonian forest to atmospheric change. *Philosophical Transactions of the Royal Society, London (B)* **359**, 539–547.

Crafts-Brandner, S. J. and Salvucci, M. E. (2000) Rubisco activase constrains the photosynthetic potential of leaves at high temperature and CO_2. *Proceedings of the National Academy of Sciences (U.S.A)* **97**, 13430–13435.

Ehleringer, J. and Bjorkman, O. (1977) Quantum yields for CO_2 uptake in C_3 and C_4 plants—dependence on temperature, CO_2, and O_2 concentration. *Plant Physiology* **59**, 86–90.

Ehleringer, J., Sage, R., Flanagan, L., and Pearcy, R. (1991). Climate change and the evolution of C_4 photosynthesis. *Trends in Ecology and Evolution* **6**, 95–99.

Ehleringer, J., Cerling, T., and Helliker, B. (1997). C_4 photosynthesis, atmospheric CO_2 and climate. *Oecologia* **112**, 285–299.

Elenga, H., Peyron, O., Bonnefille, R., Jolly, D., Cheddadi, R., Guiot, J., Andrieu, V., Bottema, S., Buchet, G. de Beaulieu, J. L. *et al.* (2000) Pollen-based biome reconstruction for southern Europe and Africa 18,000 yr BP. *Journal of Biogeography* **27**, 621–634.

Farquhar, G. and Sharky, T. (1982). Stomatal conductance and photosynthesis. *Annual Review of Plant Physiology* **33**, 317–345.

Farquhar, G. D., Ehleringer, J. R., and Hubrick, K. T. (1989) Carbon isotope discrimination and photosynthesis. *Annual Review of Plant Physiology and Plant Molecular Biology* **40**, 503–537.

Faure, H., Walter, R. C., and Grant, D. R. (2002) The coastal oasis: Ice age springs on emerged continental shelves. *Global and Planetary Change* **33**, 47–56.

Friedlingstein, P., Delire, C., Muller, J. F., and Gerard, J. C. (1992) The climate induced variation of the continental biosphere: A model simulation of the last glacial maximum. *Geophysical Research Letters* **19**, 897–900.

Gill, R., Polley, H., Johnson, H., Anderson, L., Maherali, H., and Jackson, R. (2002). Nonlinear grassland responses to past and future atmospheric CO_2. *Nature* **417**, 279–282.

Goulden, M. L., Miller, S. D., da Rocha, H. R., Menton, M. C., de Freitas, H. C., Figueira, A. M. E. S., and de Sousa, C. A. D. (2004) Diel and seasonal patterns of tropical forest CO_2 exchange. *Ecological Applications* **14**, S42–S54.

Guilderson, T. P., Fairbanks, R. G., and Rubenstone, J. L. (1994) Tropical temperature variations since 20 000 years ago: Modulating inter-hemispheric climate change. *Science* **263**, 663–665.

Haberle, S. G. and Maslin, M. A. (1999) Late Quaternary vegetation and climate change in the Amazon basin based on a 50 000 year pollen record from the Amazon fan, ODP site 932. *Quaternary Research* **51**, 27–38.

Harrison, S. P. and Prentice, I. C. (2003) Climate and CO_2 controls on global vegetation distribution at the last glacial maximum: Analysis based on palaeovegetation data, biome modeling and palaeoclimate simulations. *Global Change Biology* **9**, 983–1004.

Indermuhle, I., Stocker, T., Joos, F., Fischer, H., Smith, H., Wahlen, M., Deck, B., Mastro-
    ianni, D., Tschumi, J., Blunier, T. *et al.* (1999). Holocene carbon-cycle dynamics based on
    CO_2 trapped in ice at Taylor Dome, Antarctica. *Nature* **398**, 121–126.
Johnson, H., Polley, H., and Mayeaux, H. (1993). Increasing CO_2 and plant–plant interactions:
    Effects on natural vegetation. *Vegetatio* **104–105**, 157–170.
Kaplan, J. O., Prentice, I. C., Knorr, W., and Valdes, P. J. (2002), Modeling the dynamics of
    terrestrial carbon storage since the Last Glacial Maximum. *Geophysical Research Letters*
    29(22), article #2074.
Kastner, T. P. and Goni, M. A. (2003). Constancy in the vegetation of the Amazon Basin during
    the late Pleistocene: Evidence from the organic matter composition of Amazon deep sea
    fan sediments. *Geology* **31**, 291–294.
Kirschbaum, M. U. F. (2000) Will changes in soil organic carbon act as a positive or negative
    feedback on global warming? *Biogeochemistry* **49**, 21–51.
Kubatzki, C. and Claussen, M. (1999) Simulation of the global biogeophysical interactions
    during the Last Glacial Maximum. *Climate Dynamics* **14**, 461–471.
Kutzbach, J. E., Gallimore, R., Harrison, S. P., Behling, P., Selin, R., and Laarif, F. (1998)
    Climate and biomes simulations for the past 21,000 years. *Quaternary Science Reviews* **17**,
    473–506.
Lalic, B. and Mihailovic, D. T. (2004) An empirical relation describing leaf area density inside
    the forest for environmental modeling. *Journal of Applied Meteorology* **43**, 641–645.
Laurance, W. F. and Williamson, G. B. (2001). Positive feedbacks among forest fragmentation,
    drought, and climate change in the Amazon. *Conservation Biology* **17**, 771–785.
Laurance, W. F., Williamson, G. B., Delamonica, P., and Olivera, P. (2001) Effects of a strong
    drought on Amazonian forest fragments and edges. *Journal of Tropical Ecology* **17**, 771–
    785.
Lenton, T. M. and Huntingford, C. (2003) Global terrestrial carbon storage and uncertainties in
    its temperature sensitivity examined with a simple model. *Global Change Biology* **9**, 1333–
    1352.
Leopoldo, P. R., Chaves, J. G., and Franken, W. K. (1993). Solar energy budgets in central
    Amazonian ecosystems: A comparison between natural forest and bare soil areas. *Forest
    Ecology and Management* **59**, 313–328.
Levis, S., Foley, J. A., and Pollard, D. (1999) CO_2, climate, and vegetation feedbacks at the Last
    Glacial Maximum. *Journal of Geophysical Research: Atmospheres* **104**, 31191–31198.
Liu, B., Phillips, F., and Campbell, A. (1996). Stable carbon and oxygen isotopes of pedogenic
    carbonate, Ajo Mountain, southern Arizona: Implications for palaeoenvironmental
    change. *Palaeoceanography, Palaeclimatology, Palaeoecology* **124**, 233–246.
Liu, W. G., Feng, X. H., Ning, Y. F., Zhang, Q. L., Cao, Y. N., and An, Z. S. (2005) Delta C-13
    variation of C_3 and C_4 plants across an Asian monsoon rainfall gradient in arid north-
    western China. *Global Change Biology* **11**, 1094–1100.
Lloyd, J. and Farquhar, G. D. (1994) ^{13}C discrimination during CO_2 assimilation by the
    terrestrial biosphere. *Oecologia* **99**, 201–215.
Marod, D., Kutintara, U., Tanaka, H., and Nakashizuka, T. (2004) Effects of drought and fire
    on seedling survival and growth under contrasting light conditions in a seasonal tropical
    forest. *Journal of Vegetation Science* **15**, 691–700.
Mayeux, H., Johnson, H., Polley, H., and Malone, S. (1997). Yield of wheat across a sub-
    ambient carbon dioxide gradient. *Global Change Biology* **3**, 269–278.
Mayle, F. E. (2004) Assessment of the Neotropical dry forest refugia hypothesis in the light of
    palaeoecological data and vegetation model simulations. *Journal of Quaternary Science* **19**,
    713–720.

Mayle, F. E. and Beerling, D. J. (2004) Late Quaternary changes in Amazonian ecosystems and their implications for global carbon cycling. *Palaeogeography, Palaeoclimatology, Palaeoecology* **214**, 11–25.

Mayle, F. E., Burbridge, R., and Killeen, T. J. (2000). Millennial-scale dynamics of southern Amazonian rain forests. *Science* **290**, 2291–2294.

Medlyn, B. E., Dreyer, E., Ellsworth, D., Forstreuter, M., Harley, P. C., Kirschbaum, M. U. F., Le Roux, X., Montpied, P., Strassemeyer, J., Walcroft, A. *et al.* (2002) Temperature response of parameters of a biochemically based model of photosynthesis, II: A review of experimental data. *Plant Cell and Environment* **25**, 1167–1179.

Meir, P., Grace, J., and Miranda, A. C. (2000) Photographic method to measure the vertical distribution of leaf area density in forests. *Agricultural and Forest Meteorology* **102**, 105–111.

Mesquita, R. D., Workman, S. W., and Neely, C. L. (1998) Slow litter decomposition in a *Cecropia*-dominated secondary forest of central Amazonia. *Soil Biology and Biochemistry* **30**, 167–175.

Mix, A. C., Morey, A. E., Pisias, N. G., and Gosteetler, S. W. (1999) Foraminiferal faunal estimates of paleotemperature: Circumventing the no-analog problem yields cool ice age tropics. *Paleoceanography* **14**, 350–359.

Monnin, E., Indermuhle, A., Dallenbach, A., Fluckiger, J., Stauffer, B., Stocker, T., Raynaud, D., and Barnola, J. (2001). Atmospheric CO_2 concentrations over the last glacial termination. *Science* **291**, 112–114.

Montgomery, R. A. (2004) Effects of understory foliage on patterns of light attenuation near the forest floor. *Biotropica* **36**, 33–39.

Mora, G. and Pratt, L. M. (2001) Isotopic evidence for cooler and drier conditions in the tropical Andes during the last glacial stage. *Geology* **29**, 519–522.

Nepstad, D., Lefebvre, P., Da Silva, U. L., Tomasella, J., Schlesinger, P., Solorzano, L., Moutinho, P., Ray, D., and Benito, J. G. (2004) Amazon drought and its implications for forest flammability and tree growth: A basin-wide analysis. *Global Change Biology* **10**, 704–717.

Olago, D. O., Street-Perrott, F. A., Perrott, R. A., Ivanovich, M., and Harkness, D. D. (1999) Late Quaternary glacial–interglacial cycle of climatic and environmental change on Mount Kenya, Kenya. *Journal of African Earth Sciences* **29**, 593–618.

Pearcy, R. W. and Ehleringer, J. (1984) Comparative ecophysiology of C_3 and C_4 plants. *Plant Cell and Environment* **7**, 1–13.

Pennington, R. T., Prado, D. E., and Pendry, C. A. (2000) Neotropical seasonally dry forest and Quaternary vegetation changes. *Journal of Biogeography* **27**, 261–273.

Polley, H., Johnson, H., Marino, B., and Mayeux, H. (1993) Increases in C_3 plant water-use efficiency and biomass over glacial to present CO_2 concentrations. *Nature* **361**, 61–64.

Polley, H., Johnson, H., and Mayeux, H. (1995) Nitrogen and water requirements of C_3 plants grown at glacial to present carbon dioxide concentrations. *Functional Ecology* **9**, 86–96.

Potter, C., Klooster, S., de Carvalho, C. R., Genovese, V. B., Torregrosa, A., Dungan, J., Bobo, M., and Coughlan, J. (2001a) Modeling seasonal and interannual variability in ecosystem carbon cycling for the Brazilian Amazon. *Journal of Geophysical Research: Atmospheres* **106**, 10423–10446.

Potter, B. E., Teclaw, R. M., and Zasada, J. C. (2001b) The impact of forest structure on near-ground temperatures during two years of contrasting temperature extremes. *Agricultural and Forest Meteorology* **106**, 331–336.

Prentice, I. C., Sykes, M. T., Lautenschlager, M., Harrison, S. P., Denissenko, O., and Bartlein, P. J. (1993) Modeling global vegetation patterns and terrestrial carbon storage at the last glacial maximum. *Global Ecology and Biogeography: Letters* **3**, 67–76.

Robinson, J. M. (1994) Speculations on carbon dioxide starvation, late Tertiary evolution of stomatal regulation and floristic modernization. *Plant Cell and Environment* **17**, 345–354.

Ross, D. J., Tate, K. R., Newton, P. C. D., and Clark, H. (2002) Decomposability of C_3 and C_4 grass litter sampled under different concentrations of atmospheric carbon dioxide at a natural CO_2 spring. *Plant and Soil* **240**, 275–286.

Sage, R. F. (1995) Was low atmospheric CO_2 during the Pleistocene a limiting factor for the origin of agriculture? *Global Change Biology* **1**, 93–106.

Sage, R. and Coleman, J. (2001) Effects of low atmospheric CO_2 on plants: More than a thing of the past. *Trends in Plant Sciences* **6**, 18–24.

Sage, R. and Cowling, S. (1999) Implications of stress in low CO_2 atmospheres of the past: Are today's plants too conservative for a high CO_2 world? In: H. M. Y Luo (ed.), *Carbon Dioxide and Environmental Stress* (pp. 289–308). Academic Press, San Diego.

Sage, R. and Reid, C. (1992) Photosynthetic acclimation to sub-ambient CO_2 (20 Pa) in the C_3 annual *Phaseolus vulgaris*. *Photosynthetica* **27**, 605–617.

Sage, R. F., Wedin, D. A., and Meirong, L. (1999) The biogeography of C_4 photosynthesis: Patterns and controlling factors. C_4 *Plant Biology* (pp. 313–373). Academic Press, San Diego.

Saleska, S. R., Miller, S. D., Matross, K. M., Goulden, M. L., Wofsy, S. C., da Rocha, H. R., de Camargo, P. B., Crill, P., Daule, B. C., de Freitas, H. C. *et al.* (2003) Carbon in Amazon forests: Unexpected seasonal fluxes and disturbance-induced losses. *Science* **302**, 1554–1557.

Salvucci, M. E. and Crafts-Brandner, S. J. (2004) Inhibition of photosynthesis by heat stress: The activation state of Rubisco as a limiting factor in photosynthesis. *Physiologia Plantarum* **120**, 179–186.

Scott, L. (2002) Grassland development under glacial and interglacial conditions in southern Africa: Review of pollen, phytolith, and isotope evidence. *Palaeogeography, Palaeoclimatology, Palaeoecology* **177**, 47–57.

Sotta, E. D., Meir, P., Malhi, Y., Nobre, A. D., Hodnett, M., and Grace, J. (2004) Soil CO_2 efflux in a tropical forest in the central Amazon. *Global Change Biology* **10**, 601–617.

Street-Perrott, F. A., Huang, Y. S., Perrott, R. A., Eglinton, G., Barker, P., Benkhelifa, L., Harkness, D. D., and Olago, D. O. (1997) Impact of lower atmospheric carbon dioxide on tropical mountain ecosystems. *Science* **278**, 1422–1426.

Street-Perrott, F. A., Ficken, K. J., Huang, Y. S., and Eglinton, G. (2004) Late Quaternary changes in carbon cycling on Mt. Kenya, East Africa: An overview of the delta C-13 record in lacustrine organic matter. *Quaternary Science Reviews* **23**, 861–879.

Stute, M., Forster, M., Frischkorn, H., Serejo, A., Clark, J. F., Schlosser, P., Broecker, W. S., and Bonani, G. (1995) Cooling of tropical Brazil (5°C) during the last glacial maximum. *Science* **269**, 379–383.

Sukumar, R., Suresh, H. S., and Ramesh, R. (1995) Climate change and its impact on tropical montane ecosystems in southern India. *Journal of Biogeography* **22**, 533–536.

Thery, M. (2001) Forest light and its influence on habitat selection. *Plant Ecology* **153**, 251–261.

Tian, H., Melillo, J. M., Kicklighter, D. W., McGuire, A. D., Helfrich, J., Moore, B., and Vörösmarty, C. J. (2000) Climatic and biotic controls on annual carbon storage in Amazonian ecosystems. *Global Ecology and Biogeography* **9**, 315–335.

Tissue, D., Griffin, K., Thomas, R., and Strain, B. (1995) Effects of low and elevated CO_2 on C_3 and C_4 annuals, II: Photosynthesis and leaf biochemistry. *Oecologia* **101**, 21–28.

Trumbore, S. (2006) Carbon respired by terrestrial ecosystems: Recent progress and challenges. *Global Change Biology* **12**, 141–153.

Urrego, D. H., Silman, M. R., and Bush, M. B. (2005) The Last Glacial Maximum: Stability and change in a western Amazonian cloud forest. *Journal of Quaternary Science* **20**, 693–701.

Vieira, S., de Camargo, P. B., Selhorst, D., da Silva, R., Hutyra, L., Chambers, J. Q., Brown, I. F., Higuchi, N., dos Santos, J., Wofsy, S. C. *et al.* (2004) Forest structure and carbon dynamics in Amazonian tropical rain forest. *Oecologia* **140**, 468–479.

Ward, J. and Strain, B. (1997) Effects of low and elevated CO_2 partial pressure on growth and reproduction of *Arabidopsis thaliana* from different elevations. *Plant Cell and Environment* **20**, 254–260.

Ward, J., Antonovics, J., Thomas, R., and Strain, B. (2000) Is atmospheric CO_2 a selective agent on model C_3 annuals? *Oecologia* **123**, 330–341.

Ward, J. K., Harris, J. M., Cerling, T. E., Wiedenhoeft, A., Lott, M. J., Dearing, M. D., Coltrain, J. B., and Ehleringer, J. R. (2005) Carbon starvation in glacial trees recovered from the La Brea tar pits, southern California. *Proceedings of the National Academy of Sciences (U.S.A)* **102**, 690–694.

Williamson, G. B., Laurance, W. F., Oliveira, A. A., Delamonica, P., Gascon, C., Lovejoy, T. E., and Pohl, L. (2000) Amazonian tree mortality during the 1997 El Niño drought. *Conservation Biology* **14**, 1538–1542.

Zeng, N. (1999) Seasonal cycle and interannual variability in the Amazon hydrologic cycle. *Journal of Geophysical Research: Atmospheres* **104**, 9097–9106.

14

Modeling future effects of climate change on tropical forests

L. Hannah, R. A. Betts, and H. H. Shugart

14.1 INTRODUCTION

Alterations in climate (or even the natural variation within the current climate) can affect forest communities by altering the internal processes or by altering the proportions of different species in the forests. Experience in assessing the consequences of major climate change is based mostly on paleo-reconstructions for northern hemisphere forests responding to the climate warming that followed the last ice age. These reconstructions demonstrate that climate change also can alter forests by tearing them apart and re-assembling them in novel combinations of species. This process is dramatic in temperate zones and less well-documented but no less certain in the tropics. While the evidence from the past is clear on these points, it is not abundant worldwide. Forming a complete picture of the past is elusive in many tropical regions, even those as prominent as the Amazon, and future climate change may lack past analogs. Our ability to understand the future based on our understanding of what has happened so far in tropical forests therefore faces serious limitations.

Computer modeling of forests can bridge some parts of this gap in understanding. It can be used to explore sets of future climatic conditions that do not currently exist or which have never existed in the history of the Earth. Currently, a wide range of models have been applied to predicting changes in vegetation in response to climate. These models have different data demands and are likely to have rather different domains of applicability. Some have been tested under novel conditions; others draw their credibility by their synthesis of the current best knowledge of ecosystem processes. The models are all flavored by the intended applications and the interests of their developers.

The most frequently applied modeling techniques have their theoretical underpinnings in ecology and ecophysiology. Ecological niche theory provides a framework for understanding two important types of models: species distribution models and gap models. Ecophysiology is relevant to understanding dynamic vegetation models, including the recently developed Earth System Models.

Through an evolving ecological literature, two concepts of niche have emerged. The first involves the factors that control the geographic distributions of species. This concept originated with Joseph Grinnell (1917) and finds substantial application among management-oriented ecologists today (see Shugart, 1998 for a review). It refers to the environmental requirements of species—for example, the range of temperatures a species can withstand—and may be termed *environmental niche* (Guisan and Thuiller, 2005). An alternate concept of the niche was introduced by Charles Elton (1927) and defined species niches based on feeding relations. This *trophic niche* concept refers to the way a species obtains and uses resources, especially with respect to other species—for example, a species role within a food web. Over the years, the initial "who eats whom" trophic niche concept was developed to emphasize competitive (as opposed to the predator/prey relations implied by Elton's trophic definition). Hutchinson (1957) attempted a synthesis of the Eltonian and Grinellian niche concepts relating the overlap in environmental requirements (Grinellian niche) with the likelihood of strong competitive interactions (a post-Eltonian niche concept). In retrospect, this latter development was very important in motivating theory but somewhat less successful at reconciling two rather different concepts of the niche.

Environmental niche theory provides a framework for models that aim to describe species distributions with respect to current and future climate. These models are therefore sometimes referred to as "niche models" (e.g., Peterson *et al.*, 2002) but are also widely known as "climate envelope", "bioclimatic", or "species distribution" models (Guisan and Zimmermann, 2000; Guisan and Thuiller, 2005). In these models, a direct or statistical relationship is established between a species known distribution and current climate, and rules to describe this relationship are developed and applied to future climates. Theory would suggest that each species has a total range of environmental tolerance, or *fundamental niche*, that is greater than the range it actually occupies, or *realized niche*, due to competition, dispersal history, and other factors—but this distinction is little formalized in models of environmental niche.

Trophic niche theory suggests that species compete for resources and evolve specific resource competition strategies relative to other species. An analogy is sometimes given that the trophic niche is a species' "job" while the environmental niche is its "address". Forest gap models have been developed to simulate resource competition at a small site approximately the size of a tree-fall gap in the forest, using quantitative information about species growth rates and other parameters. Qualitative models including trophic niche interactions have also been developed (e.g., the FATE model; Moore and Noble, 1990). If resource competition determines species survival at a site, it then also influences its distribution at a landscape scale, and these models have begun to be applied on broader scales as computing power has improved.

Ecophysiology theory allows construction of models that describe how plants fix and partition carbon, including the cycling of carbon in soil pools. For some well-studied crop plants, it allows building models of how individual species will respond to specific growth conditions. Tropical forest species are not sufficiently well-studied to be amenable to species-based ecophysiological modeling. However, broad models of carbon partitioning are built from first principles and can be applied in all regions of the world. These models provide information on plant growth forms (or "plant-functional type") that may dominate at a particular site under specified climatic conditions and disturbance regimes. These models are known as "dynamic global vegetation models" (DGVMs). They can be important in simulating distributions of tropical biomes—such as savanna, dry forest, and wet forest. Simplified versions of these models may be integrated into models of global climate. The resulting Earth System Models provide important insight into the interactions of global vegetation changes on climate. Changes in tropical forests have proven to be especially important in this regard.

There are at least two outstanding challenges in modeling tropical forest ecosystems under changing conditions. One challenge involves best including what we know about the photosynthesis process at a cellular level or leaf level in a model. Models constructed to be tested against micrometeorological "flux tower" measurements are likely to include fairly detailed scaled-up leaf processes and attempt to simulate the carbon dioxide, heat, and water fluxes of a forest canopy at a fine temporal resolution. Typically, models that represent the interactions of the planet's surface with global models of climate (DGVMs and Earth System Models) are strongly oriented toward this first challenge.

The second challenge involves predicting the change in structure and composition of vegetation in response to climate changes. In models oriented toward this challenge, the life history attributes of species expressed as parameters for birth, death, and success of species is a central focus. There is also an explicit recognition of the differences in structure of the forest as an overall feature influencing the long-term dynamics of vegetation. Because these models seek to represent the structure of the forest, they better estimate longer-term processes including decomposition and death aspects of forests than models emphasizing the first challenge.

At this time, the problem of simultaneously meeting both of the challenges to model under altered conditions is far from solved—at least with a universally accepted solution (Shugart, 1998). This may stem from the rather large differences between the fine temporal and spatial scales over which photosynthesis is studied compared with the much coarser scale of understanding compositional and structural forest dynamics (Woodward, 1987). A prudent way to proceed is to better understand how different models of forest systems are composed, tested, and applied. Such knowledge is key to interpreting the results from any of the different modeling approaches. Promotion of this understanding is the objective of the review that we present here.

14.2 BIOCLIMATIC MODELS

Bioclimatic or niche models of individual species provide insight into possible range shifts due to climate change. Examples of this modeling approach include BIOCLIM, DOMAIN, generalized additive modeling (GAM), generalized linear modeling (GLM), artificial neural networks (ANNs), genetic algorithms, genetic algorithm for rule-set prediction (GARP), and several others (Guisan and Thuiller, 2005). Each of these techniques uses information about a species' present range and present climate to simulate its present or future range.

Potentially, bioclimatic projections can be very useful in prioritizing areas for field surveys to search for the species or for conservation purposes. A model constructed of the present relationship between a species' distribution and climate may be projected into the future using a model of future climate—such as the results of a global climate model (e.g., general circulation model, GCM). Such model projections may be useful in assessing the impact of climate change on individual species, or in designing conservation strategies for suites of species (Hannah *et al.*, 2002).

In its simplest form, a bioclimatic model of a species' range can be constructed by measuring the range of variation in a number of environmental variables from points at which a species has been observed. For each grid-cell, the model calculates whether each environmental variable is within the range observed for the species (Box, 1981). If it is within the observed range for every variable, the species is modeled as present in that grid-cell. If any variable is outside the observed range for the species, the species is modeled as absent in that grid cell. This type of model was proposed by Box (1981) and is sometimes referred to as a "Box model". Box models have limitations. Box's original formulation was for what he termed "life forms", based on combinations of whole plant form and morphology, size, and features of the leaves. A Box (1981) life form might be a "tropical evergreen sclerophyll tree", a "xeric tuft-treelet", or a "leaf-succulent evergreen shrub". Each of Box's 100 or so life forms was taken to have a distribution in a seven-dimensional climate statistical space. Box tested his model in predicting the life forms expected at several hundred locations with wildly contrasting climatic conditions.

More complex models build on this concept, using statistical or other measures to improve the sophistication of prediction. BIOCLIM (Nix, 1986) creates one-tailed percentile distributions for each variable, while DOMAIN calculates a distance statistic between the environmental values in a cell and the environmental values in occurrence points from the observational record. GAM and GLM are statistical modeling techniques that can be used to construct a model of species presence/absence, based on known occurrences (Hastie and Tibshirani, 1990). Other approaches—artificial neural networks (ANNs) and genetic algorithms—may be applied to get a similar result. The genetic algorithm for rule-set prediction (GARP) is noteworthy in that it uses multiple algorithms ("rule sets") to fit the data in different parts of a species' range, and thus potentially captures the benefits of multiple modeling approaches (Stockwell and Peters, 1999).

One test of models using these methods has been geographic applications involving the estimation of species' current ranges. For example, a bioclimatic model may be used to estimate a climatically suitable range for a species that is currently only known from a handful of observations. To test such an application one obvious approach is to determine whether the modeling predictions match known locations of the species that have been omitted from the model calibration step. One such test was developed by Webb (1988), who calibrated the modern climatic niche occupied by tree species using current distributions and then predicted prehistoric spatial distributions of spruce (*Picea* sp.) and other major tree taxa across North America in response to climate changes and evidenced in pollen samples from lake cores. This reconstruction was developed at three millennial intervals over the past 18,000 years.

Others have used essentially the same approach to describe the niches of individual plant species and then used these data to simulate ranges under altered climate. An important example of this approach is by Iverson and Prasad (1998, 2001) who mapped the distributions of 80 tree speces over eastern North America. This application was based on data from millions of trees sampled systematically in the National Forest Inventory overlain with spatial climatic data sets. Such massive systematic data collections do not exist for most tropical countries (a notable exception is Malaysia) and tropical applications are likely to have several data limitations. Since many species have limited (<100) numbers of observational records, the range of any one variable from the observations is unlikely to capture the full range of variation the species is actually able to tolerate. Conversely, for variables which have no role in controlling the limits of a species' distribution, the range of the variable within the observational records is actually irrelevant (unless those variables become limiting under altered climate regimes).

The strength of bioclimatic models is that they yield results for individual species, and thus are directly relevant to measuring impacts on biodiversity at the community level of ecological integration. However, the complexity of bioclimatic models varies greatly and they have numerous assumptions that may not be relevant in the real world (Pearson and Dawson, 2003). The most often cited is the assumption that current species' ranges are in equilibrium with present climate or that at least the current range of the species expresses the full environmental range of the species. Many species may in fact be responding to minor climatic cycles (e.g., the Little Ice Age) or in some cases still coming into equilibrium with the transition from the last glacial period.

A second important assumption is that a species' range is limited by climate, rather than other factors—such as competition or dispersal. This is not a trivial assumption. Many species may have distributions limited by biogeographic accident or other factors not related to climate. A prominent cautionary example is the Monterey pine (*Pinus radiata*) which has a very limited range in California but has become one of the most widely planted plantation–forestry species in the international subtropics.

Competition, which is assumed to be unimportant in correlational models, in fact probably plays a significant role in the distribution of most species. Whether or not this factor is important enough to badly distort statistical (correlational) models is problematic. Box's (1988) original formulation of bioclimatic models included an algorithm (based on an environmentally calculated potential canopy cover) that accounted for competition among life forms. However, most subsequent models do not directly address questions of competitive interaction. They may also lack a direct mechanistic link between species' physiology and the climatic limits suggested in the modeling. While this may not be a problem for certain applications, one issue involves whether or not the bioclimatic approach takes into account the complicating factor of the highly elevated carbon dioxide levels in the atmosphere. If, for example, the direct effect of elevated CO_2 is increased water-use efficiency, then how should one recalibrate the bioclimatic indices for futuristic applications (see Körner, 1993). Future projections vary substantially depending on the model type (Thuiller, 2004). Other types of models are useful alternatives for addressing competition, physiological limits, and carbon-cycle interactions.

Bioclimatic modeling has been applied to relatively few tropical tree species. Hughes *et al.* (1996) used the climatology of BIOCLIM to examine the climatic range of species in the *Eucalyptus* L'Herit genus in Australia. Over 31 tropical species in the genus had mean annual temperature ranges of less than 1°C, spanning both wet and dry habitats across Australia. Ferreira de Siquiera and Peterson (2003) examined the Cerrado tree species of tropical Brazil, finding major range shifts and consequent changes in patterns of species richness. More recently, Miles *et al.* (2003) assessed the effects of future climate change on 69 Amazonian tree species using GLM and life history simulations to 2095 under Hadley Centre climate projections. In this simulation 43% of all species became non-viable by 2095 (Miles *et al.*, 2004). These studies suggest a relatively high vulnerability of tropical tree species to climate change, but these early results must be viewed with caution. Tropical trees may occupy the upper limits of current temperatures, and thus bioclimatic models are not presented with the full range of conditions under which the species can exist. There may be warmer climates with no current analog, in which tropical species are perfectly capable of surviving but which models project as unsuitable.

Bioclimatic modeling has been conducted for more vertebrate species than tree species, but the number of independent studies is still small. Williams *et al.* (2004) modeled range changes of mammals, reptiles, amphibians, and birds in the Queensland tropical rainforests of Australia. These primarily montane species showed strong range reductions with temperature increases as species' ranges migrated upslope (Williams *et al.*, 2004). Peterson *et al.* (2002) used GARP to model the climate change response of large numbers of Mexican vertebrates, many of which were tropical. Range reduction was pronounced in some of these species, but large reductions were less widespread than in the species examined in the Australian study, perhaps due to the more varied topography of Mexico (Peterson *et al.*, 2002).

The issue of no-analog climates is of strong relevance for the interpretation of bioclimatic models for the tropics. Saxon *et al.* (2005) have projected that over 60% of the United States will be occupied by no-analog climates by 2100. Temperature increases in the tropics may result in even higher proportions of no-analog climate space, but the physiological implications of these no-analog situations are not clear. Many tropical species may be well able to withstand higher temperature than those in which they currently exist. On the other hand, no-analog conditions in moisture balance or moisture balance combined with elevated CO_2 may be much more difficult for tropical trees to withstand. Bioclimatic models probably poorly reflect both positive and negative effects of no-analog climates on species' ranges.

14.3   GAP MODELS

In the late 1960s, several forestry schools in the United States developed a forest modeling approach based on simulating a forest by computing the birth (or planting), growth, and death (thinning or harvest) of plantation forest. Applications of these models included determining plant strategies for genetically improved tree crops and ascertaining the appropriate spacings for growing trees to produce metric-dimensioned lumber (see Shugart, 1998 for review). This emphasis on tracking individual trees was soon taken up by forest ecologists interested in the simulation of forest-structural and forest-compositional changes in forest succession and in response to environmental gradients (see Shugart, 1998, ch. 8, for a review of several of these applications and model tests).

One of the features of these early individual-based forest simulators (Huston *et al.*, 1988) was geometrically elaborate computation of the effect of each tree on the others through shading or the use of nutrient and/or water resources. One simplification of the rather laborious calculations was to assume the competitive interactions were primarily occurring on a plot of land that was roughly the size of a very large canopy tree. This is the space scale of a gap in a forest canopy associated with a large tree's death. Hence, this class of models was termed "gap models" (Shugart and West, 1980). Modern computers have lifted the computational limitations that caused gap model competition simplification. Nowadays, "gap model" refers to individual-based models of naturally regenerating multi-species forests.

A variety of gap models now exist—for example, the FORET model of Tennessee forests (Shugart, 1984) and of European mountain forests, and FORCLIM (Bugmann and Solomon, 1995). Gap models utilize the demography and natural history of plant species, including physiological trade-offs, plant growth, and reproductive processes, to simulate the compositional dynamics at the scale of a tree-fall gap. These parameters are used to simulate the response of individual plants of differing species within the gap, as well as the way the environment is modified by those individuals. Each individual tree is modeled both in spatial extent and in vertical structure and its consequent effects (such as shading of trees below). Gap modeling is therefore more suitable for assessing competitive interactions and small-scale impacts on biodiversity than is bioclimatic modeling.

One of the earliest modes of testing gap models was to simulate the altitudinal zonation of forests along montane climate gradients. This leads naturally to an interest in their ability to construct expected forests under altered climates. Consequently, there has been a proliferation of this class of models with reviews of their application in assessing climate change (Shugart *et al.*, 1992; along with two special issues of *Climatic Change*, with introductory papers: Bugmann *et al.*, 2001 and Shugart and Smith, 1996). The initial entry of gap models into the task of assessing the effects of climate change on forest ecosystems sprang from applications to understand expected forest responses to past climates. These palaeo-reconstructions (Overpeck *et al.*, 1990; Solomon *et al.*, 1980; Solomon and Webb, 1985) lead naturally to the assessment of expected forest composition under future climate scenarios.

Because gap models require knowledge of the growth rates of trees along with the fundamental silvicultural features of trees—such as regeneration habit, size, growth rates, and heights—the lack of such information across the diverse array of species composing many tropical forests have made applications to tropical systems difficult. The lack of growth rate information from tropical forests has been a particularly limiting constraint. There is an associated difficulty in knowing how tree growth might change with climate. Nevertheless, there are several tropical forest gap models. The Kiambram model simulates 125 species in Australian montane rainforest (Shugart *et al.*, 1980). The Outenqua model (van Daalen and Shugart, 1989) similarly projects forest dynamics for African subtropical rainforests. Because of the lack of information on the climate response of tropical forest species, there have been relatively few climate-related applications. Doyle (1981) simulated the effects of altered hurricane disturbance frequencies for Puerto Rican montane rainforest with a gap model and O'Brien *et al.* (1992) explored landscape-scale interactions with hurricanes for the same site.

There have been applications of gap models in tropical settings using a gap model framework to produce the dynamics of forest structures over time. Bossell and Krieger (1994) have implemented a functional type (rather than species-based) model of tropical forests, basically on a gap model framework. Another important tropical application of gap models to tropical rainforests is the approach of Moorcroft *et al.* (2001), who parameterized a gap model for Amazonian rainforest using a postulated relationship between growth rate, photosynthesis, and wood density. This model was used to parameterize a statistical model of the size distribution of trees across the Amazon Basin using an approach pioneered by Kohyama (1993). This was then driven by a photosynthesis/production model to incorporate the effects of temperature and moisture.

14.4 DYNAMIC GLOBAL VEGETATION MODELS

It has long been recognized that climate exerts a general control on the vegetation zones of the Earth (Woodward and Williams, 1987). An early approach to modeling this effect was analogous to bioclimatic modeling of individual species. Models that correlated vegetation types with certain climatic conditions were shown to reproduce global vegetation zonation with reasonable accuracy. Such correlative models could also be used to project future vegetation types using projections of future climates from GCMs.

More recently, correlative models have been largely replaced by mechanistic models using plant physiology to simulate vegetation patterns. These models are known as "dynamic global vegetation models" (DGVMs). These models use equations describing basic plant physiological processes—such as photosynthesis and respiration—to determine net amounts of carbon available for plant growth and the allocation of that carbon (Prentice *et al.*, 1992; Woodward and Beerling, 1997). The results of these calculations are expressed as "plant-functional types"—for example, "evergreen needleleaf forest", "grass savanna", or "deciduous broadleaf forest". Plant-functional types in these applications are direct analogs to Box's (1981) "life forms".

Numerous authors have contributed to the development of over half a dozen DGVMs that are being actively tested and refined. SDGVM (Woodward and Lomas, 2004), TRIFFID (Cox 2001), IBIS (Foley *et al.*, 1996), LPJ (Sitch *et al.*, 2003), VECODE (Brovkin *et al.*, 1997), MC1 (Bachelet *et al.*, 2001), and HYBRID (Friend *et al.*, 1997) are examples. Widespread increases or decreases in forest cover projected in response to CO_2 rise and climate change may therefore indirectly contribute to additional contributions to the regional and global climate changes through alterations to land surface properties. Furthermore, the net changes in terrestrial carbon stocks in the tropics and elsewhere may influence the rise in CO_2 itself. Ecosystems may therefore have different impacts on climate change, both at the regional and global scale.

Cramer *et al.* (2001) present the results of a model intercomparison effort conducted with several of these major DGVMs. DGVMs have great utility in assessing the impacts of climate change, fire, and other environmental drivers on gross vegetation structure and physiognomy, especially at large scales. They are less useful for studies on individual species or biodiversity assessments at small scale. DGVMs may also be integrated into climate models to assess influences of vegetation on carbon cycles and global climatic change.

Responses to climate change and elevated CO_2 modeled by DGVMs show broad similarities but also substantial inter-model variation. Cramer *et al.* (2001) used an ensemble of six DGVMs to make projections of global vegetation responses to transient climate change simulated with the HadCM2 GCM under the IS92a greenhouse gas and sulphate aerosol concentration scenario. The DGVMs generally overestimated the amount of tropical forest for current climates, especially in Africa but also in South America and Southeast Asia (Cramer *et al.*, 2001). Tropical dry forest tends to be over-predicted in Southeast Asia and under-predicted in Africa. TRIFFID and HYBRID particularly over-predict tropical wet forest, while results from VECODE and SDGVM more closely approximate tropical moist and dry deciduous vegetation classifications derived from satellite images.

A particular feature of this study was projection for a major reduction in forest cover in the eastern half of Amazonia, due to significantly reduced precipitation and increased temperature. A drying of the Amazonian climate emerges in a number of GCMs, generally associated with an El Niño like pattern of global climate warming, but it is important to note that not all GCMs show this response. All six DGVMs showed a tendency toward reduction in forest cover due to drier conditions, and a drop in Amazonian biomass by 2100 (Figure 14.1; Cox *et al.*, 2004). Variability in outcomes is influenced by assumptions about photosynthetic response and more efficient water use by the vegatation due to increased CO_2 concentrations. While all models project a reduction in forest cover in northeastern Amazonia, considerable variability is evident in the models for other Amazonian outcomes.

DGVMs are rather difficult to test against independent data, in no small part due to their scales in time and space. The Foley *et al.* (1996) IBIS model has been tested globally against the flows of the major rivers of the world and regionally for the flows of the Amazon and its tributaries. The rationale is that the models compute evapotranspiration as one of its dynamic variables and water flow of a basin can be taken as the difference between rainfall and evapotranspiration when corrected for soil and ground-water storage. This regional- and global-scale DGVM testing is a healthy development particularly in that it uses data that are independent of model development and parameterization. Most DVGM model testing has been for consistency with overall patterns in the parameterization data or in the form of model comparisons (as opposed to independent data comparisons).

14.5 EARTH SYSTEM MODELS

Tropical ecosystems respond to climate changes, but the ecosystems themselves also exert influences on climate. For example, a number of studies have suggested that removal of the Amazonian forests may cause a warming of surface temperature and

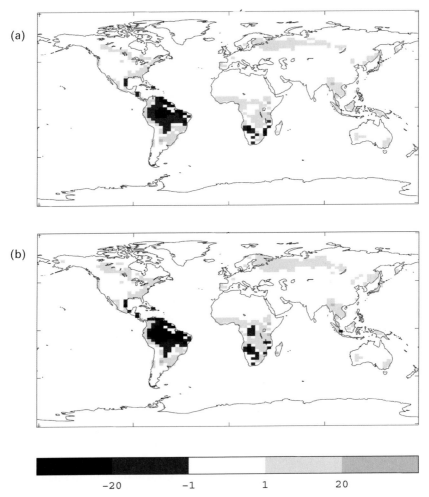

Figure 14.1. Global changes in broadleaf tree cover illustrating importance of tropical forests in CO_2 feedback effects. (a) Change in cover simulated by the HadCM3LC coupled climate–carbon cycle model (Cox *et al.*, 2000) from 1860 to 2100 without CO_2 climate feedbacks. (b) Additional changes with inclusion of CO_2 climate feedbacks (© British Crown Copyright 2003, by kind permission of the Met Office).

reduction in precipitation, due to a reduced level of transpiration from the deforested landscape (see, e.g., Lean and Rowntree, 1997 for summary). Such effects may be crucial in maintaining local climates in a state amenable to the forests themselves (Betts, 1999). Changes in forest cover may also influence the climate through changes in the production of aerosol particles, which affect cloud formation and rainfall production. As well as influencing local climate, tropical ecosystem changes may also exert more far-reaching effects. For example, changes in carbon stocks affecting the rate of CO_2 rise and changes in the near-surface energy balance and cloud processes may modify atmospheric circulatory (Hadley) cells near the equator. Gedney and

Valdes (2000) have used robust atmospheric models to show that such changes in atmospheric circulation may have influences felt across the globe.

Given the potential for major feedbacks from ecosystems, it is clear that predictions of future climate change should consider ecosystem responses and their effect on climate. This has led to the development of "Earth System models" which couple models of the atmosphere and oceans (GCMs) to models of the terrestrial and marine biosphere (DGVMs) (Foley *et al.*, 1996; Cox *et al.*, 2000; Ganapolski *et al.*, 2001). Physical and biological models interact via biogeochemical cycles and through the impact of life on the physical properties of the Earth's surface. A number of such models have been developed with a wide range of spatial and temporal resolutions, attempting to trade off model complexity and detail against computational efficiency. The models used to study the interactions between climate change and tropical forests typically feature a DGVM and/or an interactive carbon cycle included within a GCM (Cox *et al.*, 2000; Betts *et al.*, 2004).

The inclusion of DGVMs in GCMs allows climate prediction simulations to include feedbacks from ecosystems responding to climatic changes at global and regional scales (Cox *et al.*, 2000). Coupled GCM–DGVMs are therefore potentially valuable for understanding and predicting synergistic responses of ecosystems to climate change over timescales of centuries and spatial scales of hundreds of kilometers (Betts *et al.*, 2004).

Early applications of Earth System models show that widespread increases or decreases in forest cover projected in response to CO_2 rise and climate change may indirectly contribute to regional and global climate changes through alterations to land surface properties. Furthermore, net changes in terrestrial carbon stocks in the tropics and elsewhere may influence the rise in CO_2 itself. Ecosystems may therefore exert a number of feedbacks on climate change, both at the regional and global scale.

In simulations using the Hadley Centre coupled climate–ecosystem model—HadCM3LC—the forests of Amazonia showed a very large reduction in tree cover as a result of decreased rainfall (Betts *et al.*, 2004; Cox *et al.*, 2004). Some signs of the beginning of this process were already simulated by 2000, with broadleaf tree cover reducing in the northeast of Amazonia in response to a drier climate than that simulated for 1860 (Figure 14.2). The reduction in rainfall spreads towards the southwest through the 21st century, and the tree cover reduces until it is less than 1% in the northeast quarter of Amazonia by 2100. Almost all of the Amazon Basin loses at least 50% of its tree cover by the end of the simulation, to be replaced mainly by C_4 grass but also with large areas of bare soil. The general character of the region fundamentally changes from dense evergreen broadleaf forest to savanna, grassland, or even semi-desert.

The changes in tropical forest ecosystems in these simulations had significant impacts on regional climates through changes in the physical properties of the land surface. Although the drying climate in Amazonia emerged even when vegetation was fixed at the present day state, regional climate changes were significantly affected by vegetation feedbacks. In particular, precipitation reduction over Amazonia was found to be enhanced by 25% by feedbacks from the loss of forest cover. In the western part of the basin, the feedback was greater still because of the greater dependency of rainfall on recycling through evapotranspiration in the continental interior. Here

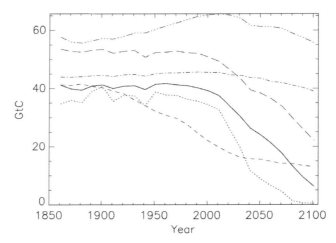

Figure 14.2. Decline of Amazon forest biomass in six different DGVMs, under a climate projection from the HadCM2 climate model (Cox *et al.*, 2004). These simulations do not account for the direct effect of rising atmospheric CO_2 in fertilization; other simulations including CO_2 fertilization using the same models showed smaller losses of biomass or small increases (© British Crown Copyright 2003, by kind permission of the Met Office).

precipitation reduction was increased by over 30% as a result of drought-induced dieback of the forests, particularly to the east. Forest loss also increased surface albedo which reduced convection and moisture convergence, providing a further positive feedback on rainfall reduction (Charney, 1975).

The model simulations of Friedlingstein *et al.* (2001) also found a reduction in precipitation to be simulated in Amazonia, but the model did not include dynamic vegetation so there was no feedback on climate through biogeophysical effects. The model used by Thompson *et al.* (submitted) included the IBIS2 dynamic vegetation model (Foley *et al.*, 1996; Kucharik *et al.*, 2000), but this model did *not* produce a drying in Amazonia.

In order to estimate some constraints on sustainable conditions for the Amazonian forests with reference to changes that have occurred in the past, Cowling *et al.* (2003) used HadCM3LC to simulate the coupled paleoclimate–vegetation state in Amazonia at the last glacial maximum (LGM) 25,000 years ago. At the LGM, forest cover was maintained but was less productive, consistent with proxy data from the paleorecord. This was despite a drier climate and lower CO_2 concentrations, both of which are less favorable for forest cover. Cowling *et al.* suggested that the variation in forest structure (leaf area index) at the LGM might have acted to drive speciation and diversity in Amazonian forests, through mechanisms somewhat analogous to those that have been proposed for forest refugia, without loss of continuous forest cover. The critical aspect of the climate at the LGM was cooler temperatures, which helped to reduce both photorespiration and evapotranspiration, leading to decreased loss of carbon and water from the vegetation. Cowling *et al.* noted that—for the future state—warmer conditions are likely to amplify the effects of any drying of the regional climate that may occur despite the likely effects of elevated CO_2.

14.6   CONCLUSION

Bioclimatic models suggest major impacts of climate change on tropical forests; however, these are early results. Several biases, not the least of which being the unknown physical tolerances of many tropical species to warmer climates, suggest that these results be viewed with caution. Gap models have seen relatively limited application to tropical forests, and still less use in assessing impacts of climate change on these forests. The difficulties of obtaining gap model parameters for tropical forests will probably continue to constrain their use, but ED-type models (Moorcroft *et al.*, 2001) allow these results to be applied regionally (notably for the Amazon Basin). DGVM results indicate possible losses in tropical forest cover in some regions, which may support the findings of bioclimatic models. The Amazon is of particular concern, as many—but not all—DGVMs indicate loss of forest cover in the Amazon under future climate projections. This effect has been further explored in coupled GCM–DGVM simulations, and the enhanced forest loss observed in these coupled simulations provides one of the most remarkable modeling results for tropical forests under climate change.

The extreme 21st century precipitation decrease and forest dieback simulated in Amazonia by HadCM3LC is the result of a complex, coupled process emerging from interactions between the atmosphere, the oceans, and the land ecosystems of the Amazon and elsewhere. The outcome is a major change in forest cover which has significant implications for both the ecosystem itself and for global climate change. These model results are still subject to considerable uncertainties. Nevertheless, while these results should not be viewed as a prediction, the analysis to date suggests that under increasing concentrations of CO_2 and other greenhouse gases, a transition to a drier, less forested Amazon cannot be ruled out. Given our rudimentary understanding of climate change impacts in the tropics, these results for the Amazon provide reason for concern for impacts on other tropical forests.

The key role of feedbacks from forest loss has major implications for the importance of the sensitivity of Amazonian forests to climate change. While the true sensitivity of the forests is still uncertain, the results presented here show that, in principle, forests could be a significant component of the sensitivity of regional and global climates to radiative forcing. This has implications for the long-term effects of human activity in Amazonia. Activities such as road-building, partial deforestation and selective logging have been shown to increase the climatic sensitivity of parts of the forest not directly affected by the activity. The exposure of a new forest "edge" can increase susceptibility to fire, and may therefore increase the sensitivity of the forest to climate change. If forest sensitivity is increased in this way, this could enhance the feedback on both regional and global climates.

14.7   REFERENCES

Bachelet, D., Neilson, R. P., Lenihan, J. M, and Drapek, R. J. (2001) Climate change effects on vegetation distribution and carbon budget in the US. *Ecosystems* **4**(3), 164–185.

Betts, R. A. (1999) Self-beneficial effects of vegetation on climate in an ocean–atmosphere general circulation model. *Geophysical Research Letters* **26**(10), 1457–1460.

Betts, R. A., Cox, P. M., Collins, M., Harris, P. P., Huntingford, C., and Jones, C. D. (2004) The role of ecosystem–atmosphere interactions in simulated Amazonian precipitation decrease and forest dieback under global climate warming. *Theoretical and Applied Climatology* **78**, 157–175.

Bossell, H. and Krieger, H. (1994) Simulation of multi-species tropical forest dynamics using a vertically and horizontally structured model. *Forest Ecology and Management* **69**, 123–144.

Box, E. O. (1981) *Macroclimate and Plant Forms: An Introduction to Predictive Modelling in Phytogeography*. Dr. W. Junk, The Hague.

Brovkin, V., Ganapolski, A., and Svirezhev, Y. (1997). A continuous climate–vegetation classification for use in climate–biosphere studies. *Ecological Modelling* **101**, 251–261.

Bugmann, H. K.M. and Solomon, A. M. (1995) The use of a European forest model in North America: A study of ecosystem response to climate gradients. *Journal of Biogeography* **22**, 477–484.

Bugmann, H., Reynolds, J. F., and Pitelka, L. F. (2001) How much physiology is needed in forest gap models for simulating long-term vegetation response to global change. *Climatic Change* **51**, 249–250.

Cowling, S. A., Betts, R. A., Cox, P. M., Ettwein, V. J., Jones, C. D., Maslin, M. A., and Spall, S. A. (2003) Constrasting simulated past and future responses of the Amazonian forest to atmospheric change. *Philosophical Transactions of the Royal Society of London, B* doi:10.1098/rstb.2003.1427.

Cox, P. M. (2001) *Description of the TRIFFID Dynamic Global Vegetation Model* (Technical Note 24). Hadley Centre, Met Office, Bracknell, U.K.

Cox, P. M., Betts, R. A., Jones, C. D., Spall, S. A., and Totterdell, I. J. (2000) Acceleration of global warming due to carbon-cycle feedbacks in a coupled climate model. *Nature* **408**, 184–187.

Cox, P. M., Betts, R. A., Collins, M., Harris, P. P., Huntingford, C., and Jones, C. D. (2004) Amazonian forest dieback under climate–carbon cycle projections for the 21st century. *Theoretical and Applied Climatology* **78**, 137–156.

Cramer, W., Bondeau, A., Woodward, F. I., Prentice, I. C., Betts, R. A., Brovkin, V., Cox, P. M., Fisher, V., Foley, J. A., Friend, A. D. *et al.* (2001) Global response of terrestrial ecosystem structure and function to CO_2 and climate change: Results from six dynamic global vegetation models. *Global Change Biology* **7**, 357–373.

Doyle, T. W. (1981) The role of disturbance in the gap dynamics of a montane rain forest: An application of a tropical forest succession model. In: D. C. West, H. H. Shugart, and D. B. Botkin (eds.), *Forest Succession: Concepts and Application* (pp. 56–73). Springer-Verlag, New York.

Elton, C. (1927) *Animal Ecology*. Macmillan, New York.

Ferreira de Siqueira, M. and Peterson, A. T. (2003) Global climate change consequences for Cerrado tree species. *Biota Neotropica* **3**, 1–14.

Foley, J. A., Prentice, I. C., Ramankutty, N., Levis, S., Pollard, D., Sitch, S., and Haxeltine, A. (1996) An integrated biosphere model of land surface processes, terrestrial carbon balance and vegetation dynamics. *Global Biogeochemical Cycles* **10**(4), 603–628

Friedlingstein, P., Bopp, L., Ciais, P., Dufresne, J.-L., Fairhead, L., and co-authors (2001) Positive feedback between future climate change and the carbon cycle. *Geophysical Research Letters* **28**, 1543–1546.

Friend, A. D., Stevens, A. K., Knox, R. G., and Cannell, M. G. R. (1997) A process-based, terrestrial biosphere model of ecosystem dynamics (HYBRID v3.0). *Ecological Modelling* **95**, 249–287.

Ganapolski, A., Petoukhov, V., Rahmstorf, S., Brovkin, V., Claussen, M., Eliseev, A., and Kubatzki, C. (2001) CLIMBER-2: A climate system model of intermediate complexity. Part II. Model sensitivity. *Climate Dynamics* **17**, 735–751.

Gedney, N. and Valdes, P. J. (2000) The effect of Amazonian deforestation on northern hemisphere circulation and climate. *Geophysical Research Letters* **27**(19), 3053–3056.

Grinnell, J. (1917) The niche relations of the California thrasher. *Auk* **34**, 364–382.

Guisan, A. and Thuiller, W. (2005) Predicting species distribution: Offering more than simple habitat models. *Ecology Letters* **8**(9), 993–1009.

Guisan, A. and Zimmermann, N. E. (2000) Predictive habitat distribution models in ecology. *Ecological Modelling* **135**(2–3), 147–186.

Hannah, L., Midgley, G. F., Lovejoy, T. E., Bond, W. J., Bush, M., Lovett, J. C., Scott, D., and Woodward, F. I. (2002) Conservation of biodiversity in a changing climate. *Conservation Biology* **16**, 264–268.

Hastie, T. and Tibshirani, R. (1990) *Generalized Additive Models*. Chapman & Hall, New York.

Hughes, L., Cawsey, E. M., and Westoby, M. (1996) Climatic range sizes of *Eucalyptus* species in relation to future climate change. *Global Ecology and Biogeography Letters* **5**, 23–29.

Huston, M., DeAngelis, D. L., and Post, W. M. (1988) New computer models unify ecological theory. *BioScience* **38**, 682–691.

Hutchinson, G. E. (1957) Concluding remarks. *Cold Spring Harbor Symposium on Quantitative Biology* **22**, 415–427.

Iverson, L. R. and Prasad, A. M. (1998) Predicting abundance of 80 tree species following climate change in the eastern United States. *Ecological Monographs* **68**, 465–485.

Iverson, L. R. and Prasad, A. M. (2001) Potential changes in tree species richness and forest community types following climate change. *Ecosystems* **4**, 186–199.

Kohyama, T. (1993) Size-structured tree populations in gap dynamic forests: The forest architecture hypothesis for stable coexistence of species. *Journal of Ecology* **81**, 131–143.

Körner, C. (1993) CO_2 fertilization: The great uncertainty in future vegetation development. In: A. M. Solomon and H. H. Shugart (eds.), *Vegetation Dynamics and Global Change* (pp. 53–70). Chapman & Hall, New York.

Kucharik, C. J., Foley, J. A., Delire, C., Fisher, V. A., Coe, M. T., and co-authors (2000) Testing the performance of a dynamic global ecosystem model: Water balance, carbon balance, and vegetation structure. *Global Biogeochemical Cycles* **14**(3), 795–825.

Lean, J. and Rowntree, P. R. (1997) Understanding the sensitivity of a GCM simulation of Amazonian deforestation to the specification of vegetation and soil characteristics. *Journal of Climate* **10**(6), 1216–1235.

Miles, L., Grainger, A., and Phillips, O. (2004) The impact of global climate change on tropical forest diversity in Amazonia. *Global Ecology and Biogeography* **13**, 553–565.

Moorcroft, P. R., Hurtt, G. C., and Pacala, S. W. (2001) A method for scaling vegetation dynamics: The ecosystem demography model (ED). *Ecological Monographs* **74**, 557–586.

Moore, A. D. and Noble, I. R. (1990) An individualistic model of vegetation stand dynamics. *Journal of Environmental Management* **31**(1), 61–81.

Nix, H. A. (1986) A biogeographic analysis of Australian elapid snakes. In: R. Longmore (ed..), *Atlas of Elapid Snakes of Australia* (Australian Flora and Fauna Series No. 7, pp. 4–15). Australian Government Publishing Service, Canberra.

O'Brien, S. T., Hayden, B. P., and Shugart, H. H. (1992) Global change, hurricanes and a tropical forest. *Climatic Change* **22**, 175–190.

Overpeck, J. T., Rind, D., and Goldberg, R. (1990) Climate-induced changes in forest disturbance and vegetation. *Nature* **343**, 51–53.

Pearson, R. G. and Dawson, T. P. (2003) Prediciting the impacts of climate change on the distribution of species: Are bioclimate envelope models useful? *Global Ecology and Biogeography* **12**, 361–371.

Peterson, A. T., Ortega-Huerta, M. A., Bartley, J., Sanchez-Cordero, V., Soberon, J., Buddemeier, R. H., and Stockwell, D. R. (2002) Future projections for Mexican faunas under global climate change scenarios. *Nature* **416**, 626–629.

Peterson, A. T., Tian, H., Martinez-Meyer, E., Soberon, J., Sanchez-Cordero, V., and Huntley, B. (2005) Modeling distributional shifts of individual species and biomes. In: T. E. Lovejoy and L. Hannah (eds.), *Climate Change and Biodiversity*. Yale University Press, New Haven, CT.

Prentice, I. C., Cramer, W., Harrison, S. P., Leemans, R., Monserud, R. A., and Solomon, A. M. (1992) A global biome model based on plant physiology and dominance, soil properties and climate. *Journal of Biogeography* **19**, 117–134.

Saxon, E., Baker, B., Hargrove, W., Hoffman, F., and Zganjar, C. (2005) Mapping environments at risk under different global climate change scenarios. *Ecology Letters* **8**, 53–60.

Sitch, S., Smith, B., Prentice, I. C., Arneth, A., Bondeau, A., Cramer, W., Kaplan, J. O., Levis, S., Lucht, W., Sykes, M. T. *et al.* (2003) Evaluation of ecosystem dynamics, plant geography and terrestrial carbon cycling in the LPJ dynamic global vegetation model. *Global Change Biology* **9**(2), 161–185.

Shugart, H. H. (1984) *A Theory of Forest Dynamics: The Ecological Implications of Forest Succession Models* (278 pp.). Springer-Verlag, New York.

Shugart, H. H. (1998) *Terrestrial Ecosystems in Changing Environments* (537 pp.). Cambridge University Press, Cambridge, U.K.

Shugart, H. H. and Smith, T. M. (1996) A review of forest patch models and their application to global change research. *Climatic Change* **34**, 131–153.

Shugart, H. H. and West, D. C. (1980) Forest succession models. *BioScience* **30**, 308–313.

Shugart, H. H., Hopkins, M. S., Burgess, I. P., and Mortlock, A. T. (1980) The development of a succession model for subtropical rain forest and its application to assess the effects of timber harvest at Wiangaree State Forest, New South Wales. *Journal of Environmental Management* **11**, 243–265.

Shugart, H. H., Smith, T. M., and Post, W. M. (1992) The potential for application of individual-based simulation models for assessing the effects of global change. *Annual Reviews of Ecology and Systematics* **23**, 15–38.

Smith, T. M., Shugart, H. H., Bonan, G. B., and Smith, J. B. (1992) Modeling the potential response of vegetation to global climate change. *Advances in Ecological Research* **22**, 93–116.

Solomon, A. M. (1986) Transient response of forests to CO_2-induced climate change: Simulation experiments in eastern North America. *Oecologia* **68**, 567–579.

Solomon, A. M. and Webb III, T. (1985) Computer-aided reconstruction of late Quaternary landscape dynamics. *Annual Reviews of Ecology and Systematics* **16**, 63–84.

Solomon, A. M., Delcourt, H. R., West, D. C., and Blasing, T. J. (1980) Testing a simulation model for reconstruction of prehistoric forest-stand dynamics. *Quaternary Research* **14**, 275–293.

Stockwell, D. and Peters, D. (1999) The GARP modelling system: Problems and solutions to automated spatial prediction. *International Journal of Geographical Information Science* **13**, 143–158.

Thuiller, W. (2004) Patterns and uncertainties of species' range shifts under climate change. *Global Change Biology* **10**, 2020–2027.

van Daalen, J. C. and Shugart, H. H. (1989) Outeniqua: A computer model to simulate succession in the mixed evergreen forests of the southern Cape, South Africa. *Landscape Ecology* **2**, 255–269.

Webb III, T. (1988) Glacial and Holocene vegetation history: Eastern North America. In: B. Huntley and T. Webb III (eds.), *Vegetation History* (pp. 385–414). Kluwer Academic, Dordrecht, The Netherlands.

Williams, S. E., Bolitho, E. E., and Fox, S. (2004) Climate change in Australian tropical rainforests: An impending environmental catastrophe. *Proceedings of the Royal Society of London, Series B* **270**, 1887–1892.

Woodward, F. I. (1987) *Climate and Plant Distribution*. Cambridge University Press, Cambridge, U.K.

Woodward, F. I. and Beerling, D. J. (1997) The dynamics of vegetation change: Health warnings for equilibrium "dodo" models. *Global Ecology and Biogeography Letters* **6**, 413–418.

Woodward, F. I. and Lomas, M. R. (2004) *Biological Reviews* **79**(3), 643–670.

Woodward, F. I. and Williams, B. G. (1987) Climate and plant distribution at global and local scales. *Vegetatio* **69**, 189–197.

15

Conservation, climate change, and tropical forests

L. Hannah and T. Lovejoy

15.1   INTRODUCTION

Conservation of tropical forests in the face of climate change is an immense task, because of the huge losses already suffered to habitat loss and because of our still rudimentary knowledge of the biology of these systems. For example, in Southeast Asia, most moist tropical forest has already been lost, resulting in the severe imperilment of hundreds of well-known species, as well as the probable extinction of thousands of species before they are described by science (Brooks *et al.*, 1997). In this setting, climate change will alter the abundance and distribution of many species whose continued existence is already precarious, in a landscape that permits little if any scope for range dynamics.

While conservationists struggle against habitat loss, we seemingly lack even a strategy for coping with climate change. But that strategy is not completely elusive. It is clear that a two-pronged response is needed to effectively cope with climate change (Hannah *et al.*, 2002a). First, on-the-ground conservation strategies must begin to consider climate change. Expanding planning horizons, modeling and assessing possible climate change effects, and monitoring potentially sensitive species are all elements of climate change integrated conservation strategies that are within easy reach. The second, and more difficult, element of conservation response is that of constraining greenhouse gas levels in the atmosphere (Hannah *et al.*, 2002a). It is clear that no conservation strategy can be successful on the ground in the face of ever mounting climate change. Greenhouse gas levels must be stabilized in the atmosphere to limit climate change, implying a huge transition in the energy economy away from fossil fuels (Lackner, 2003).

In this chapter we will describe the challenges climate change poses to tropical forest conservation, followed by an analysis of the appropriate responses and their potential scope. We will then explore greenhouse gas stabilization in the atmosphere. Taking into account the possible scope of coping with climate change in conservation strategies, what level of atmospheric greenhouse gases is "safe" and what would be required to reach those targets?

15.2   CONSERVATION CHALLENGES

Several distinctive characteristics of tropical forest response to climate change pose significant challenges for conservation. Among these are the climate-making role of tropical forests, uncertainties about past responses, the introduction of warmth on an already warm inter-glacial climate, the critical role of precipitation in tropical forest eco-physiology and the synergies of climate change with ongoing habitat loss. This section will describe each of these challenges briefly.

Tropical moist forests have interactions with regional and global climate that have profound implications for their conservation. The Amazon Basin in particular plays meso- and global-scale climatic roles (Cox *et al.*, 2000, 2004; Betts *et al.*, 2004; Marengo, 2004). The basis of this effect is the influence tropical forests have on moisture-cycling and the regional water balance. At a plot level, removing forest changes the radiative properties of the surface and reduces moisture release by evapotranspiration (Pitman *et al.*, 2000; Chapter 2 in this book). These effects result in increased convection over the cleared parcel and may result in increased precipitation if the clearing is small and isolated. As the amount of clearing increases, the effect changes to one of reduced precipitation as convection has increasingly less surrounding moisture from evapotranspiration on which to draw.

At the scale of large forested areas—such as the Amazon Basin—the net effect of forest moisture turnover is to cycle moisture entering the system from the tropical Atlantic, making the western parts of the basin significantly moister than would be the case in the absence of forest (Chapter 2 of this book; Bush, 1996). Clearing of a substantial fraction of the basin may therefore lead to additional forest loss due to loss of moisture-cycling and regional drying. This positive feedback appears to have repercussions at a global scale as well (Cox *et al.*, 2002). When carbon and surface vegetation models are incorporated into GCMs, climate change drying in eastern Amazonia leads to forest loss, progressive drying in western Amazonia, and accelerated global warming due to massive releases of CO_2 from the Amazon. Avoiding these effects may require maintaining a substantial fraction of the tropical forest cover of the basin as a whole.

A second feature complicating the conservation of tropical forests in the face of climate change is the limited knowledge of these systems' responses to past change (Flenley, 1998). This can be a key limitation is assessing the possible natural precedents for response to rapid climate change. Paleoecological evidence from the flank of the Andes suggests that forests responded to glacial–interglacial cycles, but not to the rapid climate "flickers" that appear to characterize North Atlantic climate and vegetation responses (Bush *et al.*, 2004). The very rapid millennial or shorter timespan climate "flickers" observed in Greenland ice cores are not reflected in the pollen record obtained from lakes on the flanks of the Andes. This could indicate either that these climate flickers did not occur in the tropics or that vegetation did not respond to them.

Other aspects of past response are still debated. While it is clear that tropical forests around the world have responded to climate change in the past (Flenley, 1998), the exact pattern is not well worked out for areas as significant as the Amazon (Bush, 1994). The retreat of Amazonian forests into "Pleistocene refugia" has been discounted based on a number of lines of evidence (Willis and Whittaker, 2000). However, deeper time refugia have been proposed (Haffer, 1997). It has also been suggested, based on modeling, that Amazonian forest cover may have been maintained in glacial–interglacial cycles, but that forest structure (as indicated by leaf area index in the models) may have shifted significantly, driving speciation in the absence of "refugia" (Cowling et al., 2001). This suggests that direct CO_2 effects might impact evolutionary processes in ways that would be very difficult to control or modify through conservation actions.

The relative lack of evidence about past biotic change is aggravated by the lack of climatic precedents for the speed and nature of expected future change (Overpeck et al., 2003). Warming is projected to be rapid and will occur in the context of a warm interglacial climate (IPCC, 2001). Most rapid warming over the past 2 million years has occurred in transitions out of glacial conditions. While some interglacials were warmer than the current climate, they then cycled into cooling towards a glacial period. Physical and biotic analogs to the expected warming on a warm climate are largely absent.

While lack of information is an obstacle, we have abundant data that demonstrate that all ecosystems, including tropical forests, experience climate change on a species-by-species basis. This individualistic response to climate change is reflected in numerous temperate records and in the limited tropical record. A Gleasonian view of communities as ephemeral collections of species with like climatic and biophysical tolerances is supported by this evidence. The challenge posed to conservation is how to deal with transitory communities. If climate change is to tear contemporary communities apart, as component species respond individualistically, there is no absolute baseline reference. Pre-European contact or pre-disturbance ecosystem conditions are peculiar to one point in history. Trying to replicate these conditions under future climates that have no exact past analog has no precise scientific foundation.

As no-analog communities emerge under climate change, another conservation problem surfaces. We have no precedent for managing these communities. So, just as their composition poses problems for the definition of conservation goals and endpoints, the processes of these communities pose problems for the definition of appropriate management practices. Without objective goals or management points of reference, conservation becomes relative at best and subjective at worst. Responding to the challenge faced by loss of reference points is common to all ecosystems facing climate change, but may be particularly acute in the poorly understood and mega-complex tropical forests.

Finally, the imposition of dynamics on an already severely depleted and fragmented natural system is one of the great challenges faced by tropical forest conservationists confronting climate change (Peters and Darling, 1985; Hannah et al. 2002a). Plant communities have responded in step with remarkably rapid climate changes in the past (Markgraf and Kenny, 1995). But these responses have taken place in fully natural landscapes, in which mechanisms such as micropockets of vegetation change could persist and serve as expansion fronts for subsequent change (McGlone, 1995; McGlone and Clark, 2005). Current patterns of human land disturbance indicate that most areas of the planet are now fragmented (Hannah *et al.*, 1994; Sanderson *et al.*, 2002), obscuring or obliterating many of the mechanisms for rapid response to climate change.

Tropical forests are certainly not immune to heavy fragmentation. Most of the global biodiversity hotspots fall in tropical forests, and the hotspots by definition have lost 70% or more of their primary habitat (in addition to the more widely appreciated criterion of high endemism) (Myers *et al.*, 2000). And even large forested areas such as the Amazon have undergone highly publicized fragmentation.

Yet, the amplitude of dynamics relative to fragmentation in the tropics remains poorly understood. In one of the best studies of tropical dynamics, Bush *et al.* (2004) have demonstrated that vegetation responses on the Andean flank are quite different from the records described for more temperate forests. In these Andean forests, directional change, though present, is indistinguishable from background change at any particular point in time. Does this mean that climate "flickers" were less pronounced or absent in the tropics, that vegetation response to flickers was muted, or that the records obtained to date cannot resolve the response (e.g., taxonomically)? There will be no hard answer to these questions until new data are literally dredged up from the lakes of the tropics. For now, ecologists can only be concerned that fragmentation may be a serious constraint relative to amplitudes of even background change in these complex systems.

Addressing all of the challenges discussed here will be complicated by the massive uncertainty in climate models about the magnitude and even sign of possible precipitation changes (IPCC, 2001). While there is much greater agreement about warming, consensus on precipitation change, which is critical in determining water balance, remains elusive. Water balance may be a more critical limiting factor than temperature for both tropical moist and tropical dry forests (Pacheco, 2001). Until the uncertainties associated with precipitation projections are reduced, it may be very difficult to assess possible impacts on tropical forest and appropriate conservation responses.

15.3   CONSERVATION RESPONSES

The regional feedback between tropical forests and climate is one of the conservation challenges most specific to tropical forests. The effect is expected to be greater for tropical moist forest than for tropical dry forest (Pitman *et al.*, 2000) and greater for the tropics in general than for temperate areas (Woodwell *et al.*, 1998), although the effect can play very important roles in higher latitudes as well (Pielke, 2001).

The spatial dependence of the forest–rainfall effect has been tested in the Atlantic Forest of Brazil (Webb *et al.*, 2005). In that study the relationship between forest cover and rainfall was found to be greatest at large spatial scales. The authors compared the scale of the forest–rainfall effect with areas needed to conserve mammals with large-range sizes, concluding that both area-demanding species and the forest–rainfall relationship required large reserves.

Perhaps the ultimate forest–rainfall system is the Amazon Basin. Here rainfall in the east is recycled many times through the forests of the basin, and, in fact, much of western Amazonia might not be moist forest without the rainfall generated (Betts *et al.*, 2004). Bush (1996) has suggested that preserving this moisture-recycling is probably more critical in setting conservation goals for the Amazon than are species-based concerns. Bush's suggestion is supported by the results of Webb *et al.* (2005) in the Atlantic Forest, a system with much less pronounced moisture-recycling than the Amazon. If the Amazon has a stronger forest–rainfall effect, it seems likely that the correlation between rainfall effect and scale will be even stronger, requiring even larger reserves.

The type of reserve needed to maintain forest cover may be very different from that needed for conservation of biodiversity, however. It is forest cover and physical properties that are important in the forest–rainfall effect, rather than functioning native ecosystems, so multiple-purpose reserves or even some types of plantation tree cover may be effective in maintaining moisture-recycling. At the same time, native ecosystems provide many other benefits to human society and biodiversity conservation, so the moisture-recycling properties of forest cover provide an additional strong reason for large protected areas in the Amazon. Indigenous reserves, multiple-use forest reserves and nature reserves may all depend one on the other for sufficient forest cover to maintain moisture-recycling in the basin and the future of the region's forests.

Our limited knowledge of the paleoecology of the tropics suggests that research is a critical component of conservation of tropical forest systems. Paleoecological pictures of Asian, African, and South American forests have emerged over the past 30 years (Maley, 1996; Flenley, 1998; Colinvaux and De, 2001). Yet many chapters remain to be written. Quantum improvements in spatial, taxonomic, and temporal resolution are all possible for most regions of the tropics. Some of the most celebrated of tropical forests, such as the Amazon, are among those about which the least is understood concerning past responses to climate change. The lesson for conservation is to recognize, and be open to, major research advances that will require rethinking and readjustment of conservation strategies.

The unprecedented speed and magnitude of coming change is an issue that is not unique to the tropics. Indeed, dramatic changes in the high latitudes in the early part of this century may overshadow or obscure huge tropical changes. The complexity of tropical forests and lower magnitude of change (at least in warming) will serve to make the tropical changes less obvious and slower to be documented, yet the sum impact on biodiversity as measured by species extinction may in the long run be much greater.

The recently documented amphibian extinctions in tropical forests of South America belie the idea that tropical extinctions will be slower or less dramatic. In these extinctions, synergy between climate change and chytrid fungal disease has resulted in a dramatic spate of extinctions that would not be predicted based on models of range shifts with warming. If such synergistic, threshold-linked extinctions turn out to be common, the tropical impacts of rapid large climate change may outshadow high-latitude change. It is difficult to suggest conservation responses to such unexpected effects. However, now that one such wave of extinctions has been documented, it is clear that two priorities are monitoring for rapid population crashes and capacity for rapid institution of captive breeding programs where such crashes are observed.

Individualistic species response and no-analog communities present parallel challenges to conservation. Each implies lack of historic or paleoecological precedents for acquisition and management, respectively. Resilience has been suggested as a principle to guide both acquisition and management, and for coral reef communities there is emerging evidence that properties associated with resilience can be defined (Salm et al., 2001). However, in the more physically and biologically complex tropical forest systems, properties that may convey resilience may prove more elusive. It is far from clear that resilient forests would be the most diverse, suggesting a possible loss of biodiversity to attain resilience. Nonetheless, one principle that is clear is that removing current stressors is good for forests now and maintains biodiversity, at the same time that it makes forests more resilient to climate change (Hansen et al., 2005). The resources for even this first step are far from secured, as will be discussed below.

Responding to dynamics in fragmented landscapes depends heavily on regional context and the relative scale of the two phenomena. The scale of the minimum dynamic unit in the Amazon may be nearly the entire basin owing to moisture-recycling, while in Central African forests—where moisture-recycling is less pronounced—the minimum dynamic unit may be much smaller and defined by the area demands of large species, rather than by the forest–rainfall effect. The only answer from a conservation viewpoint is to be aware of these effects and craft conservation strategies which incorporate careful consideration of scale and process. There is no substitute for intelligent design.

Uncertainty will continue to be high in impact assessments of climate change on biodiversity, yet we have done so little in our conservation strategies to get ready for climate change that there are many steps that can be taken with certainty. The following section describes some of the conservation strategies that can be employed to take these early steps.

15.4   CONSERVATION STRATEGIES

Conservation responses to climate change are drawn from the existing mainstays of conservation strategies—protected areas, conservation in multiple-use lands, connectivity between conservation areas—with new emphases and drawing on new elements necessary to respond to the challenges of a dynamic climate. In this section we will discuss these new and existing tools, and outline their application to the conservation challenges identified above.

Perhaps the greatest impediment to sound conservation in the face of climate change is the fact that present conservation systems are incomplete. The current global network of protected areas does not represent all species and is vastly underfunded, particularly in the tropics (James and Green, 1999; Rodrigues et al., 2004). Responding to climate change would be a great challenge even in the presence of a comprehensive and well-funded system of parks and conservation measures. The problem of dealing with climate change is magnified when species range dynamics, alterations in phenology, and other changes must be addressed at the same time as completing species representation and meeting basic conservation management needs. Worse, the actions needed to confront the climate change challenge must compete for resources with these other, fundamental and often more urgent needs.

Therefore, completing representation of protected areas and adequate funding for basic management of parks and other conservation measures is the number one priority for addressing climate change. The representation and funding deficiencies are greatest in tropical forests. Over 1,400 species are under-represented in current protected areas considering vertebrates alone, virtually all of which are in the tropics (Rodrigues *et al.*, 2004). The global shortfall in protected area funding alone is estimated at $1.5 billion annually, with the majority of the shortfall occurring in the tropics (James and Green, 1999).

Connectivity between conservation areas is much less well-developed than the protected areas network, and is often presumed to be crucial in responses to climate change (Hannah *et al.*, 2002b). In principle, connectivity is an advantage as climate change dynamics become more pronounced. Not all connectivity is equally good for climate change response, however. Connecting forests along spines of mountain chains or ridgetops may not be as effective as connecting lowlands and uplands. Species will move upslope with warming, so connecting ridgetop forests has relatively less benefit than connecting lowlands and uplands. Similarly, a unit of connectivity in lowlands may be less relevant in climate change strategies than a unit of connectivity in uplands. This is because lowland species experience range shifts over relatively great distances as climate changes, while montane species are more numerous and experience range adjustments on smaller scales. Montane connectivity may therefore be more effective in species conservation on a per-unit basis. Thus, not all connectivity is equal from the perspective of climate change, and large investments in connectivity in the name of climate change should be qualitatively and quantitatively weighed against other conservation options.

For example, perhaps the most cost-effective action in a climate change conservation strategy is to invest in protected areas that harbor both species present range and their projected future ranges. Such investments offer the opportunity to improve both species current representation in protected areas and their potential future representation. In contrast to strategies to connect disjunct present and future ranges through connectivity, this approach is relatively robust to model uncertainty and it is a "no regrets" action (Williams *et al.*, 2005).

Beyond protected areas and connectivity, several new or modified conservation mechanisms will be important in dealing with climate change. These include most prominently vertical and lateral coordination of conservation planning and action. Vertical coordination is needed to ensure that national, regional, and local strategies work in concert in response to climate dynamics. Lateral coordination is needed between agencies to ensure that sectoral strategies are similarly aligned. This coordination currently exists, but will require conscious and systematic development as climate change intensifies. For example, strategies to promote transitions to new vegetation types must be coordinated across regions and between management agencies to ensure consistent management strategies and outcomes.

Creatively applied, these tools can make a substantial contribution to meeting the challenges posed by climate change (Hannah *et al.*, 2002b). Protected area systems can be expanded to compensate for range shifts resulting from climate change. Corridors can be designed specifically for climate change where range translocations for multiple species are anticipated. Conservation planning can adopt longer timeframes and emphasize vertical and lateral coordination to help improve the resilience of management strategies to climate change. Each of these tools can be fit to the climate change biology of individual regions.

For instance, for Amazonian tropical forests, protected area strategies must be sized and located with both species conservation and climate maintenance in mind (Bush, 1996a). The forest area required for maintaining the internal moisture-recycling of the basin may be larger than that required for representing all species. The location of the conserved forest is important for both biodiversity conservation and climate maintenance, but the optimal geographic configurations of the two may not exactly overlap. Simultaneous consideration of both biodiversity representation and climate maintenance may be important for other moist tropical forests as well. This is a politically sensitive issue. For instance, the Brazilian government has been implementing conservations units in some parts of Amazonia, that have generated reactions among soybean producers and the timber industry.

Management in the face of the uncertainties surrounding tropical forest response to climate change will require patience and adaptability. If tropical forests have not faced rapid climate flickers in the past, they may be poorly adapted to cope with rapid future change (Bush *et al.*, 2004). Yet, their physiology may be relatively robust to warming, in comparison with temperate and boreal species. By the time climate change provides a practical demonstration of which of these factors may prevail, it will be far too late to address the source cause of that change. It is therefore prudent to consider how, and at what levels, atmospheric greenhouse gases (GHGs) could be constrained.

15.5   GREENHOUSE GAS STABILIZATION

The global instrument for dealing with climate change—United Nations Framework Convention on Climate Change (UNFCCC)—is designed to avoid dangerous interference in agriculture, economies, and ecosystems (Schneider, 2001). Since coming into existence at the time of the 1992 Earth Summit, it is becoming increasingly apparent that ecosystems are the most sensitive of the three (O'Neill and Oppenheimer, 2002).

There is statistically sound evidence of responses in nature to the climate change that has already taken place: changes in flowering and nesting times, changes in distribution of birds, butterflies, and some marine organisms (Parmesan and Yohe, 2003; Root *et al.*, 2003). More disturbing is the first extinction associated with climate change (in conservation-conscious Costa Rica) and the widespread and massive bleaching of coral reefs from warmer seas added to other stresses (Walther *et al.*, 2002). Ecosystem failures—like those of corals and the 3.5 million acres of Alaskan spruce weakened by over 15 years of above-average temperature and dying from insect attack—can be considered a preview of more such events to come.

As scientists look ahead at the impacts of additional climate change on biodiversity, a consistent pattern is emerging, no matter how imperfect, of serious biological degradation and species loss. Compounding the problem of climate change *per se* are the ubiquitous human-modified landscapes that create an obstacle course to the movement of organisms and survival of species: the normal response in past climatic changes—such as the glacial–interglacial swings dominant in the recent geological past of the northern hemisphere.

The convention specifically addresses rapidity of climate change, citing the need not to exceed rates at which species can adapt naturally. Ignoring the distinct possibility that it is a mistake to assume climate change will only be gradual and never have abrupt episodes, it is nonetheless clear that some species and ecosystems will not be able to adapt above certain levels of climate change no matter how leisurely the rate of change. Ecosystems of low-lying islands will succumb to sea level rise and those on mountain tops will simply have nowhere to go at higher altitudes as it becomes too warm for them to survive where they are. Safe levels of climate change would avoid such ecosystem disruption and the associated wave of extinctions.

So what might constitute safe levels? Where we are right now is probably safe even with 0.8°C of average global warming plus whatever additional warming would take place because of the lag between increase in gases and temperature rise. But, it is impossible to stop at this level because of rates of emissions from current energy use (IPCC, 2001).

There seems to be a growing consensus that a safe level would be at carbon dioxide concentration of 450 parts per million or less (the pre-industrial level was 280 p.p.m.; today we are at 379 p.p.m.). That roughly translates into an average global warming of 2°C. While hard to achieve, and complicated by the need to take other greenhouse gases into account, the sooner such a target is agreed upon, the easier it is to achieve. So, somewhere between 379 and 450 p.p.m. may well be the safe zone.

This may mean more than a 2°C change for tropical forests, since change over land is higher than the global mean (because change over ocean is considerably less), and is regionally variable. Even 2°C is a very ambitious goal given the social/energy restructuring implied (Lackner, 2003). Hitting a greenhouse gas target of 450 p.p.m. implies a total transition from fossil fuels to renewable energy in the next several decades. Given that renewables currently account for about 13% of energy consumption (and 80% of that is fuelwood use that may not be sustainable as currently practiced) and increase in renewables is rising less quickly than rise in overall demand, the change required is far from incremental.

Yet, it is a change that may be of critical importance to tropical forests. Early modeling results indicate major range changes in tropical species due to future climate change (Ferreira de Siqueira and Peterson, 2003; Miles *et al.*, 2004). Other studies indicate that changes in the past may have been muted (Bush *et al.*, 2004), and there is great uncertainty about past change and no analog from the past for future magnitude and speed of changes expected in the future.

15.6   CONCLUSION

International agreements have the right targets in place to take the first, most important steps towards protecting tropical forests from climate change. The Convention on Biological Diversity (CBD) has targeted a measurable reduction in global biodiversity loss by 2010, which implies completion of the global protected areas network, its adequate funding, and significant reduction of destructive practices outside of protected areas. The Kyoto Protocol of the UNFCCC is now in place, which sets a framework for international cooperation in emission reduction. The UNFCCC itself targets avoiding climate change that would impair ecosystems' ability to adapt naturally. Even though this formulation is technically awkward, its intent is clear.

But, reality clashes very strongly with these goals—in tropical forests and many other systems: habitat loss continues; evidence is mounting that climate change is compounding the damage of habitat loss; and some systems may already be past natural ability to adapt (corals). What can biologists do? In tropical forests we can work to understand critical clues to the possible future effects of climate change. These include better understanding of past responses, better understanding of current species distributions and ecology, and analysis and modeling of responses to future climate change. Above all, we can work to rapidly incorporate the results of that research into improved conservation strategies, and to advocate the lowest possible atmospheric greenhouse gas stabilization levels.

15.7   REFERENCES

Betts, R. A., Cox, P. M., Collins, M., Harris, P. P., Huntingford, C., and Jones, C. D. (2004) The role of ecosystem–atmosphere interactions in simulated Amazonian precipitation decrease and forest dieback under global climate warming. *Theoretical and Applied Climatology* **78**, 157–175.

Brooks, T. M., Pimm, S. L., and Collar, N. J. (1997) Deforestation predicts the number of threatened birds in insular Southeast Asia. *Conservation Biology* **11**, 382–394.

Bush, M. B. (1994) Amazonian speciation: A necessarily complex model. *Journal of Biogeography* **21**, 5–17.

Bush, M. B. (1996) Amazonian conservation in a changing world. *Biological Conservation* **76**, 219–228.

Bush, M. B., Silman, M. R., and Urrego, D. H. (2004) 48,000 years of climate and forest change in a biodiversity hot spot. *Science* **303**, 827–829.

Colinvaux, P. A. and De, O. P. E. (2001) Amazon plant diversity and climate through the Cenozoic. *Palaeogeography, Palaeoclimatology, Palaeoecology* **166**, 51–63.

Cowling, S. A., Maslin, M. A., and Sykes, M. T. (2001) Paleovegetation simulations of lowland Amazonia and implications for neotropical allopatry and speciation. *Quaternary Research* **55**, 140–149.

Cox, P. M., Betts, R. A., Jones, C. D., Spall, S. A., and Totterdell, I. J. (2000) Acceleration of global warming due to carbon-cycle feedbacks in a coupled climate model. *Nature* **408**, 184–187.

Cox, P. M., Betts, R. A., Jones, C. D., Spall, S. A., and Totterdell, I. J. (2002) Acceleration of global warming due to carbon-cycle feedbacks in a coupled climate model. *Nature* **408**, 184–187.

Cox, P. M., Betts, R. A., Collins, M., Harris, P. P., Huntingford, C., and Jones, C. D. (2004) Amazonian forest dieback under climate–carbon cycle projections for the 21st century. *Theoretical and Applied Climatology* **78**, 137–156.

Ferreira de Siqueira, M. and Peterson, A. T. (2003) Global climate change consequences for Cerrado tree species. *Biota Neotropica* **3**, 1–14.

Flenley, J. R. (1998) Tropical forests under the climates of the last 30,000 years. *Climatic Change* **39**, 177–197.

Haffer, J. (1997) Alternative models of vertebrate speciation in Amazonia: An overview. *Biodiversity and Conservation* **6**, 451–476.

Hannah, L., Lohse, D., Hutchinson, C., Carr, J. L., and Lankerani. A. (1994) A preliminary inventory of human disturbance of world ecosystems. *Ambio* **23**, 246.

Hannah, L., Midgley, G. F., Lovejoy, T., Bond, W. J., Bush, M. L. J. C., Scott, D., and Woodward, F. I. (2002a) Conservation of biodiversity in a changing climate. *Conservation Biology* **16**, 11–15.

Hannah, L., Midgley, G. F., and Millar, D. (2002b) Climate change-integrated conservation strategies. *Global Ecology and Biogeography* **11**, 485–495.

Hansen, L., Biringer, J., and Hoffman, J. (2003) *Buying Time: A User's Manual for Building Resistance and Resilience to Climate Change in Natural Systems*. World Wildlife Fund, Washington, D.C.

IPCC (2001) *Climate Change 2001: The Scientific Basis* (contribution of Working Group I to the Third Assessment Report of the Intergovernmental Panel on Climate Change). Cambridge University Press, Port Chester, NY.

James, A. N. and Green, M. J. B. (1999) *Global Review of Protected Area Budgets and Staff* (pp. 1–35). World Conservation Monitoring Center, Cambridge, U.K,

Lackner, K. S. (2003) A guide to CO_2 sequestration. *Science* **300**, 1677–1678.

Maley, J. (1996) The African rain forest: Main characteristics of changes in vegetation and climate from the Upper Cretaceous to the Quaternary. *Proceedings of the Royal Society of Edinburgh* **104B**, 31–73.

Marengo, J. A. (2004) Interdecadal variability and trends of rainfall across the Amazon Basin. *Theoretical and Applied Climatology* **78**, 79–96.

Markgraf, V. and Kenny, R. (1995) Character of rapid vegetation and climate change during the late-glacial in southernmost South America. In: B. Huntley, W. Cramer, A. V. Morgan, H. C. Prentice, and J. R. M. Allen (eds.), *Past and Future Rapid Environmental Changes: The Spatial and Evolutionary Responses of Terrestrial Biota* (pp. 81–102). Springer-Verlag, Berlin.

McGlone, M. S. (1995) The responses of New Zealand forest diversity to Quaternary climates. In: B. Huntley, W. Cramer, A. V. Morgan, H. C. Prentice, and J. R. M. Allen (eds.), *Past and Future Rapid Environmental Changes: The Spatial and Evolutionary Responses of Terrestrial Biota* (pp. 73–80). Springer-Verlag, Berlin.

McGlone, M. and Clark, J. S. (2005) Microrefugia and macroecology. In: T. E. Lovejoy and
   L. Hannah (eds.), *Climate Change and Biodiversity* (pp. 157–160). Yale University Press,
   New Haven, CT.

Miles, L., Grainger, A., and Phillips, O. (2004) The impact of global climate change on tropical
   forest diversity in Amazonia. *Global Ecology and Biogeography* **13**, 553–565.

Myers, N., Mittermeier, R. A., Mittermeier, C. G., Da Fonseca, G. A. B., and Kent, J. (2000)
   Biodiversity hotspots for conservation priorities. *Nature* **403**, 853–858.

O'Neill, B. C. and Oppenheimer, M. (2002) Climate change: Dangerous climate impacts and the
   Kyoto Protocol. *Science* **296**, 1971–1972.

Overpeck, J., Whitlock, C., and Huntley, B. (2003) Terrestrial biosphere dynamics in the climate
   system: Past and future. In: K. D. Alverson, R. S. Bradley, and T. F. Pederson (eds.),
   Paleoclimate, Global Change, and the Future (pp. 81–109). Springer-Verlag, Berlin.

Pacheco, M. (2001) Impacts of climate change on tropical forest plants. Unpublished work.

Parmesan, C. and Yohe, G. (2003) A globally coherent fingerprint of climate change impacts
   across natural systems. *Nature* **421**, 37–42.

Peters, R. L. and Darling, J. D. S. (1985) The greenhouse effect and nature reserves. *BioScience*
   35, 707–717.

Pielke, R. A. (2001) Influence of the spatial distribution of vegetation and soils on the prediction
   of cumulus convective rainfall. *Reviews of Goephysics* **39**, 151–177.

Pitman, A., Pielke, R., Avissar, R., Claussen, M., Gash, J., and Dolman, H. (2000) The role of
   land surface in weather and climate: Does the land surface matter. *IGBP Newsletter* **39**,
   4–24.

Rodrigues, A. S. L., Andelman, S. J., Bakarr, M. I., Boitani, L., Brooks, T. M., Cowling, R. M.,
   Fishpool, L. D. C., da Fonseca, G. A. B., Gaston, K. J., Hoffmann, M. *et al.* (2004)
   Effectiveness of the global protected area network in representing species diversity. *Nature*
   428, 640–643.

Root, T., Price, J. T., Hall, K. R., Schneider, S. H., Rosenzweig, C., and Pounds, J. A. (2003)
   Fingerprints of global warming on wild animals and plants. *Nature* **421**, 57–60.

Salm, R. V., Coles, S. L., West, J. M., Done, L. G. T., Causey, B. D., Glynn, P. W., Heyman,
   W., Jokiel, P., Obura, D., and Oliver, J. (2001) *Coral Bleaching and Marine Protected Areas*
   (p. 102). Nature Conservancy, Honolulu, HI.

Sanderson, E., Jaiteh, M., Levy, M. A., Redford, K. H., Wannebo, A., and Woolmer, G. (2002)
   The human footprint and the last of the wild. *BioScience* **52**, 891–904.

Schneider, S. H. (2001) What is "dangerous" climate change? *Nature* **411**, 17–19.

Walther, G., Post, E., Convey, P., Menzel, A., Parmesan, C., Beebee, T. J. C., Fromentin, J.,
   Hoegh-Guldberg, O.. and Bairlein, F. (2002) Ecological responses to recent climate
   change. *Nature* **416**, 389–395.

Webb, T., Woodward, F. I., Hannah, L., and Gaston, K. J. (2005) Forest cover–rainfall
   relationships in a biodiversity hotspot: the Atlantic forest of Brazil. *Ecological Applications*
   15, 1968–1983.

Williams, P., Hannah, L., Andelman, S., Midgely, G. F., Araujo, M. B., Hughes, G., Manne, L.,
   Martinez-Meyer, E., and Pearson, R. G. (2005) Planning for climate change: Identifying
   minimum-dispersal corridors for the cape Proteaceae. *Conservation Biology* **19**, 1063–1074.

Willis, K. J. and Whittaker, R. J. (2000) Perspectives: paleoecology. The refugial debate. *Science*
   287, 1406–1407.

Woodwell, G. M., MacKenzie, F. T., Houghton, R. A., Apps, M., Gorham, E., and Davidson,
   E. (1998) Biotic feedbacks in the warming of the earth. *Climatic Change* **40**, 495–518.

Subject index

Species index

Abies, 13
Acaena, 65
Afzelia, 120
Agathis, 82
Agelaea, 151
Aidia, 134
Albizia, 127
Alchornea, 43, 126, 210
Allanblachia, 132
Alnus, 13, 35, 43, 57
Alsophila, 43
Altingia, 86, 91, 100
Anacardiaceae, 127
Aningeria, 121
Annonaceae, 6, 34, 123
Anonidium, 134
Anopyxis, 152
Anthonata, 119
Antidesma, 127
Apocynaceae, 123
Aquifoliaceae, 5
Arachis, 204
Araucaria, 92
Arecaceae, 6
Artemisia, 89, 91
Astelia, 102
Asteraceae, 36
Aucoumea 161

Bactris, 203
Baillonella toxisperma, 119
Baphia, 126
Berlinia, 119
Bocconia, 65
Bosqueia, 159
Brachystegia, 119, 146
Brosimum, 43
Burseraceae, 128

Caesalpiniaceae, 119, 139
Calamus, 121
Calathea, 201
Calpocalyx, 132
Canarium, 119, 156
Canavalia, 203
Canna, 204
Capsicum, 204
Cassiopourea, 151
Castanopsis/Lithocarpus, 81, 83, 91–92,
    100–101
Casuarina, 13, 81–83
Cecropia, 57, 210
Cedrela, 65
Celtis, 86
Chaetachme, 127
Chenopodiaceae/Amaranthaceae, 94
Chloranthaceae, 5
Chlorophora, 154
Chrysobalanaceae, 123

Printing: Mercedes-Druck, Berlin
Binding: Stein+Lehmann, Berlin